Lecture Notes in Electrical Engineering

Volume 757

The book series *Lecture Notes in Electrical Engineering* (LNEE) publishes the latest developments in Electrical Engineering - quickly, informally and in high quality. While original research reported in proceedings and monographs has traditionally formed the core of LNEE, we also encourage authors to submit books devoted to supporting student education and professional training in the various fields and applications areas of electrical engineering. The series cover classical and emerging topics concerning:

- Communication Engineering, Information Theory and Networks
- Electronics Engineering and Microelectronics
- Signal, Image and Speech Processing
- Wireless and Mobile Communication
- Circuits and Systems
- Energy Systems, Power Electronics and Electrical Machines
- Electro-optical Engineering
- Instrumentation Engineering
- Avionics Engineering
- Control Systems
- Internet-of-Things and Cybersecurity
- Biomedical Devices, MEMS and NEMS

For general information about this book series, comments or suggestions, please contact leontina.dicecco@springer.com.

To submit a proposal or request further information, please contact the Publishing Editor in your country:

China

Jasmine Dou, Editor (jasmine.dou@springer.com)

India, Japan, Rest of Asia

Swati Meherishi, Editorial Director (Swati.Meherishi@springer.com)

Southeast Asia, Australia, New Zealand

Ramesh Nath Premnath, Editor (ramesh.premnath@springernature.com)

USA, Canada:

Michael Luby, Senior Editor (michael.luby@springer.com)

All other Countries:

Leontina Di Cecco, Senior Editor (leontina.dicecco@springer.com)

**** This series is indexed by EI Compendex and Scopus databases. ****

More information about this series at https://link.springer.com/bookseries/7818

Liheng Wang · Yirong Wu · Jianya Gong
Editors

Proceedings of the 7th China High Resolution Earth Observation Conference (CHREOC 2020)

A Decade of Integrated Aerospace Exploration

Springer

Editors
Liheng Wang
China Aerospace Science and Technology Corporation
Beijing, China

Yirong Wu
Aerospace Information Research Institute
Chinese Academy of Sciences
Beijing, China

Jianya Gong
Wuhan University
Wuhan, Hubei, China

ISSN 1876-1100 ISSN 1876-1119 (electronic)
Lecture Notes in Electrical Engineering
ISBN 978-981-16-5734-4 ISBN 978-981-16-5735-1 (eBook)
https://doi.org/10.1007/978-981-16-5735-1

This Springer imprint is published by the registered company Springer Nature Singapore Pte Ltd.
The registered company address is: 152 Beach Road, #21-01/04 Gateway East, Singapore 189721, Singapore

Contents

Weight Factor Determination of Reverse Distance Weighting Method in Computation of Geomagnetic Diurnal Variation Data 1
Liu Xiaogang, Liu Qian, Wang Yunpeng, and Song Ying

Optimal Rescheduling for OOS Mission Based on Squeaky-Wheel Optimization 17
Jing Yu, Dong Hao, and Xiaoqian Chen

Method of the Stability of Soil Landslide Under the Condition of Continuous Rainfall 37
Min Yang and Yihan Xie

Processing and Application of AMS-3000 Large Field of View Triple Line Array Camera in Photogrammetry 49
Shuping Xiong, Yansong Duan, and Guoqin Yuan

Evaluation and Analysis of Geometric Accuracy of the ZY-1 02D VNIC Image 57
Liping Zhao, Xingke Fu, and Xianhui Dou

Band Registration Analysis of the ZY-1 02D VNIC Image 77
Liping Zhao, Hongzhou Li, and Huai Wang

Analysis and Optimization of Geometric Accuracy of ZY-1 02C HRC Image 97
Liping Zhao

River Shoreline Recognition from High Resolution Remote Sensing Imagery Based on Water Index and Artificial Neural Network 115
Yuan Yao and Linyi Li

Collaborative Classification of Hyperspectral and LiDAR Data Based on Gram Matrices Constrained Fusion Net 125
Mengmeng Zhang, Wei Li, and Ran Tao

Joint Classification of Hyperspectral and LiDAR Data Using Improved Local Contain Profile 137
Dandan Cao, Wei Li, Lu Li, Qiong Ran, Mengmeng Zhang, and Ran Tao

A High-Resolution Spotlight SAR Imaging Method Based on Two-Step Processing Approach 151
Hao Zheng and Hongmeng Chen

Research of Industrial Heat Sources Detection Method Based on Thermal Infrared Satellite Data 163
Jie Wang, Yu Liu, Hongzhong Lu, Jianwei Qi, and Ping Ai

Robust Detection and Tracking Algorithm for Pedestrian Targets with Posture Change 173
Tao Zhao, Jin-Sheng Xiao, Wen Wang, Hai-Gang Sui, Jing-Jin Ma, and Chuan Xu

Comparative Analysis of Chinese High-Resolution Satellite Data for Sugarcane Classification Based on U-Net Model 185
Chen Chen, Linjiang Lou, Xinyuan Gao, and Yu Liu

A Differential Flat Approach for Attitude Manoeuver Control of Agile Satellite 193
Deting Li, Yikang He, Shuyu Lin, Miao Li, and Kang Shi

A Voxel-Based Fusing Method for Aerial Laser Scanning and Oblique Image Point Cloud Via Noise-and-Occupancy-Aware 205
Shiming Li, Qing Zhu, Han Hu, Xuming Ge, and Chuncheng Zhu

A Fast and Robust UAV Images Mosaic Method 229
Guangrui Yu, Changliang Ha, Chunlin Shi, Lianbing Gong, and Lili Yu

Comparative Research on Water Body Extraction Methods Based on SPOT Data 247
Linjiang Lou, Chen Chen, Xinyuan Gao, Kun Liu, Minmin Li, and Yajie Fu

Research on National Land Resource Accounting Method 257
Xinyan Zheng, Yu Zhou, Tao Cheng, Jin Liu, Chen Chen, and Lu Yi

A Configuration Approach and Effectiveness Evaluation of Modular Reconfigurable Satellite 265
Fu Xin, Mai Xiamei, Zhang Haolong, and Xue Fengtong

FD-LinkNet: A Encoder-decoder Structure Network for High Resolution Satellite Imagery Rural Road Extraction 285
Lin Yi, Miao Yang, Liang Shuang, Peng Xiangyang, and Song Wentao

Geolocation Accuracy of Bistatic InSAR Configuration with Geostationary Transmitter and LEO Receivers 295
Yuekun Wang, Zhonghao Wei, Feng Tian, and Long Zhuang

Preliminary Evaluation of the Stereo Mapping Accuracy of the Gaofen-7 303
Liping Zhao, Zhi Fang, and Qifeng Chu

Research on Cloud Detection Method of GaoFen-6 Wide Camera Data 321
Shiyun Ke, Mi Wang, Jinshan Cao, and En Long

Research on Space-Ground Integrated Control Technology of Earth Observation Spacecraft 341
Zhang Fajia, Yang Tongzhi, Zhu Jia, Xing Nan, and Dang Jiancheng

An Improved Ionospheric Estimation Algorithm for Low-Frequency SAR Interferometry 357
Beixin Qin, Yongsheng Zhang, Zaoyu Sun, and Xiaoxiang Zhu

The Research of Multi-scale Effect on Remote Sensing Image Object Detection 367
Guangkuo Ma, Yifan Dong, Yun Su, Wei Xu, and Pingping Huang

Remote Sensing Road Segmentation Based on Feature Extraction Optimization and Skeleton Detection Optimization 377
Sijun Dong, Yu Zhao, and Zhengchao Chen

Capability Requirement Analysis of Space-Based Ocean Reconnaissance and Surveillance System Based on Improved QFD Method 387
Min Wan and Yan Hou

Remote Sensing Scene Classification with Multi-task Learning 403
Tengfei Gong, Xiangtao Zheng, and Xiaoqiang Lu

Top-of-Atmosphere Radiance Image Simulation from the Visible to Thermal Infrared 419
Dong Liu, Yanbing Dong, and Hongxia Mao

Application of Improved RBF Neural Network in Remote Sensing Image Restoration 437
Yunsen Wang, Yong Wang, Daqiang Feng, Jing Yang, and Ke An

Distributed HRWS MIMO-SAR Linear Vertical Baseline Estimation and Compensation 451
Pengcheng Li, Zaoyu Sun, Yongsheng Zhang, Feng He, Shenjing Wang, and Zhen Dong

Fast Three-Dimensional Imaging Method of Aerospace Targets Based on Linear Array Radar 465
Zhang Yi, Yu Anxi, Jin Guanghu, and Dong Zhen

Edge Detection of SAR Images Based on Shearlet 483
Zengguo Sun, Guodong Zhao, Weirong Chen, Robertas Damaševičius, and Marcin Woźniak

Research on Hysteresis Control Method of Fast Steering Mirror in Satellite Laser Communication 495
Yu Zhiliang, Li Xue, Yang Mingliang, and Fang Zhiyi

Construction of Remote Sensing User Growth System Based on Hooked Mode 511
Caiping Li, Xiaoming Zhou, Jisheng Zhang, and Lei Chang

Ship Azimuth Velocity Estimation in Spaceborne SAR Based on Minimum-Entropy Criterion and Newton's Method 523
Yating Huang, Dexin Li, and Zhen Dong

Mapping 30-m Resolution Land Cover of China Based on Full Convolutional Neural Network 535
Yinhe Liu and Yanfei Zhong

Research on Brightness Compensation Algorithm for GF-3 SAR Images 547
Zengguo Sun, Xiaopeng Yan, Qianfu Chen, and Weirong Chen

The Spaceborne SAR 3D Imaging Method Based on 2D Compressed Sensing 565
He Tian, Chunzhu Dong, and Hongcheng Yin

Weight Factor Determination of Reverse Distance Weighting Method in Computation of Geomagnetic Diurnal Variation Data

Liu Xiaogang, Liu Qian, Wang Yunpeng, and Song Ying

Abstract A virtual geomagnetic diurnal station has been used to calculate the diurnal variation correction data in the regions where the ground magnetic surveying is difficult to be implemented, and this will provide diurnal variation correction data for the airborne geomagnetic measurement. The weight factors of reverse distance weighting (RDW) method usually use the empirical values, which have limitation in the application of different testing areas. The correlation coefficients are proposed and introduced to calculate the weight factors iteratively, which are more practical. The numerical experimental results show that the diurnal variation correction data, whose weight factors are calculated by correlation coefficients, have a higher precision. The precision of northern component X and horizontal component H are superior to 2.2 nT, eastern component Y and geomagnetic total intensity F are superior to 1.0 nT, vertical component Z is superior to 0.3 nT, and geomagnetic declination D and geomagnetic inclination I are superior to 10″. Therefore, the diurnal variation correction data in higher precision can be provided for the airborne scalar and vector geomagnetic measurement in the regions where the ground magnetic surveying is difficult.

Keywords Aeromagnetic measurement · Geomagnetic observatory · Geomagnetic diurnal variation (GDV) · Reverse distance weighting (RDW) method · Correlation coefficient

---

Foundation Support: National Key R&D Program of China (2018YFC1503806); The National Natural Science Foundation of China (42174001, 41774018); Open Foundation of State Key Laboratory of Geo-information Engineering (SKLGIE2021-ZZ-4)

L. Xiaogang (✉) · L. Qian · W. Yunpeng · S. Ying
State Key Laboratory of Geo-Information Engineering, Xi'an 710054, China
e-mail: liuxiaogang_1949@163.com

L. Xiaogang · L. Qian · W. Yunpeng · S. Ying
Xi'an Research Institute of Surveying and Mapping, Xi'an 710054, China

L. Wang et al. (eds.), *Proceedings of the 7th China High Resolution Earth Observation Conference (CHREOC 2020)*, Lecture Notes in Electrical Engineering 757, https://doi.org/10.1007/978-981-16-5735-1_1

1 Introduction

The aeromagnetic measurement has many merits, such as high surveying efficiency, big quantity information, large coverage area, symmetrical distribution, good precision, and so on. Therefore, it is an advanced technology to obtain geomagnetic data in high precision and resolution effectively and quickly [1–7].

The GDV correction must be considered in processing of aeromagnetic measurement data. The most efficient technological approach to implement the GDV correction is setting the GDV station in aeromagnetic testing area, which can survey the real-time variation of Earth's geomagnetic field, and the GDV correction can be computed after the processing of diurnal variation data. This strategy is in high precision and of which the results are reliable. However, for some testing areas, such as high mountain, glacier, ancient forest, desert, marsh, lake, land-sea interface, et al., where are unreachable for people and vehicles. So setting the GDV station in these places has great difficulties. Therefore, the technology of virtual GDV station has been used to solve the questions mentioned above. The fundamental of virtual GDV station is that the measurement data of surrounding geomagnetic observatories or the GDV stations are used to estimate the GDV data of some site inner the testing area, and then the aeromagnetic measurement data can be corrected by GDV correction data.

In the most commonly used technologies for the construction of virtual GDV station, application of RDW method is the most extensive. This method, which is closely related to the space changing, determines the weight using the k-th power of the reciprocal of the distance between every GDV station and computational point. Meanwhile, this method has little computational quantity and good precision, so it has great application prospect in the computation of GDV correction data in the difficult areas that ground measurement can't be complemented [8–12]. However, there is no better way to determine the weight factor k, the convention practice is giving an experience value, but the problem is that this experience value can't be available in all the testing areas, so the GDV data determined will be in low accuracy.

According to the correlation analysis of space changing of geomagnetic field [13], the geomagnetic measurement data basically reflects the physical attributes of the geomagnetic field where the station is, so it has some correlation with the measurement data of the neighboring stations. The correlation coefficient is introduced in this paper to determine the weight factor k of RDW method. When the correlation coefficient between the computational point and adjacent stations is more bigger, the results of computational point is much more closer to measurement data. Therefore, the weight factor k, which is not confined to a constant value when computing the GDV data in different testing areas, can be determined flexibly using correlation coefficient between the computational point and neighboring stations, and this method is more practical.

2 RDW Model

The basic principle of RDW method can be described as follows: the computational point is taken as the central point, the number of geomagnetic observatories within a certain searching radius is determined, the distances between computational point and geomagnetic observatories are calculated, and *k*-th power of the reciprocal of the distances is taken as weight, then the GDV data of computational point can be obtained. The computational formula is [11, 12]

$$M = \frac{\sum_{i=1}^{n}\left[\left(\frac{1}{d_i^k}\right) \times T_i\right]}{\sum_{i=1}^{n}\left(\frac{1}{d_i^k}\right)} \quad k \geq 0 \tag{1}$$

where T_{i} is the measurement data of geomagnetic observatories, and there are seven geomagnetic elements, including geomagnetic northern component *X*, geomagnetic eastern component *Y*, geomagnetic vertical component *Z*, geomagnetic total intensity *F*, geomagnetic horizontal component *H*, geomagnetic declination *D*, and geomagnetic inclination *I*, *n* is the number of geomagnetic observatories, d_i is the distance between computational point and geomagnetic observatories, and *M* is the calculation results of computational point.

3 Correlation Coefficient

The correlation coefficients of the GDV data between the computational point and geomagnetic observatories are calculated, when the correlation coefficient is more closer to numerical value 1.0, the precision of the results will be more higher. The computational formulae are [14, 15]

$$\text{cov}(x, y) = E\{[x - E(x)][y - E(y)]\} \tag{2}$$

$$cof(x, y) = \frac{\text{cov}(x, y)}{\sigma(x)\sigma(y)} \tag{3}$$

where, *E*(*x*) and *E*(*y*), which are also called average value, are mathematical expectation of variables *x* and *y*, Cov(*x, y*) is the covariance between variables *x* and *y*, *cof*(*x, y*) is the correlation coefficient between variables *x* and *y*, $\sigma(x)$ and $\sigma(x)$ are standard deviation (STD) of variables *x* and *y*.

$$E(x) = \frac{1}{n}\sum_{i=1}^{n} x_i \tag{4}$$

$$\sigma(x) = \sqrt{\frac{1}{n-1}\sum_{i=1}^{n} (x_i - E(x))^2} \tag{5}$$

4 Precision Evaluation

The standard deviation (STD) and root-mean-square error (RMSE) are used for the precision evaluation of GDV data determined in this paper.

(1) STD
The measurement data of computational point are used to calculate the STD of GDV data. More smaller the STD, more higher the precision. The computational formula is

$$R = \sqrt{\frac{\sum_{i=1}^{n} (u_i - \overline{u})^2}{n-1}} \tag{6}$$

where u_i are the differences between computational results and measurement data, of which $\overline{u}$ is the average, n is the number of measurement data, and R is the STD of computational results.

(2) RMSE
The measurement data of computational point are used to calculate the RMSE of GDV data. More smaller the RMSE, more higher the precision. The computational formula is

$$S = \sqrt{\frac{\sum_{i=1}^{n} u_i^2}{n}} \tag{7}$$

where S is the RMSE of computational results, the meanings of other quantities are the same as Eq. (6).

5 Numerical Experiment and Results Analysis

5.1 Experimentation Data Explanation

At present, the number of global geomagnetic observatories are more than one thousand, and the measurement data of 150 geomagnetic observatories, including the geomagnetic elements *F*, *H*, *D*, *Z* or *X*, *Y*, *Z*, *F*, can be download from Intermagnet website [16]. The distribution of 150 geomagnetic observatories is shown in Fig. 1.

The geomagnetic observatories participated in data communication of InterMagnet are mainly distributed in Europe. Therefore, some geomagnetic observatories in Europe are chosen in this experiment. Picking the geomagnetic observatory NCK of Hungary as the central point, the basic information of geomagnetic observatories within 10° radius of NCK are listed in Table 1.

The distances between NCK and other geomagnetic observatories are listed in Table 2.

The GDV data of 13 geomagnetic observatories in May 1, 2018 are used to testify the validity of the proposed method in this paper. The specific implementation steps are as follows:

(1) A large range of the weight factors is given firstly, and the GDV data of the computational point is calculated with different weight factors varying in a big step. Then the correlation coefficients of the GDV data between the computational point and other geomagnetic observatories are calculated. The weight factor, which is corresponding to the maximum of correlation coefficients, is defined as initial weight factor.

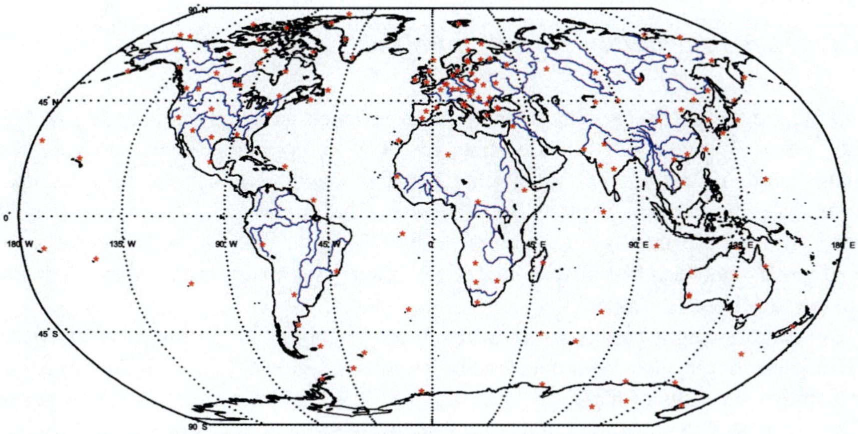

Fig. 1 Distribution of global geomagnetic observatories of InterMagnet

Table 1 The basic information of European geomagnetic observatories

Station code	Station name	Country	Latitude	Longitude
BDV	Budkov	Czech Republic	49.08°	14.02°
BEL	Belsk	Poland	51.84°	20.79°
BFO	Black Forest	Germany	48.331°	8.325°
FUR	Furstenfeldbruck	Germany	48.17°	11.28°
HLP	Hel	Poland	54.61°	18.82°
HRB	Hurbanovo	Slovakia	47.86°	18.19°
LON	Lonjsko Polje	Croatia	45.4081°	16.6592°
NCK	Nagycenk	Hungary	47.63°	16.72°
PAG	Panagjurishte	Bulgaria	42.5°	24.2°
SUA	Surlari	Romania	44.68°	26.25°
THY	Tihany	Hungary	46.9°	17.89°
WIC	Conrad Observatory	Austria	47.9305°	15.8657°
WNG	Wingst	Germany	53.74°	9.07°

(2) Approximating the initial weight factor determined above, a small range of the weight factors is then given, the GDV data of the computational point is calculated with different weight factors vary in a much smaller step. And the correlation coefficients of the GDV data between the computational point and other geomagnetic observatories are sequentially calculated. The weight factor, which is corresponding to the maximum of correlation coefficients, is sequentially defined as another initial weight factor.

(3) The weight factor is refined iteratively with above implementation steps until the difference of the weight factors between the first and second iteration fulfills a given threshold value.

5.2 *Experimentation Results Analysis*

Firstly, the weight factor k is given some experience values, such as 0.5, 1.0, 1.5, 2.0, 2.5, 3.0 and 3.5. The GDV data of NCK station are calculated using the geomagnetic measurement data of other 12 neighboring geomagnetic observatories, and then comparing the computational results of NCK station with its measurement data, the statistic results are listed in Tables 3, 4, 5, 6, 7, 8, 9 and correlation coefficients between the determined GDV data and measurement data of NCK station are listed in Table 11.

When the weight factors are in different empirical values, comparisons between GDV data and measurement data are shown in Figs. 2, 3, 4, 5, 6, 7, 8. According to the statistics results of Tables 3, 4, 5, 6, 7, 8, 9, 11 and the comparison results of Figs. 2, 3, 4, 5, 6, 7, 8, the precision of determined GDV data is the highest when the empirical value of weight factor k equals to 2.5.

Table 2 The distance between NCK and other geomagnetic observatories (km)

Station code	SUA-NCK	WNG-NCK	PAG-NCK	BFO-NCK	HLP-NCK	BEL-NCK	FUR-NCK	BDV-NCK	LON-NCK	HRB-NCK	THY-NCK	WIC-NCK
Station distance	1111	1090	1010	938	811	652	609	341	247	166	154	101

Table 3 Precision statistics of RDW method ($k = 0.5$)

Geomagnetic elements	Maximum	Minimum	Average	STD	RMSE
X (nT)	2.94	−8.40	−0.89	2.77	2.91
Y (nT)	2.41	−2.11	−0.43	0.92	1.01
Z (nT)	1.66	−0.78	0.40	0.57	0.70
F (nT)	2.54	−2.59	0.09	1.10	1.10
H (nT)	2.73	−8.24	−0.92	2.69	2.84
D (′)	0.42	−0.34	−0.04	0.17	0.18
I (′)	0.57	−0.14	0.07	0.18	0.19

Table 4 Precision statistics of RDW method ($k = 1.0$)

Geomagnetic elements	Maximum	Minimum	Average	STD	RMSE
X (nT)	3.26	−6.97	−0.63	2.45	2.53
Y (nT)	2.40	−1.95	−0.49	0.93	1.05
Z (nT)	1.19	−0.82	0.19	0.45	0.49
F (nT)	2.36	−2.04	0.01	0.97	0.97
H (nT)	3.07	−6.86	−0.66	2.38	2.47
D (′)	0.40	−0.32	−0.05	0.16	0.17
I (′)	0.46	−0.17	0.05	0.15	0.16

Table 5 Precision statistics of RDW method ($k = 1.5$)

Geomagnetic elements	Maximum	Minimum	Average	STD	RMSE
X (nT)	3.44	−6.06	−0.47	2.26	2.31
Y (nT)	2.38	−1.84	−0.51	0.94	1.07
Z (nT)	0.94	−0.85	0.06	0.38	0.38
F (nT)	2.23	−1.82	−0.04	0.90	0.90
H (nT)	3.27	−5.97	−0.50	2.20	2.26
D (′)	0.39	−0.33	−0.05	0.16	0.17
I (′)	0.03	−0.19	0.03	0.14	0.15

Table 6 Precision statistics of RDW method ($k = 2.0$)

Geomagnetic elements	Maximum	Minimum	Average	STD	RMSE
X (nT)	3.51	−5.68	−0.41	2.19	2.22
Y (nT)	2.39	−1.89	−0.51	0.96	1.09
Z (nT)	0.77	−0.87	−0.02	0.34	0.34
F (nT)	2.12	−1.76	−0.09	0.87	0.88
H (nT)	3.36	−5.60	−0.44	2.14	2.18
D (′)	0.39	−0.33	−0.05	0.16	0.17
I (′)	0.37	−0.21	0.03	0.14	0.14

Table 7 Precision statistics of RDW method ($k = 2.5$)

Geomagnetic elements	Maximum	Minimum	Average	STD	RMSE
X (nT)	3.54	−5.60	−0.41	2.18	2.21
Y (nT)	2.46	−1.90	−0.48	0.99	1.10
Z (nT)	0.74	−0.87	−0.06	0.31	0.32
F (nT)	2.02	−1.83	−0.13	0.86	0.88
H (nT)	3.40	−5.54	−0.43	2.13	2.17
D (′)	0.40	−0.33	−0.05	0.16	0.17
I (′)	0.36	−0.21	0.02	0.14	0.14

Table 8 Precision statistics of RDW method ($k = 3.0$)

Geomagnetic elements	Maximum	Minimum	Average	STD	RMSE
X (nT)	3.55	−5.68	−0.44	2.20	2.24
Y (nT)	2.56	−1.90	−0.45	1.02	1.10
Z (nT)	0.75	−0.87	−0.09	0.30	0.32
F (nT)	1.93	−2.01	−0.17	0.89	0.88
H (nT)	3.41	−5.61	−0.46	2.15	2.17
D (′)	0.41	−0.32	−0.04	0.17	0.17
I (′)	0.36	−0.21	0.03	0.14	0.14

Table 9 Precision statistics of RDW method ($k = 3.5$)

Geomagnetic elements	Maximum	Minimum	Average	STD	RMSE
X (nT)	3.55	−5.81	−0.47	2.23	2.28
Y (nT)	2.68	−1.90	−0.41	1.05	1.13
Z (nT)	0.78	−0.86	−0.10	0.29	0.31
F (nT)	1.86	−2.20	−0.20	0.91	0.93
H (nT)	3.41	−5.73	−0.49	2.18	2.23
D (′)	0.43	−0.32	−0.04	0.17	0.18
I (′)	0.37	−0.21	0.03	0.14	0.14

Secondly, the optimal weight factor (OWF) k is determined using the correlation coefficients between the computational data of NCK station and the measurement data of other 12 neighboring geomagnetic observatories. Then the GDV data of NCK station are calculated using the OWF. Comparing the computational data of NCK station with its measurement data, the statistic results are listed in Table 10, and correlation coefficients between the determined GDV data and measurement data of NCK station are listed in Table 11.

Fig. 2 Geomagnetic northern component *X*

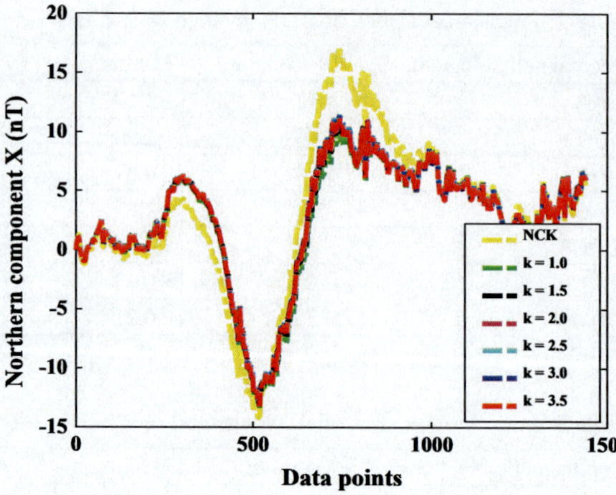

Fig. 3 Geomagnetic eastern component *Y*

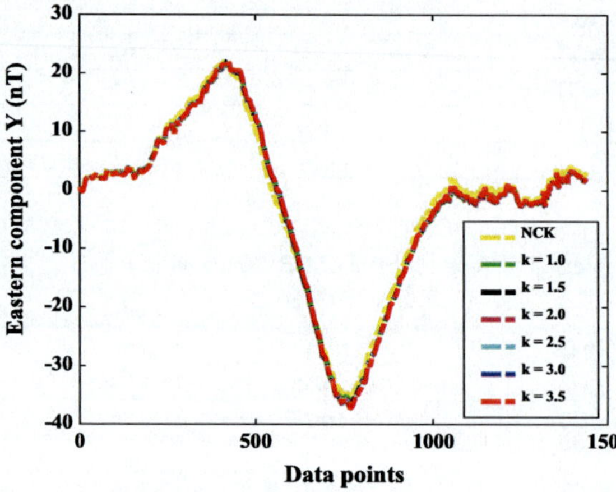

From Table 10 we can see that when using correlation coefficient to determine the OWF, the precision of determined GDV data of RDW method can be concluded as follows: the geomagnetic northern component *X* and horizontal component *H* are superior to 2.2 nT, the geomagnetic eastern component *Y* and geomagnetic total intensity *F* are superior to 1.0 nT, the geomagnetic vertical component *Z* is superior to 0.3 nT, the geomagnetic declination *D* and geomagnetic inclination *I* are superior to 10″.

When the weight factors are in different empirical values and OWF, correlation coefficients between the determined GDV data and measurement data are shown in Fig. 9.

According to the statistics results of Tables 3, 4, 5, 6, 7, 8, 9, 10 and 11 and the comparison results of Fig. 9, the precision of determined GDV data is basically the highest when the weight factor *k* equals to OWF.

Fig. 4 Geomagnetic vertical component Z

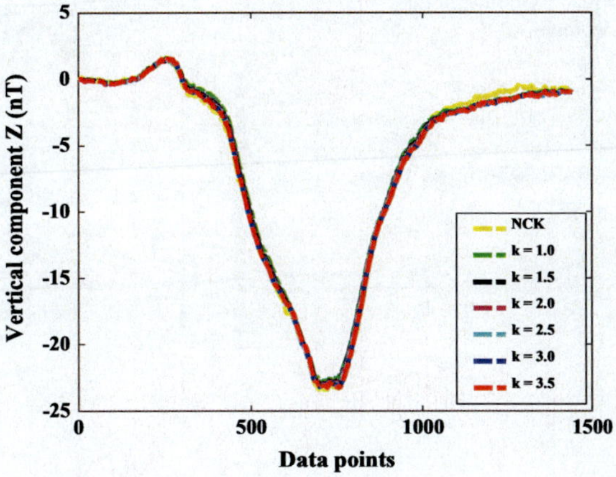

Fig. 5 Geomagnetic total intensity F

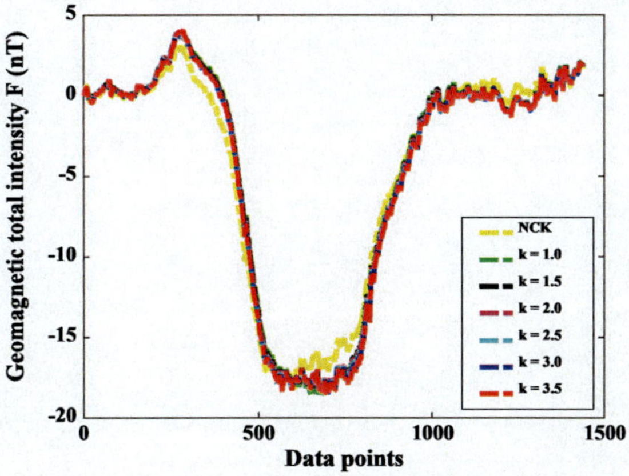

Fig. 6 Geomagnetic horizontal component H

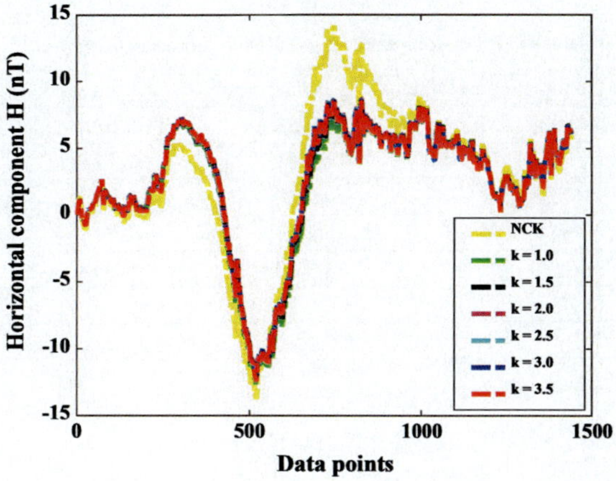

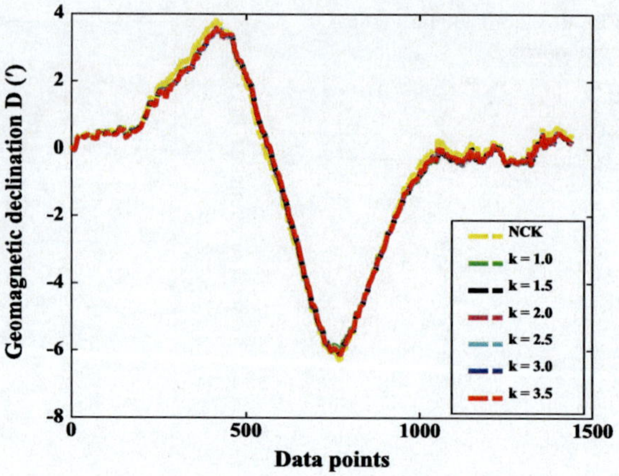

Fig. 7 Geomagnetic declination *D*

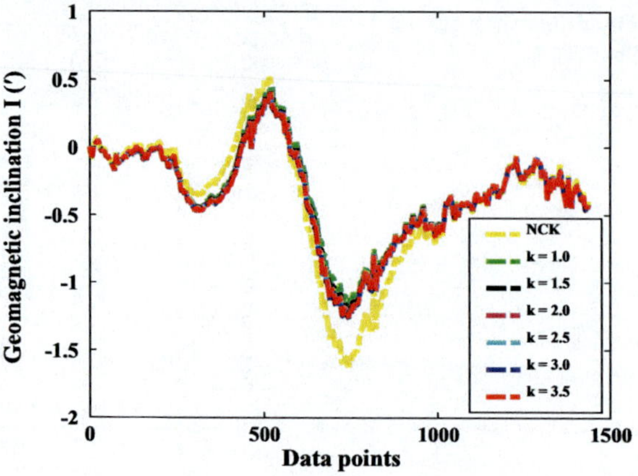

Fig. 8 Geomagnetic inclination *I*

Table 10 Precision statistics of RDW method (OWF)

Geomagnetic elements	Maximum	Minimum	Average	STD	RMSE
X (nT)	3.53	−5.60	−0.40	2.17	2.21
Y (nT)	2.41	−2.11	−0.43	0.92	1.01
Z (nT)	0.80	−0.85	−0.11	0.29	0.31
F (nT)	2.07	−1.77	−0.11	0.87	0.88
H (nT)	3.39	−5.54	−0.43	2.13	2.17
D (′)	0.39	−0.33	−0.05	0.16	0.17
I (′)	0.36	−0.21	0.02	0.14	0.14

Table 11 Correlation coefficients of RDW method

Weight factors	Geomagnetic elements						
	X	*Y*	*Z*	*F*	*H*	*D*	*I*
$k = 0.5$	0.91268	0.99735	0.99713	0.98956	0.88492	0.99727	0.96354
$k = 1.0$	0.93807	0.99750	0.99801	0.99146	0.91451	0.99754	0.97844
$k = 1.5$	0.95125	0.99757	0.99839	0.99244	0.93049	0.99767	0.98397
$k = 2.0$	0.95608	0.99755	0.99855	0.99286	0.93644	0.99768	0.98538
$k = 2.5$	0.95663	0.99745	0.99864	0.99299	0.93714	0.99759	0.98537
$k = 3.0$	0.95520	0.99729	0.99868	0.99297	0.93542	0.99744	0.98483
$k = 3.5$	0.95293	0.99710	0.99869	0.99288	0.93274	0.99726	0.98405
$k = \text{OWF}$	0.95675	0.99734	0.99869	0.99294	0.93728	0.99769	0.98544

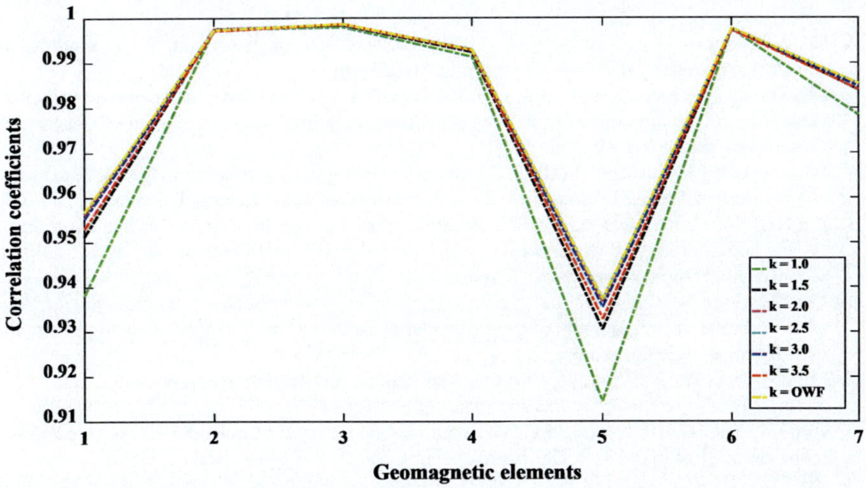

Fig. 9 Correlation coefficients between the determined GDV data and measurement data of NCK station when the weight factors are in different values

6 Conclusion

(1) When the weight factor is determined by correlation coefficients, the precision of determined GDV data are summarized as follows: the geomagnetic northern component *X* and horizontal component *H* are superior to 2.2 nT, the geomagnetic eastern component *Y* and geomagnetic total intensity *F* are superior to 1.0 nT, the geomagnetic vertical component *Z* is superior to 0.3 nT, the geomagnetic declination *D* and geomagnetic inclination *I* are superior to 10″, and these results can directly be used in diurnal variation correction of airborne scalar and vector geomagnetic measurement.

(2) The weight factor calculated iteratively by correlation coefficients is more practical, and has no application limitation in different testing areas. It is highly correlated between the calculation results and measurement data of the computational point, and it basically reflects the real physical attributes of the measurement data. Therefore, the proposed method in this paper is reliable and practicable.

Acknowledgements Thanks Intermagnet website for the geomagnetic observatory data.

References

1. Xiong SQ (2009) The present situation and development of airborne gravity and magnetic survey techniques in China. Progress Geophys 24(1):113–117
2. Chen B, Xiong SQ, Zhao BBM (2010) The preliminary study on flight altitude of aero-magnetic survey. Progress Geophys 25(3):957–961
3. Liu XG, Li YC, Xiao Y et al (2014) Study of optimal regularization parameter determination method in downward continuation of gravimetric and geomagnetic data. Acta Geodaetica et Cartographica Sinica 43(9):881–887
4. Liu XG, Li YC, Xiao Y et al (2015) Downward continuation of airborne magnetic data based on two iterative regularization methods in frequency domain. Geodesy Geodyn 6(1):34–40
5. Liu XG, Xiao Y, Guan B et al (2017) Precision evaluation of the flying experimentation data of the initial prototype of airborne vector geomagnetic measurement. In: 13th National security geophysical symposium, Tengchong
6. Liu XG, Guan B, Duan WC et al (2018a) Precision evaluation of flying experimentation data of the airborne vector geomagnetic measurement based on discrepancy of crossing points. In: 14th National security geophysical symposium, Zhangye
7. Liu XG, Sun ZM, Ji JF et al (2018b) Precision evaluation of flying experimentation data of the airborne vector geomagnetic measurement. EORSA2018, Xi'an
8. Guo JH, Xue DJ (1999) The application and method research of magnetic diurnal calibration based on multi-station. Acta Geoscientia Sinica 20(sup.):932–937
9. Bian G, Liu YC, Bian GL et al (2009) Research on computation method of multi-station diurnal variation correction in marine magnetic surveys. Chinese J Geophys 52(10):1284–1289
10. Gu CL, Zhang Y, Xu RG et al (2013) The method of geomagnetic diurnal correction to vector geomagnetic data by virtual diurnal variation station. Chinese J Geophys 56(3):834–841
11. Liu XG, Xu JL, Guan B et al (2020a) Bifactor weight determination method based on direction and distance in geomagnetic data assimilation. J Beijing Univ Aeronaut Astronaut 46(1):29–37
12. Liu XG, Xu JL, Zhang SQ et al (2020b) Bifactor weight determination method considering the influence of latitude and longitude in the calculation of diurnal variation of geomagnetic data. Geomatics and Information Science of Wuhan University. https://doi.org/10.13203/j.whugis20180454
13. Liu XG, Wu J, Wang K et al (2018) Space changing and correlation analysis of the geomagnetic field. Geomatics Sci Eng 38(1):1–7
14. Sheng Z, Xie SQ, Pan CY (1979) Probability and mathematical statistics. Higher Education Press, Beijing

15. Wu Y, Li YL, Hu QJ (1995) Application mathematical statistics. National University of Defense Technology Press, Changsha
16. Xin CJ, Li QH (2001) Brief introduction of geomagnetic observatory of intermagnet. Seismol Geomagn Obs Res 22(3):48–52

Optimal Rescheduling for OOS Mission Based on Squeaky-Wheel Optimization

Jing Yu, Dong Hao, and Xiaoqian Chen

Abstract On-Orbit Servicing (OOS) is the process of improving a space-based capability through a combination of in-orbit activities. With the successful completion of the OOS enabling technologies, the mission planning problem is issued, optimizing mission sequences and trajectories for Servicing Spacecraft (SSc) servicing multiple targets. During the mission process, even though the adequate preparations were done, some accidents may still occur, e.g. the user adds some unassigned targets in emergency, or one of the SScs malfunctions and the un-completed targets should be re-allocated. Then it is necessary to reschedule the existing mission scheme, regenerating a new mission scheme to deal with the new-arrival targets, minimizing the disturbances on the initial mission scheme, minimizing the total rendezvous cost increment, and at the same time designing the rendezvous trajectories. The OOS mission rescheduling problem for coplanar circular constellation is studied in this paper. The reschedule model is developed first. Then base on the Squeaky-Wheel Optimization framework, a general optimization framework for combinatorial optimization problems, the solution method is proposed. Numerical simulations are carried out to demonstrate the effectiveness of the model and rescheduling method.

Keywords On-orbit servicing · Squeaky-wheel optimization · Space mission · Mission planning · Mission rescheduling

J. Yu · D. Hao (✉)
China Aerodynamics Research and Development Center, Mianyang 621000, Sichuan Province, China

X. Chen
National Innovation Institute of Defense Technology, Chinese Academic of Military Science, Beijing, China

L. Wang et al. (eds.), *Proceedings of the 7th China High Resolution Earth Observation Conference (CHREOC 2020)*, Lecture Notes in Electrical Engineering 757, https://doi.org/10.1007/978-981-16-5735-1_2

1 Introduction

On-Orbit Servicing (OOS) is the process of improving a space-based capability through a combination of in-orbit activities which may include inspection; rendezvous and docking; and value-added modifications to a satellite's position, orientation, and operational status [1]. Typical OOS missions are satellite rescue, refueling, upgrading, assembly, repair operations, debris removal and so on. In the past decades, some of the most extraordinary successes in space exploration have emphasized the growing importance of on-orbit servicing [2].

With the successful completion of a series of Hubble Space Telescope repairs, as well as the assembly of the International Space Station, the automated rendezvous of Orbital Express [3], we can look forward with confidence to plan such a future that the Servicing Spacecraft (SSc) serve multiple target satellites in one mission, improving the economic returns of OOS missions [4, 5]. Then the mission planning problem is issued, which optimizes mission sequences and trajectories for SSc servicing multiple targets, aiming at reducing the mission cost, the time duration or some other merits.

So far, there have been number of reported work devoted in OOS mission planning. In the application of on-orbit refueling, Shen and Tsiotras [6], Zhang et al. [7], Bo and Feng [8], Zhou et al. [9, 10], Yu et al. [11] paid attention to the problem of one single SSc servicing multiple satellites; Ouyang et al. [12], Zhang et al. [13], Zhou et al. [14], Yu et al. [15, 16] focused on the mission scenario of multiple SScs servicing multiple satellites in parallel; Zhao et al. [17], Du et al. [18], and Dutta et al. [19–24] studied the OOS mission with cooperative maneuver; while Chen and Jing [25], and Tsiotras et al. [26] furtherly studied the OOS mission of mixed strategy. They presented different optimization models and solutions for various mission scenarios, considering different constraints, and mission merits. In the application of debris removal, Braun et al. [27], Yu et al. [28–30], Liu et al. [31], Madakat et al. [32], Missel and Mortari [33], and Cerf [34] made contributions on how to select the debris, how to evaluate the merits and how to provide a path optimization strategy for the whole mission of different orbits.

Despite the achievements mentioned above, there is a non-negligible pre-condition and assumption of these researches, that once the mission starts, there would be no change on mission targets and any other related OOS resources. As a matter of fact, even though the adequate preparations were done, some accidents may still occur during the mission process, e.g. the user adds some unassigned targets in emergency, or one of the SScs malfunctions and the un-completed targets should be re-allocated. Then it is necessary to adjust the existing mission scheme in short time. There are two ways to address this problem: (1) cancel the whole mission, and re-plan it according to the new requirements using the existing mission planning method; (2) based on the existing mission scheme, insert the new targets to the initial sequences according to some principles. We name the former one re-planning, and the latter one rescheduling.

So far, there were little reported researches focusing on the OOS mission rescheduling problem. One reported work is presented by Yu et al. [35], focusing on the rescheduling problem of debris removal in a coplanar GEO orbit, and the solution method is only suitable for GEO debris removal mission with special space distribution. However, there has been some reported work devoted in rescheduling of other missions, e.g. Earth Observation Rescheduling problem [36]. Squeaky-Wheel Optimization (SWO) is one of the conventional rescheduling methods.

SWO is firstly developed by Joslin and Clements [37]. It is a general optimization framework for combinatorial optimization problems. It is fast, allowing for rescheduling without significant changes to the initial schedule. It has already been used for several optimization areas, such as coloring problem, ground station re-scheduling problem, and satellites mission re-scheduling problem etc. [37–39].

In this paper, the OOS mission rescheduling problem for coplanar circular constellation is studied. The goal of the rescheduling problem is to regenerate a new mission scheme quickly and robustly, dealing with the variation of the targets and resources, minimizing the disturbances on the initial mission scheme, and minimizing the whole mission cost.

This paper is organized as follows. Section 2 details the mission scenario studied in this paper, as well as analyzing the potential accidents. Section 3 presents the optimization model. Section 4 introduces the SWO framework, and then gives the solution method based on this framework. Finally, in Section 5, numerical examples are carried out to demonstrate the effectiveness of the model and solution method, and different simulation cases are compared and discussed.

2 Mission Description

Before an OOS mission starting, numerous preparations should be done, involving mission planning, resource assignment, as well as presenting schemes for the operations of all control centers, communication infrastructures etc. Afterwards, the SSc would be launched, and serve the targets in order, repeatedly rendezvousing with them, until the whole mission is completed.

During the whole process, some accidents may occur, e.g. the user adds new targets in emergency (which is named target variation), or some other un-completed targets from malfunction SScs require re-allocation (which is named resource variation). Then, mission rescheduling is required. The goal of the rescheduling problem is to regenerate a new mission scheme quickly and robustly, dealing with the variation of the targets and resources, minimizing the disturbances on the initial mission scheme, as well as minimizing the mission cost. Fig. 1 shows a typical rescheduling mission.

Target variation and resource variation are two main accidents considered in this paper. Whatever accidents occur, the main requirement is to insert all new targets to the existing mission scheme with lowest possible cost. Definitely, the whole rescheduling problem could be transformed into a new mission planning problem,

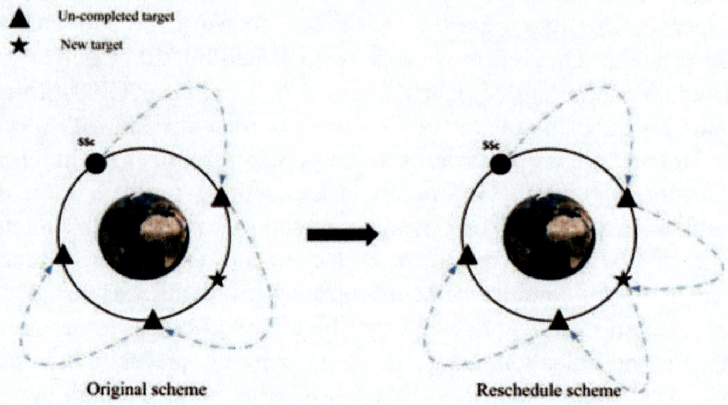

Fig. 1 Mission scenario

reallocating all un-completed targets using existing mission planning algorithms. Most mission planning methods make use of intelligent algorithms, which are of random nature. So after re-planning, the mission sequences may be changed a lot, compared with the initial ones. As mentioned above, before one OOS mission starting, schemes for the operations of all control centers, communication infrastructures etc. should be prepared. If the OOS mission sequence is modified or even redesigned, the corresponding operations for other infrastructures should also be adjusted or re-arranged. The cost is not high for changing the OOS mission sequences, but the adjustment cost for other infrastructures would be incredible large. So, there is an important demand in mission re-scheduling that maintain the initial mission scheme as much as possible. Taking Earth Observation Rescheduling problem as an instance, in [36], it is defined that all new targets should be inserted under the premise that all existing targets in the initial mission scheme should be satisfied and unchanged.

Despite the increasingly mature mission planning technologies, it is hard to demand the intelligent algorithms presenting optimized results meeting large number of strict constraints. Mission re-planning methods are not suitable for mission rescheduling.

Summing up, the OOS rescheduling problem studied in this paper could be stated as: the SScs are conducting OOS mission on schedule, and the initial mission sequences are known. Then, some accidents occur, and new/un-completed targets are required to insert to the existing mission scheme. A new mission sequence with new targets and un-completed targets is required. The goal of the rescheduling is to work out such a new scheme for the SSc, minimizing the disturbances on the initial mission scheme, minimizing the total rendezvous cost, and at the same time designing the rendezvous trajectories.

3 Mathematical Description

Key factors of the optimization problem, i.e. objective functions, constraints and design variables are detailed in this section.

3.1 Design Variables

Suppose that the existing mission sequence is $Q_0 = [q_1, q_2, \ldots q_n]$, which indicates that n targets have been assigned to the SSc. Starting from the initial position, the SSc serves targets q_1 to q_n in order. And after q_n is completed, the SSc should return to its initial position. That is to say the actual mission sequence is $[q_0, q_1, q_2, \ldots q_n, q_0]$. Corresponding to each rendezvous, the transfer path is $P = \left[p_{0,1}, p_{1,2}, \ldots, p_{n-1,n}, p_{n,0}\right]$, in which $p_{i,j} = \left[tf_{i,j}, dv1_{i,j}, dv2_{i,j}\right]$. It indicates that all trajectories in the OOS mission are carried out by two-impulse maneuver.

Assume that the new targets arrive at the time that the SSc is serving the i-th target. Then before re-scheduling, the existing/initial mission scheme is $Q = [q_i, q_{i+1}, \ldots, q_n, q_0]$, and $P = \left[p_{i,i+1}, \ldots, p_{n-1,n}, p_{n,0}\right]$. Denote the new targets as $INS = \{ins_1, ins_2, \ldots, ins_m\}$. After rescheduling the new mission is assumed to be Q_{new}, and $\{Q_{new}\} = INS \cup Q$.

Q_{new}, as well as the corresponding transfer path P_{new} is the design variable. Q_{new} indicates the new mission sequence after re-scheduling, while P_{new} gives the transfer path of each rendezvous.

3.2 Constraints

C_1: The targets order in the initial mission sequence should not be changed.
C_2: The targets in the initial mission sequence should not be canceled.
C_3: The servicing time for the old targets should not be changed. After rendezvous, complex operations should be done for serving, including attitude adjustment, ground/space monitor, and some other in-orbit activities. Once the servicing time is determined, the corresponding scheme for other infrastructures should be planned. In order to minimize the disturbance on the other infrastructures scheme, the designed servicing time should not be changed.
C_4: Suppose the minimum servicing duration for each service is defined to be t_{oos}, the mission sequence is $Q = [q_0, q_1, q_2, \ldots, q_n, q_0]$, the transfer path is $P = \left[p_{0,1}, p_{1,2}, \ldots, p_{n-1,n}, p_{n,0}\right]$, and the latest time of the SSc arriving at each target is $\left[t_0, t_1, \ldots, t_n, t_f\right]$ (t_0 and t_f are mission starting and ending time, respectively).

Then, the time distribution should satisfy the constraint: $t_i + t_{oos} + tf_{i,i+1} \leq t_{i+1}$ $(0 < i \leq n)$.
C_5: It is defined that between any two neighboring targets in the initial mission sequence, only one new target could be inserted.

Here gives an example to further clarify these constraints. Suppose the initial mission sequence is $Q = [q_0, q_1, q_2, q_0]$, the corresponding transfer path is $P = [p_{0,1}, p_{1,2}, p_{2,0}]$, and the new target is q_x. Due to constraints C_1 and C_2, the rescheduled mission sequence can only be $Q_1 = [q_0, q_x, q_1, q_2, q_0]$, $Q_2 = [q_0, q_1, q_x, q_2, q_0]$, or $Q_3 = [q_0, q_1, q_2, q_x, q_0]$. Taking Q_2 as an instance, due to constraints C_4 and C_3, the time distribution should satisfy $t_1 + t_{oos} + tf_{1,x} + t_{oos} + tf_{x,2} \leq t_2$, while t_1 and t_2 are the same as the value in the initial mission scheme.

3.3 *Objective Functions*

In this paper, two-impulse maneuver including Phasing Maneuver and Lambert Transfer, is employed for each rendezvous, and the rendezvous cost is evaluated by the velocity increment.

Suppose the initial mission sequence is Q, and the re-scheduling mission sequence is Q_{new}. Then the objective function could be stated as $\min(\sum \Delta v(Q_{new}) - \sum \Delta v(Q))$.

3.4 *Optimization Model*

Summing up, the re-scheduling problem studied in this paper can be stated as:

$$\mathit{find} \quad Q_{new}, P_{new} \tag{1}$$

$$\text{minimize} \quad \Delta Jv = \left(\sum \Delta v(Q_{new}) - \sum \Delta v(Q)\right) \tag{2}$$

$$s.t. \quad C_1 \cup C_2 \cup C_3 \cup C_4 \tag{3}$$

4 Solution Based on SWO Algorithm

4.1 *Squeaky-Wheel Optimization*

The core of SWO is a Constructor/Analyzer/Prioritizer cycle, as shown in Fig. 2. Each task is analyzed by analyzer, estimating its blame value. According to the

Fig. 2 SWO cycle

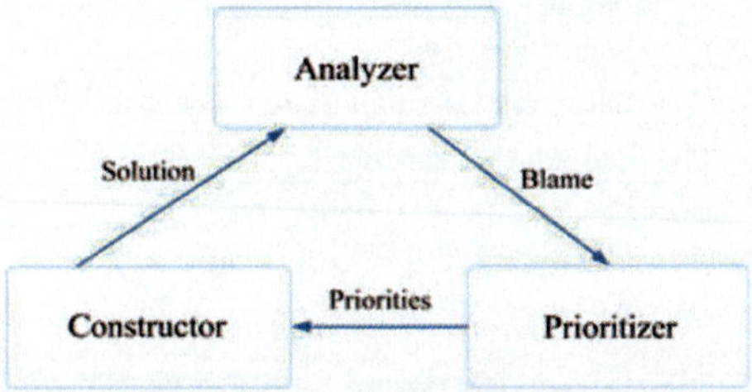

blame value, tasks to be scheduled are ordered in a priority queue. Then each task would be assigned in order, and a solution is generated by the constructor, using a greedy algorithm. This solution is then analyzed by analyzer again. This cycle repeats until a termination condition occurs. In essence, SWO finds good quality solutions quickly by searching two spaces simultaneously, the traditional solution space and the priority space.

4.2 SWO Design

In the problem studied in this paper, targets are divided into two classes: the assigned ones and the unassigned ones. At the beginning, the order of the assigned targets is the same as the initial schedule, while the unassigned targets are the new targets that required to be inserted, and they are in disorder. In SWO process, analyzer firstly analyzes the unassigned targets, presenting their blame value. Then, the unassigned targets are sorted in prioritizer. And in constructor, it always assigns the unassigned targets according to the order determined by the prioritizer. After iterations, a termination condition occurs and a new mission scheme is generated (see Fig. 3).

Different from the basic SWO operator, in this paper, analyzer and prioritizer only work on unassigned targets, while constructor works on both assigned and unassigned targets. Here details each operator.

(1) Analyzer

The goal of the analyzer is to estimate the minimum rendezvous cost increment of inserting the target to the existing mission sequence. For example, if the existing mission sequence is $Q = [q_0, q_1, q_2, q_0]$, and the new target is q_x, then the candidate mission sequences are $Q_1 = [q_0, q_x, q_1, q_2, q_0]$, $Q_2 = [q_0, q_1, q_x, q_2, q_0]$, and $Q_3 = [q_0, q_1, q_2, q_x, q_0]$ (Fig. 4). The blame value of q_x is

$$blame(q_x) = \min_{i=1,2,3}([\Delta v(Q_i) - \Delta v(Q)]) \tag{4}$$

```
Algorithm: SWO
Input: the existing mission sequence Q, and the new targets set INS
Output: the recheduling plan Q_new
-----------------------------------------------------------------
Set Assigned = Q, and UnAssigned = INS
(1) Blame = Analyzer(UnAssigned)
(2) Prioritizer(UnAssigned, Blame)
(3) (Assigned, UnAssigned) = Constructor(Assigned, UnAssigned)
If meet the termination condition then
    Exit
else
    goto (1)
end if
```

Fig. 3 Pseudo code for SWO algorithm

```
Algorithm: SWO_Analyzer
Input: the existing mission sequence Q, and the unassigned targets set
UnAssigned
Output: the Blame value and Rec(Q)
-----------------------------------------------------------------
For each target q in UnAssigned
    For each possible inserting location in Q
        Path planning and time distribution (see section 4.2);
    Update the optimal rendezvous cost;
    End for
Record Rec(q) containing the rendezvous cost and the transferring path for each
possible inserting location.
Record the optimal cost as Blame, the inserting location and the corresponding
transferring path for q.
End for
```

Fig. 4 Pseudo code for SWO_Analyzer

(2) Prioritizer

In Prioritizer, all unassigned targets are sorted in ascending order of blame value. Suppose that *Sort* is the sorted queue (see Fig. 5).

```
Algorithm: SWO_Prioritizer
Input: the Blame value, Rec(Q), and the unassigned targets set UnAssigned
Output: the sorted queue  Sort
--------------------------------------------------------------------------
For each target  q  in UnAssigned
    Sort the inserting location in ascending order of rendezvous cost in Rec(q);
End for
Sort the targets in UnAssigned in ascending order of blame value and get  Sort
```

Fig. 5 Pseudo code for SWO_Prioritizer

```
Algorithm: SWO_ Constructor
Input: the existing mission sequence  Q , the sorted queue  Sort  and the sorted
Rec(Q).
Output: the updated mission plan  Q_new  and the unassigned targets set
UnAssigned_new
--------------------------------------------------------------------------
Set  Q_new = Q, and  UnAssigned_new = ∅
For each target  q  in  Sort
    For each location in Rec(q)
        If it is possible to insert, insert and  Q_new = Q ∪ q
        Else if there were an inserted new target  q'  at the same location,
    check if it is possible to swap  q  with the inserted target  q' , making no
    increment on the objective. If possible, swap, update  Q_new  and
    UnAssigned_new = UnAssigned_new ∪ q'
        End if
    End for
    If there are no inserting and no swapping,
        UnAssigned_new = UnAssigned_new ∪ q
    End if
End for
```

Fig. 6 Pseudo code for SWO_Constructor

(3) Constructor

The goal of the constructor is to regenerate a mission scheme without significant change on the initial mission sequence. For each target in *Sort*, the constructor tries to insert it to the location with minimum blame value. If it is possible, update the mission sequence and record the corresponding information. If it is impossible, it means that some other target has already been inserted to this location. Then check that if it is preferable to swap these two targets. If preferable, swap (insert the new one to the location, cancel the inserted target and put it to the unassigned target set), and if not, make no change, and the target is put into the unassigned target set (see Fig. 6).

4.3 Path Planning and Time Distribution

The mission re-scheduling problem is combined by an outer-loop and an inner-loop optimal problem. The outer one determines the optimal mission sequence, while the inner one finds the optimal rendezvous trajectory with minimum cost. In the re-scheduling problem studied in this paper, the SWO framework drives the outer-loop, while the constructor drives the inner-loop for given mission sequence. These two loops interact with each other. To determine the new sequence, the cost of the modified rendezvous should be determined first. In this section, the path planning and the time distribution methods for inner-loop are detailed.

(1) Path planning

When the location of the chaser and the target is given, e.g. (q_i, q_{i+1}), as well as the permitted maximum flight duration tf_m, the path planning problem could be stated as:

$$\begin{array}{ll} find & [tf, dv_1, dv_2] \\ \min & |dv_1| + |dv_2| \\ s.t. & tf \le tf_m \end{array} \tag{5}$$

In this paper, two-impulse maneuver including Phasing Maneuver and Lambert Transfer, is employed for each rendezvous, and the rendezvous cost is evaluated by the velocity increment. The resulting transfer path can be stated as $[tf, dv_1, dv_2]$, in which tf denotes the transferring duration, dv_1 describes the velocity change of the first maneuver, while dv_2 describes the second.

Phasing Maneuver helps the Chaser transferring from one point to another in a circular orbit, with the velocity changes occurring at the same point in the transfer orbit. The following formulations are used to calculate the velocity increment Δv_p.

$$\Delta v_p = \sqrt{\mu}\left|\sqrt{\frac{2}{r} - \frac{1}{a}} - \sqrt{\frac{1}{r}}\right| \tag{6}$$

$$a = r\left(\frac{2\pi n_t + \Delta\theta}{2\pi n_s}\right)^{2/3} \tag{7}$$

where r is the orbit radius of the satellite, a is the semi-major axis of the transfer orbit, n_t and n_s is the number of revolutions of the target satellite and SSc during phasing, $\Delta\theta$ is the phase angle which is measured from the target to the SSc with the direction of the target's motion defined as the positive direction, and $\Delta\theta \in [-\pi, \pi]$.

Lambert Transfer is the problem of determining the transfer orbit given a starting position $R1$, an arrival position $R2$, and a transfer duration tf. Phasing Maneuver is the exceptional case of lambert transfer.

In this paper, only spacecrafts in coplanar circular constellation are considered. It has been shown in [6] that, for any two satellites on a circular orbit, the cost for Lambert Transfer depends on the initial separation angle and the given time-of-flight. That is to say, as long as the initial separation angle between the chaser and the target is determined, the rendezvous cost for any flight duration could be calculated, having nothing to do with the beginning and the ending states of the chaser and the target. Typically, for GEO satellites with a particular initial separation angle of 30°, the plot of the cost as a function of the time-of-flight is shown in Fig. 7.

In Fig. 7, the blue dashed line describes the results of Multi-revolution Lambert Transfer, while the red solid line gives the minimum-fuel of the blue line, considering coasting. At point P, for blue dashed line the velocity increment is 364.08 m/s, and the transferring time is 1.1 *Tr*, but for red solid line, the velocity increment is 186.26 m/s, and the transferring time is 0.9 *Tr* not 1.1 *Tr*, since some time is used to coast in orbit, in order to achieve the minimum-fuel cost. *Tr* here denotes the orbital period of GEO.

In this paper, each rendezvous path in the initial scheme is generated by Phasing Maneuver, while in the re-scheduling scheme, Lambert Transfer with coasting is employed.

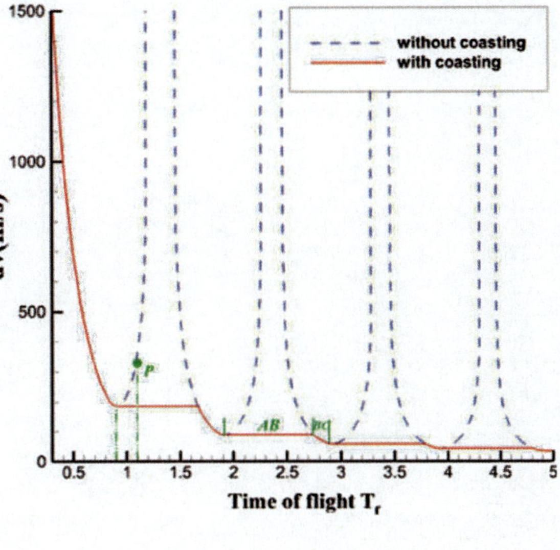

Fig. 7 The plot of the transfer cost as a function of the time-of-flight

(2) Time distribution

Before path planning, the permitted maximum flight duration tf_m should be pre-defined. And this is another optimization problem for time distribution. Suppose the mission sequence is given as (q_i, q_x, q_j), where q_x is the inserted target and (q_i, q_j) is the initial mission sequence. The transferring duration $tf_{i,j}$ from q_i to q_j can be obtained from the initial transfer path set P (see Sect. 3.1). Then the time distribution problem can be stated as:

$$\begin{array}{ll} find & (tf_{i,x}, tf_{x,j}) \\ \min & \Delta v(q_i, q_x) + \Delta v(q_x, q_j) \\ s.t. & tf_{i,x} + tf_{x,j} + toos \le tf_{i,j} \end{array} \quad (8)$$

When orbital coasting is considered, the optimal rendezvous cost curve consists of several constant segments and smoothing curves. As depicted in Fig. 7, AB is a constant segment while BC is a smoothing curve segment. In this paper, to simplify the question, we change the rendezvous cost function line as Fig. 8, in which the cost line consists of constant segments except for the first smoothing curve.

Disregarding the first curve, the rendezvous cost function only consists of constant segments. Then the time distribution problem can be translated into the combinatoric search problem, matching constant segments, and the solution is detailed in Fig. 9.

It would spend much computational cost for path planning in time distribution. In this paper, only coplanar circular orbits are taken into consideration, and the rendezvous cost can be determined only by the initial phase angle, having nothing to do with the orbital state before maneuver. So, before optimization, all rendezvous

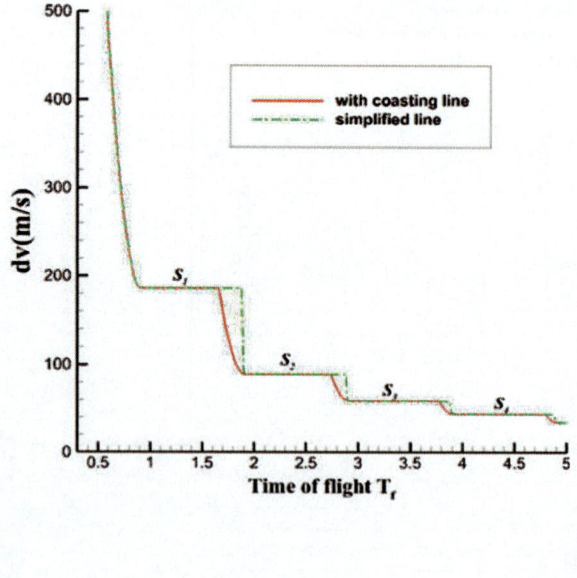

Fig. 8 Simplified plot of the transfer cost as a function of the time-of-flight

```
Algorithm: Time distribution
Input: related information about mission sequence (q_i, q_x, q_j)
Output: (tf_{i,x}, tf_{x,j})
----------------------------------------------------------------
Path planning for (q_i, q_x) and obtain the rendezvous cost function line L_{i,x}
Path planning for (q_x, q_j) and obtain the rendezvous cost function line L_{x,j}
Seg(L_{i,x}) = {S_1,...,S_n} is the segment set for L_{i,x};
Seg(L_{x,j}) = {S_1,...,S_m} is the segment set for L_{x,j};
For each segment tmpS_i in Seg(L_{i,x})
    For each segment tmpS_j in Seg(L_{x,j})
        If min t(tmpS_i) + min t(tmpS_j) + t_OOS <= tf_{i,j}
            Record the combination and the corresponding cost
        End if
    End for
End for
Record the combination with the optimal cost, and return the corresponding
(tf_{i,x}, tf_{x,j})
```

Fig. 9 Pseudo code for time distribution

cost v.s. different flight duration for any two targets can be pre-calculated and stored. In the process of rescheduling, the rendezvous cost can be called at any time, reducing the computational cost.

5 Numerical Simulation

Suppose that all objects initially run on a coplanar circular GEO orbit with the altitude of 35,798 km. The path planning for the initial scheme is generated by Phasing Maneuver, and the relevant constraints are given in Table 1.

Three methods are employed to carry out the simulations:

SWO: the method developed in this paper.

TSA: it has been pointed out in [6] that for servicing multiple satellites in a circular constellation, the best rendezvous sequence is always among the group of sequences that assume the minimum TSA. TSA is the abbreviation for Total Sweep Angle, and it gives the sum of the sweep angles of all the rendezvous segments in the given mission sequence. Sweep angle describes the smaller angle along the circular arc between the SSc and the satellite to be serviced. That is to say the

Table 1 Constraints for phasing maneuver

Parameters	n_t	n_s	t_{OOS}
Value	≤ 10	≤ 10	1.5 day

optimal sequence can be determined independently by the space distribution of the targets and the SSc. In TSA method, the mission sequence is re-scheduled by the principle that the TSA of the new scheme is the smallest. After the new sequence is obtained, the corresponding trajectories would be modified.
SA: Simulated Annealing algorithm, an intelligent algorithm.

5.1 *Simulation for Inserting Single Target*

In this section, only one new target will be inserted. The initial slots of the objects are randomly given in Table 2, where 0 denotes the SSc, number 1–5 describe the targets and "new" presents the new target. Table 3 gives the initial mission scheme after optimization, while Table 6 gives the initial mission sequence by random generation. Suppose that the new target arrives at the time that the SSc is servicing the second target in mission sequence. Then before re-scheduling, the existing mission schemes are (4, 3, 2, 1, 0) and (3, 2, 1, 4, 0), respectively.

It can be seen from Tables 4, 5, 7, 8, the optimization results of SWO and SA are better than those of TSA. There is a significant premise of TSA algorithm in [6], but the authors have not pointed out explicitly. That is all transfer path should be

Table 2 Space distribution for all objects

Objects	0	1	2	3	4	5	new
Slot (°)	0	30	100	178	240	330	50

Table 3 The initial mission scheme after optimization

Variable	Value	Description
Sequence	(0, 5, **4**, **3**, **2**, **1**, **0**)	Mission sequence
Jv (m/s)	223.01	Total rendezvous cost
Tseq (Tr)	(9.08, 9.25, 9.17, 9.22, 9.19, 9.08)	Time distribution for each rendezvous segment
DVseq (m/s)	(18.80, 55.39, 38.48, 48.18, 43.34, 18.80)	Transfer cost for each rendezvous segment

Table 4 The rescheduling results of SWO and SA for optimized initial scheme

Variables	Value	Description
New scheme	(4, 3, new, 2, 1, 0)	Mission sequence
ΔJv (m/s)	85.96	The growth of the velocity increments
Tseq (Tr)	(9.17, **3.8**, **3**, 9.19, 9.08)	Time distribution for each rendezvous segment
DVseq (m/s)	(38.48, **107.10**, **27.04**, 43.34, 18.80)	Transfer cost for each rendezvous segment

Table 5 The rescheduling results of TSA for optimized initial scheme

Variables	Value	Description
New scheme	(4, 3, 2, new, 1, 0)	Mission sequence
ΔJv (m/s)	177.71	The growth of the velocity increments
ΔTSA (°)	0	The growth of the TSA
Tseq (Tr)	(9.17, 9.22, **2**, **5.5**, 9.08)	Time distribution for each rendezvous segment
DVseq (m/s)	(38.48, 48.18, **42.40**, **178.65**, 18.80)	Transfer cost for each rendezvous segment

Table 6 The initial mission scheme by random generation

Variables	Value	Description
Sequence	(0, 5, **3**, **2**, **1**, **4**, **0**)	Mission sequence
Jv (m/s)	371.72	Total rendezvous cost
Tseq (Tr)	(9.08, 9.42, 9.22, 9.19, 9.42, 8.67)	Time distribution for each rendezvous segment
DVseq (m/s)	(18.80, 91.85, 48.18, 43.3440, 90.70, 78.83)	Transfer cost for each rendezvous segment

Table 7 The rescheduling results of SWO and SA for random initial scheme

Variables	Value	Description
Sequence	(3, new, 2, 1, 4, 0)	Mission sequence
ΔJv (m/s)	85.96	The growth of the velocity increments
Tseq (Tr)	(**3.80**, **3**, 9.19, 9.42, 8.67)	Time distribution for each rendezvous segment
DVseq (m/s)	(**107.10**, **27.04**, 90.70, 78.83)	Transfer cost for each rendezvous segment

Table 8 The rescheduling results of TSA for random initial scheme

Variables	Value	Description
Sequence	(3, 2, new, 1, 4, 0)	Mission sequence
ΔJv (m/s)	177.71	The growth of the velocity increments
ΔTSA (°)	0	The growth of the TSA
Tseq (Tr)	(9.22, **2.0**, **5.5**, 9.42, 8.67)	Time distribution for each rendezvous segment
DVseq (m/s)	(48.18, **42.40**, **178.65**, 90.70, 78.83)	Transfer cost for each rendezvous segment

globally optimized. Only when globally optimal the time distribution and path planning are, the TSA principle can hold true. So, in this paper, when the time distribution is constrained, the conclusion derived by [6] is no long of use. Taking

Table 6 as an example, the ΔJv for inserting new target to different location of the initial scheme are separately 85.96 m/s, 177.71 m/s, 435.98 m/s, and 307.10 m/s. It can be seen that though TSA algorithm would not necessarily give the optimal result, it still can provide a sub-optimal one.

5.2 *Simulation for Inserting Multiple Targets*

In this section, algorithms SWO and SA would be compared. The optimization results of 12 cases are compared. The number of the initial targets and the new targets for each case is given in Table 9. The initial locations for all targets are randomly given. The initial true anomaly of the SSc is always 0 degree. And the initial mission sequences are always randomly generated, while the transfer paths are optimized by Eqs. (6) and (7), considering the constraints given in Table 1. It is assumed that for all cases, the re-scheduling requirements are issued at the time the SSc is servicing the first target in the initial mission sequence. The comparisons of the optimization results and time consumption are depicted in Figs. 10 and 11.

Some conclusions can be conducted from Fig. 10. (1) When dealing with small number of target (cases 1–6), the optimization results of SWO and SA are almost the same; To some degree, it demonstrates the efficiency of the SWO algorithm presented in this paper; (2) In some cases (e.g. cases 3, 5 and 6), the optimization results of SA are better than those of SWO, and this is mainly due to the iteration

Table 9 Configuration for all simulation cases

Case	1	2	3	4	5–6	7–8	9–10		11–12
Initial	5	10	10	20	20	50	50	50	
New	3	3	6	3	10	10	20	30	

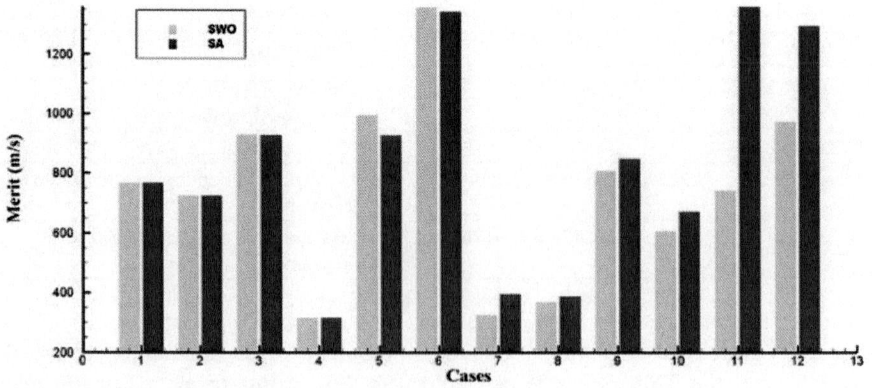

Fig. 10 Comparison for optimization results

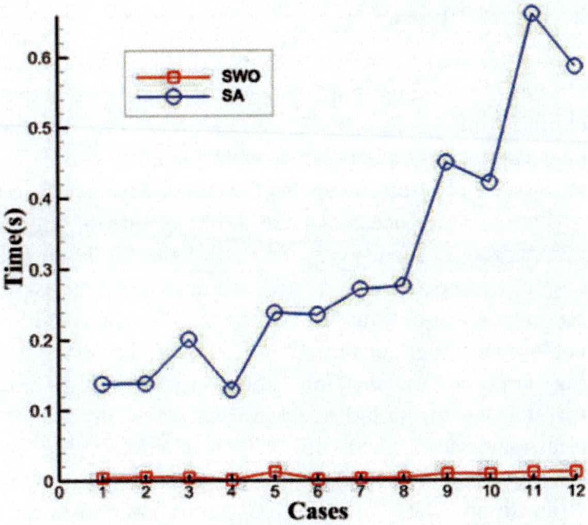

Fig. 11 Comparison for time consumption

mechanism of SWO. It always greedily tries to insert the new target to the current-optimal location, and this may miss the global-optimal solution. (3) When the number of new target is fixed, with the growth of the initial target number, the merit value is usually decreasing (e.g. cases 1–2 and cases 5–8). As the increment of the initial target number, more locations are available for selection, and then more chances are presented to the new targets to find a location with better performance. (4) When dealing with large number of target (cases 7–12), the optimization results of SWO are better than those of SA. With the increment of the target number, the search space is growing and the performance of the SA is decreasing. As is known, one of the shortcomings of the intelligent algorithms is the weakness at dealing with the problems with large number of design variables.

Figure 11 illustrates that both algorithms could rapidly obtain the optimization results, and the time consumed by SWO is shorter than that of SA. In the whole process, the most time-consuming step is the Lambert Transfer in path planning. Due to the characteristics of the problem studied in this paper, the Lambert Transfer for different initial separation angles and different flight duration could be computed and stored beforehand. Then in the optimization cycle, the time consumed by path planning and time distribution is largely reduced. So the time difference between SA and SWO is not apparent. Additionally, SWO is easier to implement compared with SA.

To sum up, the SWO presented in this paper is effective and efficient in dealing with the re-scheduling problem studied in this paper, especially for the cases with numerous targets.

6 Conclusions

On-Orbit Servicing is the process of improving a space-based capability through a combination of in-orbit activities. With the successful completion of the OOS enabling technologies, the mission planning problem is issued, optimizing mission sequences and trajectories for Servicing Spacecraft (SSc) servicing multiple targets.

During the process of OOS, some accidents may occur, e.g. the user adds some new targets in emergency, or one of the SScs malfunctions and the un-completed targets should be re-allocated. Then it is necessary to adjust the existing mission scheme in short time. In this paper, the OOS mission rescheduling problem for coplanar circular constellation is studied. The goal of the rescheduling problem is to regenerate a new mission scheme quickly and robustly, dealing with accidents, maintaining the initial mission scheme as much as possible, and at the same time minimizing the mission cost. The reschedule model is developed first. Then base on the Squeaky-Wheel Optimization framework, the solution method is proposed. Algorithms SWO, SA and TSA are compared in simulations. It is concluded that (1) TSA algorithm would not necessarily give the optimal result, but it still can provide a sub-optimal one; (2) when dealing with small number of target, the optimization results of SWO and SA are almost the same; (3) when dealing with large number of target, the optimization results of SWO are better than those of SA; (4) the time difference between SA and SWO is not apparent due to the pre-processing in path planning. Additionally, SWO is easier to implement compared with SA. Summing up, the SWO presented in this paper is effective and efficient in dealing with the re-scheduling problem studied in this paper, especially for the cases with numerous targets.

Acknowledgements This work was supported by National Natural Science Foundation of China under Grant No. 71601179.

References

1. Richards MG (2006) On-orbit serviceability of space system architectures. Massachusetts Institute of Technology
2. On-orbit satellite servicing study, NASA satellite servicing project report 0511, 2010
3. Friend RB (2008) Orbital express program summary and mission overview. Proc SPIE. https://doi.org/10.1117/12.783792
4. Yao W et al (2013) On-orbit servicing system assessment and optimization methods based on lifecycle simulation under mixed aleatory and epistemic uncertainties. Acta Astronaut 87:107–126
5. Chen X, Yu J (2017) Optimal mission planning of GEO on-orbit refueling in mixed strategy. Acta Astronaut 133:63–72
6. Shen H, Tsiotras P (2002) Optimal scheduling for servicing multiple satellites in a circular constellation. In: AIAA/AAS Astrodynamics specialists conference and exhibit, (Monterey, CA), Aug 2002

7. Zhang J, Luo YZ, Tang GJ (2012) Hybrid planning for LEO long-duration multi-spacecraft rendezvous mission. Sci China Tech Sci 55:233–243
8. Bo X, Feng Q (2011) Research on constellation refueling based on formation flying. Acta Astronaut 68:1987–1995
9. Zhou Y et al (2017) Multi-objective planning of a multiple geostationary spacecraft refueling mission. Eng Optim 49(3):531–548
10. Zhou Y et al (2015) Mission planning optimization for multiple geosynchronous satellites refueling. Adv Space Res 56:2612–2625
11. Yu J, Yu Y-G, Huang J-T, Chen X-Q, Liu H-Y (2017) Optimal scheduling of GEO on-orbit refuelling with uncertain object satellites. In: 2017 International conference on mechanical, material and aerospace engineering, Beijing, China
12. Ouyang Q, Yao W, Chen XQ (2010) Mission programming of on-orbit re-fueling for geosynchronous satellites (in Chinese). J Astronaut 31:2629–2634
13. Zhang J et al (2014) Multispacecraft refueling optimization considering the J2 perturbation and window constraints. J Guid Control Dyn 37(1):111–122
14. Zhou Y et al (2015) Optimal scheduling of multiple geosynchronous satellites refueling based on a hybrid particle swarm optimizer. Aerosp Sci Technol 47:125–134
15. Yu J, Yu Y-G, Hao D, Chen X-Q, Liu H-Y (in press) Biobjective mission planning for geosynchronous satellites on-orbit refueling. Proc IMechE Part G: Journal of Aerospace Engineering. https://doi.org/10.1177/0954410017740383
16. Yu J, Ouyang Q, Chen Z-Q, Chen, L-H (2012) Optimal peer-to-peer maneuvers for refueling satellites in circular constellations. In: 2012 International conference on spacecraft structures, materials and mechanical testing, Xiamen, China, Dec 2012
17. Zhao Z et al (2016) LEO cooperative multi-spacecraft refueling mission optimization considering J2 perturbation and target's surplus propellant constraint. Adv Space Res 59:252–262
18. Du B et al (2015) Optimal scheduling of multi-spacecraft refueling based on cooperative maneuver. Adv Space Res 55:2808–2819
19. Dutta A (2009) Optimal cooperative and non-cooperative peer-to-peer maneuvers for refueling satellites in circular constellation. Georgia Institute of Technology
20. Dutta A, Tsiotras P (2009) A cooperative egalitarian peer-to-peer strategy for refueling satellites in circular constellations. Am Inst Aeronaut Astronaut 134:103–119
21. Dutta A, Tsiotras P (2007) A greedy random adaptive search procedure for optimal scheduling of P2P satellite refueling. Adv Astronaut Sci 127:749–764
22. Dutta A, Tsiotras P (2008) An egalitarian peer-to-peer satellite refueling strategy. J Spacecr Rocket 45(3):608–618
23. Dutta A, Tsiotras P (2010) Network flow formulation for cooperative peer-to-peer refueling strategies. J Guid Control Dyn 33(5):1539–1549
24. Dutta A, Arora N, Russell RP (2012) Peer-to-peer refueling strategy using low-thrust propulsion. J Spacecr Rocket 49(5):944–954
25. Chen X-Q, Jing Y (2017) Optimal mission planning of GEO on-orbit refueling in mixed strategy. Acta Astronaut 133:63–72
26. Tsiotras P, Nailly A (2005) Comparison between peer-to-peer and single spacecraft refueling strategies for spacecraft in circular orbits, Info tech at Aerospace Conference, AIAA Paper 05-7115 Sept 2005
27. Braun V, Pken AL, Flegel S et al (2013) Active debris removal of multiple priority targets. Adv Space Res 51:1638–1648
28. Yu J, Chen XQ, Chen LH, Hao D (2014) Optimal scheduling of GEO debris removing based on hybrid optimal control theory. Acta Astronaut 93:400–409
29. Jing Y, Chen X-Q, Chen L-H (2014) Biobjective planning of GEO debris removal mission with multiple servicing spacecrafts. Acta Astronaut 105(1):311–320
30. Jing Y, Chen X-Q, Chen L-H (2015) Optimal planning of LEO active debris removal based on hybrid optimal control theory. Adv Space Res 55(11):2628–2640

31. Liu Y et al (2017) Multi-objective optimal preliminary planning of multi-debris active removal mission in LEO. Sci China Inf Sci 60(7):072202
32. Madakat D, Morio J, Vanderpooten D (2013) Biobjective planning of an active debris removal mission. Acta Astronaut 84:182–188
33. Missel J, Mortari D (2013) Path optimization for space sweeper with sling-sat: a method of active space debris removal. Adv Space Res 52:1339–1348
34. Cerf M (2013) Multiple space debris collecting mission—debris selection and trajectory optimization. J Optim Theory Appl 156:761–796
35. Yu J, Hao D, Liu H, Chen X (2018) Mission rescheduling for GEO active debris removal. In: 4th IAA conference on dynamics and control of space systems (DyCoSS2018), Changsha, China, May 21–23
36. Verfaillie G, Schiex T (1994) Solution reuse in dynamic constraint satisfaction problems. In: Proceedings of the 12th conference of the American association of artificial intelligence, Seattle, USA
37. Joslin ED, Clements PD (1998) "Squeaky Wheel" optimization. In: Proceedings of the 15th national conference on artificial intelligence, Madison, USA
38. Barbulescu L, Whitley LD, Howe AE (2004) Leap before you look: an effective strategy in an oversubscribed scheduling problem. In: Proceedings of the 9th national conference on artificial intelligence, Anaheim, USA
39. Globus A, Crawford J, Lohn J (2004) A comparison of techniques for scheduling earth observing satellites. In: Proceedings of the 16th conference on innovative applications of artificial intelligence, San Jose, USA

Method of the Stability of Soil Landslide Under the Condition of Continuous Rainfall

Min Yang and Yihan Xie

Abstract In recent years, landslide disasters occur frequently, making the evaluation of the susceptibility of landslide hazards a major difficulty and hot topic in current research. At present, the research mainly focuses on using statistical model or information model to analyze landslide disaster, but the accuracy is not high. In this study, the combination of Limit Equilibrium Slice Method and Finite Element Method is used to model the landslide and analyze the stability of soil landslide under complex conditions. On this basis, taking a typical soil landslide in a village of Shaanxi Province as the research area, combined with multi-source geological data, the possible landslide mass and slip surface under the condition of continuous rainfall are obtained, which provides effective technical support for landslide disaster prediction and prevention planning.

Keywords Rainfall · Landslide · Stability

1 Introduction

The term landslide is defined as the sliding of rocks, cuttings or rocks along the landslide [1]. It has caused serious harm to natural resources, human life and property. Historical records of landslides have caused great losses to buildings, dams, reservoirs, highways, railways, rivers and lifelines. For example, on February 17, 2006, a huge landslide in the Philippines resulted in more than 1800 deaths and

National Key R&D Program of China (2016YFC0803107).

M. Yang (✉)
School of Environment and Spatial Informatics, University of Mining and Technology,
No. 1 Daxue Road, Xuzhou, China
e-mail: cumtym@126.com

Y. Xie
School of Remote Sensing and Information Engineering, Wuhan University,
No. 129 Luoyu Road, Wuhan, China

L. Wang et al. (eds.), *Proceedings of the 7th China High Resolution Earth Observation Conference (CHREOC 2020)*, Lecture Notes in Electrical Engineering 757, https://doi.org/10.1007/978-981-16-5735-1_3

disappearances; on May 12, 2008, the Wenchuan earthquake in China caused more than 1600 deaths and disappearances in Beichuan landslide (Fig. 1); in 2010, a huge mudslide in Zhouqu resulted in 127 deaths and more than 2000 disappearances. With the change of environment in the development of human society, the disaster loss caused by landslides continues to increase. Therefore, the research on landslides is particularly important [2].

Although there are many researches on landslide disaster at home and abroad, there is not a more perfect model to predict landslide more accurately. Therefore, this paper puts forward a new method of disaster prediction for landslide points based on geological data, precipitation data and other multi-year and multi-source data, and establishes a multi-element comprehensive landslide prediction model. First of all, this paper collects the data of a landslide in Shaanxi Province in recent years, including rainfall, local geological conditions, etc.; then, this paper compares the geological data and precipitation data of landslide area and non landslide area, finds out the main differences, and focuses on these factors. After consulting a large number of geological and rainfall data, this paper adopts a new modeling method, that is, the combination of Limit Equilibrium Slice Method and Finite Element Method to model and analyze the stability of soil landslide under complex conditions. Taking the existing landslide data as samples, this paper establishes the landslide model in geo-studio 2007, and obtains the possible landslide mass and slip surface according to the calculation model. Compared with the existing data, the method in this paper is more feasible, and it also provides a new way for later scholars to study the stability of landslide.

Fig. 1 Large scale landslide in Beichuan city

2 Method of Landslide Stability Analysis

2.1 Overview

There are many theoretical methods to evaluate the stability of soil landslide, especially the Limit Equilibrium Slice Method and numerical analysis method of quantitative analysis method have been widely used. In this study, the combination of Limit Equilibrium Slice Method and Finite Element Method is used, and (Landslide/W) module and (Scep/W) module of Canadian computing software geo-studio 2007 are used to analyze the stability of soil landslide under complex conditions.

Among the two methods of quantitative analysis of soil landslide stability, Limit Equilibrium Slice Method is widely used because of its advantages of simple calculation and clear physical concept. However, due to the complex characteristics of rock and soil, such as discontinuity, heterogeneity, anisotropy, natural initial ground stress, groundwater influence and complex boundary condition treatment, numerical analysis method has also become another relatively rational method The choice you want.

Nowadays, due to the advantages and disadvantages of various methods, it has gradually become an idea of landslide stability evaluation. The comprehensive application of various methods is also the development trend of landslide stability evaluation in the future.

2.2 Limit Equilibrium Slice Method

The Limit Equilibrium Slice Method is to divide the sliding soil into several slices, analyze the force on each slice, and then use the static equilibrium condition of force and moment to calculate the stability coefficient of the landslide. According to whether the static and moment equilibrium conditions are fully satisfied in the stability calculation, the Limit Equilibrium Slice Method can be divided into two categories: accurate slice method and semi accurate slice method.

The limit equilibrium strip method is based on the Mohr Coulomb strength criterion. Its characteristic is that the stress state of every point on the sliding surface is required to be on the Mohr circle (reaching the limit equilibrium), but it does not require that the stress state of every point in the sliding body is on the Mohr circle (in the limit equilibrium state).

There are many kinds of Limit Equilibrium Slice Methods, which can be divided into the following three categories according to the different assumptions of the inter slice forces.

One is to assume the magnitude of the inter strip force or its distribution function, such as Sarma method;

The second is to assume the action direction of the inter slice force, such as Swedish slice method, simplified Bishop method, Morgenstern price method, Spencer method; The third is to assume the position of the action point of the inter bar force, such as Janbu method.

In this study, the Morgenstern price method is used to solve the safety factor FS by geo-studio 2007 software (Figs. 2 and 3). The stress diagram of landslide profile and single block when using this method is as follows.

2.3 Finite Element Method

Now the theory of finite element analysis has reached the mature stage. It can conveniently consider the constitutive model, boundary condition, yield criterion, positive (negative) pore water pressure, soil permeability, seismic force and so on of soil material, and can reflect the real stress-strain state, seepage field and coupling of landslide, which overcomes all kinds of assumptions and rationality of rigid body Limit Equilibrium Slice Method It can simulate the real situation of landslide very well.

(1) Stress analysis of landslide under initial stress

In stress analysis, the stress equation of finite element under the action of initial stress is incremental analysis method, and the specific expression is as follows:

$$\int_V [B]^T [C][B] dv\{a\} = b \int_V \langle N \rangle^T dv + p \int_A \langle N \rangle^T dA + \{F_n\} \tag{1}$$

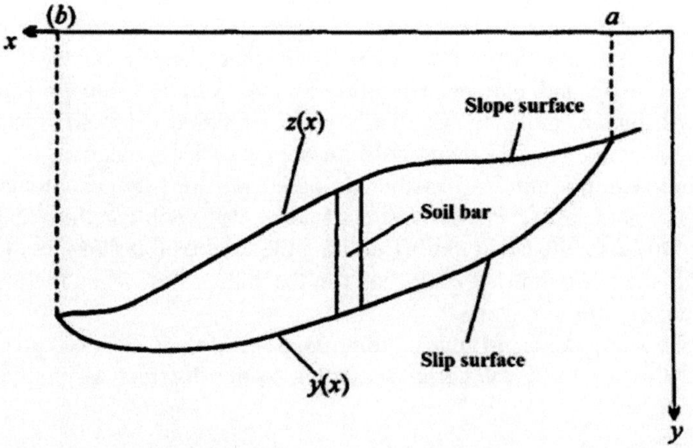

Fig. 2 Geological profile of landslide [3]

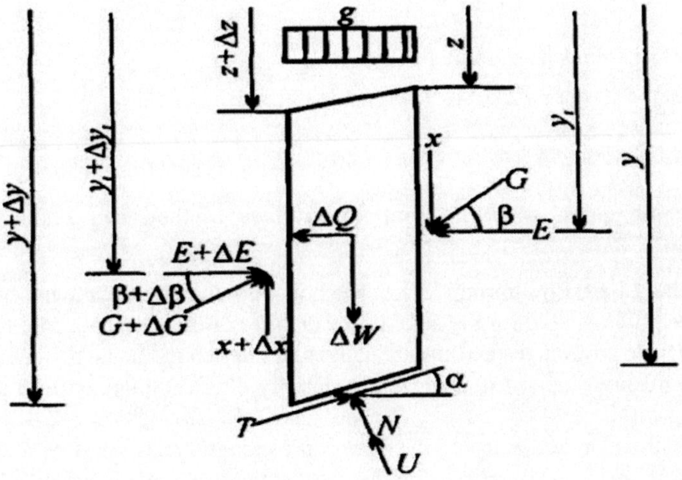

Fig. 3 Forces acting on the column [3, 4]

where

[B]	Strain displacement matrix;
[K]	Stiffness matrix;
{a}	Additional X and Y displacement node column vectors;
$\langle N\rangle^T$	Interpolation function row vector;
A	Boundary area of element;
V	Unit volume;
B	Unit load strength;
P	Increased surface pressure;
$\{F_n\}$	Increased force of concentrated node.

(2) Stress analysis of landslide under precipitation

The influence of precipitation on landslide body is complex. Only the change of pore water pressure is considered in this paper. Pore water pressure is produced by the change of two parts of stress. The specific expression is as follows:

$$\Delta u = \Delta u_\beta + \Delta u_\alpha \tag{2}$$

$$\Delta u_\beta = \frac{\beta}{3}(\Delta\sigma_x + \Delta\sigma_y + \Delta\sigma_z) \tag{3}$$

$$\Delta u_\alpha = \alpha\sqrt{(\Delta\sigma_x - \Delta\sigma_y)^2 + (\Delta\sigma_y - \Delta\sigma_z)^2 + (\Delta\sigma_z - \Delta\sigma_x)^2 + 6\Delta\tau_{xy}^2} \tag{4}$$

where

Δu_β Functions related to normal stress;
Δu_α Functions related to deviator stress;

(3) The combination of Limit Equilibrium Slice Method and Finite Element Method

Because the Limit Equilibrium Slice Method and the finite element numerical method are both suitable for the stability evaluation of the soil landslide, therefore, considering the advantages and disadvantages of the two methods, the combination of the two methods is used to analyze the stability of the soil landslide in the study area.

In the finite element analysis, the stress of element Gauss point or node in the finite element mesh can be given directly, that is, the stress at any point of landslide body. According to the coordinate value of node center, the normal stress and shear stress of any node of landslide body can be obtained by stress tensor transformation, and then the total stress at any sliding surface can be obtained by integration or summation.

According to the definition method of stability coefficient of limit equilibrium theory (the ratio of total sliding force to total resistance) and Mohr Coulomb yield criterion, the finite element stability coefficient K is obtained.

$$K = \frac{\sum_{i=1}^{m} (c_i + \sigma_i \tan \varphi_i)}{\sum_{i=1}^{n} \{F\}_i} \tag{5}$$

where

m Number of nodes on sliding surface;
n Number of nodes on the whole landslide mass;
c Cohesive force;
φ Internal friction angle;
F Load vector of strip;
σ Element stress vector.

3 Research Process of Soil Landslide Stability

3.1 Research Area

The landslide in the study area (Fig. 4) is located in a village of Shaanxi Province. The landslide is an accumulation landslide with a volume of about 19×10^4 m^3,

Fig. 4 UAV image of research area

endangering 521 villagers in 115 households. After the "5.12" earthquake in 2008, the landslide body suffered from local collapse and cracks, with poor stability. The material composition of the sliding mass is mainly composed of silty clay mixed with gravel. It is grayish yellow, yellowish brown, slightly wet, plastic, with great thickness change (1.0–10.5 m). The content of silty clay can reach 65–85%, and the content of gravel can reach 25–50%, in angular shape. The particle size is mostly 1.5–5.0 cm, with the maximum particle size of 8–10 cm. The structure is loose and the permeability is good. According to the survey, no groundwater is found in the landslide.

3.2 Establishment of Basic Model

According to the landslide data in the study area, the basic model is established in geo-studio 2007 software. The landslide mass material is divided into two layers, the corresponding input parameters: material model: Mohr Coulomb model; unit weight of upper and lower layer materials are 15 kN/m^3 and 18 kN/m^3 respectively; cohesion of upper and lower layer materials are 5 kpa and 10 kpa respectively; internal friction angle phi of upper and lower layer materials are 20 and 25° (Fig. 5).

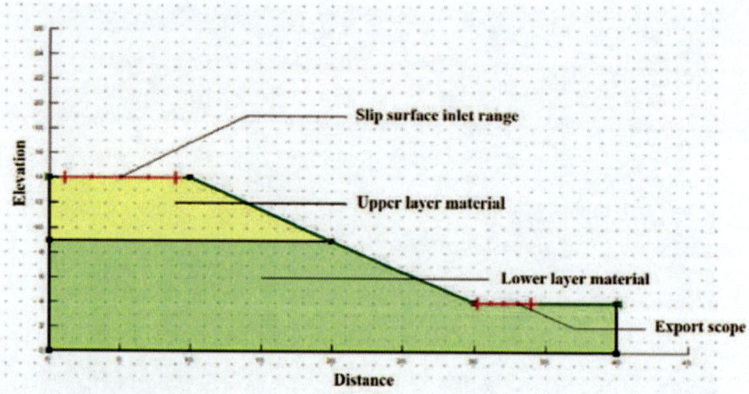

Fig. 5 Model in initial state

3.3 *Determination of Sliding Surface Under the Condition of Continuous Rainfall*

Rainfall will lead to the change of pore water pressure, which will affect the stability of landslide. In this paper, the corresponding parameters are set according to the actual situation, the total rainfall duration is 48 h, and other basic parameters are set as follows (Figs. 6 and 7):

The Table 1 shows the distribution of pore water pressure of landslide model in the first, 16th, 32nd and 48th hours of precipitation:

Fig. 6 Set volume moisture content function

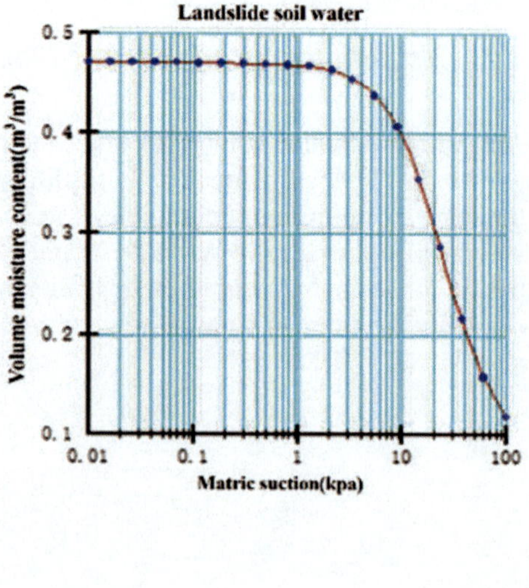

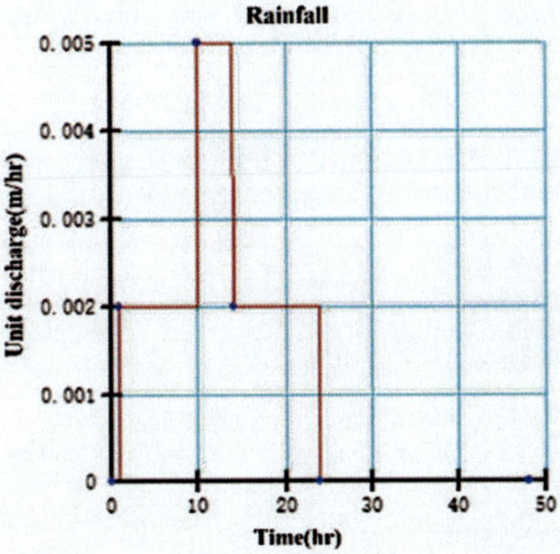

Fig. 7 Set hydraulic boundary function

3.4 *Figures Analysis of the Results*

The Figure 8 shows the stability of the landslide in the 48th hour. The stability of the slider is expressed by the safety factor. The smaller the safety factor is, the lower the stability of the slider is. The green part in the figure is the critical slide, that is, the slide with the minimum safety factor. The grey line indicates the possible slip surface.

4 Conclusions

In this paper, the stability of a landslide in Shaanxi Province is studied. The combination of Limit Equilibrium Slice Method and Finite Element Method is used to analyze the stability of soil landslide under complex conditions by using (Slope/W) module and (Scep/W) module of geo-studio 2007. This technology can show the deformation, displacement state and trend of geological disasters, provide quantitative data for the prevention and control of geological disasters, and inform relevant departments to take timely protective measures to reduce or avoid the loss of life and property caused by landslide disasters. This study was supported by National Key R&D Program of China (2016YFC0803107).

Table 1 Comparison of pore water pressure distribution in landslide model (1–48 h)

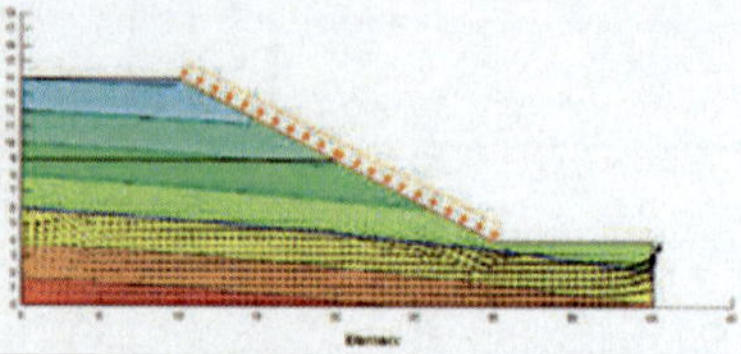
Comparison of pore water pressure distribution in landslide model(1h)
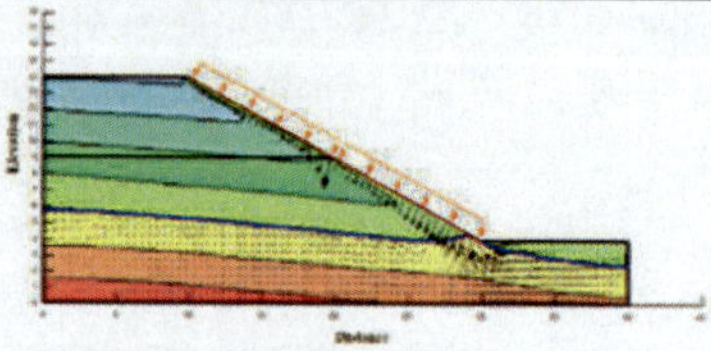
Comparison of pore water pressure distribution in landslide model(16h)
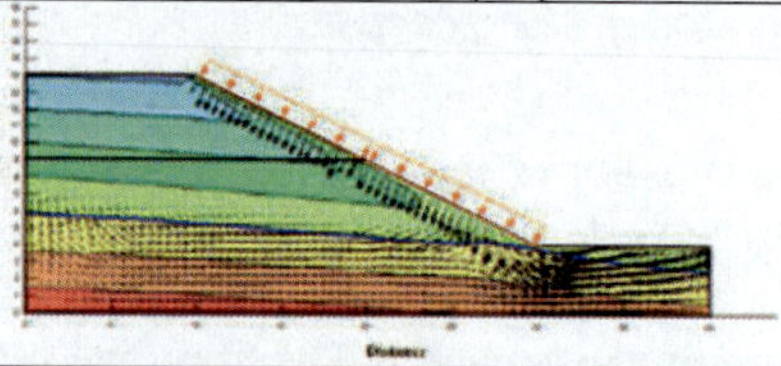
Comparison of pore water pressure distribution in landslide model(32h)
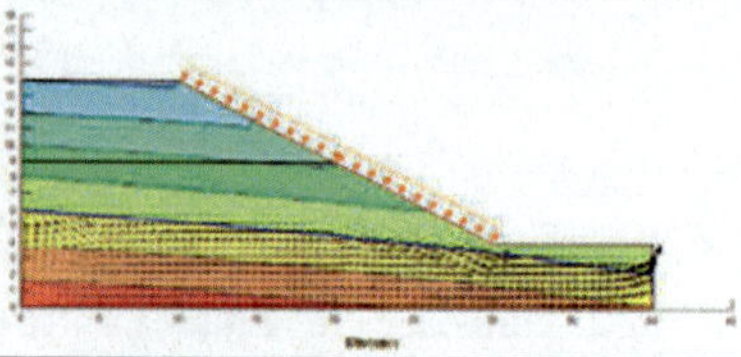
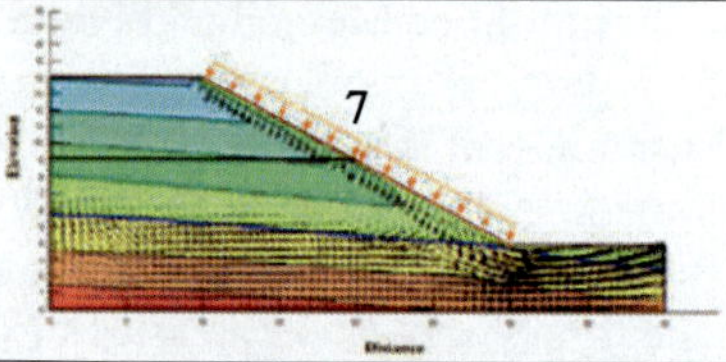
Comparison of pore water pressure distribution in landslide model(48h)

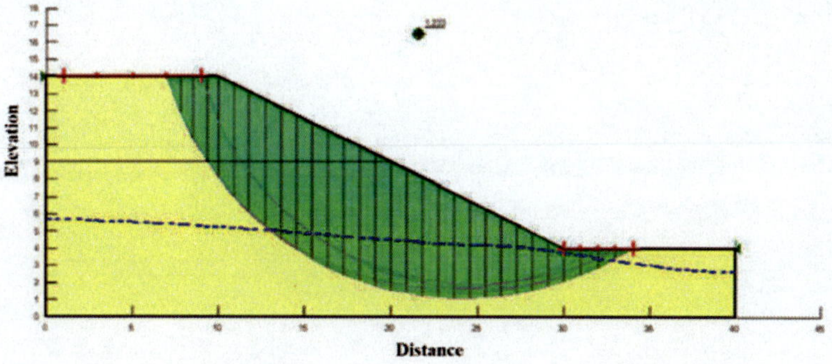

Fig. 8 Stability of the landslide (48 h)

References

1. Cruden DM (1991) A simple definition of a landslide. Bull Int Assoc Eng Geol 43:27–29
2. Changchun L (2014) Risk assessment of landslides in Wanzhou city of three gorges reservoir area. China University of Geosciences
3. Longyuan H, Jihong L, Laishan C, Hui X (2002) Application of Morgenstern price improved method in landslide stability calculation. J Anshan Iron Steel Inst 02:129–133
4. Mao L (2011) Sensitivity study on influencing factors of soil landslide stability. Chengdu University of technology

Processing and Application of AMS-3000 Large Field of View Triple Line Array Camera in Photogrammetry

Shuping Xiong, Yansong Duan, and Guoqin Yuan

Abstract AMS-3000 is a three-line stereo aerial camera with large field of view and high resolution. It can not only acquire panchromatic and RGB high-resolution images, but also has high accuracy. Compared with the world's most advanced ADS40/80/100 camera, this type of camera has higher resolution and higher working efficiency. This article mainly analyzes the imaging principle, working principle, main technical parameters and working efficiency of AMS-3000. By comparing the focal length, the resolution, and coverage width at the same resolution with ADS40/80/100, the superiority of AMS-3000 has emerged. Then the processing flow of the AMS-3000 system is proposed. Next, take Yangjiang in China as the survey area, the good working efficiency and high imaging performance have been confirmed. Finally, the future development and prospect of the AMS-3000 camera are forecasted.

Keywords Sensor · Three-line array · Large field of view · CCD · Imaging principle

1 Introduction

The traditional method of photogrammetry is to obtain digital images by scanning photographic film, which has a large workload and low accuracy. In recent years, with the continuous improvement of sensor imaging technology, the use of digital aerial cameras has developed rapidly. Compared with traditional cameras, the advantages are obvious: (1) Large-format aerial photography can be achieved, and high-quality image information can be obtained; (2) Combined with the position

S. Xiong (✉) · Y. Duan
School of Remote Sensing, China Wuhan University, Luonan Street, Wuhan 430079, China
e-mail: xiongsp@stu.pku.edu.cn

G. Yuan
Changchun Institute of Optics, Fine Mechanics and Physics, Chinese Academy of Sciences, China Wuhan University, No. 3888 Dong Nanhu Road, Changchun 130033, China

L. Wang et al. (eds.), *Proceedings of the 7th China High Resolution Earth Observation Conference (CHREOC 2020)*, Lecture Notes in Electrical Engineering 757, https://doi.org/10.1007/978-981-16-5735-1_4

and attitude measurement system, it guarantees high accuracy and improves the working efficiency of photogrammetry; (3) Digital aerial cameras reduce weather requirements, improve data acquisition efficiency, and reduce time costs.

The imaging methods of digital aerial cameras can be divided into two categories: linear array sweep and area array imaging. The area array aerial camera data processing method is close to that of traditional film cameras, but the number of pixels of a single-format CCD chip is not high enough, which results in low aerial photography efficiency and a small base height, resulting in poor elevation accuracy [1]. Linear array aerial camera adopts CCD array push-broom imaging method, which has the advantages of large viewing angle, high accuracy, low power consumption, and can obtain panchromatic images and multispectral images at the same time.

Internationally, the representative products of line scan aerial cameras are Leica ADS40/80 cameras, which have been widely used due to their high resolution and large imaging angle of view. In contrast, China lags in the research and development of Airborne line-array sensors. AMS-3000 (Airborne measurement system), developed by DPGrid research group of Wuhan university of China and Changchun Opto-Mechanics, is a three-dimensional wide-field array aerial survey camera with high-resolution and high-precision. It can obtain high resolution images of panchromatic and single wave bands of RGB at the same time, and integrates high precision POS and stable platform to adapt to a variety of carrier operations. After testing, the AMS-3000 camera has good working efficiency and imaging performance. What's more, the precision of this product's plane and elevation is high.

2 Working Principle of Three-Line Stereo Camera

The three-line stereo aerial camera pushes and scans from three angles of forward view, front view, and rear view. The projection method is center projection, which can obtain panchromatic, multi-spectral high-precision images. As the loading platform flies forward, three CCD arrays image different ground surfaces at the same time. The forward-looking detector *F* obtains the forward-looking image *Fs* through the center projection, the front-viewing detector *N* is vertically projected to form the front view *Ns*, and the rear-viewing detector tilts backwards. A rear-view image *Bs* is formed. Three lines intersect on the same ground to achieve the purpose of stereo imaging. If the outer azimuth element and the inner azimuth element of the camera are known, the geographic coordinates of the image point can be calculated (Figs. 1 and 2).

The construction equation of the pusher sensor with CCD line array is as follows [2]:

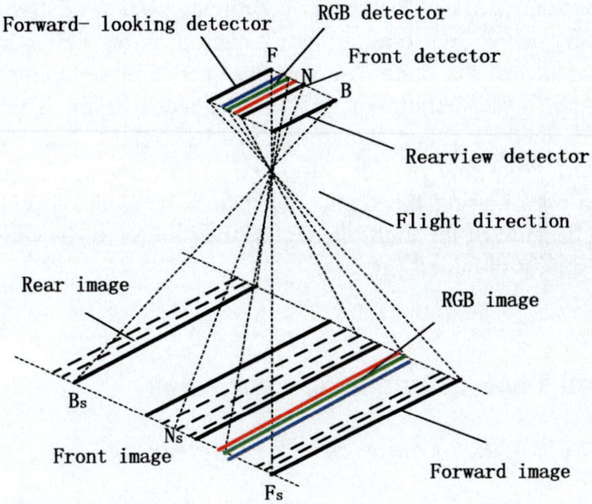

Fig. 1 Principle of three—line array stereo aerial survey camera

Fig. 2 Working principle diagram of large field-of-view three-line array airborne mapping camera

$$\begin{bmatrix} X \\ Y \\ Z \end{bmatrix}_P = \begin{bmatrix} X \\ Y \\ Z \end{bmatrix}_S + \lambda A_t R_\theta \begin{bmatrix} 0 \\ Y \\ -f \end{bmatrix} \quad (1)$$

where $(X,Y,Z)_p$ = the coordinates of the ground point P in the ground coordinate system, $(X,Y,Z)_S$ = the geocentric coordinates of the sensor projection center S,$(0, y, -f)$ = at time t, the image point coordinates of line array imaging, λ = scale size, A_t= the rotation matrix of the sensor coordinate system relative to the ground coordinate system, and R_θ= a rotation matrix caused by an inclination Angle θ.

The characteristics of the three—dimensional aerial survey camera is high resolution, large field of view, multi-spectral, and stereo imaging. There are 6 linear array detectors in the focal plane, including three R, G and B linear array detectors and three forward, front and rear detectors, so as to achieve high resolution and multi-spectral effects. During the flight, the system scans the ground at a certain frequency. And because of the multiple angles, the stereo images with nearly 100% overlap can be finally obtained [3].

3 AMS-3000 Photogrammetric Processing

3.1 AMS-3000 Main Technical Parameters

AMS-3000 large field-of-view three-line aerial camera adopts high-resolution line array CMOS element as the detector, the lens adopts the center vertical projection design, each row has 32 K pixels [4], the pixel size is 5 μm × 5 μm, the focal length is 130 mm, The base height ratio is 0.89, according to the 1: 1000 (2400 m aerial height) mapping photography operation. Each CCD scan line has 32,768 pixels with high precision and high data acquisition efficiency. After the RGB information obtained by the multi-spectral sensor and the high-resolution panchromatic image are fused, a panchromatic orthographic image and a high-precision DEM model can be obtained in the mapping area. The specific technical parameters are shown in Table 1, and the sensor structure is shown in Figure 3.

Table 1 Main technical parameters of the AMS-3000

Spectrum band	Panchromatic and RGB
Pixel	32/16 k
Focal length	130 mm
Base to height ratio	0.89
Map scale	1:1000 (2400 m flying height)
Transverse field of view angle	64°
Intersection angle	21°, 27°
Storage capacity	6 TB
Volume	500 mm × 500 mm × 915 mm
Weight	72 kg
Operation temperature	−20–60 °C

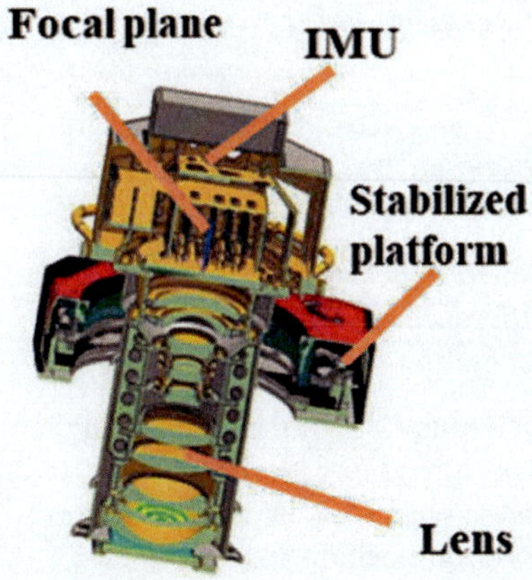

Fig. 3 Structure diagram of AMS-3000

3.2 AMS-3000 System Processing Flow

According to the characteristics of the AMD-3000 camera, based on the original ADS40 processing software, related processing algorithms have been added and the processing process has been redesigned, as shown in Figure 4.

We tailor the ground processing system DPGridAMS specifically for the characteristics of the AMS-3000 digital aerial camera. DPGridAMS directly inputs the raw data obtained by AMS-3000 digital aerial camera, and can process DEM, DOM, DLG, 3D data and other surveying and mapping products through automatic air triangulation, automatic DEM production, and large-scale image toning.

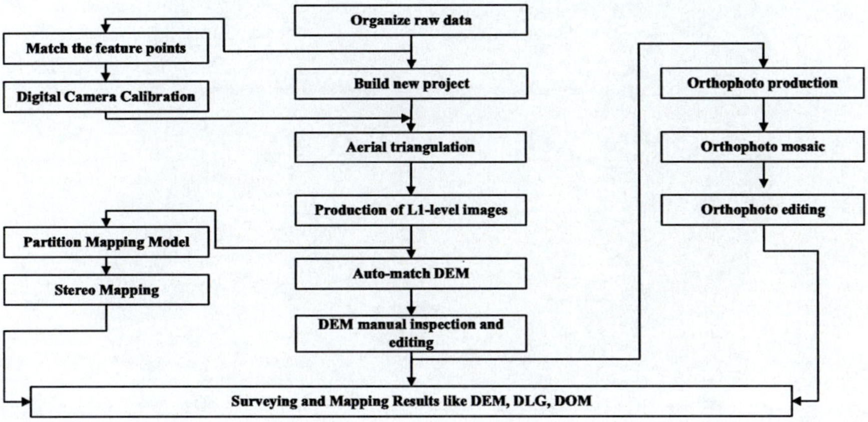

Fig. 4 Photogrammetric processing flow of AMS-3000 camera

3.3 *AMS-3000 Processing Effect*

At present, AMS-3000 has completed several test flights. A total of 16TB of raw data were obtained, covering an area of 7000 km^2. Each piece of data contains:

1. Panchromatic image of front, bottom and rear view.
2. RGB color image located between the bottom view and the back view.
3. Pos data after post-difference processing.
4. Parameter files corresponding to image line and POS.

Figure 5 is a true color orthophoto of the entire Yangjiang survey area, obtained by AMS-3000 flight. It was filmed on December 18, 2018. There are a total of nine air belts, and the ground coverage area is 264.21 km^2. The mapping scale is 1: 1000, and the ground decomposition rate is 0.1.

4 The Advantages of AMS-3000

Compared with foreign advanced ADS40/80 sensors, AMS-3000 has the advantages of high resolution and large field of view, especially the resolution, which is almost 2.7 times that of ADS40/80. The base height ratio, the number of pixels, and the pixel size are all better than ADS40/80, and the working field angles are equivalent, all being 64° (Table 2).

ADS100 is a new aerial camera launched by Leica in 2013. It adopts a single main optical axis design, with small eccentricity error and high geometric stability. Compared with this camera, although AMS-3000 is heavier, it has more advantages in resolution and environmental adaptability (Table 3).

Fig. 5 True color orthophoto of the whole Yangjiang survey area

Table 2 Technical parameters of the AMS-3000 are compared with the ADS40/80

Parameters	ADS40/80	AMS-3000
Focal length (m)	62.77	130
Base to height ratio	0.78	0.89
Pixel elements	12,000	32,768
Pixel size (μm)	6.5 × 6.5	5 × 5
Field angle	64°	64°
2000 m resolution ratio (m)	0.207	0.077
Coverage width (m) with same resolution (0.1 m)	1206.9	3249.3

Table 3 Technical parameters of the AMS-3000 are compared with the ADS100

Key indicators	AMS-3000	ADS100
Focal length (m)	130 mm	62.5 mm
Base to height ratio	0.89	0.78
Intersection angle	21°, 27°	14°, 27°
Pixel elements	32,768	16,000
2000 m GSD	0.08 m	0.16 m
Transverse field of view angle	64°	60°
Operation temperature	−20 to 60° C	−20 to 45° C
Weight	72.5 kg	50 kg

Compared with ADS40/80/100, the main features of domestic AMS-3000 airborne three-line camera are as follows:

(1) Acquire high-resolution images with panchromatic and RGB at the same time;
(2) 32 k long linear array detector, high operating efficiency;
(3) Integrated high-precision POS and stable platform to adapt to various carrier operations;
(4) Large base-to-height ratio and strong stereo imaging capability;
(5) Airtight and active temperature control design, good environmental adaptability;
(6) Automated data processing system to process data more efficiently;
(7) Large width, larger field of view and greatly reduced flight time.

5 Application Prospects of AMS-3000

At present, the AMS-3000 aerial survey camera has passed environmental tests and 9 field missions. The camera's key technical indicators have reached the advanced level of similar products in the world. The successful research and development of

AMS-3000 has to a certain extent filled the gap in China's airborne stereo array aerial survey cameras, and can meet China's requirements for large-scale, high-resolution, high-speed and efficient aerial mapping. AMS-3000 can be applied in the following areas:

(1) Digitalization of cities. Different from the traditional frame-type camera, the large-field-of-view three-line array sensor has the characteristics of high resolution, digital mapping, and strong environmental adaptability. At this stage, the topographic maps on the scale of cities in China are insufficient, and the urban spatial information is outdated and incomplete. With the rapid development of the economy, China urgently needs a large number of digital large-scale topographic maps of cities. The AMS-3000 can not only obtain high-resolution aerial mapping images, but also process data more efficiently, which is very suitable as a data source for digital cities.
(2) Remote sensing of resources. By mounting the AMS-3000 camera on a satellite on the ground or a moon, it can quickly obtain the spatial information of the ground or the surface of the moon, which is beneficial to the dynamic monitoring of various resources.
(3) Three-dimensional mapping. The three-line camera realizes the establishment of a three-dimensional model through the stereo image pairs in the front-view, forward-view, and rear-view push-broom images. The characteristics of AMS-3000 stereo mapping provide a new solution for 3D mapping.

References

1. Fang Y, Hu H (2013) Development status and technical system analysis of digital aerial camera. Surveying and Mapping Science and Engineering, pp 47–52
2. Zhang Z, Zhang J (1997) Digital photogrammetry. Wuhan University Press
3. Yao Y, Xu Y, Ding Y (2018) Optical system design of large field of view three-line array aerial mapping camera. Opt Precis Eng 26(9):208–217
4. AMS-3000 large field of view triple line camera production test achieved good results [OL]. http://www.sohu.com/a/308911803_650579

Evaluation and Analysis of Geometric Accuracy of the ZY-1 02D VNIC Image

Liping Zhao, Xingke Fu, and Xianhui Dou

Abstract The ZY-1 02D reached orbit in September 12, 2019. It is equipped with a 9-band visible near-infrared camera (VNIC) and a 166-band hyperspectral camera. The VNIC can acquire a panchromatic/multi-spectral image with a spatial resolution of 2.5/10 m and a width of 115 km. The geometric accuracy of ZY-1 02D VNIC is evaluated using multiple revisited data in Northeast and Northwest China. Based on the ground control points (GCPs) extracted from the ZY-3 standard stereo triplet products, the accuracy of the bundle block adjustment with the affine correction in image space is analyzed, and the root mean square error (RMSE) of the panchromatic image is better than 0.5 pixel in most cases. However, through the static and dynamic residual analysis of GCP, it is found that the vibration of the satellite platform will cause significant dynamic imaging errors, especially in the push-broom direction along the orbit, an effect of approximately 1.0 pixels in size can be detected. In general, the panchromatic images of ZY-1 02D VNIC have good geometric accuracy and meet the requirements of design specifications. However, the accuracy of panchromatic image is adversely affected by platform vibration.

Keywords ZY-1 02D · VNIC · Geometric accuracy · Vibration · Analysis

1 Introduction

The ZY-1 02D is an important part of China's space-based infrastructure, and reached orbit in September 12, 2019. The primary objective is to provide high-resolution panchromatic, multispectral and hyperspectral data for natural resources monitoring and management, ecological and forestry monitoring, disaster prevention and control, environmental protection, etc. The instruments

L. Zhao (✉) · X. Fu · X. Dou
Land Satellite Remote Sensing Application Center, Ministry of Nature Resource, Beijing, China
e-mail: zhaolpwww@163.com

L. Wang et al. (eds.), *Proceedings of the 7th China High Resolution Earth Observation Conference (CHREOC 2020)*, Lecture Notes in Electrical Engineering 757, https://doi.org/10.1007/978-981-16-5735-1_5

onboard the spacecraft are the Visible Near-Infrared Camera (VNIC) and the Advanced Hyper-Spectral Imager (AHSI). The VNIC uses long linear detector arrays aligned on the focal planes collect imagery in a push-broom method, and is able to acquire 10-meter data for visible and near infrared spectral bands and provide for a 2.5-meter panchromatic (PAN) band with a width of 115 km (see Tables 1 and 2).

The geometric performance is an important indicator of image quality and a key factor affecting data applications. Generally speaking, the geolocation accuracy, the internal geometric accuracy and the band registration accuracy are the most critical geometric performance indicators, which are the main content of geometric calibration activities. In order to ensure the stability of the geometric performance of the system and verify whether the data product meets the specification, these geometric characterization need to be continuously verified, monitored and improved throughout the life of the mission.

In this paper, the geolocation accuracy and internal regularity accuracy of VNIC image product located in Northeast and Northwest China are assessed, and through residual analysis, the geometric distortion is studied, and the satellite platform vibration problem is discussed.

Table 1 Main characteristics of the ZY-1 02D

Launch date	September 12, 2019
Design life	5 years
Altitude	778 km
Repeat cycle	55 days
Revisit time	2–3 days, depending on latitude
Pointing capability in cross-track	±26°

Table 2 Specifications of the VNIC

Imager type	Push-broom with TDI capability
Swath width	>115 km
GSD at nadir	PAN: <2.5 m MS: <10 m
Pixel size	PAN: 7 μm MS: 28 μm
Sample	PAN: >47,000 MS: >11,800
Data quantization	12bits (4096)
FOV (Field of view)	>8.5°

2 Geometric Model

2.1 Overview

In this study, the geolocation accuracy is the accuracy assessment in the object space, generally expressed in ground units, such as meters, inches, etc., and can also be converted into image coordinate system. According to whether GCP is used, it is divided into absolute and relative geolocation accuracy. The absolute geolocation accuracy does not use GCP, which is also called direct geolocation accuracy. It is mainly affected by ephemeris errors, attitude errors, and alignment errors between spacecraft and payloads. The relationship between them is often given in the form of an installation matrix. The relative geolocation accuracy is the accuracy after adjustment calculation with GCP, which is mainly affected by the geometric accuracy inside the image.

The internal geometric accuracy is the accuracy in the image space, usually expressed in pixels (image coordinate system) or micrometer/angle (camera focal plane coordinate system). It generally refers to the internal geometric accuracy within the scene, which include static geometric distortion of the camera and dynamic imaging error. The static error is the geometric calibration residual error of the camera, that is, the error of the Line of Sight (LOS) for the detectors on the focal plane, and the dynamic error refers to the error caused by attitude vibration during the push-broom imaging process.

The initial installation matrix of the camera and the LOSs for each detector are both meticulously calibrated in prelaunch testing. The calibration parameters are not expected to change significantly through launch, but they may require minor refinement during the on-orbit commissioning period, as well as continuous monitoring and optimization in routine operational use.

2.2 Focal Plane Geometry

Using a focal plane array comprised of 2 × 4 Sensor Chip Assemblies (SCA), the VNIC achieves the imaging function of 9 spectral bands with a 115 km swath width. Among them, four SCAs are aligned in the cross-track direction to realize a push-broom width of 115 km, five-spectral and four-spectral SCAs are respectively deployed in two channels along the track direction to collect data in one panchromatic band and 8 multi-spectral bands. Based on the focal plane integration scheme, the field of view (FOV) of the telescope is about approximately 8.5° in the across-track direction and 1.6° in the along-track direction.

2.3 RPC Model

The RPC (Rational Polynomial Coefficients) model is a general geometric model that establishes the mapping relationship between ground objects and image pixels. Generally, the geometric accuracy of the RPC model is equivalent to the strict physical model. In this paper, the RPC model attached to the image data is used to study the absolute geolocation accuracy, and it is also used to perform the bundle adjustment with GCP to optimize geometry performance.

2.4 Simulation

In order to obtain the prior knowledge of ZY-1 02D VNIC, according to the imaging geometric model, for PAN data, Figs. 1, 2, 3 show the simulation results of the influence of ground reference data elevation error on the accuracy of image adjustment.

It can be seen from Figs. 1, 2 and 3 that the accuracy in the along-track direction is hardly affected by the roll angle and terrain-induced errors, while in the across track direction, as the roll angle increases, the adverse effect of the elevation error becomes more and more obvious. Analysis Figs. 1, 2 and 3 show that the elevation error of the ground reference data has a significant influence on the image accuracy at different FOV. This is due to the fact that the FOV of the telescope is approximately 8.5° in the across-track direction. Therefore, in general, it is desirable to have an elevation accuracy better than 2 m in order to ensure that the terrain induced error at the FOV edge is less than 0.3 PAN pixel.

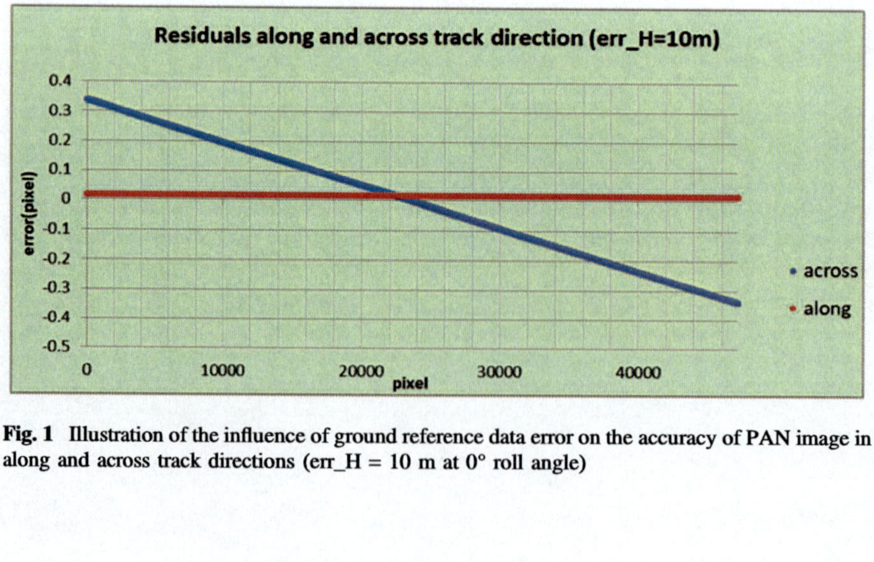

Fig. 1 Illustration of the influence of ground reference data error on the accuracy of PAN image in along and across track directions (err_H = 10 m at 0° roll angle)

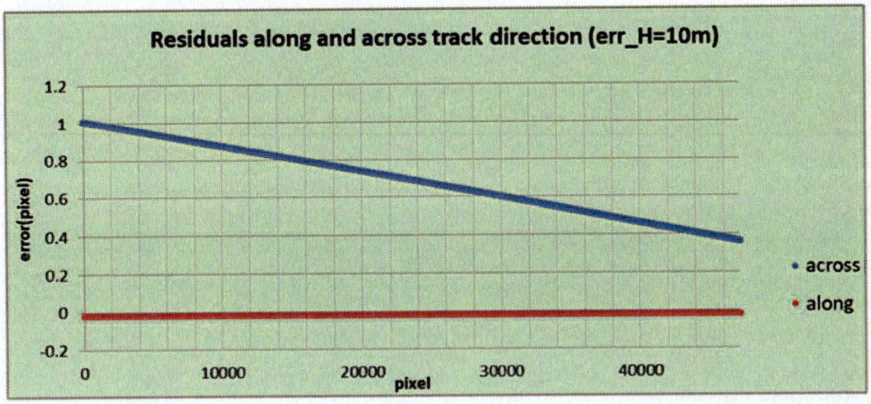

Fig. 2 Illustration of the influence of ground reference data error on the accuracy of PAN image in along and across track directions (err_H = 10 m at −8.8° roll angle)

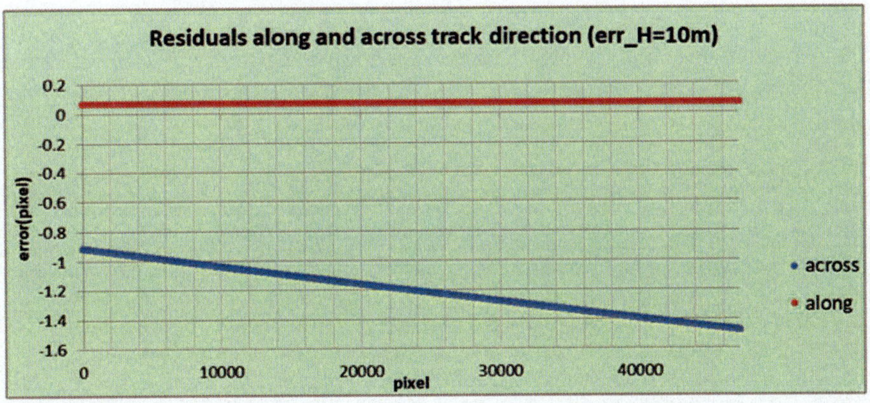

Fig. 3 Illustration of the influence of ground reference data error on the accuracy of PAN image in along and across track directions (err_H = 10 m at +16° roll angle)

3 Datasets

Harbin in northeastern China and Hami in northwestern China are selected as experimental areas for accuracy evaluation. Both the ADS80 reference data and the ZY-3 stereo data in the Harbin experimental area are available, while only the ZY3 data can be used as a reference in the Hami experimental area.

3.1 Study Area

The Harbin area has the advantages of good weather conditions, moderate intensity of human activities, and more available data. The land cover in this region is mainly farmland and grassland, and the main topographic features of the test area are flat with a height difference of less than 60 m, which is conducive to analyzing the geometric properties of the image itself, eliminating terrain effects as much as possible.

The terrain of the Hami test area has large undulations, with an elevation range of approximately 800m–1800 m above the WGS84 ellipsoid, which can be used to explore the influence of elevation errors on geometric accuracy. This area not only has good weather, but also has simple land cover and small seasonal changes, which is conducive to extracting GCP/ICP (Independent Check Points) from multi-temporal images.

3.2 ZY-1 02D Data

The test image data and its main characteristics are summarized in Table 3. In the Harbin test area, the data of 4 repeat cycles were collected, some of which are

Table 3 Main characteristics of the ZY-1 02D test data

No	Area	Date	Scene ID	Terrain (m)	GSD sample	GSD line	Roll angle	Reference data
1	Harbin	2020-04-14	70,981	<100	2.464	2.452	0.0	ADS80_2014 and ZY-3 stereo
2		2020-04-17	72,008	<100	2.464	2.453	0.0	ZY-3 stereo
3		2020-04-28	76,149	<100	2.560	2.661	16.0	ZY-3 stereo
4		2020-05-13	82,208	<100	2.478	2.477	5.6	ADS80_2014
5		2020-05-16	83,122	<100	2.465	2.449	0.0	ADS80_2014
6		2020-06-08	92,196	<100	2.464	2.453	0.0	ADS80_2014
7		2020-06-11	93,215	<100	2.464	2.452	0.0	ADS80_2014
8		2020-06-17	96,093	<100	2.513	2.557	−10.7	ADS80_2014
9		2020-07-10	105,954	<100	2.463	2.452	0.0	ADS80_2014
10		2020-07-13	106,716	<100	2.464	2.452	0.0	ADS80_2014
11	Hami	2020-05-30	88,816	>1000	2.461	2.448	−1.0	ZY-3 stereo
12		2020-05-30	88,817	>1000	2.460	2.447	−1.0	ZY-3 stereo
13		2020-06-02	89,891	>1000	2.478	2.483	−6.4	ZY-3 stereo
14		2020-06-02	89,892	>1000	2.477	2.483	−6.4	ZY-3 stereo
15		2020-06-05	91,978	>1000	2.512	2.558	−11.0	ZY-3 stereo
16		2020-06-05	91,979	>1000	2.511	2.557	−11.0	ZY-3 stereo
17		2020-06-08	92,242	>1000	2.558	2.660	−15.0	ZY-3 stereo
18		2020-06-08	92,243	>1000	2.556	2.658	−15.0	ZY-3 stereo

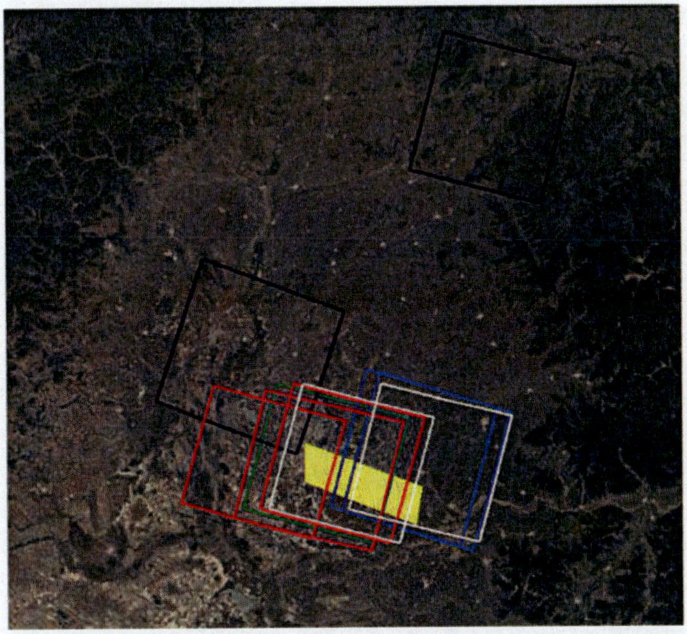

Fig. 4 Illustration of ZY-1 02D image distribution in the Harbin test area. Yellow: ADS80_2014 reference data; Green: 2020-04-14; Black: left: 2020-04-17, right: 2020-04-28; Blue: left: 2020-05-16, right: 2020-05-13; Red: left: 2020-06-08, middle:2020-06-17, right: 2020-06-11; White: left: 2020-07-10, right: 2020-07-13

covered by a small amount of cloud, which does not affect the accuracy assessment. In the Hami test area, we obtained 4 revisit data that overlapped almost completely with different off-nadir angles, so the relationship between the off-nadir view and geometric accuracy could be studied. Figs. 4 and 5 respectively give illustration of ZY-1 02D image distribution in the Harbin and Hami test area. Figure 6 shows the typical image data of the two test areas.

3.3 ADS80 Reference Data

There is a reference data set in Harbin area, which is comprised of ortho-image with a GSD of 0.5 m and an accuracy around 0.5 m and digital surface model with a grid of 2 m and an accuracy around 1.0 m. The panchromatic ortho-image and DSM are produced from ADS80 aerial imageries acquired in October 2014 using software such as X-Pro and INPHO. This set of data is called ADS80_2014 reference dataset in this article.

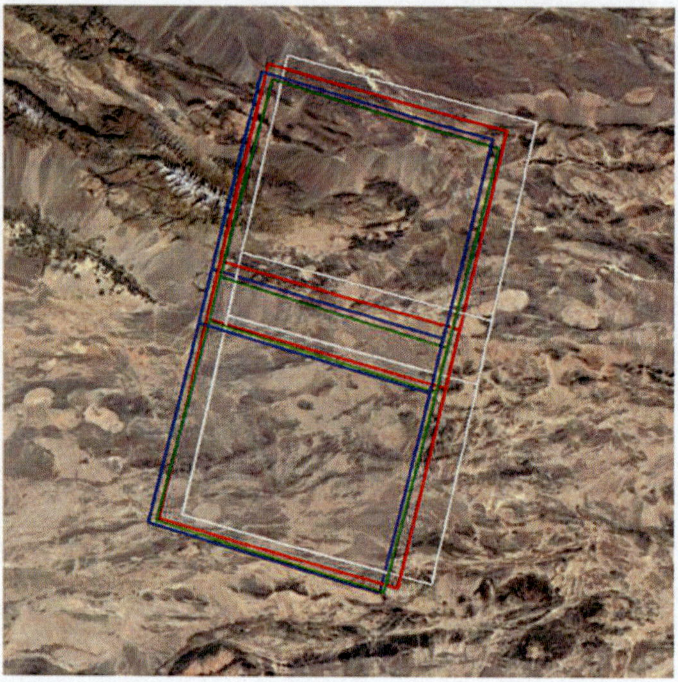

Fig. 5 Illustration of ZY-1 02D image distribution in the Hami test area. Green: 2020-05-30 (upper: 88,816, lower: 88,817); Blue: 2020-06-02 (upper: 89,891, lower: 89,892); Red: 2020-06-05 (upper: 91,978, lower: 91,979); White: 2020-06-08 (upper: 92,242, lower: 92,243)

Fig. 6 ZY-1 02D thumbnail images of test data. Left: Harbin test image (2020-04-14_70981); Right: Hami test image (2020-06-02_89892)

Moreover, since the range of the reference data is about 120 km × 30 km, and the size of ZY-1 02D VNIC image data reaches 115 km × 115 km, GCPs extracted based on the high-precision data on each scene could only cover part of the image. In addition, the reference data will gradually become less useful over time as the imagery becomes obsolete. Therefore, during this evaluation, these reference data are mainly used to evaluate the absolute geolocation accuracy.

3.4 ZY-3 Stereo Data

Although the absolute geolocation accuracy of the ZY-3 stereo data is sometimes slightly worse, but the internal geometric accuracy is very good, and the accuracy and reliability can be improved based on the stereo triplets [8]. According to the simulation results, it can meet the accuracy analysis requirements of VNIC image. In addition, the ZY-3 data close to the ZY-1 02D data acquisition date can also be selected to ensure that the difference in land cover between the images is small and the best tie-points extraction results can be obtained.

However, the reference data also has a shortcoming, that is, the range of ZY-3 stereo is about 50 km × 50 km, while the range of ZY-1 is 115 km × 115 km, which requires multiple ZY-3 stereo pairs to cover a scene of ZY-1. Therefore, in some cases in this test, the ZY-3 only covers part of the ZY-1, but overall does not affect the accuracy assessment and analysis. In a word, the ZY-3 stereo triplets are currently the optimal reference data that can be obtained at low cost. In this study, it is mainly used to analyze the internal geometric accuracy of VNIC PAN images.

3.5 GCP and ICP

For the VNIC test data, based on the above-mentioned high-precision ADS80 reference dataset and ZY-3 stereo triplets, a large number of evenly distributed tie points (TPs) are extracted using the phase correlation method [1, 6]. For the ADS80 reference data, the planimetric coordinates and elevation values are calculated by DOM and DSM. For the ZY-3 stereo triplets reference data, three-dimensional coordinates are directly obtained from stereo model. After removing gross errors, these TPs can be used as GCP or ICP for geometric analysis.

4 Geolocation Accuracy

The geolocation performance is one of the main performance characteristics of satellite missions, which is evaluated and discussed using Level 1 data from the Harbin and Hami experimental area.

4.1 Geolocation

The absolute geolocation accuracy of image products in the object space, also known as the direct geolocation accuracy, is evaluated using the RPC model without any GCP, that is, all automatically extracted tiepoints are used as ICP. It should be noted that since the VNIC data is a mono image, the elevation coordinates of the ICP need to be used as known values to participate in the calculation.

Table 4 statistics the absolute geolocation accuracy of the ZY-1 02D VNIC PAN images in the Harbin test area. The ICP distribution of ZY-1_02D_VNIC_2020-07-10 extracted from the high-precision ADS80_2014 reference data is shown in Fig. 7. Although the coverage of the ADS80_2014 reference data is relatively small, it does not affect the evaluation results of the geolocation accuracy.

4.2 Relative Geolocation

The relative geolocation accuracy can be obtained by using affine adjustment model with all GCPs in image space to refine the absolute accuracy [2–5, 7]. Tables 5 and 6 respectively summarize the adjustment accuracy of the Harbin and Hami test areas. Figures 8 and 9 show the typical distribution diagrams of GCP in the Harbin and Hami regions. The schematic of the GCP distribution and elevation of data ZY-1_02D_VNIC_2020-05-30_88817 is shown in Fig. 10.

In the Harbin experimental area, based on the ADS80_2014 reference data, there are differences in the accuracy of the 8 scenes, which may be caused by the different distribution ranges of these GCPs in each scene, or the accuracy of the images themselves.

Table 4 The absolute geolocation accuracy of the ZY-1 02D VNIC PAN image in the Harbin

No	Date	ICP Count	ΔPixel (pixel)	ΔLine (pixel)	ΔRadial (pixel)	ΔACT* (meter)	ΔALT* (meter)	ΔLocation (meter)
19	2020-04-14	9036	5.403	−8.628	10.195	13.313	−21.156	24.996
20	2020-05-13	5461	8.241	1.159	8.361	20.421	2.871	20.622
21	2020-05-16	8527	2.394	−0.832	2.779	5.901	−2.038	6.243
22	2020-06-08	5824	14.167	−15.545	21.039	34.907	−38.132	51.697
23	2020-06-11	6001	7.475	−24.658	25.771	18.418	−60.461	63.205
24	2020-06-17	6665	13.631	−4.207	14.295	34.255	−10.757	35.904
25	2020-07-10	7158	13.380	−4.373	14.148	32.955	−10.723	34.655
26	2020-07-13	2767	18.624	3.587	19.068	45.890	8.795	46.725
Mean			10.414	−6.687	14.457	25.758	−16.450	35.506

**ALT* Along track direction; *ACT* Across track direction

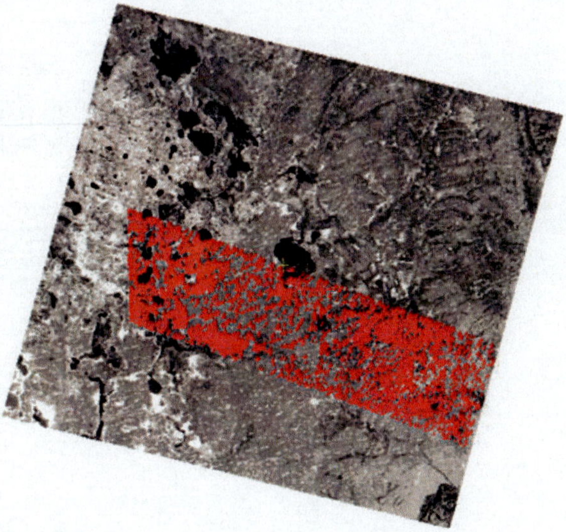

Fig. 7 The distribution diagram of ICPs extracted from the ADS80_2014 reference data (ZY-1_02D_VNIC_2020-07-10)

In the Hami test area, the 8 scenes are divided into upper and lower two groups according to the geographical location (refer to Fig. 5). The GCP of the 4 scenes in each group is extracted based on ZY-3 stereo at the same time, so the number and location of the GCP are exactly the same, that is, each GCP is tiepoints on the 4 scenes, so the accuracy between them is comparable.

On the whole, the accuracy of the Hami data is better than 0.5 pixels, and the accuracy decreases slightly with the increase of the roll angle, and the surface undulation of more than 1000 m does not cause a significant degradation of accuracy.

A comprehensive analysis of the data of the two test areas shows that:

- The relative geolocation accuracy of VNIC images is good, and some of the lower accuracy may be caused by insufficient accuracy of GCP;
- Terrain relief did not have a significant impact on the accuracy, indicating that the accuracy of the automatically extracted GCP is high enough, and the accuracy of the VNIC image is very consistent;
- Imaging by off-nadir view has little influence on accuracy;
- The accuracy in across track direction is better than along track direction for all data.
- As reference data, the accuracy obtained by ZY-3 stereo is slightly higher than that of ADS80_2014, which may be caused by the expiration of ADS80_2014.

Table 5 The relative geolocation accuracy of the ZY-1 02D VNIC PAN images in the Harbin

No	Date	Roll angle (degree)	GCP Count	ΔPixel (pixel)	ΔLine (pixel)	ΔRadial (pixel)	ΔACT* (meter)	ΔALT* (meter)	ΔLocation (meter)	Reference data
1	2020-04-14	0.0	9036	0.302	0.383	0.488	0.744	0.939	1.198	ADS80_2014
2	2020-04-14	0.0	75,910	0.172	0.224	0.282	0.424	0.549	0.694	ZY-3_2020-04-15
3	2020-04-17	0.0	4708	0.271	0.498	0.567	0.668	1.222	1.392	ZY-3_2018-05-13 ZY-3_2018-05-18
4	2020-04-28	16.0	23,548	0.227	0.328	0.398	0.581	0.873	1.049	ZY-3_2020-04-05
5	2020-05-13	5.6	5461	0.558	0.652	0.858	1.383	1.615	2.126	ADS80_2014
6	2020-05-16	0.0	8527	0.456	0.568	0.728	1.124	1.391	1.788	ADS80_2014
7	2020-06-08	0.0	5824	0.218	0.356	0.418	0.537	0.873	1.025	ADS80_2014
8	2020-06-11	0.0	6001	0.437	0.488	0.655	1.077	1.197	1.610	ADS80_2014
9	2020-06-17	−10.7	6665	0.290	0.430	0.519	0.729	1.100	1.319	ADS80_2014
10	2020-07-10	0.0	7158	0.390	0.476	0.615	0.961	1.167	1.512	ADS80_2014
11	2020-07-13	0.0	2767	0.458	0.463	0.651	1.129	1.135	1.601	ADS80_2014

ALT Along track direction; *ACT* Across track direction

Table 6 The relative geolocation accuracy of the ZY-1 02D VNIC PAN images in the Hami

No	Date	Scene ID	Roll angle (degree)	GCP Count	ΔPixel (pixel)	ΔLine (pixel)	ΔRadial (pixel)	ΔACT* (meter)	ΔALT* (meter)	ΔLocation (meter)
1	2020-05-30	88,816	−1.0	11,537	0.159	0.350	0.385	0.391	0.857	0.942
2	2020-06-02	89,891	−6.4	11,537	0.176	0.282	0.332	0.436	0.700	0.825
3	2020-06-05	91,978	−11.0	11,537	0.199	0.295	0.356	0.500	0.755	0.905
4	2020-06-08	92,242	−15.0	11,537	0.252	0.344	0.427	0.645	0.915	1.119
5	2020-05-30	88,817	−1.0	12,981	0.151	0.288	0.325	0.371	0.705	0.797
6	2020-06-02	89,892	−6.4	12,981	0.183	0.287	0.341	0.453	0.713	0.845
7	2020-06-05	91,979	−11.0	12,981	0.177	0.331	0.375	0.444	0.846	0.956
8	2020-06-08	92,243	−15.0	12,981	0.196	0.393	0.439	0.501	1.045	1.159

**ALT* Along track direction; *ACT* Across track direction

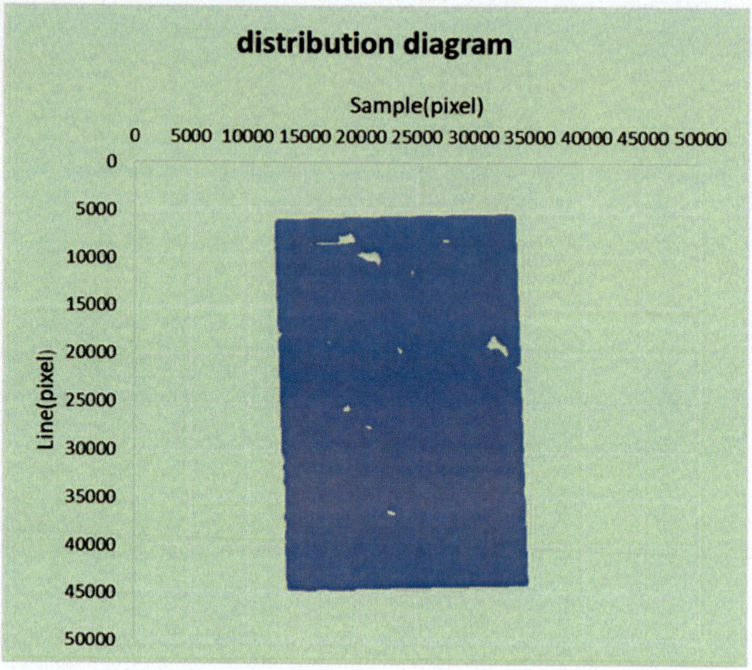

Fig. 8 The distribution diagram of GCPs extracted from the ZY-3 stereo triplets in the Harbin test area (ZY-1_02D_VNIC_2020-04-14). The acquisition dates of ZY-1 02D and ZY-3 (collected on 2020-04-15) images differ by one day, and there is almost no change in the ground surface, so the extraction accuracy of GCP is sufficiently high

5 Internal Geometric Accuracy

The internal geometric accuracy of the image product not only reflects the geometric calibration performance of the camera, but also is affected by the attitude control performance of the spacecraft, which is further analyzed and discussed in terms of attitude vibration and geometric distortion based on typical Level 1 data optimized in the Harbin and Hami test areas.

5.1 Attitude Vibration

Since ZY-1 02D obtains image data by push-broom method, the dynamic imaging error caused by the attitude vibration of the spacecraft not only exists in the push-broom direction along the track, but also in the cross-track direction. In order to reduce the influence of geometric distortion residuals, it is generally possible to analyze some GCP data along the track.

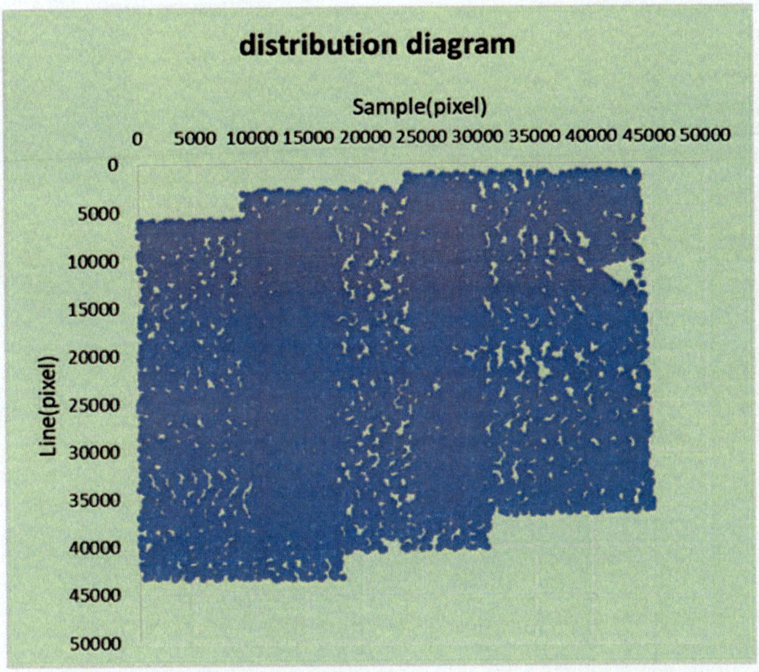

Fig. 9 The distribution diagram of GCPs extracted from the ZY-3 stereo triplets in the Hami test area (ZY-1_02D_VNIC_2020-05-30_88817). From left to right, the ZY-3 reference data were obtained on 2020-04-04, 2020-04-09, and 2019-10-15

In the Harbin area, taking data ZY-1_02D_VNIC_2020-04-14_70981 as an example, using part of the GCP data, and the vibration residuals in the CCD direction and the push-broom direction are plotted in Figs. 11 and 12, respectively. Figures 13 and 14 respectively show the residuals of the Hami data (ZY-1_02D_VNIC_2020-05-30_88817) in these two directions.

Analyzing the residual curves in Figs. 11, 12, 13, 14, it is not difficult to find:

- Both the along-track direction and the across-track direction are affected by the vibration and jitter of the platform;
- For the each scene, the interference in the along-track direction is more serious than the across-track direction.

For the degradation of image geometric accuracy caused by platform vibration, on the one hand, it can be improved by subsequent data processing, and on the other hand, it is also necessary to suppress spacecraft vibration.

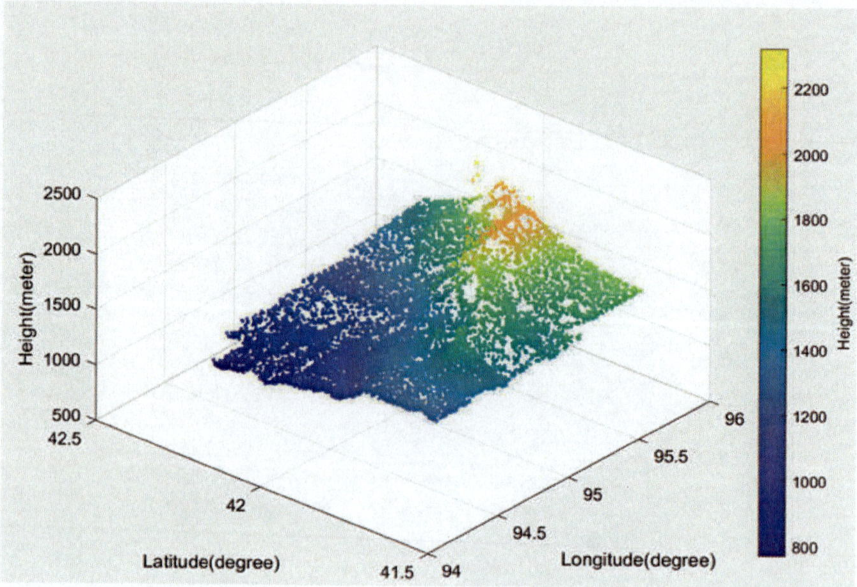

Fig. 10 The schematic of GCP/ICP distribution and elevation in the Hami test area (ZY-1_02D_VNIC_2020-05-30_88817)

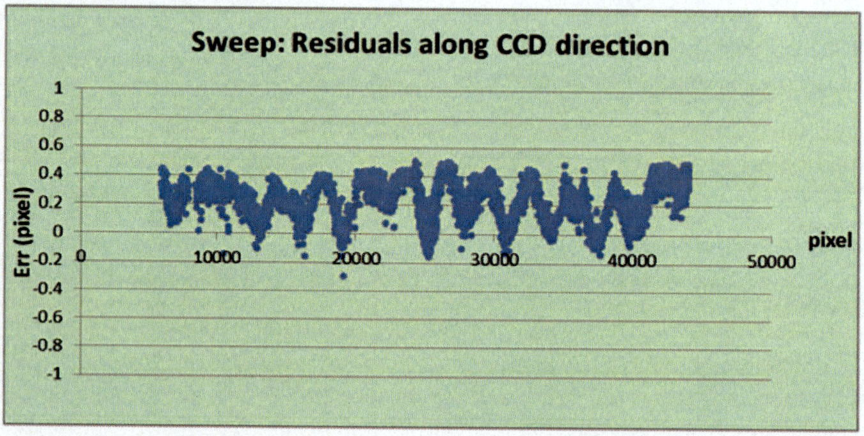

Fig. 11 The attitude residuals distribution along CCD direction (ZY-1_02D_VNIC_2020-04-14_70981)

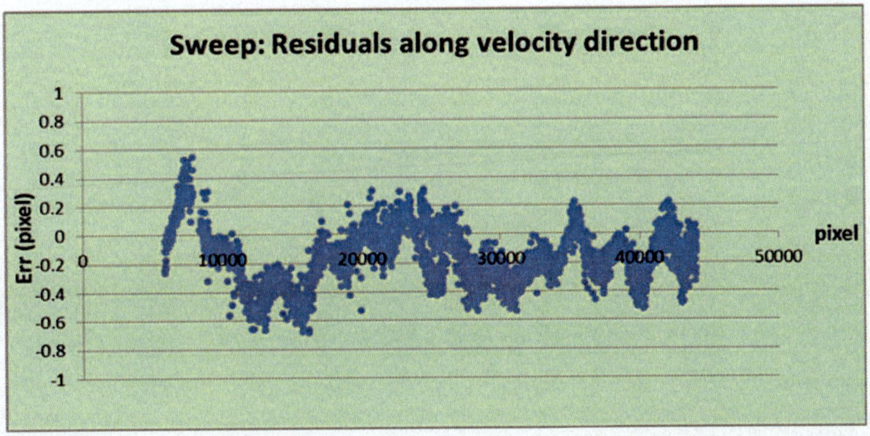

Fig. 12 The attitude residuals distribution along velocity direction (ZY-1_02D_VNIC_2020-04-14_70981)

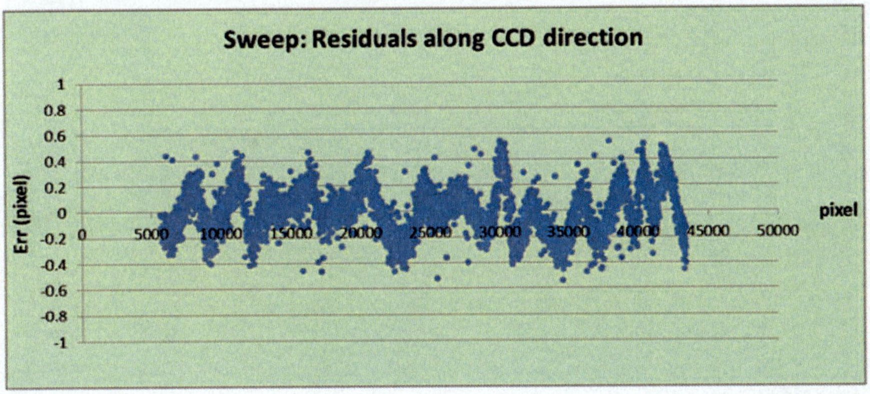

Fig. 13 The attitude residuals distribution along CCD direction (ZY-1_02D_VNIC_2020-05-30_88817)

5.2 *Geometric Distortion*

In order to overcome the influence of other factors on the analysis of geometric distortion, the interference of the vibration residual can be reduced by optimizing part of the GCP data along the CCD direction. Figures 15 and 16 show the geometric distortion residuals of the Harbin data (ZY-1_02D_VNIC_2020-04-14_70981) in the CCD direction and the track direction respectively. The residuals of the Hami data (ZY-1_02D_VNIC_2020-05-30_88817) in these two directions are plotted in Figs. 17 and 18. It seems that the residual geometric distortion is not particularly obvious, and further analysis is needed.

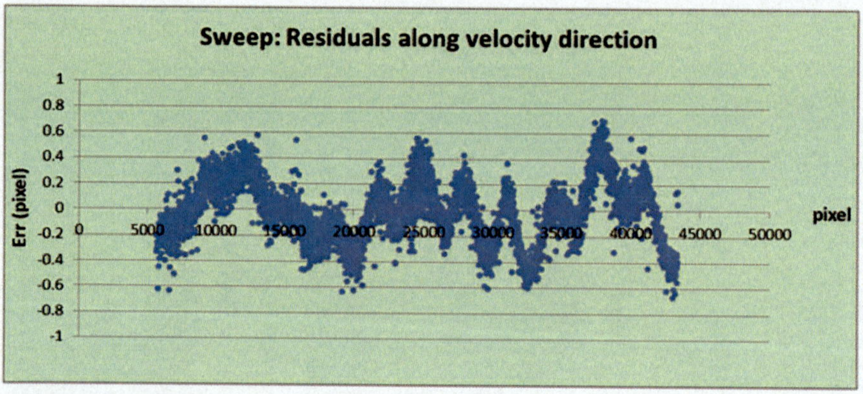

Fig. 14 The attitude residuals distribution along velocity direction (ZY-1_02D_VNIC_2020-05-30_88817)

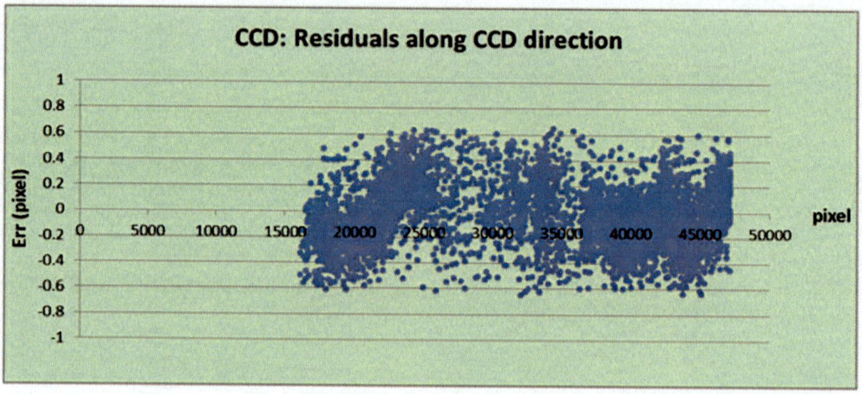

Fig. 15 The distortion residuals distribution along CCD direction (ZY-1_02D_VNIC_2020-04-14_70981)

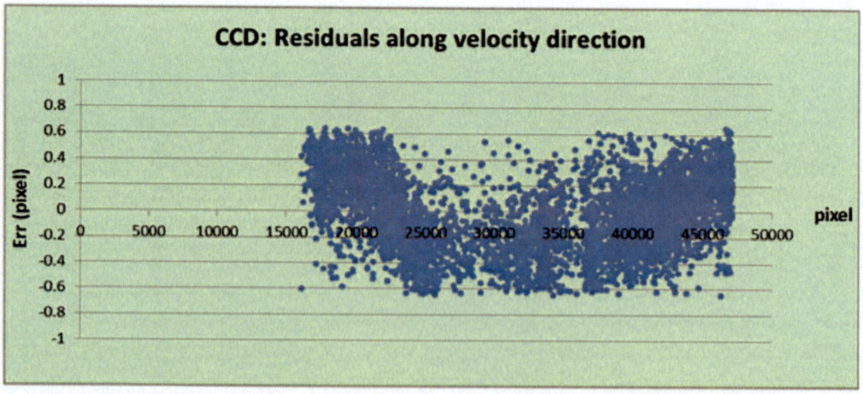

Fig. 16 The distortion residuals distribution along velocity direction (ZY-1_02D_VNIC_2020-04-14_70981)

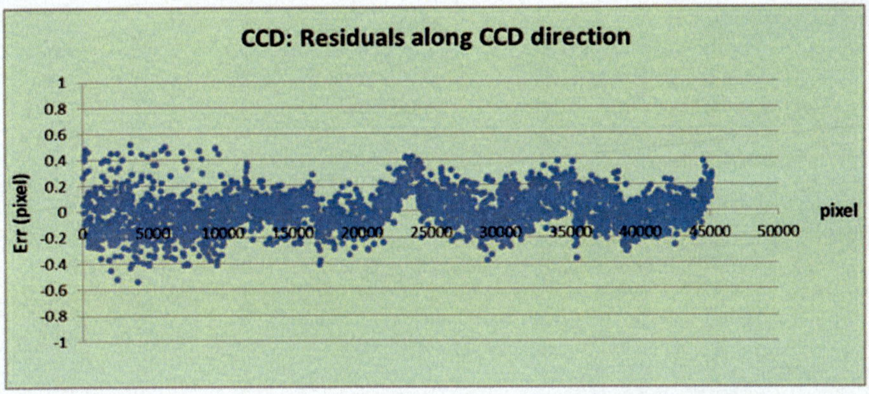

Fig. 17 The distortion residuals distribution along CCD direction (ZY-1_02D_VNIC_2020-05-30_88817)

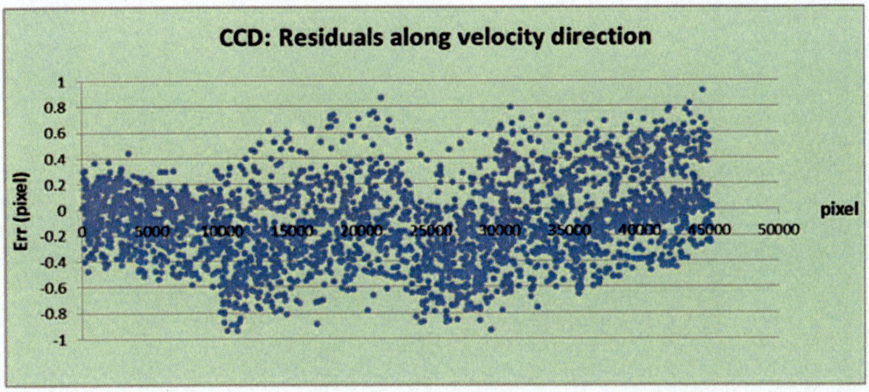

Fig. 18 The distortion residuals distribution along velocity direction (ZY-1_02D_VNIC_2020-05-30_88817)

6 Conclusion

The geometric accuracy of ZY-1 02D VNIC panchromatic image is quantitatively evaluated and analyzed. The root mean square error of the control points based on the ZY-3 reference data is better than 0.5 pixels in most cases, which meets the requirements of the design specification. However, the dynamic imaging error caused by the vibration of the spacecraft is also significant, and the impact of about 1.0 pixel size can be detected in the direction of the push-broom. It is recommended to improve this problem in the subsequent satellite design.

References

1. Foroosh H, Zerubia JB (2002) Extension of phase correlation to sub-pixel registration. IEEE Trans Image Proces 11(3):188–200
2. Grodecki J (2001) IKONOS stereo feature extraction—RPC approach. IN: ASPRS annual conference St. Louis
3. Grodecki J, Dial G (2003) Block adjustment of high-resolution satellite images described by rational functions. Photogramm Eng Remote Sens 69:59–70
4. Jacobsen K (2017) Problems and limitations of satellite image orientation for determination of height models. Int Arch Photogramm Remote Sens Spatial Inf Sci XLII-1/W1:257–264
5. Jacobsen K (2018) Systematic geometric image errors of very high resolution optical satellites. Int Arch Photogramm Remote Sens Spatial Inf Sci XLII-1:233–238
6. Kuglin CD, Hines DC (1975) The phase correlation image alignment method. Proc. Int. Conf. Cybernetics Society:163–165
7. Zhao L, Fu X, Dou X, Liu H, Fang Z (2018) Comparison and analysis of accuracy of elevation extraction based on the ZY-3 01 and 02 satellites stereoscopic images. Int Arch Photogramm Remote Sens Spatial Inf Sci XLII-3:2379–2384. https://doi.org/10.5194/isprs-archives-XLII-3-2379-2018
8. Zhao L (2020) Geometric accuracy evaluation of the ZY-3 stereo mapping satellite for 8 years. Int Arch Photogramm Remote Sens Spatial Inf Sci XLIII-B1:415–421. https://doi.org/10.5194/isprs-archives-XLIII-B1-2020-415-2020

Band Registration Analysis of the ZY-1 02D VNIC Image

Liping Zhao, Hongzhou Li, and Huai Wang

Abstract The ZY-1 02D mission is mainly devoted to the operational services of monitoring, understanding, and managing natural resources in China. The Visible Near-Infrared Camera (VNIC) carried on the ZY-1 02D is equipped with 8 multispectral bands at 10 m resolution with a 115 km swath width. The registration accuracy of the multi-spectral bands is analyzed using the multiple images cloud free from northern China. For each image, the relative registration errors among all bands are measured based on more than 10,000 evenly distributed and highly accurate tie-points extracted using the phase correlation method. According to the design characteristics of VNIC, the 8 bands are divided into 2 groups, the first 4 bands are B1234, and the last 4 bands are B5678. In most cases, the root mean square error (RMSE) of the registration is up to 0.3 pixels within group B1234 or B5678. However, between B1234 and B5678, the registration accuracy drops to about 0.4 pixels. After preliminary analysis, this degradation is believed to be caused by the slight vibration of the satellite platform and the geometric distortion between the bands.

Keywords ZY-1 02D · VNIC · Band registration · Accuracy · Analysis · Distortion · Vibration

1 Introduction

As an important part of China's space-based infrastructure, the ZY-1 02D mission is mainly devoted to the operational monitoring services of natural resource management, ecological and environmental protection, disaster prevention and control, etc. It was launched into orbit on September 12, 2019.

L. Zhao (✉) · H. Li · H. Wang
Land Satellite Remote Sensing Application Center, Ministry of Nature Resource, Beijing, China
e-mail: zhaolpwww@163.com

L. Wang et al. (eds.), *Proceedings of the 7th China High Resolution Earth Observation Conference (CHREOC 2020)*, Lecture Notes in Electrical Engineering 757, https://doi.org/10.1007/978-981-16-5735-1_6

The satellite features improved imagers compared to its predecessor, it is equipped with a 9-band Visible Near-Infrared Camera (VNIC) and a 166-band hyperspectral camera. The VNIC is a push-broom sensor with no moving parts, and it can simultaneously obtain the data consisting of one panchromatic band at 2.5 m resolution and eight multi-spectral bands at 10 m resolution with a 115 km swath width. Tables 1, 2 and 3 summarize the main characteristics of the ZY-1 02D and the specifications of the VNIC camera, respectively.

The geometric performance is an important indicator of image quality and a key factor affecting data applications. Generally speaking, the geolocation accuracy, the internal geometric accuracy and the band registration accuracy are the most critical

Table 1 Main characteristics of the ZY-1 02D

Launch date	September 12, 2019
Design life	5 years
Altitude	778 km
Repeat cycle	55 days
Revisit time	2–3 days, depending on latitude
Pointing capability in cross-track	±26°

Table 2 Specifications of the VNIC

Imager type	Push-broom with TDI capability
Swath width	>115 km
GSD at nadir	PAN: <2.5 m MS: <10 m
Pixel size	PAN: 7 μm MS: 28 μm
Sample	PAN: >47,000 MS: >11,800
Data quantization	12bits (4096)

Table 3 Spectral specifications of the VNIC

Band		Name	Wavelength (μm)	GSD (m)
Panchromatic		PAN	0.450–0.900	2.5
Multispectral	Band 1	Blue	0.450–0.520	10
	Band 2	Green	0.520–0.590	
	Band 3	Red	0.630–0.690	
	Band 4	Near-Infrared 1 (NIR1)	0.770–0.890	
	Band 5	Coastal Blue	0.400–0.450	
	Band 6	Yellow	0.590–0.625	
	Band 7	Red Edge	0.705–0.745	
	Band 8	Near-Infrared 2 (NIR2)	0.860–1.040	

geometric performance indicators, which are the main content of geometric calibration activities. These geometric characterization need to be continuously verified, monitored and improved throughout the life of the mission.

The goal of the band-to-band registration assessment is to verify the alignment accuracy of the multi-spectral bands and provide a basis for band placement calibration. In this paper, the registration accuracy of the eight multi-spectral bands of the VNIC camera is evaluated using multi-temporal image data located in northern China, and the influence of geometric distortion and platform stability is studied through residual analysis. Further using ZY-3 stereo image data as reference, the vibration of the spacecraft bus and the geometric accuracy of each multispectral band are discussed.

2 Geometric Model

In this study, the band-to-band registration assessment refers to the measurement and analysis of the relative alignment of the eight VNIC multi-spectral bands in scenes. Terrain influence is not considered, and the RPC model is not used. For a two-dimensional image, it will not only be affected by the residual geometric distortion on the focal plane of the camera, but also by the attitude vibration during the dynamic push-broom imaging process.

According to the dual-channel focal plane design of the VNIC camera, the 8 multi-spectral bands are divided into 2 groups in order, B1 to B4 (called Band1234 or B1234) are integrated into one Sensor Chip Assembly (SCA), and B5 to B8 (called Band5678 or B5678) are arranged into another SCA. In the focal plane of the camera, the 8 SCAs are arranged in two rows of four. 4 SCAs with Band1234 are stitched along one line, and the other 4 SCAs with Band5678 are splicing along another line. The field of view (FOV) of the telescope is about 1.6° in the along-track direction and approximately 8.5° in the across-track direction. The distance between the two groups of devices is much greater than the distance between the bands.

3 Datasets

The Hami area in north-western China has the advantages of good weather conditions, simple land coverage and moderate topography relief, so it is preferred as the experimental area. For 8 bands of VNIC image data, a large number of evenly distributed tie points (TPs) are extracted based on the phase correlation method, which are used for the band co-registration accuracy analysis.

3.1 Study Area

As shown in Fig. 1, the research area is located in the southeast of Hami City. The air in this region is dry, the atmosphere is transparent, and the cloud cover is little. The land cover is dominated by Gobi, vegetation is scarce, which is hardly affected by seasonal changes. Table 4 lists 4 scenes of the experimental area, which were obtained on May 30 and June 2, 2020. The two phases of data overlap almost completely, and the off-nadir angles are −1.00° and −6.38°, respectively. Figure 2 shows the thumbnails of two of them.

3.2 Tiepoints

The extraction accuracy and point distribution of the tiepoints among various bands are the key factors for analyzing the registration accuracy. Many methods have been designed for image registration and accuracy estimation. The phase correlation

Fig. 1 The schematic diagram of the study area and image coverage. Green: ZY-1 02D (2020-05-30); Blue: ZY-1 02D (2020-06-02)

Table 4 The ZY-1 02D VNIC image data of the research area

Scene ID	Date	Orbit No	Yaw (degree)	Roll (degree)	Tiepoint Count
88816	2020-05-30	003747	2.848	−1.000	10,037
88817	2020-05-30	003747	2.890	−1.000	10,041
89891	2020-06-02	003790	2.843	−6.380	10,015
89892	2020-06-02	003790	2.885	−6.380	10,028

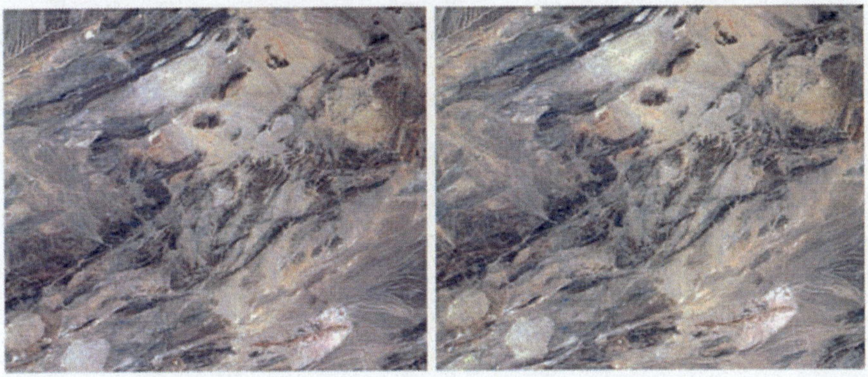

Fig. 2 The schematic diagram of the study area and image coverage. Left: ZY-1 02D (2020-05-30_88817); Right: ZY-1 02D (2020-06-02_89892)

method [1, 2] can not only obtain sub-pixel registration accuracy estimation, but also has strong robustness against noise and speckle. As shown in Figs. 3 and 4, this method can uniformly extract more than 10,000 points on the 8 bands of each scene.

4 Evaluation

4.1 Methodology

Due to the different viewing angles of the Band1234 and B5678 on the focal plane of the VNIC camera, the accuracy evaluation is divided into the following three cases:

(1) The registration accuracy of the 4 bands inside Band1234;
(2) The registration accuracy of the 4 bands inside Band5678;
(3) The registration accuracy between Band1234 and Band5678.

Fig. 3 The tie points distribution diagram of the study image

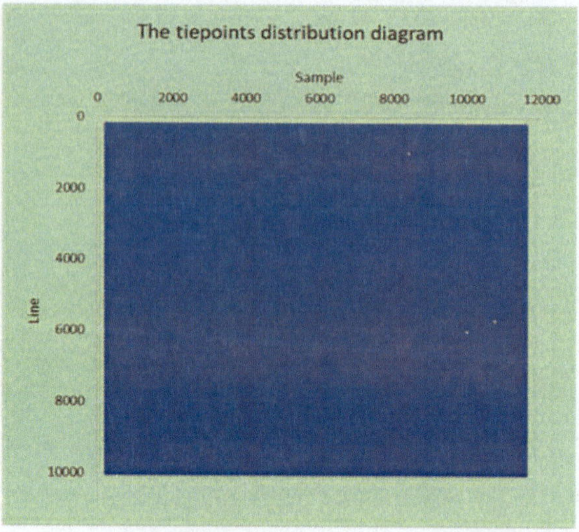

(a) ZY-1 02D 2020-05-30_88817 (tiepoint count: 10041)

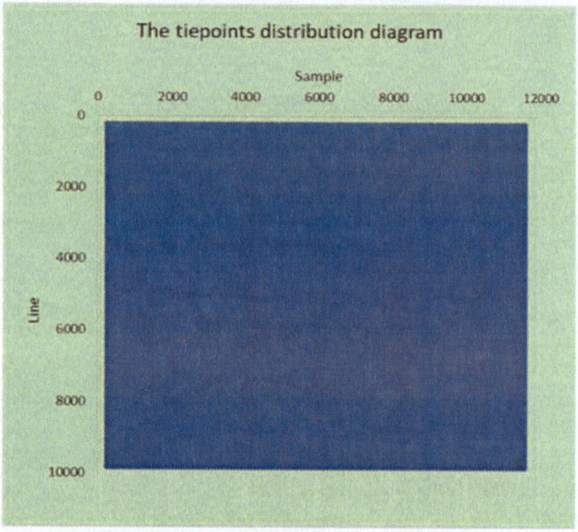

(b) ZY-1 02D 2020-06-02_89892 (tiepoint count: 10028)

In addition, due to the slight difference in the viewing angle of each band, the geometric displacement caused by topographical relief is basically negligible, so the image for band registration assessment has not been corrected for terrain.

Using the tiepoints extracted between any two bands, the registration accuracy can be calculated by the following process:

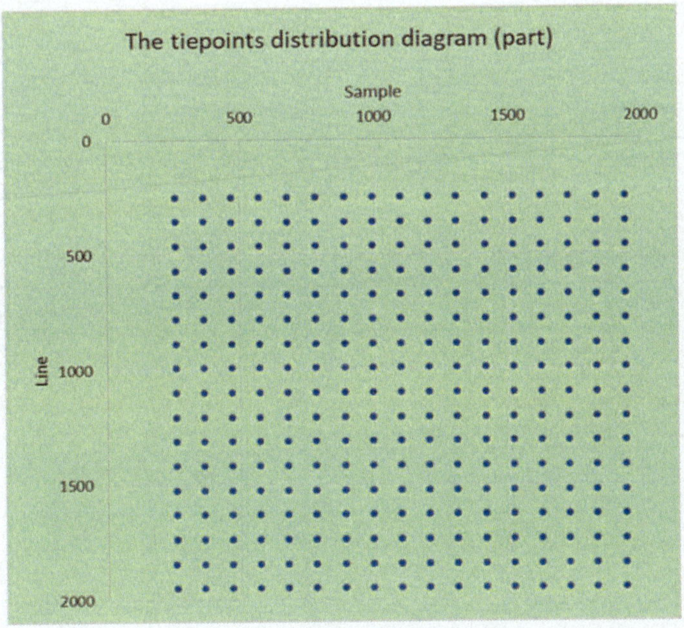

Fig. 4 The tie points distribution diagram of the test image (Part of image 2020-05-30_88817)

(1) For each tiepoint, calculate the difference in the direction of sample and line respectively, results in a list of "ΔLine" and "ΔSample" values;
(2) For each pair of bands, compute the mean, standard deviations and RMS error of "ΔLine" and "ΔSample" values of all TPs.

4.2 Results

Tables 5, 6, 7, 8 summarize the registration accuracy of the 4 scenes of the research area. In general, the registration accuracy of the Band1234 or Band5678 is higher, while the accuracy between Band1234 and Band5678 is lower, which requires further detailed analysis.

Take ZY-1_02D_VNIC_20200530_88817 as an example. For Band1234, Tables 9, 10, 11 list the mean, standard deviations and RMS error of "ΔLine", "ΔSample" and "ΔRadial" respectively. The accuracy data of Band5678 are given in Tables 12, 13, 14, and the accuracy between Band1234 and Band5678 is calculated in Tables 15, 16, 17.

Table 5 The RMSE of band registration of image (ZY-1_02D_VNIC_20200530_88816) (unit: pixel)

Band	B2	B3	B4	B5	B6	B7	B8
B1	0.141	0.141	0.363	0.413	0.376	0.464	0.449
B2		0.104	0.277	0.434	0.382	0.428	0.418
B3			0.302	0.453	0.397	0.469	0.453
B4				0.531	0.485	0.403	0.383
B5					0.125	0.294	0.296
B6						0.269	0.266
B7							0.121

Table 6 The RMSE of band registration of image (ZY-1_02D_VNIC_20200530_88817) (unit: pixel)

Band	B2	B3	B4	B5	B6	B7	B8
B1	0.097	0.119	0.309	0.338	0.321	0.335	0.371
B2		0.073	0.241	0.383	0.356	0.324	0.372
B3			0.249	0.371	0.332	0.325	0.362
B4				0.493	0.463	0.317	0.355
B5					0.112	0.292	0.304
B6						0.265	0.276
B7							0.143

Table 7 The RMSE of band registration of image (ZY-1_02D_VNIC_20200602_89891) (unit: pixel)

Band	B2	B3	B4	B5	B6	B7	B8
B1	0.128	0.136	0.353	0.304	0.282	0.408	0.409
B2		0.093	0.269	0.313	0.269	0.349	0.357
B3			0.301	0.315	0.263	0.382	0.382
B4				0.407	0.360	0.246	0.243
B5					0.136	0.326	0.328
B6						0.284	0.284
B7							0.108

Comparing the "mean" data in Tables 9, 10, 11, 12, 13, 14, 15, 16, 17, it can be seen that,

(1) Inside Band1234 or Band5678, the alignment error in the sample direction is less than 0.1 pixel, while that in the line direction is slightly larger, not exceeding 0.2 pixel;
(2) The alignment error between Band1234 and Band5678 is larger, especially in the line direction between some bands.

Table 8 The RMSE of band registration of image (ZY-1_02D_VNIC_20200602_89892) (unit: pixel)

Band	B2	B3	B4	B5	B6	B7	B8
B1	0.133	0.153	0.290	0.318	0.297	0.394	0.374
B2		0.084	0.211	0.333	0.263	0.338	0.327
B3			0.232	0.342	0.258	0.367	0.344
B4				0.356	0.288	0.244	0.233
B5					0.199	0.307	0.290
B6						0.243	0.219
B7							0.109

Table 9 The band registration accuracy in the SAMPLE direction of image (ZY-1_02D_VNIC_20200530_88817) (B1234) (unit: pixel)

Band	Mean			Std. Dev			RMSE		
	B2	B3	B4	B2	B3	B4	B2	B3	B4
B1	0.030	0.026	0.097	0.043	0.084	0.174	0.052	0.088	0.200
B2		−0.004	0.067		0.050	0.154		0.050	0.168
B3			0.071			0.148			0.164

Table 10 The band registration accuracy in the LINE direction of image (ZY-1_02D_VNIC_20200530_88817) (B1234) (unit: pixel)

Band	Mean			Std. Dev			RMSE		
	B2	B3	B4	B2	B3	B4	B2	B3	B4
B1	0.070	0.039	0.186	0.041	0.070	0.145	0.081	0.080	0.236
B2		−0.031	0.116		0.042	0.128		0.053	0.173
B3			0.148			0.115			0.187

Table 11 The band registration accuracy in the RADIAL direction of image (ZY-1_02D_VNIC_20200530_88817) (B1234) (unit: pixel)

Band	Mean			Std. Dev			RMSE		
	B2	B3	B4	B2	B3	B4	B2	B3	B4
B1	0.088	0.101	0.258	0.039	0.063	0.170	0.097	0.119	0.309
B2		0.063	0.190		0.037	0.148		0.073	0.241
B3			0.211			0.132			0.249

Table 12 The band registration accuracy in the SAMPLE direction of image (ZY-1_02D_VNIC_20200530_88817) (B5678) (unit: pixel)

Band	Mean			Std. Dev			RMSE		
	B6	B7	B8	B6	B7	B8	B6	B7	B8
B5	−0.001	0.020	−0.014	0.082	0.150	0.198	0.082	0.152	0.199
B6		0.022	−0.013		0.150	0.197		0.152	0.198
B7			−0.034			0.104			0.110

Table 13 The band registration accuracy in the LINE direction of image (ZY-1_02D_VNIC_20200530_88817) (B5678) (unit: pixel)

Band	Mean			Std. Dev			RMSE		
	B6	B7	B8	B6	B7	B8	B6	B7	B8
B5	0.015	0.204	0.153	0.074	0.143	0.172	0.076	0.249	0.230
B6		0.189	0.138		0.106	0.134		0.217	0.192
B7			−0.051			0.077			0.092

Table 14 The band registration accuracy in the RADIAL direction of image (ZY-1_02D_VNIC_20200530_88817) (B5678) (unit: pixel)

Band	Mean			Std. Dev			RMSE		
	B6	B7	B8	B6	B7	B8	B6	B7	B8
B5	0.090	0.266	0.257	0.066	0.121	0.163	0.112	0.292	0.304
B6		0.247	0.235		0.094	0.145		0.265	0.276
B7			0.097			0.105			0.143

Band alignment errors can be solved by offset correction in subsequent data processing.

Under normal circumstances, the source or expression of error can be revealed through the analysis of standard deviation data by means of visual graphic display. The distribution of registration errors for band 1 and band 2 is shown in Fig. 5. The error is very small, less than 0.1 pixel, which is very ideal.

Figure 6 illustrates the distribution of registration errors for band 1 and band 4. There are significant static errors along the detector array direction and along the push-broom direction, which are caused by the inconsistent geometric distortion correction between B1 and B4. It can be found from Fig. 7 that similar problems exist between B5 and B8. Figure 8 shows the registration error between the furthest B1 and B8; in addition to the obvious difference in geometric distortion correction, there is also a dynamic imaging error along the push-broom direction.

Table 15 The band registration accuracy in the SAMPLE direction of image (ZY-1_02D_VNIC_20200530_88817) (B1234 and B5678) (unit: pixel)

Band	Mean				Std. Dev				RMSE			
	B5	B6	B7	B8	B5	B6	B7	B8	B5	B6	B7	B8
B1	0.137	0.136	0.158	0.124	0.135	0.119	0.237	0.283	0.193	0.181	0.284	0.309
B2	0.107	0.106	0.128	0.094	0.136	0.099	0.226	0.271	0.173	0.145	0.260	0.287
B3	0.111	0.110	0.132	0.098	0.153	0.100	0.235	0.275	0.189	0.149	0.270	0.292
B4	0.040	0.039	0.060	0.026	0.150	0.113	0.150	0.168	0.155	0.119	0.161	0.170

Table 16 The band registration accuracy in the LINE direction of image (ZY-1_02D_VNIC_20200530_88817) (B1234 and B5678) (unit: pixel)

Band	Mean				Std. Dev				RMSE			
	B5	B6	B7	B8	B5	B6	B7	B8	B5	B6	B7	B8
B1	−0.236	−0.221	−0.032	−0.083	0.147	0.146	0.174	0.188	0.278	0.265	0.177	0.206
B2	−0.306	−0.291	−0.102	−0.153	0.153	0.143	0.165	0.179	0.342	0.325	0.194	0.236
B3	−0.275	−0.260	−0.071	−0.122	0.161	0.144	0.167	0.176	0.319	0.297	0.181	0.214
B4	−0.422	−0.408	−0.219	−0.269	0.202	0.184	0.164	0.155	0.468	0.447	0.273	0.311

Table 17 The band registration accuracy in the RADIAL direction of image (ZY-1_02D_VNIC_20200530_88817) (B1234 and B5678) (unit: pixel)

Band	Mean				Std. Dev				RMSE			
	B5	B6	B7	B8	B5	B6	B7	B8	B5	B6	B7	B8
B1	0.304	0.292	0.289	0.295	0.149	0.133	0.168	0.226	0.338	0.321	0.335	0.371
B2	0.354	0.330	0.286	0.302	0.147	0.132	0.153	0.217	0.383	0.356	0.324	0.372
B3	0.342	0.309	0.293	0.298	0.142	0.122	0.141	0.205	0.371	0.332	0.325	0.362
B4	0.454	0.428	0.292	0.318	0.194	0.175	0.124	0.156	0.493	0.463	0.317	0.355

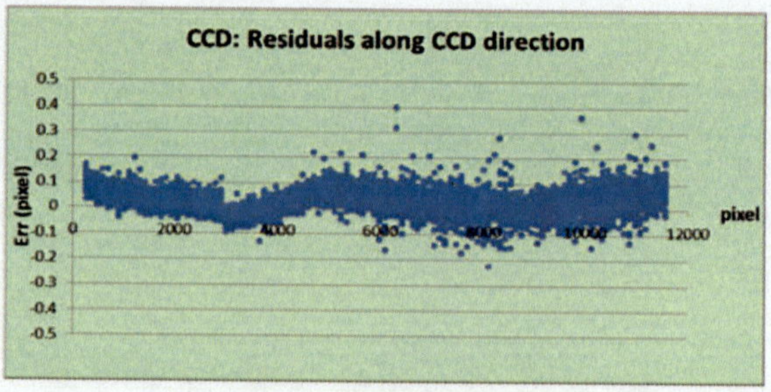

(a)

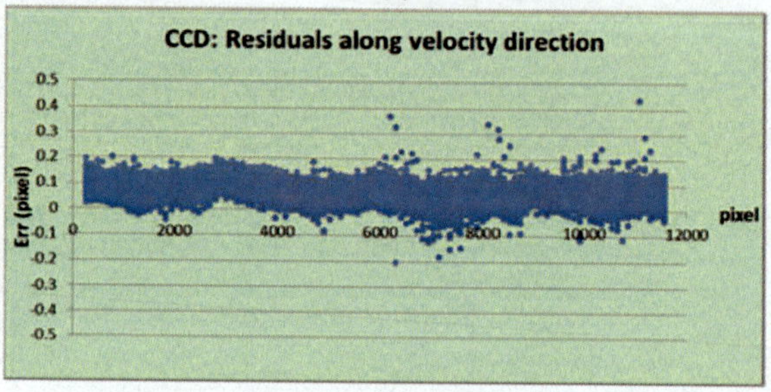

(b)

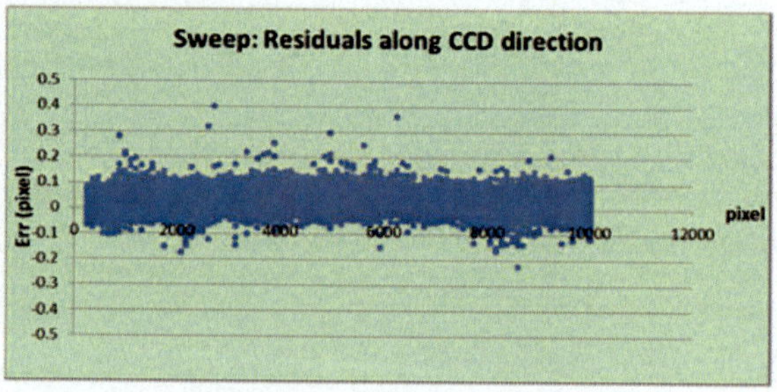

(c)

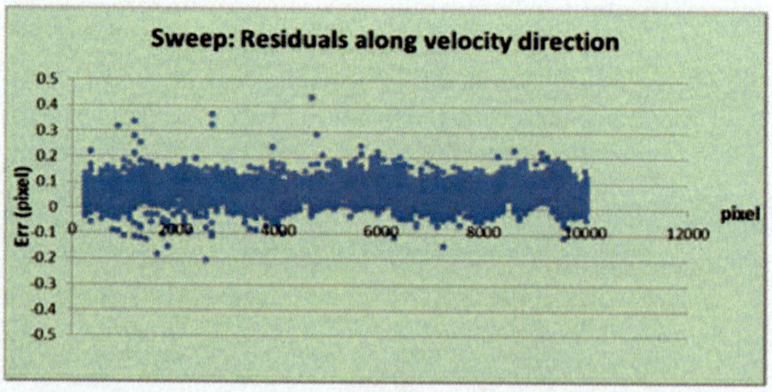

(d)

Fig. 5 The distribution of registration errors for band 1 and band 2 (ZY-1_02D_VNIC_20200530_88817). **a** Static error along the direction of the detector array; **b** Static error in the direction perpendicular to the detector array; **c** dynamic error in the cross-track direction; **d** dynamic error along the track direction

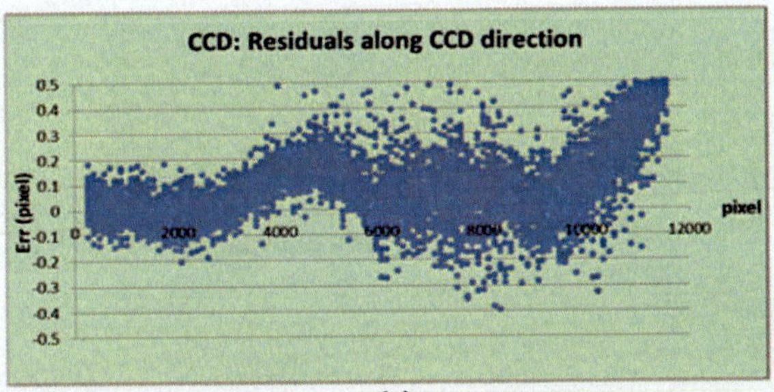

(a)

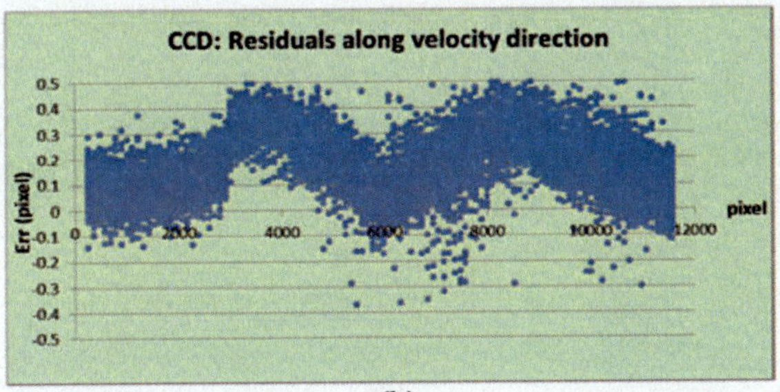

(b)

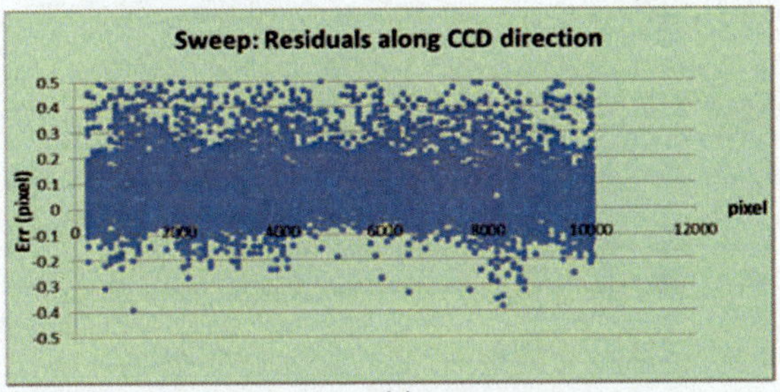

(c)

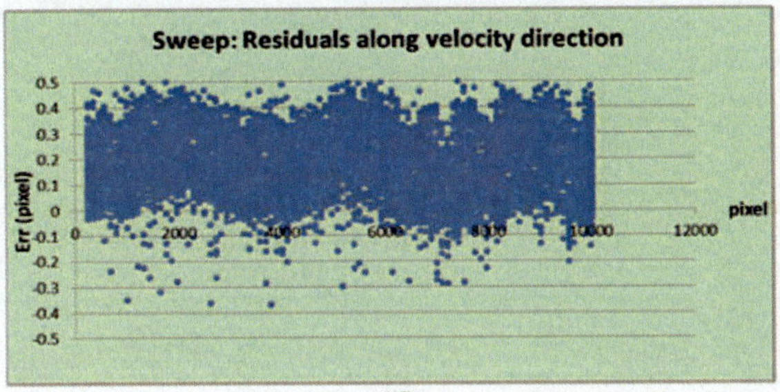

(d)

Fig. 6 The distribution of registration errors for band 1 and band 4 (ZY-1_02D_VNIC_20200530_88817). **a** Static error along the direction of the detector array; **b** Static error in the direction perpendicular to the detector array; **c** dynamic error in the cross-track direction; **d** dynamic error along the track direction

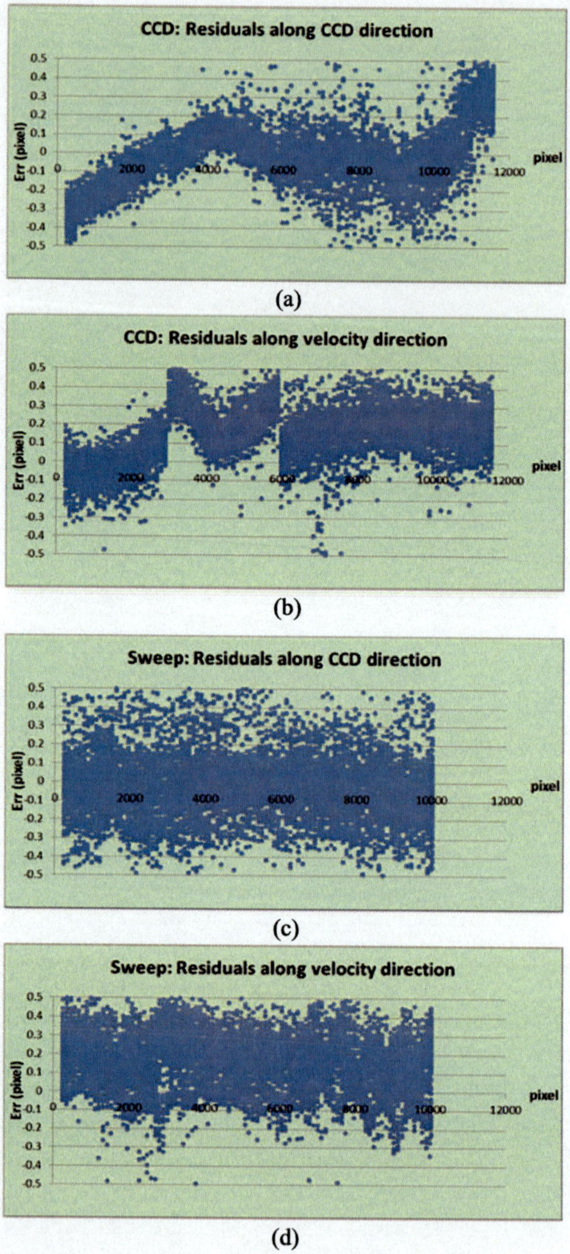

Fig. 7 The distribution of registration errors for band 5 and band 8 (ZY-1_02D_VNIC_20200530_88817). **a** Static error along the direction of the detector array; **b** static error in the direction perpendicular to the detector array; **c** dynamic error in the cross-track direction; **d** dynamic error along the track direction

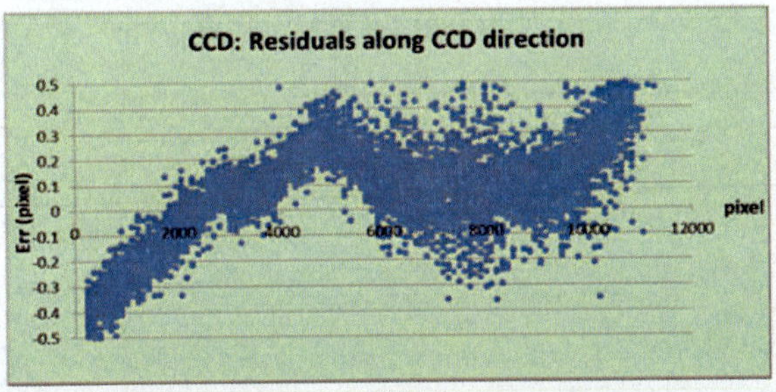

(a)

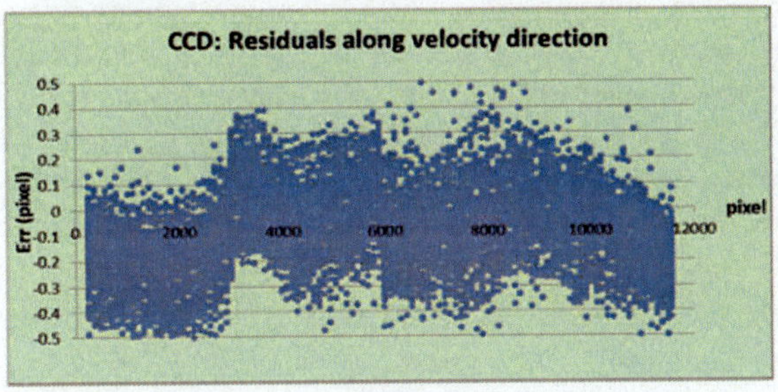

(b)

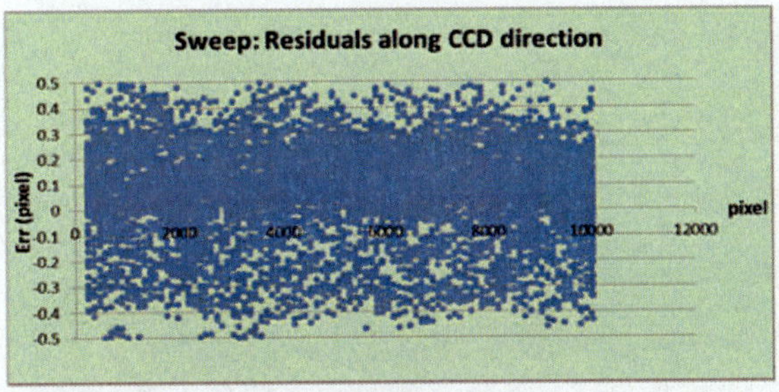

(c)

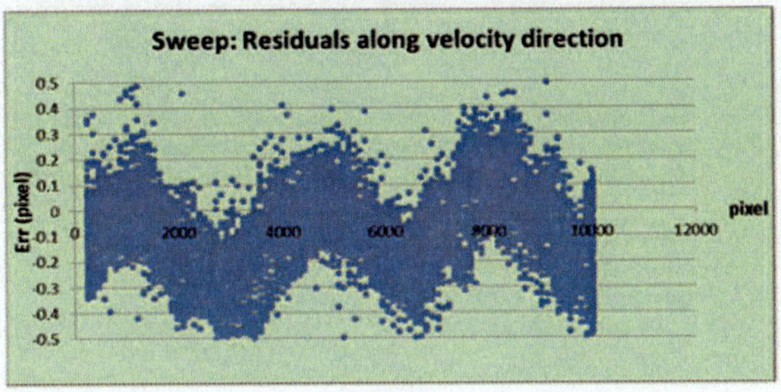

(d)

Fig. 8 The distribution of registration errors for band 1 and band 8 (ZY-1_02D_VNIC_20200530_88817). **a** Static error along the direction of the detector array; **b** static error in the direction perpendicular to the detector array; **c** dynamic error in the cross-track direction; **d** dynamic error along the track direction

Summarizing Figures 5, 6, 7, 8, two problems were detected:

(1) There are differences in geometric distortion correction between bands;
(2) There are errors caused by vibration in the dynamic imaging process.

5 Analysis

In order to further analyze the band registration error caused by static geometric distortion and dynamic imaging error. The ZY-3 stereo triplets are used as reference data [3], and the geometric accuracy of the panchromatic and multi-spectral bands of the VNIC camera has been tested. The GCP distribution is shown in Fig. 9, and the adjustment accuracy with all GCPs in image space is listed in Table 18.

The error distribution of the PAN image with 2.5 m GSD is shown in Fig. 10. The vibration during the dynamic imaging process along the push-broom direction is very obvious. It can also be seen from Fig. 11 that there are still geometric distortions in the multi-spectral bands that need to be further corrected.

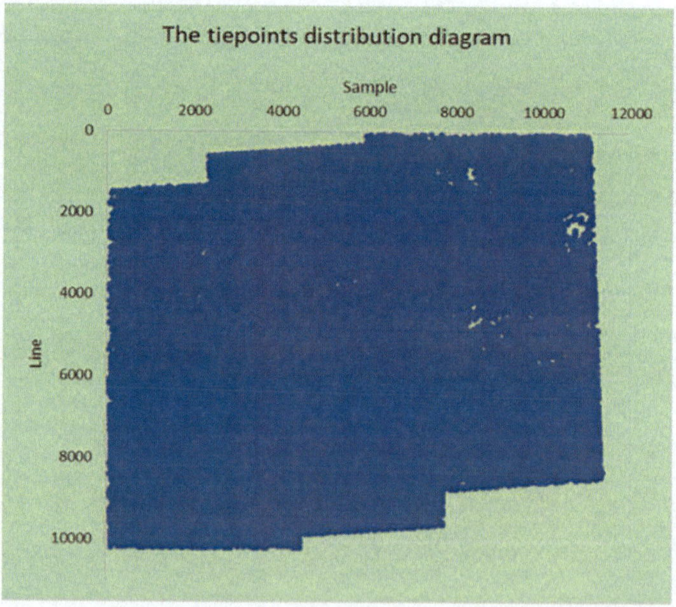

Fig. 9 The GCP distribution diagram of the study image (ZY-1 02D 2020-05-30_88817)

Table 18 The accuracy of each band of ZY-1 02D VNIC image (ZY-1 02D 2020-05-30_88817) (unit: pixel)

Bands	ΔSample	ΔLine	ΔRadial	Count
PAN	0.160	0.303	0.366	49,125
B1	0.112	0.148	0.195	49,125
B2	0.097	0.140	0.176	49,125
B3	0.092	0.134	0.167	49,125
B4	0.114	0.162	0.206	49,125
B5	0.102	0.150	0.185	49,125
B6	0.081	0.136	0.157	49,125
B7	0.098	0.163	0.196	49,125
B8	0.099	0.166	0.195	49,125

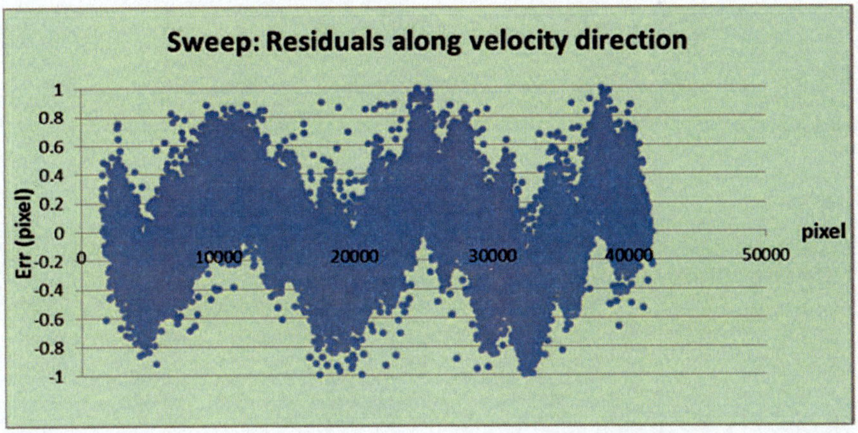

Fig. 10 The error distribution of the PAN image (ZY-1 02D 2020-05-30_88817)

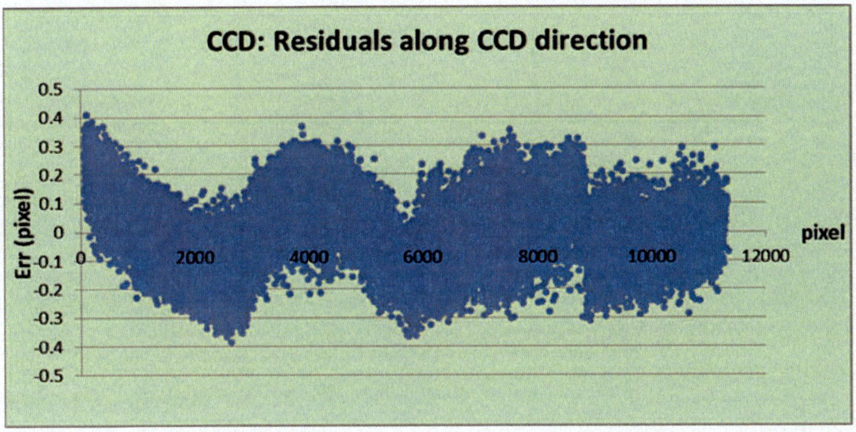

Fig. 11 The error distribution of the band 4 (ZY-1 02D 2020-05-30_88817)

6 Conclusion

The registration accuracy between the 8 multi-spectral bands of the ZY-1 02D VNIC camera is quantitatively evaluated. In most cases, the root mean square error (RMSE) of the registration inside B1234 or B5678 can reach 0.3 pixels. However, the registration accuracy between B1234 and B5678 will drop to 0.4 or even 0.5 pixels. According to the preliminary analysis, the band alignment error, the geometric distortion residuals of each band of the multi-spectrum, and the vibration error in the dynamic imaging along the push-broom direction are the three main factors that cause the registration accuracy to decrease significantly. Band alignment errors can be resolved by offset correction in subsequent data processing. The geometric distortion error is systematic and stable and should be improved in the products provided by the supplier. Suppressing the vibration of the spacecraft bus is the best solution to overcome dynamic imaging errors, and correcting the vibration error in data post-processing can also be used as a remedial measure.

References

1. Foroosh H, Zerubia JB (2002) Extension of phase correlation to sub-pixel registration. IEEE Trans Image Process 11(3):188–200
2. Kuglin CD, Hines DC (1975) The phase correlation image alignment method. Proc Int Conf Cybernetics Society:163–165
3. Zhao L (2020) Geometric accuracy evaluation of the ZY-3 stereo mapping satellite for 8 years. Int Arch Photogramm Remote Sens Spatial Inf Sci XLIII-B1:415–421. https://doi.org/10.5194/isprs-archives-XLIII-B1-2020-415-2020

Analysis and Optimization of Geometric Accuracy of ZY-1 02C HRC Image

Liping Zhao

Abstract The ZY-1 02C is China's first operational high-resolution civilian remote sensing satellite. It was launched on December 22, 2011, it has been in a continuous operation on orbit for more than 9 years. It is equipped with a panchromatic and multispectral scanner and a panchromatic High Resolution Camera (HRC) with 2.36 m spatial resolution. The geometric accuracy of ZY-1 02C HRC is evaluated using two revisited data in Northwest China. Based on the ground control points (GCPs) extracted from the ZY-3 standard stereo triplet products, the adjustment accuracy is analyzed, the root mean square error (RMSE) of panchromatic images is within 2.0 pixels. Further, through static and dynamic residual analysis of GCP, it is found that due to the influence of the vibration of the satellite platform, there is a significant dynamic imaging error in the image data, especially in the push-broom direction along the orbit, an effect of about 5.0 pixels in size is detected from the image. After correcting the vibration, the geometric accuracy of the image can be improved to within 1.0 pixels, reaching the geometric accuracy of most images, reducing the adverse effects on the application. In general, this article quantitatively analyzes the geometric accuracy of the ZY-1 02C HRC image, studies the non-negligible vibration effects of the satellite platform, optimizes the geometric accuracy through the data processing method of attitude compensation, and improves the quality of image data.

Keywords ZY-1 02C · Geometric accuracy · Vibration · Analysis · Optimization

---

L. Zhao (✉)
Land Satellite Remote Sensing Application Center, Ministry of Nature Resource, Beijing, China
e-mail: zhaolpwww@163.com

L. Wang et al. (eds.), *Proceedings of the 7th China High Resolution Earth Observation Conference (CHREOC 2020)*, Lecture Notes in Electrical Engineering 757, https://doi.org/10.1007/978-981-16-5735-1_7

1 Introduction

ZY-1 02C is China's first operational high-resolution civilian remote sensing satellite, is mainly devoted to the application services of natural resources investigation and monitoring, disaster prevention and reduction, ecology and environment protection. Although the design life of the satellite is 3 years, it has been in orbit for more than 9 years since launching on December 22, 2011. The spacecraft is equipped with a panchromatic and multispectral scanner (PMS) and two panchromatic high resolution cameras (HRC) with 2.36 m spatial resolution in a swath of 54 km. Tables 1 and 2 summarize the main characteristics of the ZY-1 02C and the specifications of the HRC camera, respectively.

The geometric accuracy of the image is an important indicator of data quality and a key factor affecting data application. In this paper, the ZY-3 stereo triplet data is used as reference data to evaluate the geometric accuracy of HRC, and through the residual analysis method, the effects of geometric calibration accuracy and spacecraft attitude stability on the image geometric accuracy are studied. Aiming at the vibration problem of the spacecraft bus, the attitude compensation method is adopted to optimize the geometric accuracy and improve the quality of the image data.

2 Datasets

The Hami area in northwestern China is preferred as the experimental area based on weather, land coverage, and topographic relief. The geometric accuracy of the ZY-1 02C HRC is evaluated and analysed using the ZY-3 stereo data obtained in this area as reference data.

Table 1 Main characteristics of the ZY-1 02C

Launch date	December 22, 2011
Design life	3 years
Altitude	780 km
Repeat cycle	55 days
Revisit time	2–3 days, depending on latitude

Table 2 Specifications of the HRC (HR1/HR2)

Focal length	3300 mm
Spectral range	500–800 nm
Swath width	Single: >27 km Double: >54 km
GSD at nadir	2.36 m
Pixel size	10 μm
Detectors	TDI CCD: 4096 pixels × 3 (single camera)
Sample	>12,200 pixels (single camera)
Pointing capability cin cross-track	±25°

2.1 Study Area

The study area is located in the southeast of the Hami City, as shown in Fig. 1. The elevation range of the experimental area is approximately 800–1800 m above the WGS84 ellipsoid, but the topography varies relatively little. The land cover in this area is simple and the change is small, which is conducive to the extraction of GCP/ICP (Independent Check Points). Table 3 lists the image data of the research area.

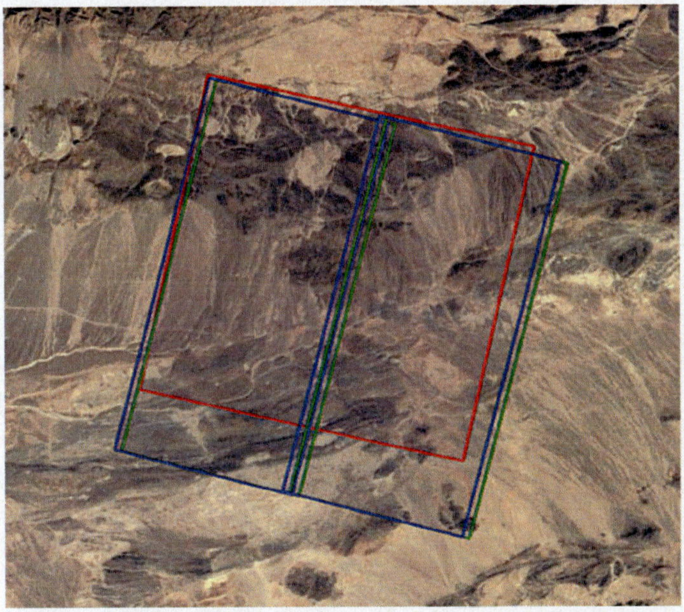

Fig. 1 The schematic diagram of the study area and image coverage. Red: ZY-3 (2020-04-09); Green: ZY-1 02C (2019-12-07); Blue: ZY-1 02C (2020-03-26)

Table 3 The image data of the research area

Satellite	Camera	Date	Orbit No	Roll angle (degree)
ZY-1 02C	HR1 HR2	2019-12-07	41704	0.005
	HR1 HR2	2020-03-26	43282	0.005
ZY-3	Stereo camera	2020-04-09	21439	−0.004

2.2 GCP and ICP

When studying geometric accuracy, high-precision reference data is generally needed, such as ground-surveyed control points, or high-precision digital orthophoto map (DOM) and digital elevation model (DEM). However, the acquisition of these data generally requires a higher cost. The ZY-3 stereo triplets, with the good internal geometric accuracy [1], can be used to extract GCP/ICP with sufficient accuracy. In particular, ZY-1 02C and ZY-3 data with similar date or season can be selected, which is helpful to extract a large number of dense and uniformly distributed reference points with high precision.

In this paper, a large number of evenly distributed tie points (TPs) are extracted based on the phase correlation method [1–3] for HR1 and HR2 camera test data. The planimetric coordinates and elevation values are calculated by the ZY-3 stereo triplets. After removing the gross error, the number of TPs retained is 17777 and 22266, respectively. Figure 2 gives the distribution diagram.

3 Evaluation

3.1 Methodology

Rational function model (RFM) is often used as an imaging geometry model to establish the mapping relationship between image and ground. The bundle block adjustment with the affine correction in image space based on RFM is used to evaluate the geometric accuracy according to the previous experience [4–8]. The general workflow is as follows:

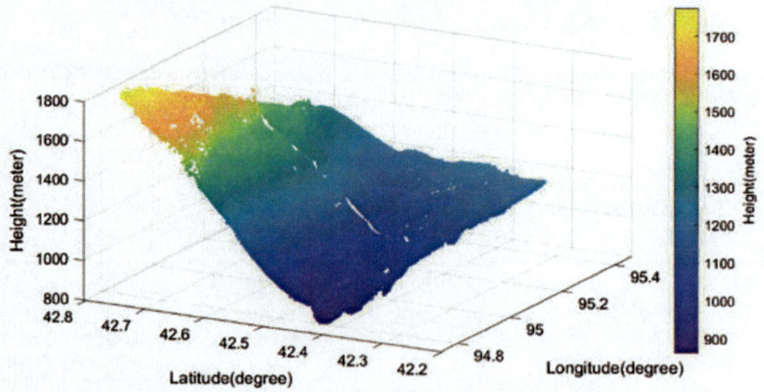

Fig. 2 The schematic of GCP/ICP distribution and elevation

(1) Bundle adjustment with the GCP to optimize geometry of ZY-1 02C image;
(2) Calculate the adjusted image coordinates of the GCP;
(3) For each GCP, subtract image reference coordinates from calculated image coordinates, results in a list of "ΔLine", and "ΔSample" values;
(4) For each image, compute the mean, standard deviations and RMS error of "ΔLine", and "ΔSample" values.

3.2 Results

Table 4 summarizes the accuracy of HR1 and HR2 image, and the overall accuracy is better than 2.0 pixels. An abnormal phenomenon can be found, the accuracy of "line" is significantly lower than that of "sample", which needs further detailed analysis.

4 Analysis

A simple analysis of the standard deviation in Table 4 shows that the accuracy of the push-broom direction is significantly greater than that of the detector array direction, which may be caused by static geometric distortion of the camera or dynamic push-broom imaging.

Comparing the error data in Table 4, it is found that ZY-1 02C HR2 20191207 has the largest error. Therefore, taking this data as an example, a three-dimensional distribution diagram of "ΔLine" error as shown in Fig. 3 is drawn. It can be clearly seen from the figure that the error distribution presents three parts. In addition, the

Table 4 The accuracy of the ZY-1 02C HR1 and HR2 image (unit: pixel)

	Image	ΔSample	ΔLine	ΔRadial	GCP count
Mean	ZY-1 02C HR1 20191207	0.000	0.000	1.445	17,777
Standard deviation		0.585	1.653	0.994	
RMSE		0.585	1.653	1.754	
Mean	ZY-1 02C HR2 20191207	0.000	0.000	1.535	22,266
Standard deviation		0.601	1.666	0.884	
RMSE		0.601	1.666	1.771	
Mean	ZY-1 02C HR1 20200326	0.000	0.000	1.210	17,777
Standard deviation		0.252	1.372	0.695	
RMSE		0.252	1.372	1.395	
Mean	ZY-1 02C HR2 20200326	0.000	0.000	1.262	22,266
Standard deviation		0.300	1.460	0.791	
RMSE		0.300	1.460	1.490	

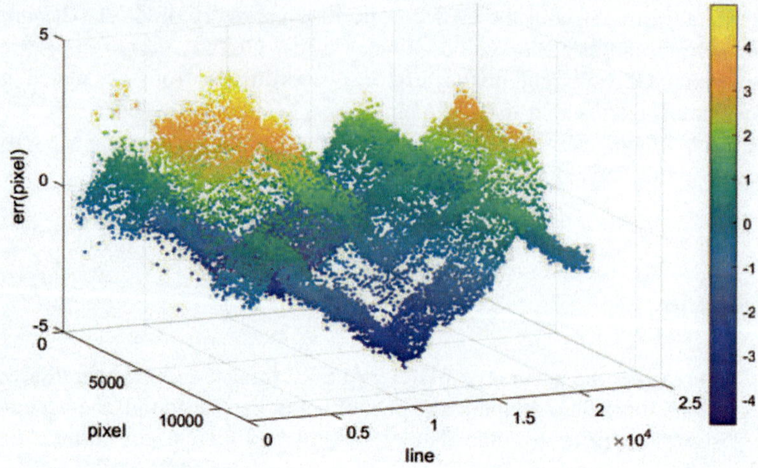

Fig. 3 Three-dimensional distribution of the "ΔLine" errors based on line coordinates (ZY-1_02C_HR2_20191207)

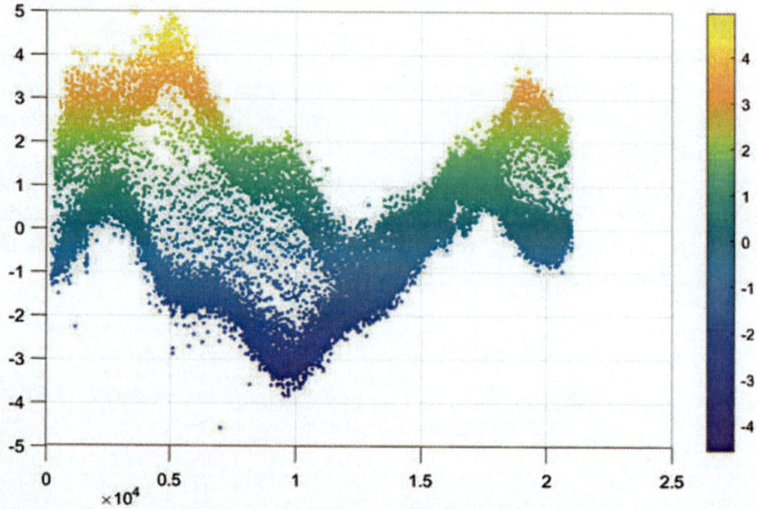

Fig. 4 Two-dimensional distribution of the "ΔLine" errors based on line coordinates (ZY-1_02C_HR2_20191207)

two-dimensional distribution diagram (as shown in Fig. 4) is also given, in which the error is divided into two parts.

In fact, HR1 and HR2 cameras are equipped with three TDICCDs in their focal planes. Figures 5, 6, and 7 show the "ΔLine" error distribution of each CCD sub-images based on line coordinates. Comparing these three figures, it is found

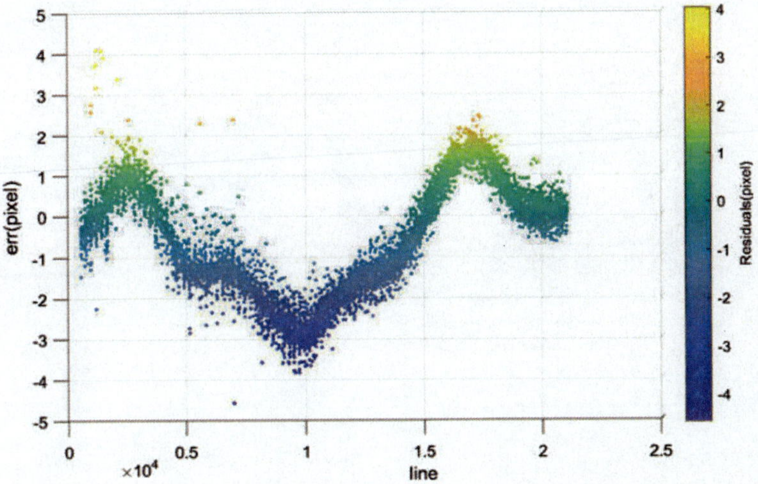

Fig. 5 The "ΔLine" error distribution of CCD1 sub-image based on line coordinates (ZY-1_02C_HR2_20191207)

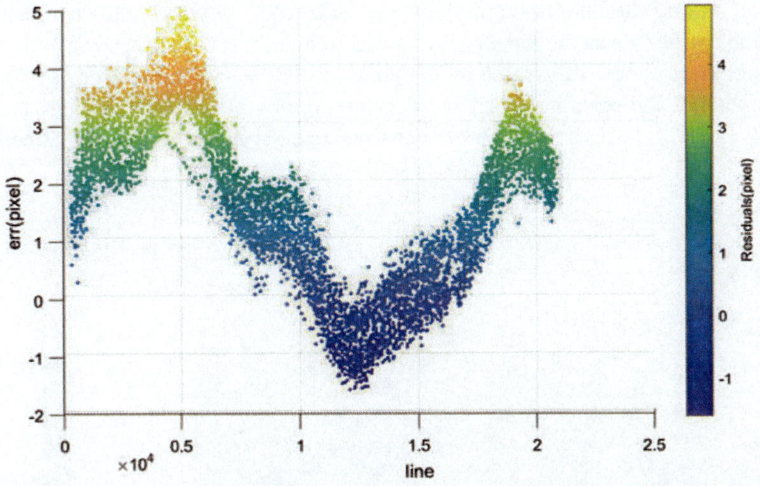

Fig. 6 The "ΔLine" error distribution of CCD2 sub-image based on line coordinates (ZY-1_02C_HR2_20191207)

that they have similar shapes and close errors. The main difference is that the CCD2 sub-image error and the CCD1/CCD3 sub-image error are offset in the line direction, which is consistent with the placement of the three CCDs of the camera on the focal plane, that is, CCD2 Not in the same line as CCD1/CCD3.

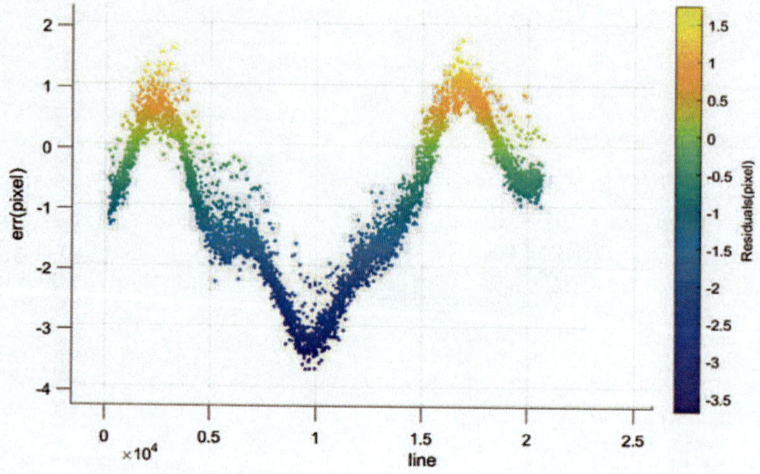

Fig. 7 The "ΔLine" error distribution of CCD3 sub-image based on line coordinates (ZY-1_02C_HR2_20191207)

Since the HRC camera uses push-broom imaging, the imaging time of each line is different, and the imaging attitude is also different. Therefore, the error statistics with line coordinate as the independent variable mainly reflects the error of dynamic imaging, that is, the error caused by the vibration of the satellite platform. Using the same method, the error in the cross-track direction is also shown in Figs. 8, 9, 10 and 11. From these figures, it can be seen that the error in the cross-track direction is

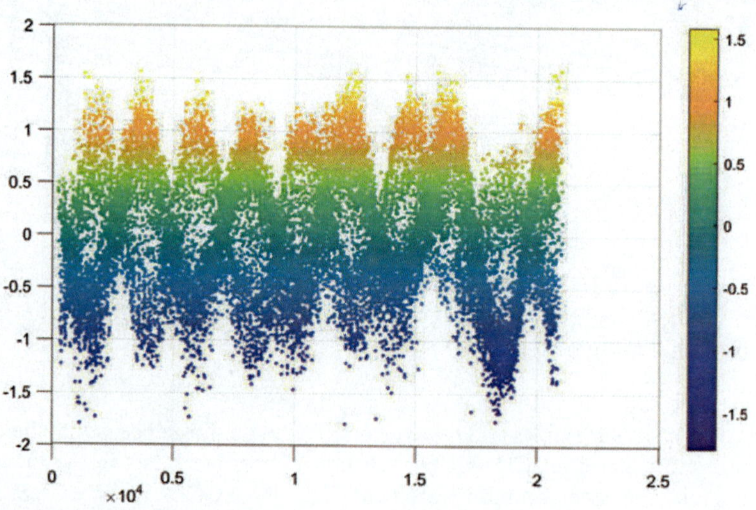

Fig. 8 Two-dimensional distribution of the "ΔSample" errors based on line coordinates (ZY-1_02C_HR2_20191207)

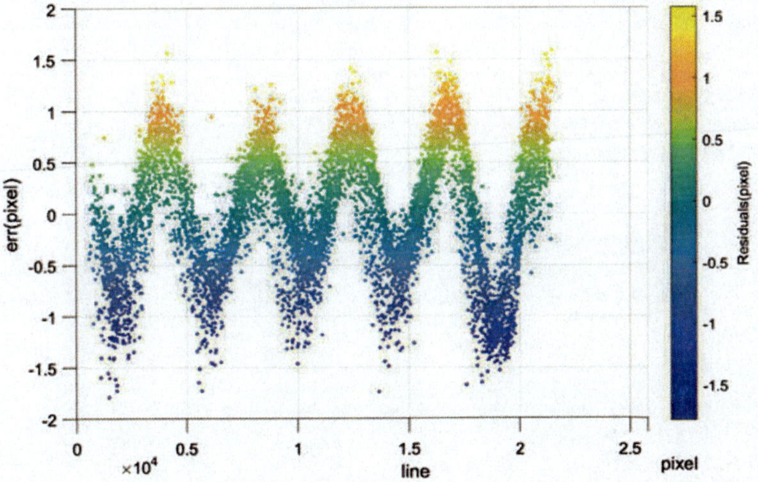

Fig. 9 The "ΔSample" error distribution of CCD1 sub-image based on line coordinates (ZY-1_02C_HR2_20191207)

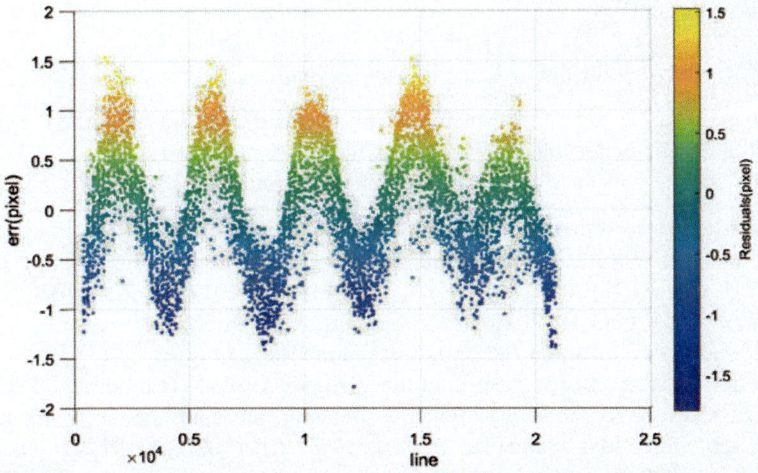

Fig. 10 The "ΔSample" error distribution of CCD2 sub-image based on line coordinates (ZY-1_02C_HR2_20191207)

small, but there are also obvious vibration phenomena. In addition, the geometric calibration residuals will also reduce the geometric accuracy of the image, but since this error is not the main error source that affects the accuracy of the HRC image, this article will not discuss it.

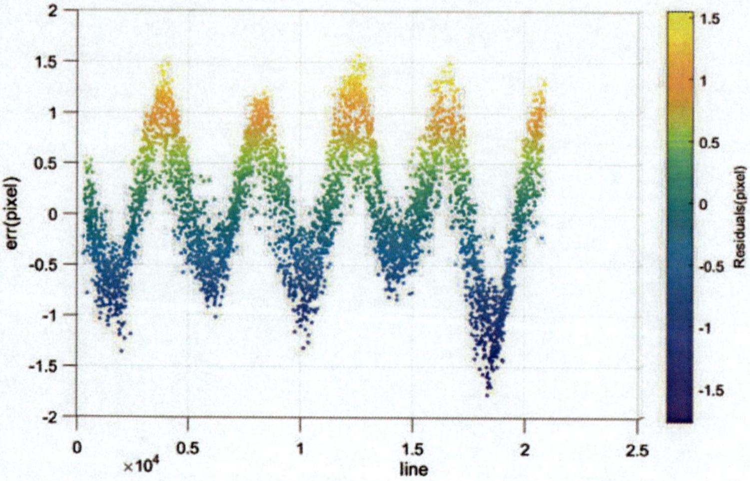

Fig. 11 The "ΔSample" error distribution of CCD3 sub-image based on line coordinates (ZY-1_02C_HR2_20191207)

5 Comparison

From the error data in Table 4, it can also be found:

(1) In the sample and line directions, the accuracy of the image ZY-1_02C_20200326 is better than that of the image ZY-1_02C_20191207;
(2) The accuracy of the HR1 and HR2 images acquired at the same time is close.

Referring to the above data analysis method, taking the CCD2 sub-image as an example, Figs. 12, 13, 14 and 15 respectively give the dynamic imaging errors of ZY-1_02C_HR1_20191207, ZY-1_02C_HR2_20191207, ZY-1_02C_HR1_2020326, ZY-1_02C_HR2_2020326 in along-track direction, Figs. 16, 17, 18 and 19 show their errors in the cross-track direction.

Further analysis and comparison of these graphics reveals that both the HR1 and HR2 cameras have the same dynamic imaging error in the along-track and cross-track directions, whether it is data ZY-1_02C_20191207 or data ZY-1_02C_2020326. Moreover, the vibration of the platform changes dynamically, each image is different, which has a significant impact on the geometric accuracy.

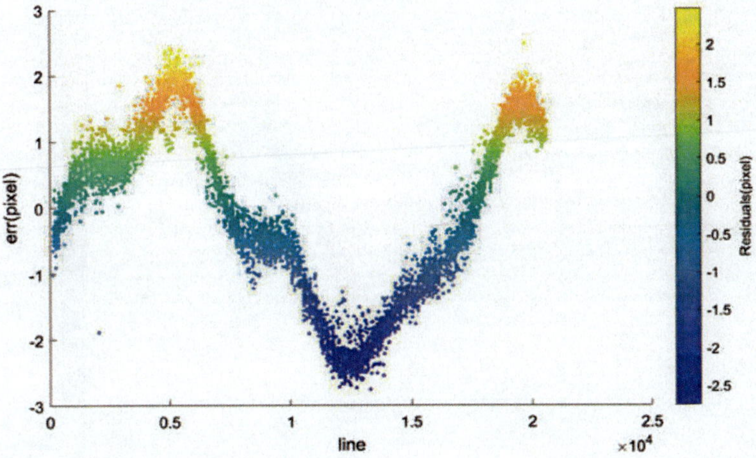

Fig. 12 The "ΔLine" error distribution of CCD2 sub-image based on line coordinates (ZY-1_02C_HR1_20191207)

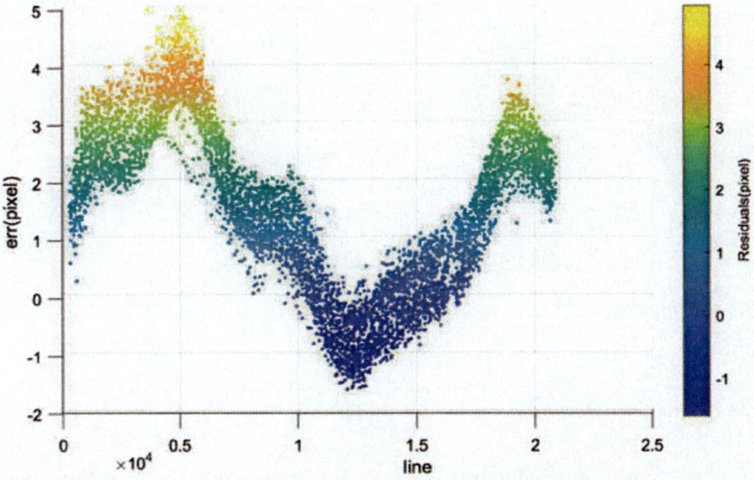

Fig. 13 The "ΔLine" error distribution of CCD2 sub-image based on line coordinates (ZY-1_02C_HR2_20191207)

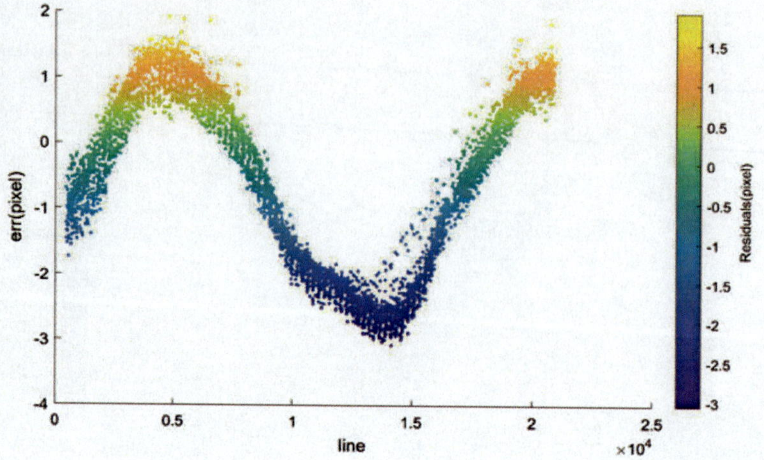

Fig. 14 The "ΔLine" error distribution of CCD2 sub-image based on line coordinates (ZY-1_02C_HR1_2020326)

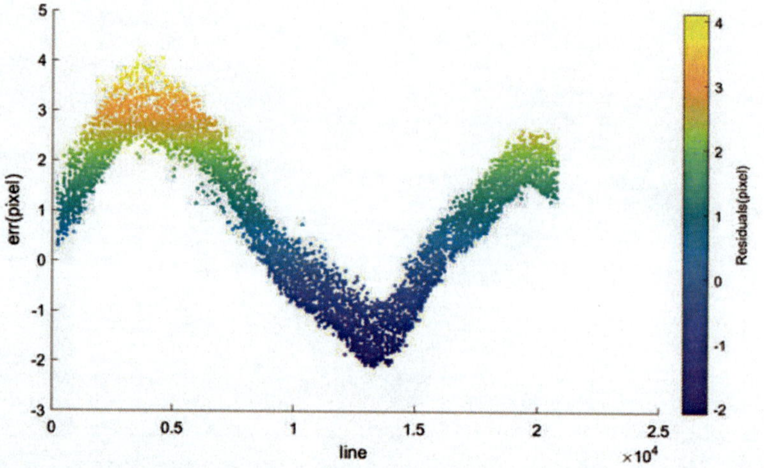

Fig. 15 The "ΔLine" error distribution of CCD2 sub-image based on line coordinates (ZY-1_02C_HR2_2020326)

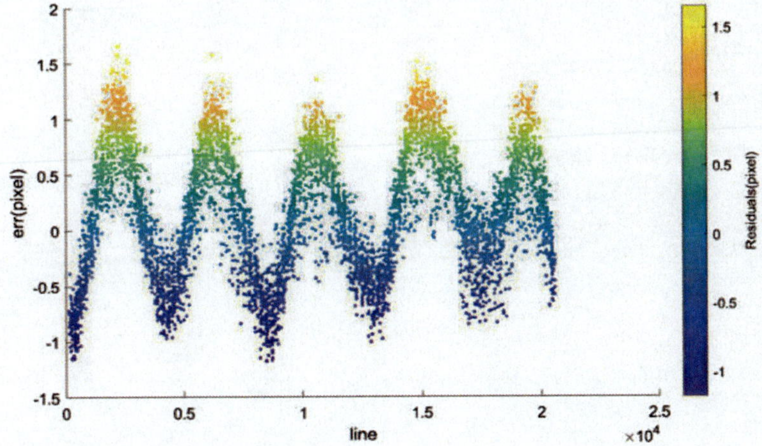

Fig. 16 The "ΔSample" error distribution of CCD2 sub-image based on line coordinates (ZY-1_02C_HR1_20191207)

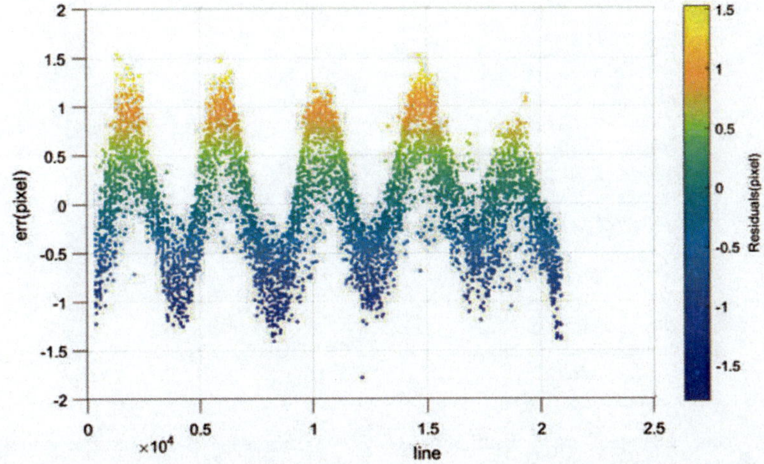

Fig. 17 The "ΔSample" error distribution of CCD2 sub-image based on line coordinates (ZY-1_02C_HR2_20191207)

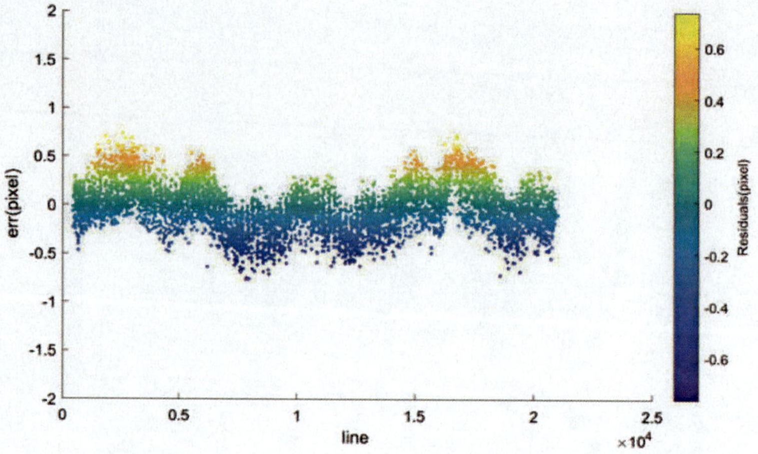

Fig. 18 The "ΔSample" error distribution of CCD2 sub-image based on line coordinates (ZY-1_02C_HR1_2020326)

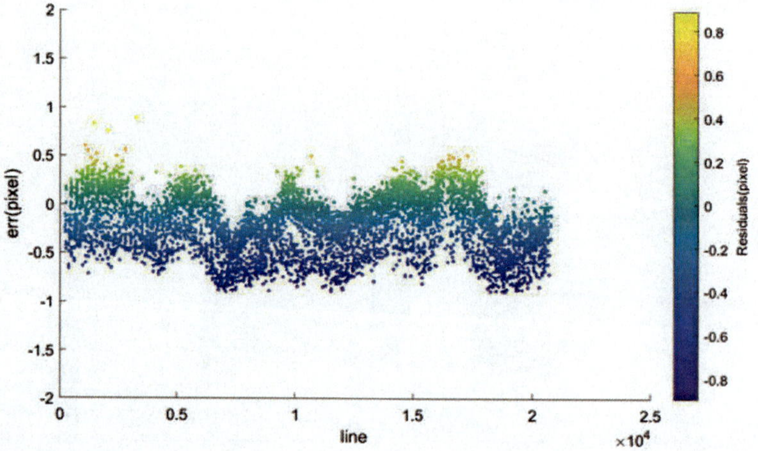

Fig. 19 The "ΔSample" error distribution of CCD2 sub-image based on line coordinates (ZY-1_02C_HR2_2020326)

6 Optimization

According to the above analysis results, it is obvious that the entire HR1 or HR2 image cannot be geometrically corrected as a whole, but the image data needs to be divided into 3 pieces of CCD sub-images, and the dynamic imaging errors are corrected for the CCD1, CCD2 and CCD3 sub-images respectively.

In general, the sine function can fit the vibration error well, and either Eq. (1) or Eq. (2) can be used.

$$f(x) = a * \sin(b * x + c). \tag{1}$$

$$f(x) = a * \sin(b * x + c) + d * \sin(e * x + f). \tag{2}$$

where, x and $f(x)$ are image coordinates and their residuals respectively, and a, b, c, d, e and f are equation coefficients.

Table 5 summarizes the accuracy of the three CCDs and the entire image data before and after sinusoidal correction. The accuracy after correction is significantly improved.

For the CCD3 sub-image of ZY-1_02C_HR2_20191207, Figures 20 and 21 show the error fitting curve and fitting residual in along-track direction and

Table 5 The accuracy of ZY-1_02C_HR2_20191207 before and after sinusoidal correction

Data	Before sinusoidal correction			After sinusoidal correction			Count
	ΔSample	ΔLine	ΔRadial	ΔSample	ΔLine	ΔRadial	
CCD1 sub-image	0.619	1.497	1.620	0.324	0.475	0.575	8167
CCD2 sub-image	0.626	1.549	1.670	0.283	0.389	0.481	5626
CCD3 sub-image	0.584	2.001	2.085	0.357	0.531	0.640	6721
Whole image	0.609	1.676	1.784	0.324	0.470	0.570	20,514

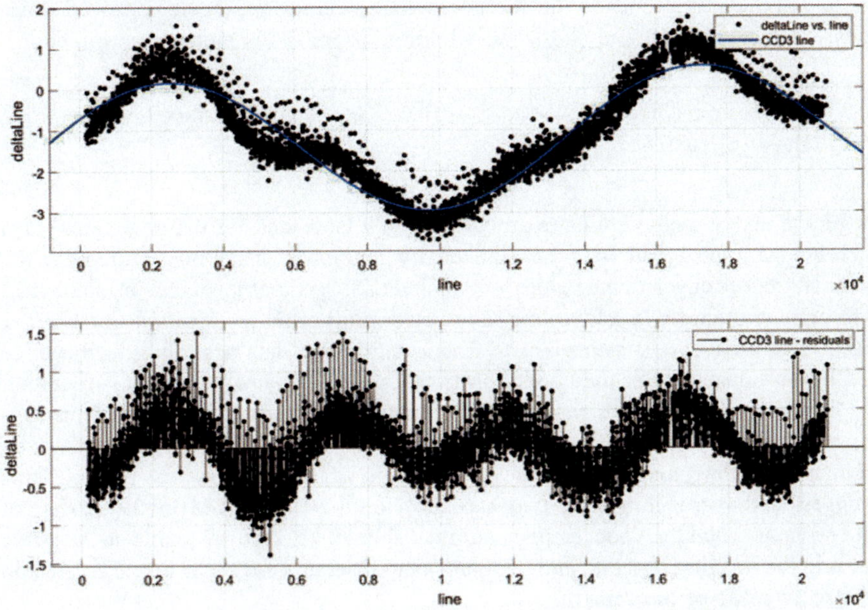

Fig. 20 The CCD3 sub-image error fitting curve and residual in along-track direction

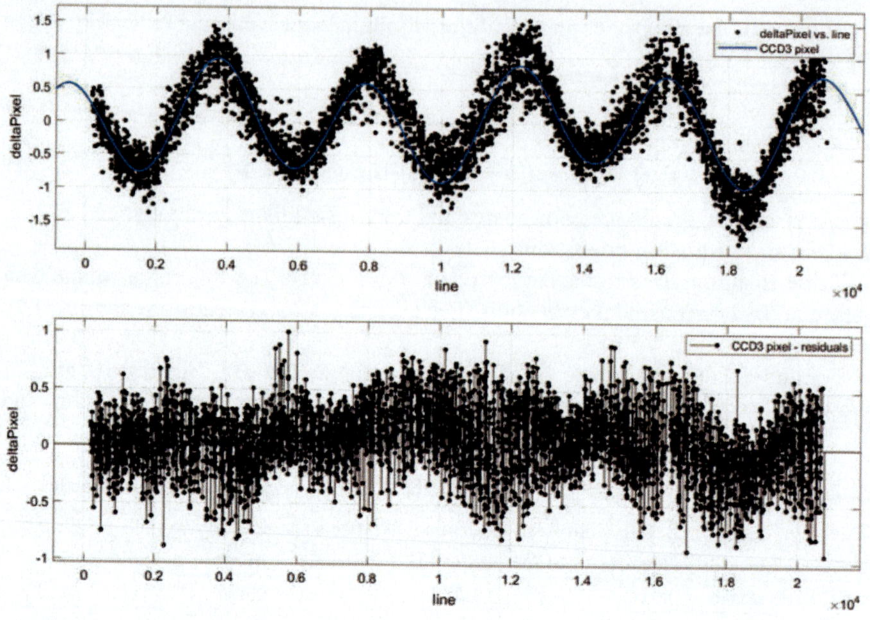

Fig. 21 The CCD3 sub-image error fitting curve and residual in across-track direction

across-track direction, respectively. On the whole, the vibration phenomenon has been significantly reduced, and the geometric accuracy has been improved obviously. However, there are still small vibration errors in the along-track direction.

7 Discussion

With the above method, a large amount of GCP is needed for the operational data processing, which can only be realized by automatic matching. Moreover, the general processing software does not include the geometric model of sinusoidal vibration correction. In addition, before performing ortho-rectification of the ZY-1 02C HRC image, it is necessary to divide the image data according to the three CCD sub-images. After the correction is completed, the image mosaic is required.

The non-collinear arrangement of the multi-chip detectors in the focal plane is more adversely affected by posture vibration during the push-broom process of dynamic imaging, and it is necessary to suppress vibration as much as possible. It is suggested that the following two aspects should be improved in the design of subsequent satellites. The camera attitude should be kept as stable as possible during the imaging process, and the multi-chip detectors are set as a straight line on the focal plane of the camera.

8 Conclusion

The geometric accuracy of ZY-1 02C HRC image is quantitatively analyzed. It is found that the vibration of the spacecraft bus is the main reason for the significant decrease of geometric accuracy. In addition, the non-collinear design of multiple TDICCDs located in the focal plane brings great challenges to image data processing. Although the sinusoidal function can be used to fit the vibration error well and improve the geometric accuracy, it is difficult to implement this method in the operational system. Therefore, it is recommended to avoid these problems in the subsequent satellite design.

References

1. Zhao L (2020) Geometric accuracy evaluation of the ZY-3 stereo mapping satellite for 8 years. In: International archives of the photogrammetry, remote sensing and spatial information sciences, XLIII-B1, pp 415–421. https://doi.org/10.5194/isprs-archives-XLIII-B1-2020-415-2020
2. Foroosh H, Zerubia JB (2002) Extension of phase correlation to sub-pixel registration. IEEE Trans Image Process 11(3):188–200
3. Kuglin CD, Hines DC (1975) The phase correlation image alignment method. In: Proceedings of international conference on cybernetics society, pp 163–165
4. Grodecki J (2001) IKONOS stereo feature extraction—rpc approach. In: ASPRS annual conference St. Louis 2001
5. Grodecki J, Dial G (2003) Block adjustment of high-resolution satellite images described by rational functions. Photogramm Eng Remote Sensing 69:59–70
6. Jacobsen K (2017) Problems and limitations of satellite image orientation for determination of height models. In: International archives of the photogrammetry, remote sensing and spatial information sciences, XLII-1/W1, pp 257–264
7. Jacobsen K (2018) Systematic geometric image errors of very high resolution optical satellites. In: International archives of the photogrammetry, remote sensing and spatial information sciences, XLII-1, pp 233–238
8. Zhao L, Fu X, Dou X, Liu H, Fang Z (2018) Comparison and analysis of accuracy of elevation extraction based on the ZY-3 01 and 02 satellites stereoscopic images. In: International archives of the photogrammetry, remote sensing and spatial information sciences, XLII-3, pp 2379–2384. https://doi.org/10.5194/isprs-archives-XLII-3-2379-2018

River Shoreline Recognition from High Resolution Remote Sensing Imagery Based on Water Index and Artificial Neural Network

Yuan Yao and Linyi Li

Abstract River shoreline recognition from high resolution remote sensing images is a hot spot in remote sensing and relevant research and application fields. Due to the complexity of river shorelines in high resolution remote sensing images, automatic recognition of the shorelines is still difficult, which needs further development. In this paper, a new intelligent method for river shoreline recognition was proposed, which was based on the integration of water index and artificial neural network. Key issues in the recognition algorithm were analyzed and the recognition algorithm was developed. Two representative areas of the Changjiang River and the Zhujiang River, were selected as the study areas. Corresponding GF-1 and GF-2 multispectral images were used as the experimental data. Visual comparison and quantitative assessments of the recognition results by the proposed method and the traditional methods were given. The experimental results show that the proposed method is effective and better than the traditional methods, which obtained most satisfactory recognition results of river shorelines.

Keywords High resolution remote sensing imagery · River shoreline · Intelligent recognition · Water index · Artificial neural network

1 Introduction

River shoreline is the boundary between the body of river and the land. Recognition of river shoreline and the acquisition of the relevant information are of great significance to the ecological environment [1–3]. The automatic recognition of river shoreline from remote sensing images is a hot spot in remote sensing and related

Y. Yao · L. Li (✉)
School of Remote Sensing and Information Engineering,
Wuhan University, Wuhan 430079, China
e-mail: lilinyi@whu.edu.cn

Y. Yao
e-mail: yaoyuan2019@whu.edu.cn

L. Wang et al. (eds.), *Proceedings of the 7th China High Resolution Earth Observation Conference (CHREOC 2020)*, Lecture Notes in Electrical Engineering 757, https://doi.org/10.1007/978-981-16-5735-1_8

research and application fields. There have been many recognition methods proposed [4–7]. However, due to the complexity of river shorelines in high resolution remote sensing images, automatic recognition of the shorelines is still difficult, which needs further development.

Normalized Difference Water Index (NDWI) is based on the green and near-infrared bands. NDWI has good effects of water extraction and is widely used in the relevant fields of water extraction [8, 9]. BP neural network belongs to the multi-layer feed-forward neural networks, which is a classic and widely used neural network using error reverse-propagation algorithm for training [10, 11]. In this study, NDWI and BP neural network are integrated and applied for the intelligent recognition of river shorelines.

This paper is organized as follows: in Sect. 2, the method of river shoreline recognition based on NDWI and BP neural network is proposed; the experimental results are presented and analyzed in Sect. 3 and conclusions are given in Sect. 4.

2 Recognition of River Shoreline Based on NDWI and BP Neural Network

2.1 Basic Principles of NDWI and BP Neural Network

NDWI is defined according to the following equation [9]:

$$NDWI = (Green - NIR)/(Green + NIR) \tag{1}$$

where Green and NIR represent the green and near-infrared bands of multi-spectral remote sensing images, respectively. The index not only uses the contrast characteristics of water bodies that gradually decrease in reflectivity in the visible light band to the near-infrared band, but also uses the increasing reflectivity of vegetation and building information in this range. By using band differences and ratios, water body information is emphasized and other information is curbed.

The algorithmic of BP neural network is divided into two main steps: forward propagation and error backward propagation. In the forward propagation process, the sample goes through the input layer, the hidden layer into the output layer. If the output result does not meet the desired result, the reverse propagation process is applied. The output error is calculated by the reverse path backward propagation, in which the error is shared to each node in the network, and the neuron threshold and weight are adjusted by gradient descent method to reduce the error to an acceptable range [10, 11].

2.2 *Key Issues in the Recognition Algorithm*

2.2.1 Water Recognition Model Based on BP Neural Network Using NDWI

BP neural network is used to classify the pixels of the remote sensing imagery into two classes: water and non-water. In order to make good use of the characteristics of water, the blue band, red band and NDWI index of the multispectral remote sensing images are selected and used as the inputs of the BP neural network in this study. The input feature vector is defined as follows:

$$X = (BLUE, RED, NDWI)^T \tag{2}$$

The outputs of the two classes are defined as follows:

$$Y_{Water} = [1\ 0]^T; \quad Y_{Non-Water} = [0\ 1]^T \tag{3}$$

The BP neural network is made of three layers. The input layer has 3 neurons, the hidden layer has 10 neurons, and the output layer has 2 neurons. The structure of BP neural network for water recognition is shown in Fig. 1.

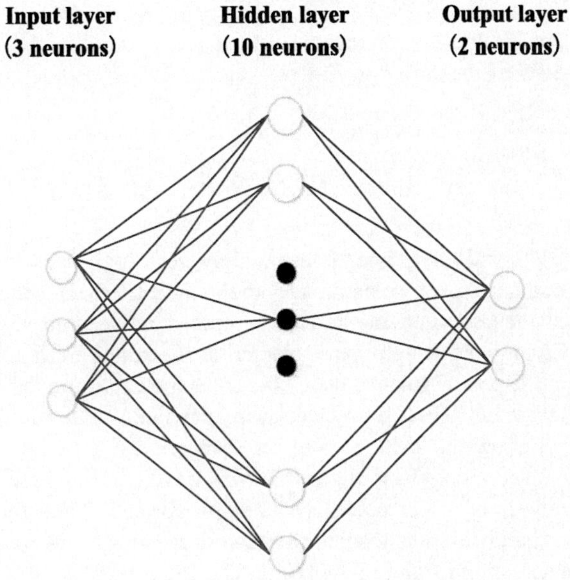

Fig. 1 Structure of BP neural network for water recognition

2.2.2 Remove the Non-target Water Body

There are non-targets in the result image by BP neural network. The non-targets are divided into two classes: (1) non-targets outside the target water body, such as paddy fields and ponds; (2) non-targets within the target water body, such as ships.

Region growing algorithm is used to deal with the first class of non-targets. This method is to select seed points artificially in the target water body, and to implement region growing algorithm to obtain the corresponding target water body, so as to retain the targets and remove the non-targets.

The ships are surrounded by water in the target water bodies. The threshold of ship size and the region growing algorithm are used to recognize ships and then remove them.

2.2.3 Shoreline Recognition Based on Spatial Relationship

The shoreline belongs to the non-water and there must be a water body pixel in the neighbourhood of the shoreline pixel. Each pixel in the image is traversed based on the above spatial neighbourhood relationship.

2.3 *The Algorithm Flowchart*

The flow chart of recognition of river shoreline based on NDWI and BP neural network is shown in Fig. 2.

3 Experimental Results and Analysis

GF-1 and GF-2 images are widely used high resolution remote sensing images in remote sensing research and application fields. In order to validate the effectiveness of the proposed method, two representative areas of the Zhujiang River and the Changjiang River, were selected as the study areas. Corresponding GF-1 and GF-2 multispectral images were chosen as the experimental data. The spatial resolution of GF-1 and GF-2 multispectral images is 8m and 4m, respectively. Two study areas are shown in Figs. 3a and 4a, respectively.

The recognition algorithm was developed in MATLAB 2015. 200 points were randomly selected and visually interpreted. Then the 200 points are used as the training samples for neural network training. L-M reverse propagation algorithm is used as the training function. The maximum iterative number of training is 1000 and the global training minimum error is set to 0.005.

Shoreline recognition results using traditional methods and proposed method are shown in Figs. 3 and 4, respectively. As can be seen from Figs. 3 and 4, compared

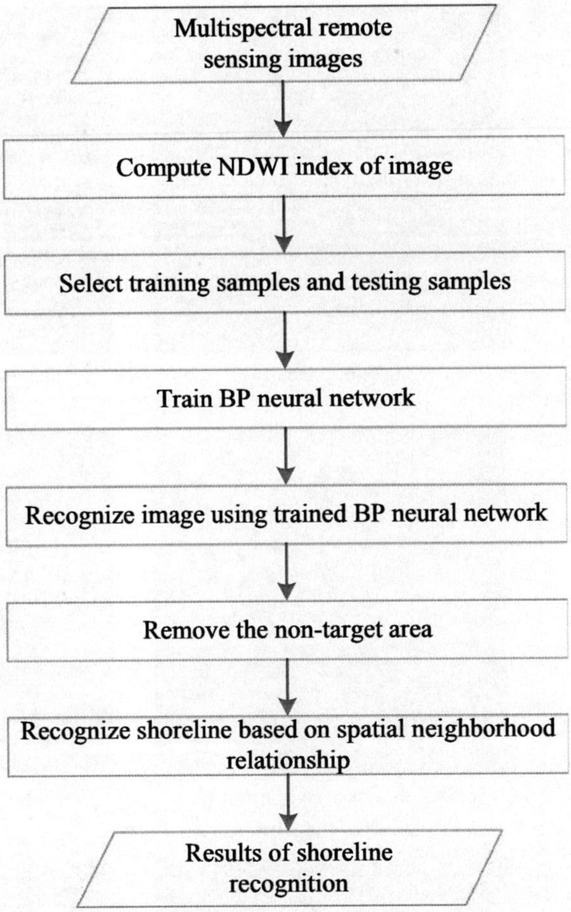

Fig. 2 Flowchart of river shoreline recognition based on NDWI and BP neural network

to the traditional methods, the proposed method has higher accuracy and obtains satisfactory shoreline recognition results. The proposed method has good adaptability when applied to different images, because this method makes good use of the characteristics of water and the strong recognition ability of BP neural network.

In order to evaluate the recognition results quantitatively, we create a buffer of the shorelines, randomly select 200 points in the buffer as the test samples, and visually interpret them into 3 categories: (1) shoreline class; (2) water body class; (3) land class. The confusion matrix is established and the evaluation can be carried out. Evaluation of shoreline recognition based on confusion matrix is shown in Table 1. From Table 1, we can see that overall accuracy, kappa coefficient, average producer accuracy and average user accuracy are 93.0%, 0.894, 92.5% and 93.0% for study area 1, 95.0%, 0.922, 95.0% and 95.3% for study area 2, respectively. The proposed method obtained high recognition accuracy.

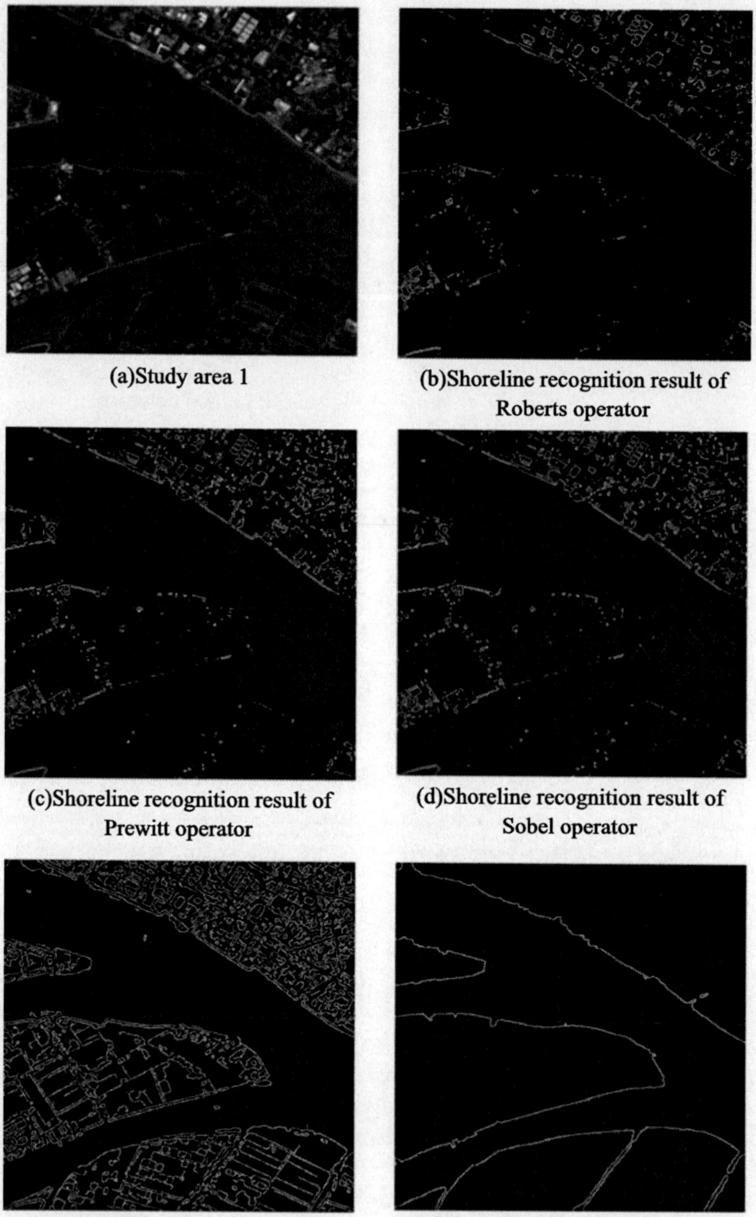

(a)Study area 1

(b)Shoreline recognition result of Roberts operator

(c)Shoreline recognition result of Prewitt operator

(d)Shoreline recognition result of Sobel operator

(e)Shoreline recognition result of Canny operator

(f)Shoreline recognition result of Proposed method

Fig. 3 Shoreline recognition result of study area 1 of the Zhujiang River using GF-1 multispectral image

(a)Study area 2

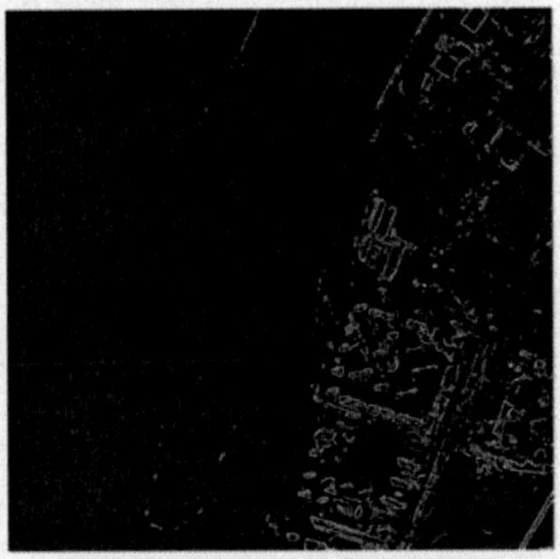

(b)Shoreline recognition result of Roberts method

(c)Shoreline recognition result of Prewitt method

(d)Shoreline recognition result of Sobel method

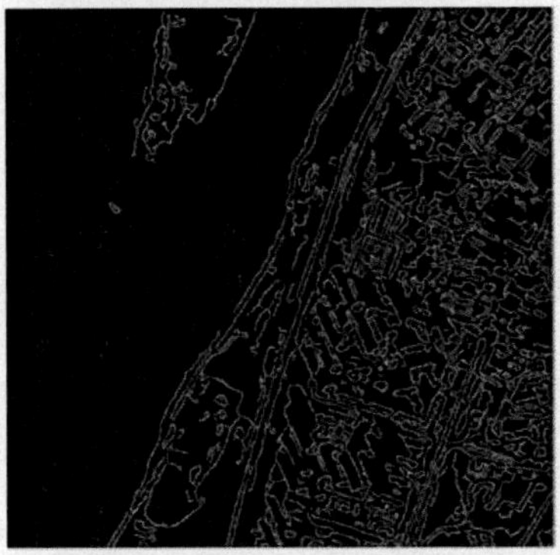

(e)Shoreline recognition result of Canny method

(f)Shoreline recognition result of proposed method

Fig. 4 Shoreline recognition result of study area 2 of the Changjiang River using GF-2 multispectral image

Table 1 Evaluation of shoreline recognition based on confusion matrix

Evaluation index	Study area 1	Study area 2
Overall accuracy (%)	93.0	95.0
Kappa coefficient	0.894	0.922
Average producer accuracy (%)	92.5	95.0
Average user accuracy (%)	93.0	95.3

4 Conclusions

In this paper, a new intelligent method for river shoreline recognition was proposed, which was based on the integration of water index and artificial neural network. Key issues in the recognition algorithm were analyzed and the recognition algorithm was developed. Two representative areas of the Changjiang River and the Zhujiang River, were selected as the study areas. Corresponding GF-1 and GF-2 multispectral images were used as the experimental data. Visual comparison and quantitative assessments of the recognition results by the proposed method and the traditional methods were given. The experimental results show that the proposed method is effective and better than the traditional methods, which obtained most satisfactory recognition results of river shorelines.

Acknowledgements This paper was funded by [the National Key Research and Development Program of China] grant number [2018YFC0407804].

References

1. Balasaraswathi P, Srinivasalu S (2019) Seasonal influence on Cuddalore shoreline change: remote sensing. GIS Statis Approach Mausam 70(2):879–894
2. Ciritci D, Turk T (2019) Automatic detection of shoreline change by geographical information system (GIS) and remote sensing in the Goksu Delta, Turkey. J Indian Soc Remote Sens 47(2):233–243
3. Yousef A, Iftekharuddin K (2014) Shoreline extraction from the fusion of LiDAR DEM data and aerial images using mutual information and genetic algrithms. IEEE Int Joint Conf Neural Netw 1007–1014
4. Demir N, Bayram B, Seker DZ, Oy S, Ince A, Bozkurt S (2019) Advanced lake shoreline extraction approach by integration of SAR image and LIDAR data. Mar Geodesy 42(2): 166–185
5. Incekara AH, Seker DZ (2018) Qualifying the LIDAR-derived intensity image as an infrared band in NDWI-based shoreline extraction. IEEE J Select Topics Appl Earth Observ Remote Sens 11(12):5053–5062
6. Tarmizi NM, Samad AM, Yusop MSM (2014) Shoreline data extraction from QuickBird satellite image using semi-automatic technique. In: IEEE 2014 IEEE 10th international colloquium on signal processing and its applications, pp 157–162

7. Widyantara IMO, Ary Esta Dewi Wirastuti NM, Asana IMDP, Adnyana IBP (2017) Gamma correction-based image enhancement and canny edge detection for shoreline extraction from coastal imagery. In: IEEE 2017 1st international conference on informatics and computational sciences (ICICoS), pp 17–22
8. Chao Z, Sheng M (2011) Study on extraction methods for water information in Nantong city, China using Landsat ETM+ data. In: IEEE international conference on remote sensing, environment and transportation engineering, pp 771–774
9. Ozelkan E (2020) Water body detection analysis using NDWI indices derived from Landsat-8 OLI. Polish J Environ Stud 29(2):1759–1769
10. Shi C, Bian D (2010) Listed logistics company performance prediction by BP neural network. IEEE Int Conf Comput Des Appl 2:V2–466-V2–469
11. Wang B (2012) A novel algorithm on BP-neural network. IEEE Int Conf Comput Sci Electron Eng 1:506–509

Collaborative Classification of Hyperspectral and LiDAR Data Based on Gram Matrices Constrained Fusion Net

Mengmeng Zhang, Wei Li, and Ran Tao

Abstract Multi-sensor information collaborative utilization has attracted considerable attention in remote sensing area. While earth observation benefits from information diversity, multi-sensor collaborative classification technique is still confronted with varied challenges, including inconsistent data volume, different data structures, and uncorrelated physical properties. In this paper, a Gram matrices constrained fusion net (GMCF-Net) is designed for controlling multi-source heterogeneous information and improving classification performance. Exploiting the capability of the Gram matrix in capturing image textures, a multi-source structure control module is constructed to simultaneously address issues involved with different data volume and data structures, which matches the Gram matrices of multi-domains in a double-interweaving pattern. Finally, classification results are obtained based on the discriminative fusion from GMCF-Net. Extensive experiments built from two benchmark remote sensing data sets are reported, and the results demonstrate that the proposed framework yields state-of-the-art performance on hyperspectral and LiDAR data collaborative classification.

Keywords Convolutional neural network · Deep learning · Hyperspectral image · LiDAR data · Collaborative classification

1 Introduction

The recent success of sensor technology has made remarkable progress in mass data acquisition [1], allowing us to measure various aspects of the objects on the surface of earth. Hyperspectral images (HSI) can provide diagnostic spectral information, thus promoting fine-grained classification of observed area [2]. While Light Detection and Ranging (LiDAR) data can provide structured information about the

M. Zhang · W. Li (✉) · R. Tao
School of Information and Electronics, Beijing Institute of Technology, Beijing 100081, China
e-mail: liwei089@ieee.org

L. Wang et al. (eds.), *Proceedings of the 7th China High Resolution Earth Observation Conference (CHREOC 2020)*, Lecture Notes in Electrical Engineering 757, https://doi.org/10.1007/978-981-16-5735-1_9

same research area with HSI, have shown useful compensation information for HSI classification tasks [3]. Collaborative classification of HSI and LiDAR data is helpful to synthesize diverse information, which further improves the performance of earth observation, and a series of approaches have been developed to perform fusion of data collected from these two remote sensing sources [4].

At the existing of data volume differences of hyperspectral and LiDAR data, one straightforward strategy towards improving the performance of collaborative classification is to take full advantage of the data source with large information capacity. Apparently, precise analysis of HSI can help raise the upper bound on collaboration performance. For HSI classification, both machine learning and deep learning based strategies were developed [5–9]. Support vector machine (SVM) [5], which can deal with the linear inseparable problem, has been extensively used in HSI classification in the early stage. Extreme learning machine (ELM) was also investigated for HSI classification [6], and ELM-based algorithms with backward propagation have become a benchmark in neural networks. Deep learning-based methods have been successfully applied in HSI domain; for example, convolutional neural network (CNN)-based pixel-pair feature framework [7], contextual CNN [8], and convolutional recurrent neural network (CRNN) [9] were recently proposed and achieved satisfactory performance for HSI classification. However, aforementioned classification methods are not specifically tailored for multi-source data analysis, and the core issues of inconsistent data structures and uncorrelated properties among multiple sources cannot be effectively handled.

With the emergence of feature and decision-fusion based researches, collaborative classification performance of HSI and LiDAR data has been greatly improved accordingly. Adaptive feature extraction and fusion becomes a popular strategy towards improving utilization of multiple sources. In [10, 11], morphological algorithms were exploited to extract both HSI and LiDAR features. By design, the features learned from morphological operators are ***source-related***; therefore, when simply stacking features with inconsistent properties and unbalanced data distributions, multi-source information (cooperative information, competition information, and complementary information) cannot be equally represented or measured [12]. Accordingly, decision-fusion based methods for HSI and LiDAR data classification have been presented [13, 14], and these methods avoid the balance control issue involved within multi-source features utilization. CNN-based methods further combined HSI and LiDAR features using multi-branch architecture [15] for efficient collaborative classification. Nevertheless, for aforementioned decision-fusion based methods and CNN-based feature-fusion methods, the multi-source information fusion and classification is ***task-related***, both heterogeneous properties and complementary information between HSI and LiDAR data cannot be properly parsed without the guidance of whole data fields of multiple sources.

In this paper, a Gram matrices constrained fusion net (GMCF-Net) is proposed for promoting the utilization of multi-source information and improving classification performance. The proposed network consists of two intersecting branches (HSI and LiDAR data reconstruction branches) and a structure control module. Through structure control module, the feature distribution in the intersection layer is quantified using the Gram matrices [16]. Specifically, Gram matrix from the source of LiDAR data can be considered as the texture-structure reference for fusion feature, and the Gram matrix from the source of HSI can be considered as the spectral-structure reference for fusion feature. While remained reconstruction setting can help to maintain the fusing information integrity of HSI and LiDAR data. Finally, the fusion layer is connected to a multi-branch CNN for obtaining final classification result.

2 Collaborative Classification Based on GMCF-Net

Differences in the representations of objects between HSI and LiDAR data cause shifts in data volume and data structure, thus it is hard for direct-classification model to acquire satisfactory collaborative classification result based on limited training data set, resulting in a decrease in classification performance. In the proposed collaborative classification method, multi-source data are firstly integrated into the Gram matrices constrained fusion net for initial integration, and the fusion layer can further be connected to multi-branch CNN to obtain classification map. The whole framework of the proposed GMCF-Net is illustrated in Fig. 1.

Let $X = [X_1, \ldots, X_C] \in R^{M \times D}$ denote HSI dataset with M samples over C classes, and D is the spectral dimension. Besides, $Y = [Y_1, \ldots, Y_C] \in R^{M \times d}$ denote LiDAR data with feature dimension of d. In the proposed GMCF-Net, $\phi(\cdot)$ is defined as a composite function of consecutive operations in one hidden layer. Accordingly, $f^{(l)}(X, Y) = \phi^{(l)}(\cdots(\phi^{(1)}(X, Y)))$ denote hidden features obtained from the lth layer of GMCF-Net. Given $\hat{d}^{(l)}$ and $m^{(l)}$ represent the channel number and vectorized size (height times width) of hidden features in the lth layer. As shown in Fig. 2, HSI and LiDAR data are fed into GMCF-Net simultaneously, and consecutive operations constitute the feed-forward mechanism, completing information fusion and data reconstruction in a double-interweaving manner.

Gram matrix has the capability of capturing image textures. For HSI, each element in Gram matrix G could be expressed as:

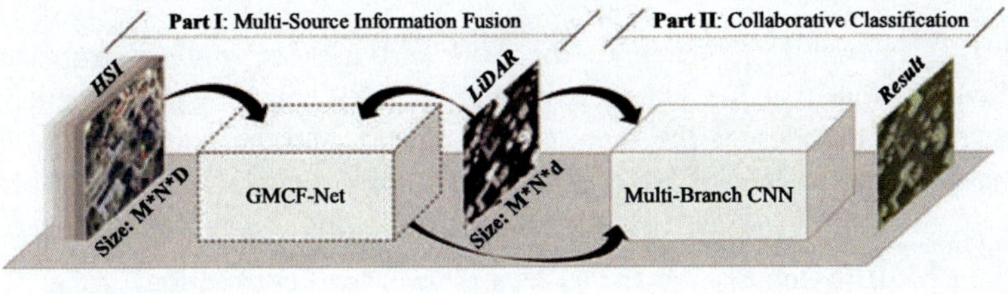

Fig. 1 Information fusion and classification framework using multi-source remote sensing data

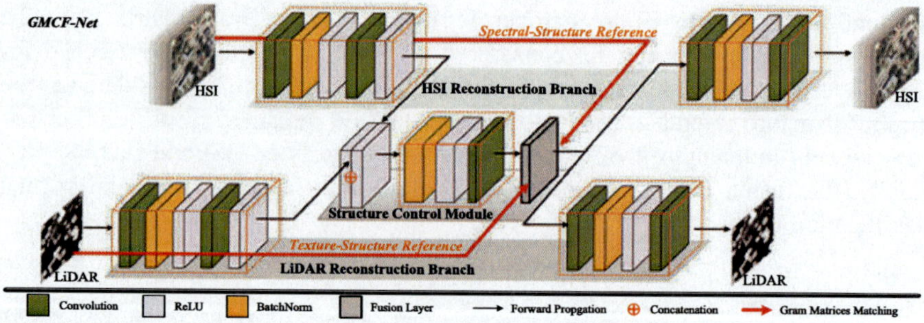

Fig. 2 Overall configuration of the proposed GMCF-Net

$$G_{i,j}(X) = \sum_{b=1}^{M} X_{i,b}X_{j,b}, \tag{1}$$

where i and j denote the ***channel index*** of HSI, and $G_{i,j}$ indicates the structure relationship of bands i and j. Specifically, $G_{i,j}$ is calculated by the inner product of the vectorized feature maps of the ith and the jth channels in HSI. Since the Gram matrix reflects information about which features activate together, spectral-structure relations within HSI can be well captured through $G_{i,j}(X)$. Similarly, for texture-structure acquisition of LiDAR data, the Gram matrix $\widehat{G}_{i,j}(Y)$ could be calculated by the inner product of the ith and the jth vectorized samples in the LiDAR data, while i and j here denote the ***index of samples***. Let $G(f^{(l)}(X,Y))$ and $\widehat{G}(f^{(l)}(X,Y))$ denote the spectral Gram matrix and spatial Gram matrix of the fusion features at the lth layer, respectively. In the structure control module, the loss between Gram matrices of multi-source inputs and the fusion features is defined as:

$$\begin{cases} L_{Gram}^{Spectral} = \frac{1}{4\hat{d}^{(l)^2} m^{(l)^2}} \sum_{i=1}^{\hat{d}^{(l)}} \sum_{j=1}^{\hat{d}^{(l)}} \left(G_{i,j}(X) - G_{i,j}(f^{(l)}(X,Y))\right)^2 \\ L_{Gram}^{Spatial} = \frac{1}{4\hat{d}^{(l)^2} m^{(l)^2}} \sum_{i=1}^{m^{(l)}} \sum_{j=1}^{m^{(l)}} \left(\widehat{G}_{i,j}(X) - \widehat{G}_{i,j}(f^{(l)}(X,Y))\right)^2 \end{cases} \tag{2}$$

where $L_{Gram}^{Spectral}$ and $L_{Gram}^{Spatial}$ denote the HSI and LiDAR guided structure loss, respectively. Note that the channel number of the feature maps generated from structure control module is the same as that of HSI, and the spatial size of feature maps is the same as that of HSI- and LiDAR-input, i.e., for the feature block of fusion layer, $\hat{d}^{(l)} = D$, $m^{(l)} = M$. Moreover, for multi-source data reconstruction task, only MSE (Mean Square Error) loss is used for optimization.

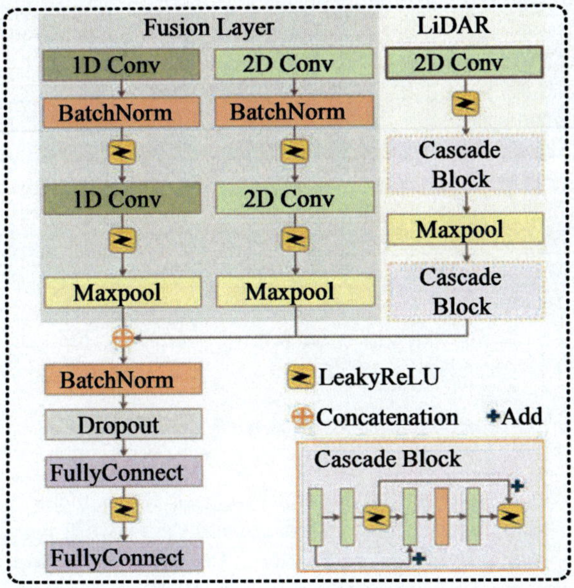

Fig. 3 Architecture of muti-branch CNN

With fully trained GMCF-Net, the fusion layer is then connected to the multi-branch CNN [15] for further classification. Since the data structure differences have been effectively analyzed and controlled based on whole data fields of multiple sources in aforementioned initial fusion stage, complementary information of HSI and LiDAR data can be well utilized in classification task even with limited training samples. Figure 3 illustrates detailed structure of the classification model proposed in [15], where the fusion layer of GMCF-Net is connected with original HSI-related branches as indicated in Fig. 1. In GMCF-Net, all the convolution operations are executed with zero padding, and the convolution stride is set as 1; the number of convolution kernels used in fusion layer of GMCF-Net is the same as the channel number of HSI data as mentioned earlier. Therefore, the channel number of fusion feature is consistent with HSI data, and the spatial dimension of fusion feature is consistent with both HSI data and LiDAR data, ensuring that the initial fusion achieved by GMCF-Net could be combined with multi-branch CNN in a seamless way. For the training setting of the proposed collaborative classification framework: (1) The learning rate of GMCF-Net is initialized at 0.001, and learning process is set follow the policy of Adam [17]. (2) Besides, learning process towards classification task is set as that in [15].

Table 1 HSI and LiDAR-based DSM data sets used for evaluation

Dataset	MUUFL [18]		Trento [19]	
Location	Long Beach, Mississippi, USA		Trento, Italy	
Sensor type	HSI	LiDAR-based DSM	HSI	LiDAR-based DSM
Image size	325 * 220	325 * 220	600 * 166	600 * 166
Spatial resolution	0.54 m * 1.0 m	0.60 m * 0.78 m	1 m	1 m
Number of bands	64	2	63	1
Wavelength range	0.38–1.05 μm	1.06 μm	0.42–0.99 μm	–
Sensor name	CASI-1500	Gemini ALTM LiDAR	AISA Eagle	Optech ALTM 3100EA

3 Experiments and Analysis

For the proposed collaborative classification method, all the programs are implemented using Python language, and the network is constructed using TensorFlow[1] with the high-level API Keras.[2] TensorFlow is an open source software library for numerical computation using data flow graphs, and Keras can be seen as a simplified interface to TensorFlow.

3.1 Experimental Data

The performance of the proposed collaborative classification method is evaluated on two multi-sensor data sets. The first scene was acquired over the University of Southern Mississippi Gulf Park Campus [18]. The data set is composed of HSI and LiDAR-based DSM, and detailed information is shown in Table 1. The second scene was acquired over a rural area in the south of the city of Trento, Italy [19]. Besides, detailed data descriptions can also be found in Table 1. Moreover, Tables 5 and 6 further list the training and testing samples for these two data sets.

3.2 Parameter Tuning and Ablation Analysis

(1) *The Scale of Input Patch*: The effect of different sizes of input patch is studied first. The performance on different window sizes of 9×9, 11×11, 13×13 and 15×15 is shown in Table 2. The overall accuracy (OA), average accuracy

[1] https://github.com/fchollet/keras.
[2] http://tensorflow.org/.

Table 2 The classification performance of the proposed GMCF-net with different patch scales

Method	Matrices	9 × 9	11 × 11	13 × 13	15 × 15
MUUFL	OA (%)	91.66	92.65	**93.41**	93.18
	AA (%)	93.03	94.27	**94.91**	94.75
	Kappa coefficient	0.8900	0.9029	**0.9129**	0.9099
Trento	OA (%)	97.77	97.69	**98.18**	96.62
	AA (%)	96.41	96.55	**97.10**	94.60
	Kappa coefficient	0.9702	0.9691	**0.9757**	0.9550

(AA), and Kappa coefficient are listed, and the results indicate that the input blocks with different sizes yield different classification performance. Also, the input size of 13 × 13 offers the best performance for both MUUFL and Trento data.

(2) *Impact of Gram Matching*: The impact of Gram matching in GMCF-Net is then evaluated, and the MUUFL data set is used for validation here. This data set contains more texture and structure information, making collaborative classification to be more challenging. As for 'Without Gram Matching' setting, we train the proposed fusion net by directly minimizing the MSE loss in HSI and LiDAR data reconstruction branches for collaborative classification. As shown in Table 3, the proposed GMCF-Net that uses the Gram matrices improves the classification performance by 1.1% in terms of overall accuracy.

(3) *Impact of Multi-source Reconstruction*: The impact of the reconstruction setting of GMCF-Net is evaluated in this subsection. For 'Without Reconstruction' setting, we train the proposed fusion net by directly minimizing the Gram matching loss in structure control module for collaborative classification. As shown in Table 4, the proposed GMCF-Net with intersecting reconstruction branches improves the classification performance by 0.75% in terms of overall accuracy.

Table 3 The classification performance of GMCF-net with or without gram matching

Method	Without gram matching	With gram matching (GMCF-net)
OA (%)	92.29	**93.41**
AA (%)	93.85	**94.91**
Kappa coefficient	0.8984	**0.9129**

Table 4 The classification performance of GMCF-net with or without multi-source reconstruction

Method	Without reconstruction	With reconstruction (GMCF-net)
OA (%)	92.66	**93.41**
AA (%)	93.96	**94.91**
Kappa coefficient	0.9030	**0.9129**

3.3 Classification Performance

To validate the effectiveness, the proposed collaborative classification framework is compared with several classifiers. Note that SVM is implemented using the LIBSVM toolbox,[3] and all the comparative methods are implemented with optimal parameters. Tables 5 and 6 list the OA, AA, and Kappa coefficient for all baselines. The collaborative classification performance with GMCF-Net is superior to other methods. Taking the MUUFL data for example, the OA of the proposed collaborative framework is 93.41%, which is 10.67% and 13% higher than that of classical SVM and ELM classifier, respectively. Besides, our approach outperforms original multi-branch CNN by approximately 3.06%. Also, when the HSI-related input is replaced by GS-merge result [20], the classification performance of multi-branch CNN is still worse than the proposed method. Accordingly, it is clear that the performance improvement of the proposed method can be attributable to the effective information fusion of data-related GMCF-Net. Similarly, due to more robust complementary information extraction of multi-source data, the proposed method can also provide obvious performance improvement in Trento data.

Figure 4 further lists the classification performance of all baselines with different numbers of training samples, evaluating the sensitivity of all methods to the training data size. Note that the percentage of training data set is changed from 20 to 100%, and for MUUFL data and Trento data, the training samples of different percentages are randomly selected from the training set listed in Tables 5 and 6, respectively. Obviously, the proposed method consistently outperforms other frameworks. Taking MUUFL data as an example, even for an extremely small training data size, such as 20%, the proposed collaborative protocol provides excellent classification performance, while the Kappa coefficient of other methods are all below 0.75. Through the comparison of classification performance among all baselines versus different training set, it can verifies that the proposed method is robust to small training sample size.

4 Summary

In this paper, a Gram matrices constrained fusion net was proposed for multi-source heterogeneous information control, further constructing the collaborative classification framework. Experimental results demonstrated that the proposed GMCF-Net in conjunction with multi-branch classification model could achieve excellent collaborative classification performance. Besides, benefiting from the effective integration of multi-source complementary information, the proposed GMCF-Net substantially improves the classification performance of the multi-branch classification model under different training-data-size condition.

---

[3] http://www.csie.ntu.edu.tw/cjlin/libsvm/.

Table 5 Comparison of the classification performance using MUUFL data

Class (train/test)	Performance						
	SVM	ELM	CNN-PPF	Multi-branch CNN	Multi-branch CNN (merge)	Context CNN	GMCF-net
Trees (150/23246)	83.22	80.10	89.07	92.35	94.28	91.29	94.95
Mostly grass (150/4270)	82.34	81.01	85.71	59.30	82.84	63.09	90.02
Mixed ground surface (150/6882)	72.87	74.09	80.15	94.47	84.58	81.84	85.29
Dirt and sand (150/1826)	89.27	88.44	93.10	93.74	97.32	93.92	97.43
Road (150/6687)	88.07	88.01	88.98	92.76	92.96	89.44	94.36
Water (150/466)	94.64	96.35	98.93	98.42	99.05	92.92	99.68
Building shadow (150/2233)	88.67	89.34	89.07	95.68	93.28	84.73	95.49
Building (150/6240)	81.67	72.87	92.15	94.01	95.42	81.22	95.93
Sidewalk (150/1385)	76.25	76.10	74.45	86.64	84.13	81.30	91.66
Yellow curb (150/183)	98.91	99.45	100.00	100.00	100.00	98.91	100.00
Cloth panels (150/269)	100.00	97.77	100.00	96.64	97.48	99.63	99.16
OA (%)	82.74	80.41	90.97	90.35	91.95	86.07	**93.41**
AA (%)	86.90	85.78	90.24	91.27	92.58	87.12	**94.91**
Kappa coefficient	0.7798	0.7518	0.8446	0.8727	0.8935	0.8189	**0.9129**

Table 6 Comparison of the classification performance using trento data

Class (train/test)	Performance						
	SVM	ELM	CNN-PPF	Multi-branch CNN	Multi-branch CNN (merge)	Context CNN	GMCF-net
Apple trees (129/3905)	88.62	95.81	90.11	98.07	98.51	99.26	98.74
Buildings (125/2778)	94.04	96.97	83.34	95.21	92.49	86.81	99.66
Ground (105/374)	93.53	96.66	71.13	93.32	100.00	87.91	98.75
Woods (154/8969)	98.90	99.39	99.04	99.93	97.32	97.31	100.00
Vineyard (184/10317)	88.96	82.24	99.37	98.78	100.00	99.82	99.71
Roads (122/3525)	91.75	86.52	89.73	89.98	92.56	84.63	85.73
OA (%)	92.77	91.32	94.76	97.92	97.46	96.11	**98.18**
AA (%)	92.63	92.93	88.97	96.19	96.80	94.29	**97.10**
Kappa coefficient	0.9585	0.9042	0.9304	0.9681	0.9661	0.9481	**0.9757**

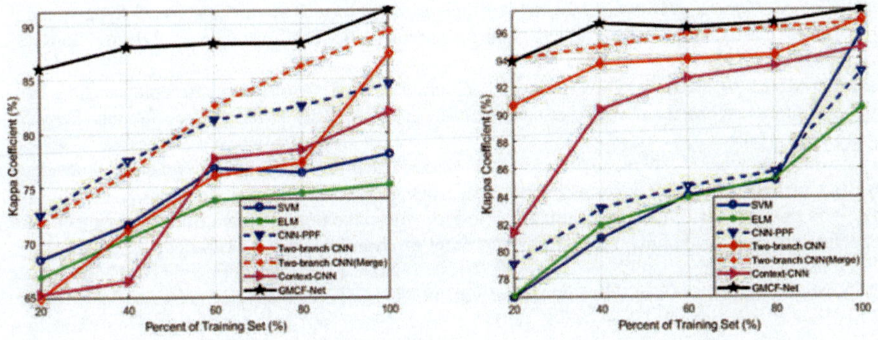

(a) (b)

Fig. 4 Classification performance of methods with different size of training samples

References

1. Zhang M, Li W, Du Q, Gao L, Zhang B (2020) Feature extraction for classification of hyperspectral and LiDAR data using patch-to-patch CNN. IEEE Trans Cybern 50(1):100–111
2. Zhang L et al (2018) Simultaneous spectral-spatial feature selection and extraction for hyperspectral images. IEEE Trans Cybern 48(1):16–28
3. Jung J, Pasolli E, Prasad S, Tilton JC, Crawford MM (2014) A framework for land cover classification using discrete return LiDAR data: adopting pseudo-waveform and hierarchical segmentation. IEEE J Sel Top Appl Earth Obs Remote Sens 7(2):491–502
4. Koetz B, Morsdorf F, Van der Linden S, Allgöwer B (2008) Multi-source land cover classification for forest fire management based on imaging spectrometry and LiDAR data. Forest Ecol Manag 256(3):263–271
5. Melgani F, Bruzzone L (2004) Classification of hyperspectral remote sensing images with support vector machines. IEEE Trans Geosci Remote Sens 42(8):1778–1790
6. Li W, Chen C, Su H, Du Q (2015) Local binary patterns and extreme learning machine for hyperspectral imagery classification. IEEE Trans Geosci Remote Sens 53(7):3681–3693
7. Li W, Wu G, Zhang F, Du Q (2017) Hyperspectral image classification using deep pixel-pair features. IEEE Trans Geosci Remote Sens 55(2):844–853
8. Lee H, Kwon H (2017) Going deeper with contextual CNN for hyperspectral image classification. IEEE Trans Image Process 26(10):4843–4855
9. Wu H, Prasad S (2017) Convolutional recurrent neural networks forhyperspectral data classification. Remote Sens 9(3)
10. Ghamisi P, Souza R, Benediktsson JA, Zhu XX (2016) Extinction profiles for the classification of remote sensing data. IEEE Trans Geosci Remote Sens 54(10):5631–5645
11. Ghamisi P, Höfle B, Zhu X (2017) Hyperspectral and LiDAR data fusion using extinction profiles and deep convolutional neural network. IEEE J Sel Top Appl Earth Obs Remote Sens 10(6):3011–3024
12. Khodadadzadeh M, Li J, Prasad S, Plaza A (2015) Fusion of hyperspectral and LiDAR remote sensing data using multiple feature learning. IEEE J Sel Top Appl Earth Obs Remote Sens 8 (6):2971–2983
13. Liao W, Bellens R, Piurica A, Gautama S, Philips W (2014) Combining feature fusion and decision fusion for classification of hyperspectral and LiDAR data, IGARSS, Quebec City, Canada, pp 1241–1244

14. Zhao C, Gao X, Wang Y, Li J (2016) Efficient multiple feature learning based hyperspectral image classification with limited training samples. IEEE Trans Geosci Remote Sens 54 (7):4052–4062
15. Xu X, Li W, Ran Q, Du Q, Gao L, Zhang B (2018) Multisource remote sensing data classification based on convolutional neural network. IEEE Trans Geosci Remote Sens 56 (2):937–949
16. Long M, Cao Y, Wang J, Jordan MI (2015) Learning transferable features with deep adaptation networks. arXiv preprint arXiv: 1502.02791
17. Kingma DP, Ba JL (2015) Adam: a method for stochastic optimization. In: Proceedings of 3rd international conference on learning represention, San Diego, CA, U.S.
18. Paul G, Alina Z, Ryan C, Jen A, Grady T (2013) MUUFL gulfport hyperspectral and LiDAR airborne data set, Univ. Florida, Gainesvill, FL, U.S., Tech. Rep. REP-2013-570
19. Rasti B, Ghamisi P, Gloaguen R (2017) Hyperspectral and LiDAR fusion using extinction profiles and total variation component analysis. IEEE Trans Geosci Remote Sens 55(7): 3997–4007
20. Zhao X, Tao R, Li W, Li HC, Du Q, Liao W, Philips W (2020) Joint classification of hyperspectral and lidar data using hierarchical random walk and deep cnn architecture. IEEE Trans Geoence Remote Sens 58(10):7355–7370

Joint Classification of Hyperspectral and LiDAR Data Using Improved Local Contain Profile

Dandan Cao, Wei Li, Lu Li, Qiong Ran, Mengmeng Zhang, and Ran Tao

Abstract Joint classification using multi-source remote sensing data has drawn increasing attention. For some complex surveyed scenes, relying on a single hyperspectral image (HSI) is not enough to meet the purpose of high-precision classification. Comparatively, light detection and ranging (LiDAR) data is rich in structure and elevation characteristics. To extract discriminative features, state-of-the-art extinction profile (EP) and local contain profile (LCP) have been investigated. However, EP is more sensitive to the external environment such as shadow occlusion, and the filtering strategy in LCP is prone to lose useful information when dealing with complex terrain scenes. Therefore, an improved LCP (ILCP) method is proposed to extract features from HSI and LiDAR data for joint classification. The proposed ILCP is more stable than EP, and the dimension of features extracted by ILCP is half of EP, which can avoid Hughes phenomenon. Compared to LCP, ILCP uses threshold-based filtering instead of extinction value, which retains more useful information for classification. Furthermore, feature-level fusion is applied to extracted features, and then the integrated features are input to the support vector machine (SVM) for final classification. In this paper, EP and LCP are also employed as comparison methods. Experimental results validated with one real multi-resource remote sensing data demonstrate that the proposed ILCP is superior to traditional methods.

Keywords Improved local contain profile (ILCP) · Hyperspectral image classification · Multi-source data · Feature extraction

---

D. Cao · Q. Ran
College of Information Science Technology, Beijing University of Chemical Technology, Beijing, China

W. Li (✉) · M. Zhang · R. Tao
School of Information and Electronics, Beijing Institute of Technology, Beijing, China
e-mail: liwei089@ieee.org

L. Li
College of Automation, Beijing Information Science and Technology University, Beijing, China

L. Wang et al. (eds.), *Proceedings of the 7th China High Resolution Earth Observation Conference (CHREOC 2020)*, Lecture Notes in Electrical Engineering 757, https://doi.org/10.1007/978-981-16-5735-1_10

1 Introduction

Single remote sensing data is no longer sufficient to deal with complex scene [1, 2]. In recent years, in order to take full advantage of different sensors, fusion of multi-sensor remote sensing data has gotten increasing attention [3, 4]. Hyperspectral image (HSI) has high spectral discrimination ability, and there is slight difference in spectral dimensions of different objects, which can be helpful for accurate objects recognition and classification. Light detection and ranging (LiDAR) data can give the elevation information of the object surface relative to the sensor, and it is rich in structure and elevation characteristics, which are useful for distinguishing homogeneous materials. Joint classification and feature extraction based on multi-source data, e.g., HSI and LiDAR data, have been investigated [5–7].

There are many ways to extract spatial information in remote sensing images. The methods based mathematical morphology are commonly used. Mathematical morphological methods include morphology profile (MP), attribute profile (AP), extinction profile (EP) and local contain profile (LCP). MP is constructed based on a series of opening/closing operations, which are applied to an image including a set of known shapes, called structuring elements (SE). The spatial features extracted by MP were used for classification of hyperspectral data from urban areas [8]. Compared with MP, attribute profile (AP) is an extension of MP, which can model the geometric characteristics of the objects rather than size. When calculating MP, a fixed SE (e.g., disk, square, length, etc.) needs to be predefined. Compared to MP, AP is more flexible, because AP can process images according to different attributes, and the attributes can generally be any shapes [9, 10]. In [11], hyperspectral data were projected to a subspace by the independent component analysis (ICA) method, and then the typical AP was applied to extract features of different subsets for classification. Extreme-oriented extinction profile (EP) and AP are both based on component tree (Max/Min-tree). Compared with threshold-oriented AP, EP has less sensitivity to image resolution, and EP is based on extinction value. Therefore, EP has better classification performance, and it is fully automatic [7, 12]. In addition to AP and EP, there is also a morphological algorithm, i.e., local contain profile (LCP). Compared with EP, LCP is based on topological tree, which is insensitive to external factors. Therefore, LCP is more stable and has better classification performance, and the dimension of features is half less than EP [13].

However, when the external factors (such as lighting, shadow occlusion) change, the component tree (Max/Min-tree) is different for the same scene. In other words, EP based on component tree is sensitive to external factors, and the dimensions of extracted features are very large. On the other hand, although LCP is more stable than EP, and the dimension of the extracted features is only half of EP; for complex scenes, the filtering methods of LCP lose useful information and reduce the classification accuracy. Based on that, we attempt to design an improved LCP (ILCP), including the construction of topological tree, threshold-based filtering, and image reconstruction.

In this paper, an efficient ILCP is proposed to extract features of HSI and LiDAR for joint classification. Compared with the original LCP, ILCP uses a filtering method based on threshold values instead of extinction values to cut connected nodes, which are not satisfied with conditions. Compared with EP, the dimension of features extracted by ILCP is less than half of EP, and it is relatively stable. For complex scenes, it can better preserve useful information for classification than LCP. The proposed ILCP can obtain higher classification accuracy than LCP and EP.

2 Proposed ILCP Feature Extraction

2.1 LCP

The process of extracting features using LCP includes three steps, (1) the construction of topological tree, (2) extinction filtering and (3) image reconstruction. Step 1 and 3 are described in detail below. For step 2, after the topological tree is built, the attribute values should be calculated to characterize each node. Extinction values represent the maximum attribute value of the branch where a leaf node is located. Attributes can be divided into increasing and non-increasing attributes. For increasing attributes, it is easy to calculate extinction values, but for non-increasing attributes, the attribute values of child nodes may be larger than parent nodes, which makes the calculation difficult. Therefore, it is necessary to build a second tree based on the established topological tree, and new extinction values need to be recalculated. More details about this process can be found in [13].

First, the principle of extinction filtering is first sorting the calculated extinction values. Then according to the set value n, the leaf nodes and their branches with the first n extinction values are retained, the remaining nodes and branches are cut.

Compared with EP, LCP can reduce the dimension of features by half, and LCP is less sensible to light and shadow. Therefore, LCP is more stable than EP [14, 15].

2.2 Improved LCP (ILCP)

As is shown in Fig. 1, the process of extracting image feature using the proposed ILCP includes three steps. The construction process is introduced as follows.

2.2.1 The Construction of Topological Tree

Same as LCP, a linear algorithm is used in the calculation of the topological tree [14], and the algorithm includes three processes: interpolation, sorting and union-find. Specifically, the Khalimskey grid is first used to interpolate. Then the pixels in the interpolated image are sorted in decreasing order, i.e., the access order.

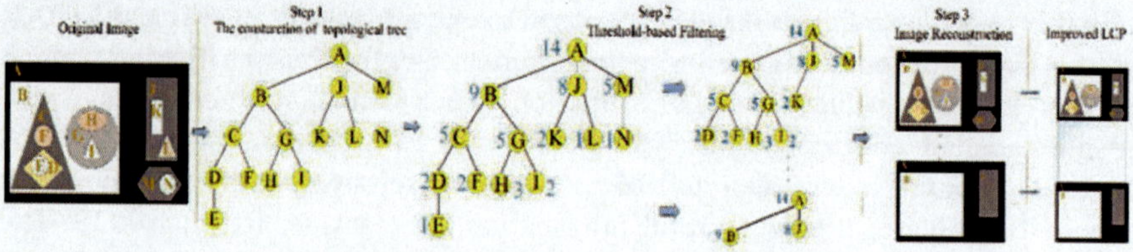

Fig. 1 The flow chart of improved LCP (ILCP) construction for feature extraction

And the hierarchical pixel queue is used to store the extracted pixel information [15]. Finally, the union-find process is used to compute the topological tree in reverse order. The step 1 in Fig. 1 gives an example of a topological tree of original image.

2.2.2 Threshold-Based Filtering

As mentioned above, the nodes and their branches that do not meet the condition will be all discarded during extinction filtering in LCP. However, for some complex scenes, the branches of the topological tree are very large, so it is easy to lose important information when the entire branch is cut. For solving this problem, this paper proposes a method that uses threshold-based filtering instead of extinction filtering. For the given threshold λ, the node (connected component) and the attribute value $att(C)$, if $att(C) > \lambda$, the node is retained, otherwise the node is covered by its parent node. The threshold setting principle is introduced below. The threshold has to be selected for guaranteeing the fulfillment of the absorption property. Compared with LCP, the improved method can preserve useful information more effectively. The step 2 in Fig. 1 reflects the specific progress of threshold-based filtering.

2.2.3 Image Reconstruction

After getting the threshold-based filtered topological tree, the pixels of the connected components represented by the cut nodes will be replaced by the ones of its parent nodes, and then the new topological tree is reconstructed into an image. The features extracted by ILCP are obtained.

2.3 *Three Useful Attributes*

In this paper, three attributes are used which are area, the diagonal of the bounding box and height. Among them, area, bounding box diagonal and height are increasing attribute. Area is to calculate the number of pixels in the connected area, the diagonal of the bounding box is to calculate the diagonal of the border of the connected area and height is to calculate the difference between the pixels of the connected area and the local pixels.

The formula for calculation of these attributes are given below:

$$\text{Area (N)} = \{\#p|p \in N\} \tag{1}$$

$$\text{Diag (N)} = \sqrt{(x_{p.\max} - x_{p.\min})^2 + (y_{p.\max} - y_{p.\min})^2} \tag{2}$$

$$\text{Height (N)} = \underset{p \in N}{\text{Max}}(f(p)) - \underset{p \in N}{\text{Min}}(f(p)) \tag{3}$$

where N denoted a connect domain, p is a pixel of connect domain. Diag is the abbreviate of the diagonal of the bounding box. $(x_{p.\max} - x_{p.\min})$ and $(y_{p.\max} - y_{p.\min})$ represent the largest length and width of connected area respectively. $x_{p.\max}$, $x_{p.\min}$, $y_{p.\max}$ and $y_{p.\min}$ represent the extreme point of the abscissa and ordinate in the connected region. f is a function for getting the value of a pixel.

2.4 *Analysis on Proposed ILCP*

Same as LCP, ILCP is also constructed based on topological tree and includes three steps. The biggest difference between the two algorithms mainly lies in the filtering process. For the filtering process of LCP, the entire branch that does not meet the extinction value will be discarded. However, as mentioned above, ILCP is an improvement on the existing LCP filtering strategy, that is, setting a series of threshold values, only nodes that are more than the threshold are retained. Therefore, compared with LCP, ILCP can extract important information much better. For example, when the value is set to 2, for the extinction filtering, the branches where the node with the first two maximum extinction values are retained, and the other branches are cut, as is shown in Fig. 2b. But for threshold filtering, only the nodes with attribute value less than 2 need to be discarded and other nodes and branches are reserved as shown in Fig. 2c.

Compared with EP, there are two main differences between ILCP and EP: (1), EP is based on the component tree (Max/Min-tree), while ILCP is based the topological tree. The latter is insensitive to the environment and relatively stable; (2), When using EP to extract features from an image, it is necessary to perform both thicking and thinning operations at the same time, so the dimension of features extracted by EP is twice as many as ILCP, which may cause Hughes phenomenon.

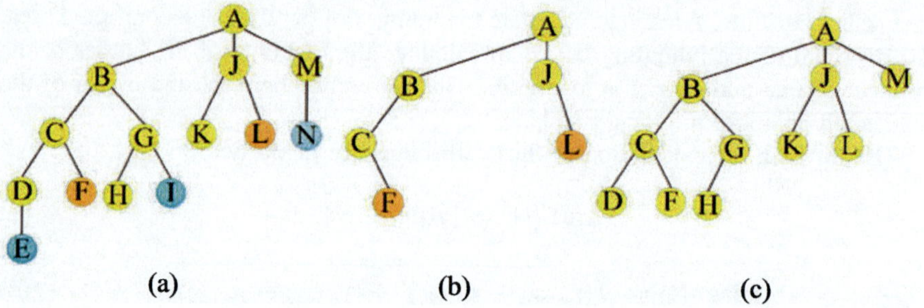

Fig. 2 The difference between LCP and the proposed ILCP. **a** Original topological tree; the orange nodes represent two nodes with the first 2 largest extinction values and the blue nodes represent nodes with attributes values which is smaller than 2. **b** Result of the extinction filtering. **c** Result of the threshold filtering

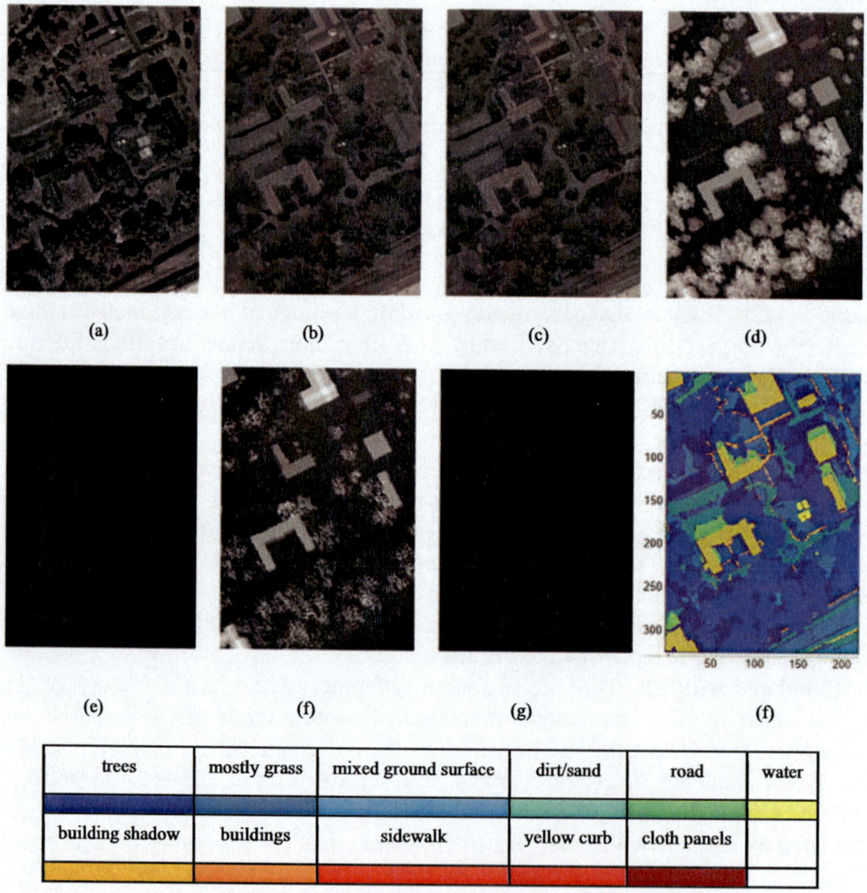

Fig. 3 MUUFL Gulfport data. The first three PCs of HSI: **a** the first component, **b** the second component, **c** the third component. The four images of LiDAR: **d** the first band of LiDAR1, **e** the second band of LiDAR1, **f** the first band of LiDAR2, **g** the second band of LiDAR2. **h** the ground-truth map

Table 1 The results of HSI using LCP, EP and ILCP respectively

No	EP				LCP				ILCP			
	a	*d*	*h*	*a* + *d* + *h*	*a*	*d*	*h*	*a* + *d* + *h*	*a*	*d*	*h*	*a* + *d* + *h*
1	59.02	68.67	80.64	81.61	48.20	56.86	75.52	77.63	69.44	76.75	81.95	84.74
2	62.31	63.33	81.70	80.40	52.83	53.18	77.29	78.25	79.73	77.57	85.35	83.95
3	51.91	55.12	74.29	68.84	43.08	53.26	67.68	66.09	73.40	78.19	77.03	78.78
4	65.18	64.49	88.75	81.17	58.48	59.63	86.36	82.14	87.04	88.18	87.86	90.73
5	67.14	66.04	84.20	76.06	59.45	64.53	77.75	77.54	73.36	82.68	85.74	85.38
6	91.82	92.57	97.81	93.44	93.61	92.90	95.35	95.25	96.99	94.59	98.36	96.56
7	78.44	76.77	90.30	89.35	85.41	78.98	89.04	86.42	78.90	87.22	87.67	89.88
8	50.50	58.76	80.49	72.47	44.38	51.86	70.35	69.00	67.55	75.13	84.84	83.19
9	47.01	45.88	65.59	62.54	43.19	43.80	66.74	99.96	56.00	54.96	67.55	75.16
10	65.06	58.56	80.72	80.24	52.05	46.50	78.31	73.25	69.88	81.20	86.99	85.30
11	92.42	90.25	98.11	96.67	83.55	91.71	97.04	98.46	98.94	95.75	95.97	98.58
OA	59.42	64.87	80.80	78.07	51.02	57.50	75.23	75.23	71.96	77.98	82.64	**84.10**
AA	44.14	47.10	63.21	62.82	38.26	43.01	58.15	58.15	57.37	61.09	64.26	**65.92**
Kappa	0.5041	0.5630	0.7548	0.7195	0.4140	0.4816	0.6879	0.6879	0.6492	0.7211	0.7781	**0.7955**

a, ***d*** and ***h*** represent the area, the diagonal of the bounding box and height. Metrics AA and OA are reported in percentages. Kappa is a coefficient what changes in the range of 0 and 1

2.5 Experimental Results and Analysis

2.5.1 Experimental Data and Parameter

The selected experimental scene is MUUFL Gulfport data set, which was collected over the Mississippi Bay Park University campus in long beach, Mississippi in 2010. It contains one HSI data and two LiDAR data. The original hyperspectral data contains 72 spectral bands and a size of 325 * 337 pixels. Due to noise, the first four and last four bands were removed, resulting in a hyperspectral image with 64 bands. The lower right corner of the original image contains invalid regions, so only the first 220 columns are used for ground-truth map, and the new hyperspectral image size is 325 * 220 * 64. The image of three PCs of HSI, four images of LiDAR (note that two LiDAR data both have two dimensions) and ground-truth map are shown in Fig. 3. The number of training sample of each class is 100, and the number of test sample is 23246, 4270, 6882, 1826, 6687, 466, 2233, 6246, 1385, 183, and 269, respectively. The specific class can be seen in Fig. 3.

In this paper, three attributes including area, the diagonal of the bounding box and height are used, and ***a***, ***d*** and ***h*** represent these three attributes respectively below. For the extinction filter, there is a value n needs to be set in advance, then after calculating the extinction value, the nodes with the first n largest extinction values and their branches will be retained. For the specific process, please refer to [13]. n is set in accordance with the method in [13], i.e., $n = b^m$, $(a = 1, 2, 3 \ldots;$ $m = 0, 1, 2, \ldots s - 1)$ where b is the basic parameter, s is the number of thresholds. In this experiment, $b = 3$, $s = 7$, i.e., the seven thresholds are {1, 3, 9, 27, 81, 243, 729}. The threshold setting of area and the diagonal of the bounding box mainly refers to the suggested values [9, 10], and the value of height is set according to its physical meaning. The specific threshold is: λ_a = {49, 169, 361, 625, 961, 1369, 1849}, λ_d = {10, 25, 50, 100, 200, 250, 350} and λ_h = {5, 10, 30, 60, 100, 150, 200}. The superiority of the proposed method over EP and LCP is compared by the overall accuracy (OA), the average accuracy (AA), and the kappa coefficient.

(a) (b) (c) (d)

Fig. 4 Classification map of HSI. **a** Original ground-truth, **b** EP, **c** LCP, **d** ILCP

Table 2 The results of LiDAR using LCP, EP and ILCP respectively

No	EP				LCP				ILCP			
	a	***d***	***h***	***a + d + h***	***a***	***d***	***h***	***a + d + h***	***a***	***d***	***h***	***a + d + h***
1	57.39	65.32	63.99	72.80	44.33	50.36	59.23	69.26	65.26	61.48	65.67	69.27
2	45.12	47.70	60.93	49.61	44.96	48.84	61.25	53.75	26.39	56.83	64.99	70.61
3	45.36	45.04	49.84	50.07	33.93	37.31	52.27	48.40	9.73	43.62	42.07	48.75
4	61.96	65.20	80.99	63.05	52.62	58.94	77.49	71.38	36.67	77.60	73.69	83.85
5	57.30	57.69	71.85	57.39	53.27	49.89	65.39	59.13	8.49	79.51	80.72	72.84
6	97.65	97.87	98.80	97.65	95.08	92.51	98.70	97.81	98.80	99.18	99.67	99.23
7	43.91	50.48	58.33	56.03	44.72	47.81	57.99	56.44	35.20	35.89	45.59	58.11
8	79.45	82.93	86.92	81.72	71.88	82.11	87.49	83.75	87.60	88.33	88.66	89.83
9	41.03	39.50	39.00	40.14	25.05	35.98	33.95	40.56	17.77	30.35	33.03	41.23
10	51.09	51.09	63.38	52.77	51.57	56.39	63.13	69.16	21.21	36.38	52.29	55.42
11	86.04	87.58	87.46	87.93	91.72	87.93	90.89	92.19	92.90	93.25	93.37	92.54
OA	57.49	61.45	65.62	65.53	47.73	52.46	62.87	64.88	47.45	63.22	66.10	**69.30**
AA	45.09	46.72	51.31	50.32	36.32	40.52	48.45	49.51	31.33	46.81	50.55	**52.63**
Kappa	0.4802	0.5216	0.5767	0.5647	0.3782	0.4302	0.5477	0.5610	0.3700	0.5524	0.5832	**0.6173**

a, ***d*** and ***h*** represent the area, the diagonal of the bounding box and height. Metrics AA and OA are reported in percentages. Kappa is a coefficient what changes in the range of 0 and 1

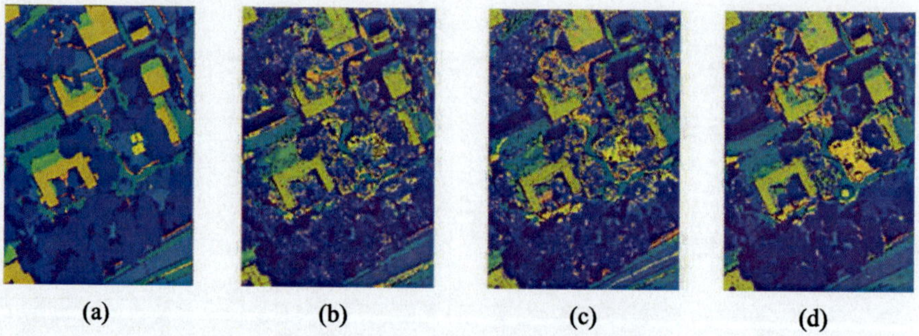

Fig. 5 Classification map of LiDAR. a Original ground-truth, b EP, c LCP, d ILCP

2.5.2 Experimental Results and Analysis

In experiment, the dimension of HSI in MUUFL Gulfport data set is reduced to 3 using principle component analysis (PCA). LCP, EP and ILCP are applied to HSI, LiDAR respectively. After the feature extraction, the features are integrated, which called feature-level fusion. And then the fused features are input into typical support vector machine (SVM) for classification.

Table 1 lists the results of HSI using EP, LCP and ILCP, respectively. From Table 1, it can be seen that for HSI, the proposed ILCP can obtain the best results whether it is single attribute or a combination of multiple attributes, and compared with EP and LCP, ILCP has obvious advantage. Furthermore, the classification accuracy of each class is also listed in Table 1. The classification map is illustrated in Fig. 4.

Table 2 provides the accuracy of each class and OA, AA and Kappa of LiDAR using three methods. Different from HSI, the results of ILCP are better than LCP on each attribute, but compared to EP, ILCP's results of area attribute is not as good as EP. However, in addition to this attribute, the effect of ILCP is still better than EP, especially after combining the three attributes, the accuracy of ILCP is 3.37% higher than that of EP. Overall, ILCP still has advantages over EP and LCP for LiDAR. The classification map is shown in Fig. 5.

The test accuracies for the combination of HSI and LiDAR using EP, LCP and ILCP are shown in Tables 3. The highest OA, AA and Kappa was achieved when ILCP was used for the combined data. Same as HSI, the results of single attribute and multi-attributes using ILCP are better than EP and LCP. The results in the last column of Table 3 are noteworthy. By comparing with Tables 1 and 2, it can be found that the classification results after data fusion are better than single data. The classification map is illustrated in Fig. 6.

Table 3 The results of the combination of HSI and LiDAR using LCP, EP and ILCP respectively

No	EP				LCP				ILCP			
	a	***d***	***h***	***a + d + h***	***a***	***d***	***h***	***a + d + h***	***a***	***d***	***h***	***a + d + h***
1	79.55	83.77	89.15	89.64	72.01	75.64	86.66	87.39	81.56	86.20	88.18	89.75
2	75.50	75.31	84.57	77.68	65.94	64.59	83.68	80.67	89.06	87.42	88.29	86.72
3	67.92	69.77	77.34	75.81	53.31	56.81	73.84	72.81	74.17	78.49	79.22	82.03
4	73.84	77.59	91.56	76.87	65.03	64.20	88.13	83.46	89.31	87.35	90.64	91.18
5	76.60	74.80	86.54	79.97	70.03	70.00	85.72	82.93	82.61	88.07	88.57	88.64
6	96.50	96.29	99.02	97.87	96.94	91.71	98.53	97.92	98.36	98.74	99.23	98.36
7	83.84	86.30	91.32	87.86	79.20	81.27	88.09	89.89	91.86	90.51	90.20	89.27
8	88.12	89.04	93.23	89.25	79.15	80.49	91.67	91.59	90.64	93.28	93.82	94.38
9	64.95	64.83	75.31	71.50	55.24	58.32	73.32	74.44	61.93	67.31	69.85	76.37
10	74.22	67.71	80.96	77.59	60.48	62.96	79.76	81.93	85.54	85.54	88.92	87.71
11	97.29	95.15	98.11	93.73	90.77	75.64	99.06	96.45	98.58	97.75	96.21	97.28
OA	78.16	80.31	87.33	84.77	69.57	72.78	85.13	84.66	82.76	86.24	87.56	**88.70**
AA	49.05	65.16	72.16	70.17	51.38	55.11	67.65	68.43	67.19	68.27	71.02	**71.98**
Kappa	0.7207	0.7470	0.8352	0.8009	0.6168	0.6542	0.8079	0.8011	0.7795	0.8224	0.8386	**0.8526**

a, d and ***h*** represent the area, the diagonal of the bounding box and height. Metrics AA and OA are reported in percentages. Kappa is a coefficient what changes in the range of 0 and 1

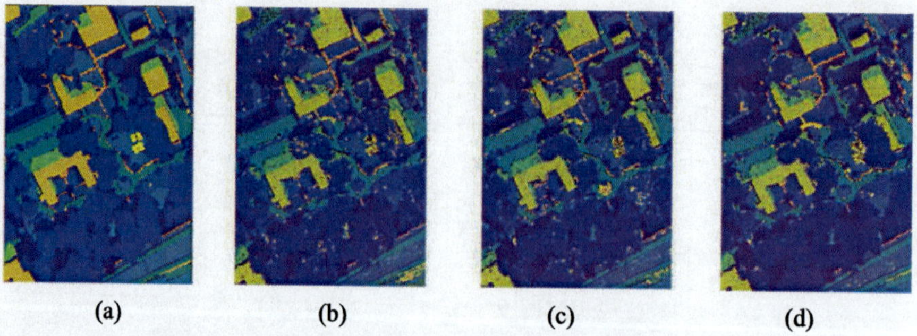

Fig. 6 Classification map of the combination of HSI and LiDAR. **a** Original ground-truth, **b** EP, **c** LCP, **d** ILCP

Furthermore, different numbers of training samples are selected to train SVM classifier. For HSI, LiDAR and the combination of two data, the classification results of each group of training samples have been illustrated in Fig. 7. It can be seen from the results that even with a very small number of training samples, performance of the proposed ILCP is better than LCP and EP, which confirms its effectiveness.

3 Conclusions

An improved LCP (ILCP) was developed for feature extraction of single data and multi-resource remote sensing data. The proposed ILCP employed a threshold-based filtering during tree-pruning stage which only discarded nodes not the whole branches that was unable to meet the requirement, and better retained important information. The proposed method performed better than the state-of-the-art EP and LCP for feature extraction. Furthermore, experimental results demonstrated that the difference and complementarity between HSI and LiDAR data can obtain superior classification performance.

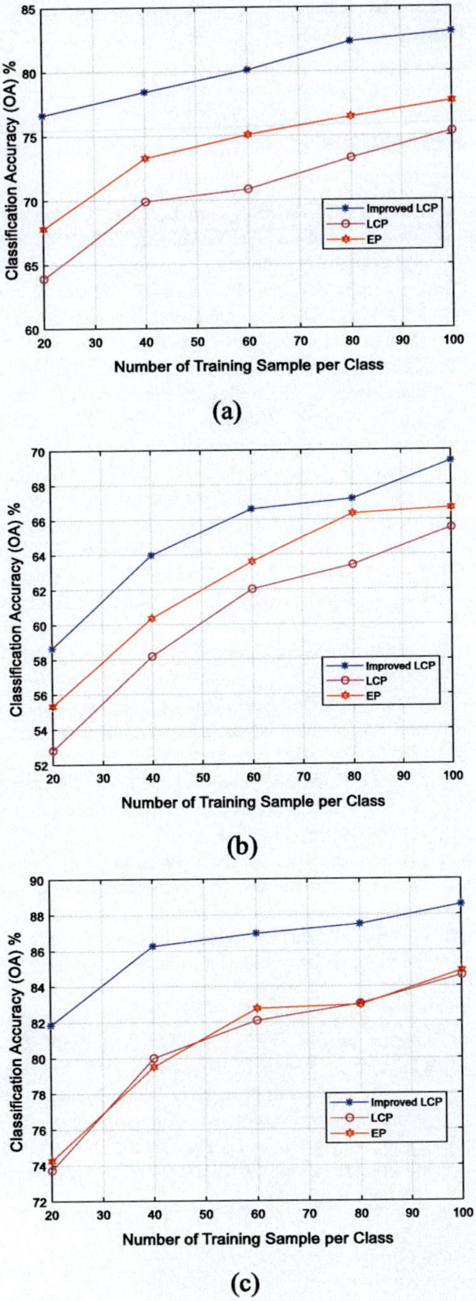

Fig. 7 Effect of different numbers of training samples on the ILCP. **a** HSI, **b** LiDAR, **c** the combination of HSI and LiDAR

Acknowledgement This work was supported by the National Natural Science Foundation of China (61922013).

References

1. Liao W, Coillie FV, Gao L, Li L, Chanussot J (2018) Deep learning for fusion of apex hyperspectral and full-waveform LiDAR remote sensing data for tree species mapping. IEEE Access 6:68716–68729
2. Ge C, Du Q, Li W, Li Y, Sun W (2019) Hyperspectral and LiDAR data classification using kernel collaborative representation based residual fusion. IEEE J Sel Top Appl Earth Obs Remote Sens 12(6):1963–1973
3. Chen Y, Li C, Ghamisi P, Jia X, Gu Y (2017) Data fusion of remote sensing data for accurate classification. IEEE Geosci Remote Sens Lett 14(8):1253–1257
4. Yu Z, Wang T, Zhang X, Zhang J, Ren P (2019) Locality preserving fusion of multisource images for sea-ice classification. Acta Oceanol Sin 38(7):129–136
5. Zhang M, Ghamisi P, Li W (2017) Classification of hyperspectral and LiDAR data using extinction profiles with feature fusion. Remote Sens Lett 8(10):957–966
6. Xu X, Li W, Ran Q, Du Q, Zhang B (2018) Multisource remote sensing data classification based on convolutional neural network. IEEE Trans Geosci Remote Sens 56(2):937–949
7. Fang L, He N, Li S, Ghamisi P, Benediktsson JA (2017) Extinction profiles fusion for hyperspectral images classification. IEEE Trans Geosci Remote Sens 56(3):1803–1815
8. Benediktsson JA, Palmason JA, Sveinsson JR (2005) Classification of hyperspectral data from urban areas based on extended morphological profiles. IEEE Trans Geosci Remote Sens 43(3):480–491
9. Dalla Mura M, Benediktsson JA, Waske B, Bruzzone L (2010) Morphological attribute profiles for the analysis of very high resolution images. IEEE Trans Geosci Remote Sens 48 (10):3747–3762
10. Song Q, Huang R, Wang K (2015) Remote sensing image retrieval based on attribute profiles. In: 2015 international conference on computer science and mechanical automation (CSMA), Hangzhou, pp 231–234
11. Mura MD, Villa A, Benediktsson JA, Chanussot J, Bruzzone L (2010) Classification of hyperspectral images by using extended morphological attribute profiles and independent component analysis. IEEE Geosci Remote Sens Lett 8(3):542–546
12. Ghamisi P, Souza R, Rittner L, Benediktsson JA, Lotufo R, Zhu X (2016) Extinction profiles for the classification of remote sensing data. IEEE Trans Geosci Remote Sens 54(10):5631–5645
13. Li W, Wang Z, Li L, Du Q (2019) Feature extraction for hyperspectral images using local contain profile. IEEE J Sel Top Appl Earth Obs Remote Sens 12(12)
14. Ghamisi P, Souza R, Rittner L, Benediktsson JA, Lotufo R, Zhu XX (2016) Extinction profiles: a novel approach for the analysis of remote sensing data. In: 2016 IEEE international geoscience and remote sensing symposium (IGARSS), Beijing, pp 5122–5125
15. Li W, Wang Z, Li L, Du Q (2018) Modified extinction profiles for hyperspectral image classification. In: 2018 10th IAPR workshop on pattern recognition in remote sensing (PRRS), Beijing, pp 1–5

A High-Resolution Spotlight SAR Imaging Method Based on Two-Step Processing Approach

Hao Zheng and Hongmeng Chen

Abstract The modern radar system not only works on the conventional stripmap synthetic aperture radar (SAR) mode, but also on the high-resolution spotlight SAR mode with refined imaging. However, due to the azimuth spectral aliasing of echo signals in the spotlight SAR, the direct stripmap SAR imaging algorithm fails. To solve this problem, a high-resolution spotlight SAR imaging method based on deramp "two-step processing" is proposed. Firstly, the motion error is compensated based on the inertial navigation system (INS) and echo data, then azimuth deramp operation is performed to eliminates the azimuth spectral aliasing, and thus obtain the 2D signal spectrum without blurring. Then the modified Range-Doppler (RD) method is used to complete the range cell migration (RCM) correction and range compression. Finally, azimuth compression is performed by means of azimuth matched filtering to acquire the focused image. In this method, the imaging framework for both the stripmap SAR and the spotlight SAR are unified by the deramp operation without interpolation. The imaging results in the spotlight SAR mode have demonstrated the effectiveness of the method.

Keywords Synthetic aperture radar · Spotlight SAR · Deramp operation

1 Introduction

With increasing applications of the SAR technology, a growing number of requirements have been put forward for such performance indicators as imaging resolution and surveying bandwidth. Therefore, advanced SAR systems now often need multiple working modes to adapt to this trend [1–3]. Multiple modes means that in addition to the basic stripmap SAR mode, the systems can also work on such wide swath modes as ScanSAR and TOPSAR, and such high-resolution modes as

H. Zheng (✉)
China Satellite Navigation Office, Beijing, China

H. Chen
Beijing Institute of Radio Measurement, Beijing, China

L. Wang et al. (eds.), *Proceedings of the 7th China High Resolution Earth Observation Conference (CHREOC 2020)*, Lecture Notes in Electrical Engineering 757, https://doi.org/10.1007/978-981-16-5735-1_11

spotlight SAR and sliding spotlight SAR. The earth observation ability of a multi-mode SAR system can be greatly improved [3, 4].

Compared with the stripmap SAR, the spotlight SAR [5–7] can continuously adjust the antenna angle during the work process so that the antenna beam always illuminates the target area. The azimuth resolution of spotlight SAR is not subject to the restriction of the actual antenna aperture length, which breaks through the azimuth resolution of the traditional stripmap SAR. Moreover, the spotlight SAR can obtain the information about different sides of the target (as the radar beam can illuminate the imaging target from different angles) and improve the Doppler bandwidth by increasing the azimuth accumulation time, so that 2D high-resolution SAR images can be acquired. Therefore, the technology is of great value in military applications.

However, the imaging methods of spotlight SAR and stripmap SAR are not interchangeable when the pulse repetition frequency (PRF) of the system is low [8–11]. This is mainly because the echo signals of spotlight SAR are aliased in the azimuth frequency domain, which brings a lot of difficulties to imaging processing. When a high PRF is used to avoid aliasing, it will cause an extremely high data volume, leading to an increase in system implementation complexity. A method to avoid azimuth spectrum aliasing is to increase the PRF to make it greater than the full-aperture Doppler bandwidth. But a higher PRF will cause range ambiguity and increase the data volume. Therefore, in case of a high full-aperture Doppler bandwidth, the azimuth spectrum aliasing cannot be avoided only by increasing the PRF. To this end, a high-resolution spotlight SAR imaging method based on Deramp "two-step processing" approach is proposed herein. This method first compensates the non-ideal motion of the carrier platform with the inertial navigation data, and then performs the Deramp operation to transform the azimuth signals into a transform domain with non-aliased azimuth spectrum by multiplying the two phases of azimuth signals in the time domain and the frequency domain, so as to de-aliase the azimuth spectrum of the sliding spotlight SAR signals. On this basis, the improved RD imaging algorithm is used for high-resolution spotlight SAR imaging, which is relatively efficient. The processing results of the data actually extracted by the airborne radar have verified the effectiveness of this method.

2 Spotlight SAR Imaging Model Under a Non-ideal Platform

Suppose that the geometric configuration of the airborne spotlight SAR is shown in Fig. 1. The carrier flies horizontally at the speed V and height H along the positive direction of the Y-axis. The squint angle of the antenna beam is θ, and the pitch angle relative to the horizontal direction is φ. Suppose that A and B are the start and end positions of the synthetic aperture respectively. The actual flight track can be expressed as $[X(t_m), Y(t_m), Z(t_m)]$, and the ideal flight track as $[Vt_m, 0, 0]$. The coordinate differences between the two tracks are the 3 error components of the

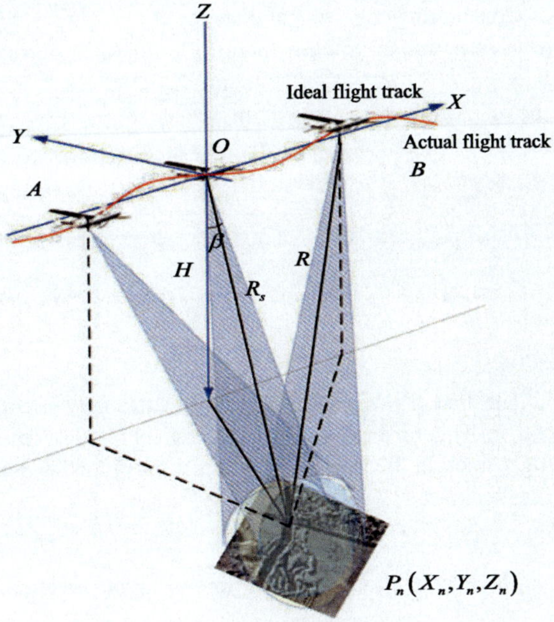

Fig. 1 Geometrical configuration of spotlight SAR mode

radar APC position. The variable of the radar APC position itself is the slow time t_m. As the radar works at the cycle T_r and t_m is discretized at integer multiples of T_r, for the "ideal flight track", the APC position interval can also be evenly arranged along the *X*-axis at an equal interval VT_r. However, compared with the ideal flight track, the error components of the actual flight track are $[X(t_m) - Vt_m, Y(t_m), Z(t_m)]$, where $Y(t_m)$ and $Z(t_m)$ are respectively the deviations from the origin in the normal plane at the time t_m. Such factors as airflow disturbance can cause sub-meter to meter-level random displacements of the carrier in the vertical direction and along the course, which change quickly. Such factors will also affect the speed of the carrier, causing the instantaneous speed to deviate from the average value. Such changes are relatively slow. If long data is processed at one time, as a certain instantaneous speed deviates from the average speed during a period of time, the APC position error $X(t_m) - Vt_m$ in the *X*-axis is sometimes relatively large, up to more than ten meters or more.

Suppose there are many targets on a straight line perpendicular to the ideal flight track at a distance of R_B in the scene and the coordinates of the *n*th scattering point P_n are (X_n, Y_n, Z_n), then the distance from the point *A* to point P_n of the actual radar APC is as follows:

$$R(t_m) = \sqrt{(X(t_m) - Vt_m - X_n)^2 + (Y(t_m) - Y_n)^2 + (Z(t_m) - Z_n)^2} \tag{1}$$

Considering the vertical distance $R_B = \sqrt{X_n^2 + Y_n^2}$ and the side viewing angle β from the radar to the point P_n in the normal plane, it is derived that $\beta = \cos^{-1}(Z_n/R_B)$, $Y_n = R_B \sin\beta$, $Z_n = R_B \cos\beta$. Based on the above relations, the above formula may be written as

$$\begin{aligned} R(t_m) &= \sqrt{(X(t_m) - Vt_m - X_n)^2 + R_B^2 + Y^2(t_m) + Z^2(t_m) - 2R_B Y(t_m)\sin\beta - 2R_B Z(t_m)\cos\beta} \\ &\approx \sqrt{(X(t_m) - Vt_m - X_n)^2 + R_B^2 - 2R_B Y(t_m)\sin\beta - 2R_B Z(t_m)\cos\beta} \\ &\approx R_B + \frac{(X(t_m) - Vt_m - X_n)^2}{2R_B} - Y(t_m)\sin\beta - Z(t_m)\cos\beta \end{aligned} \tag{2}$$

The first and second terms reflect the movement error of the carrier along the course. The third and fourth terms reflect the movement error of the carrier along the flight track in the line-of-sight direction in the normal plane, which is expressed as:

$$\Delta r_{loss} = Y(t_m)\sin\beta + Z(t_m)\cos\beta \tag{3}$$

The acceleration along the line-of-sight direction is:

$$a_{loss} = a_Y(t)\sin\beta + a_Z(t)\cos\beta \tag{4}$$

where, $a_Y(t)$ and $a_Z(t)$ represent the acceleration along the Y-axis and the Z-axis directions respectively.

From the formula above, the modulation frequency of the target point P_n is derived:

$$\gamma_n(t_m) = -\frac{2V^2(t_m)}{\lambda R_B} - \frac{2[X(t_m) - Vt_m - X_n]a(t_m)}{\lambda R_B} - \frac{2}{\lambda}a_{loss}(t_m) \tag{5}$$

where, $V(t_m)$ is the actual flight speed of the carrier. $a(t_m)$ is the acceleration along the course and $a_{loss}(t_m)$ is that along the line-of-sight direction. The second term in the above formula is related to the azimuth position of the target, because it reaches the maximum value at the edge of the synthetic aperture. If the phase error caused by the second term is less than $\pi/4$, the condition under which the effect can be ignored is:

$$L_{\max} \le \left(6\lambda R_B V^2 / a_{\max}\right)^{\frac{1}{3}} \tag{6}$$

where, $a_{\max}$ is the maximum absolute value of acceleration along the course, and $L_{\max}$ is the maximum synthetic aperture length. Due to the great inertia of the carrier, the speed change along the course is usually small, and $a_{\max}$ is so low that it can be ignored, so the above formula can be approximated as follows:

$$\begin{aligned}\gamma_n(t_m) &= -\frac{2V^2(t_m)}{\lambda R_B} - \frac{2}{\lambda}a_{loss}(t_m) \\ &= -\frac{2V^2(t_m)}{\lambda R_B} - \frac{2}{\lambda}\left[a_Y(t_m)\left(1 - \frac{H^2}{2R_B^2}\right) + a_Z(t_m)\frac{H}{R_B}\right] \\ &= -\frac{2V^2(t_m)}{\lambda R_B} - \frac{2}{\lambda}a_I(t_m) - \frac{2}{\lambda}a_{II}(t_m)\Delta r\end{aligned} \tag{7}$$

where:

$$a_I(t_m) = a_Y(t_m)\left(1 - \frac{H^2}{2R_0^2}\right) + a_Z(t_m)\frac{H}{R_0} \tag{8}$$

$$a_{II}(t_m) = a_Y(t_m)\frac{H^2}{R_0^3} - a_Z(t_m)\frac{H}{R_0^2} \tag{9}$$

R_0 is the distance from the center of the scene. $a_I(t)$ is the projection of the transceiver platform at the center of the scene in directions Y and Z under acceleration. $a_{II}(t)$ represents the spatial variation of acceleration along with the distance Δr in the line-of-sight direction. The approximation relations of $R_B \gg H$, $R_B = R_0 + \Delta r$, $\frac{1}{R_B} \approx \frac{1}{R_0}\left(1 - \frac{\Delta r}{R_0}\right)$ and $\frac{1}{R_B^2} \approx \frac{1}{R_0^2}\left(1 - \frac{2\Delta r}{R_0}\right)$ are used here. In case of a large motion error, it will seriously affect the imaging quality, and even cause the imaging effect to drop sharply, so motion error compensation must be performed.

3 A Spotlight SAR Imaging Method Based on Deramp Two-Step Processing Approach

As the total Doppler bandwidth of the azimuth in the spotlight mode is generally greater than the system PRF, the azimuth data is under-sampled and the azimuth spectrum is aliased, the traditional frequency domain imaging algorithm fails. For spotlight imaging, the azimuth spectrum aliasing must be solved first.

The demodulated baseband signal of the SAR echo signal is:

$$\begin{aligned}s(\tau, t_m) = A \cdot rect\left(\frac{\tau - \frac{2R(t_m,R)}{c}}{T_p}\right) \cdot rect\left(\frac{t_m}{T_{\text{spot}}}\right) \\ \exp\left\{-j4\pi\frac{R(t_m, R)}{\lambda} - j\pi K_r\left[\tau - \frac{2R(t_m, R)}{c}\right]^2\right\}\end{aligned} \tag{10}$$

The expression in the two-dimensional frequency domain of the echo signal is:

$$S_{2df}(f_\tau, f_\eta) = A \cdot \omega_r(f_\tau)\omega_a(f_\eta - f_{\eta c}) \\ \exp\left\{-j4\pi\frac{R(f_\tau + f_c)}{c}\sqrt{1 - \frac{c^2 f_\eta^2}{4v^2(f_\tau + f_c)^2}} + \pi\frac{f_\tau^2}{K_r} - \frac{2\pi X_n f_\eta}{v}\right\} \quad (11)$$

where $\omega_r(f_\tau)$ and $\omega_a(f_\eta)$ are the range and azimuth frequency domain envelopes respectively, t_m is the azimuth slow time, τ is the range fast time, K_r is the modulation frequency of the transmitted signal, $R(t_m, R)$ is the instantaneous slant range, f_r is the range frequency, f_η is the azimuth Doppler frequency and $f_{\eta c}$ is the azimuth Doppler center frequency.

The core of the Deramp algorithm is the two-step dechirp processing of azimuth. The essence of azimuth dechirp pre-processing is to solve the azimuth under-sampling problem of the original echo signal data domain. The basic processing flow is a time-domain convolution operation, and the purpose is to resample the original undersampled signals into new non-undersampled signals.

The azimuth deramp convolves the signals and the reference phase function, thereby realizing the azimuth scaling, which equivalently improves the azimuth sampling rate. The phase reference function is:

$$H_1 = \exp(-j\pi K_a t_m^2) \quad (12)$$

where $K_a = -\frac{2v^2\cos^2\theta}{\lambda R_{ref}}$ denotes the Doppler modulation frequency at the center of the scene, and t_a denotes the azimuth time. The function of this phase multiplication is to remove the rate of change of the Doppler center and ensure that the azimuth bandwidth is smaller than the original PRF, so that the azimuth spectrum is not aliased. The azimuth time–frequency relationship before and after Deramp is shown in Fig. 2.

$$\begin{aligned} s_a(\tau, t'_m) &= s_1(\tau, t_m) * H_1 \\ &= \exp(-j\pi K_a t'^2_m)\int s_1(\tau, t)\exp(-j\pi K_a t^2)\exp(j2\pi K_a t'_m t)\,dt \end{aligned} \quad (13)$$

Convert the above formula into a discrete form:

$$s_a(\tau, m\Delta t'_m) = \exp\left(-j\pi K_a(m\Delta t'_m)^2\right)\sum_{i=-A/2}^{A/2} s_1(\tau, i\Delta t_m) \\ \exp\left(-j\pi K_a(i\Delta t_m)^2\right)\exp(j2\pi K_a m i\Delta t'_m \Delta t_m) \quad (14)$$

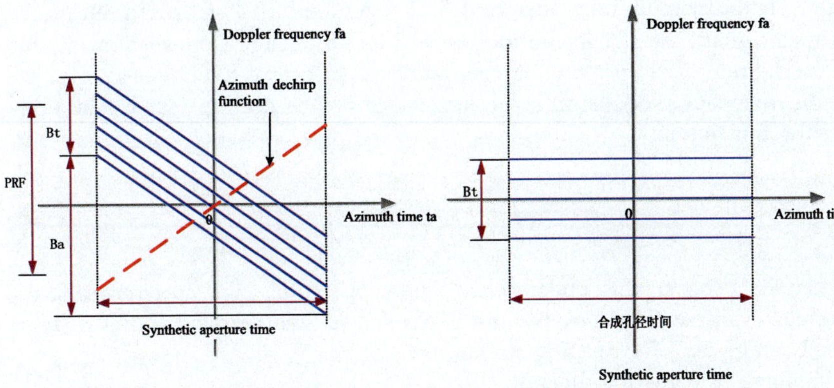

(a) Spectrogram before Deramp (b) Spectrogram after Deramp

Fig. 2 Deramp flow chart

where $m = -B/2, \ldots, B/2$, A and B are the azimuth sampling points before and after the Deramp transformation, and Δt_m and $\Delta t'_m$ are the pulse repetition intervals before and after the Deramp transformation, respectively.

$$M = \frac{1}{K_a \Delta t_m \Delta t'_m} \tag{15}$$

If the transformation satisfies Formula (8), zerofill the position signals so that the input and output samples are both M, then Formula (8) can be written as Formula (9),

$$\begin{aligned} s_a(\tau, m\Delta t'_m) = \exp\left(-j\pi K_a (m\Delta t'_m)^2\right) \sum_{i=-M/2}^{M/2} s_1(\tau, i\Delta t_m) \\ \exp\left(-j\pi K_a (i\Delta t_m)^2\right) \exp\left(j\frac{2\pi mi}{M}\right) \end{aligned} \tag{16}$$

From the above formula, the azimuth deramp is actually realized by multiplying the signal reference function, then processing with FFT and finally compensating with a secondary phase. After the above operation, the signal azimuth scale is transformed from t_m to t'_m, thus realizing the transformation of PRF. There will be no azimuth spectrum aliasing after the 2D FFT into the frequency domain.

At this time, the azimuth dechirp factor of Deramp is:

$$H_2 = \exp\left(-j\pi K_a t'^2_m\right) \tag{17}$$

where t_m' is the azimuth time after zero filling. Azimuth FFT is carried out on the dechirped signals, the signals are transformed into the range Doppler domain, and then the azimuth dechirp phase compensation is performed to compensate for the azimuth frequency modulation slope introduced by the dechirp factor. The compensation function is:

$$H_{RVP} = \exp\left(-j\pi\frac{f_a^2}{K_a}\right) \tag{18}$$

After the echo signals undergo the above operations, the spectrum aliasing phenomenon in the frequency domain is eliminated, and the high squint imaging method can be used for imaging subsequently.

The range compression function is:

$$H_3 = \exp\left(j\pi\frac{f_r^2}{K_r}\right) \tag{19}$$

The range cell migration correction amount is:

$$\Delta R = \frac{R_{ref}}{\sqrt{1 - \frac{\lambda^2 f_a^2}{4v^2}}} - R_{ref} \tag{20}$$

where R_{ref} is the reference slant range, and is selected as the scene center slant range. When range block processing is used, R_{ref} is the center slant range of each block.

The range cell migration correction can be achieved by interpolation in the time domain, and the interpolation requires a compromise between the amount of calculation and the accuracy of the interpolation. This algorithm adopts a method that realizes the range cell migration correction in the range frequency domain. According to the properties of Fourier transform, the time shift in the time domain corresponds to the linear phase in the frequency domain. Through frequency domain operation, the amount of calculation is reduced while the accuracy is improved.

Now, the range cell migration correction function is:

$$H_4 = \exp\left(j\frac{4\pi f_r R_{ref}}{C}\left(\frac{\lambda f_a}{2v}\right)^2\right) \tag{21}$$

After the range cell migration correction is completed, the echo signals of the target are distributed in the same range unit, and the azimuth matched filtering can be performed to obtain the azimuth focusing effect. Since the data is in the azimuth frequency domain after the range cell migration correction, the azimuth matched filtering is performed directly in the frequency domain.

The azimuth compression function is:

$$H_5 = \exp\left(j\frac{2\pi R_r}{v}\sqrt{f_{aM}^2 - f_a^2}\right) \tag{22}$$

4 Results and Analysis of Measured Data

In order to verify the effectiveness of this algorithm, we use a set of airborne measured spotlight data for analysis. Some radar parameters are shown in Table 1. Figure 3 is the Google map of the experimental scene.

Figure 4 shows the comparison results of different methods for spotlight imaging. In the figure, the azimuth width of the direct imaging result and the imaging result after pre-filtering is severely reduced, and there is a scene shift in the image (such as Region 1 and Region 2 in the red boxes), which is mainly due to the azimuth spectrum of the spotlight SAR. In Fig. 4a, b, the aircrafts and hangars on the aprons in Region 1 are marked on the airport runways, which indicates obvious azimuth aliasing in the imaging results. In Region 2, the airport edge scene is moved below the airport apron, which indicates azimuth aliasing in the imaging results The reason is mainly due to the azimuth spectrum aliasing of the spotlight SAR.

Table 1 System parameters

Parameters	Value	Parameters	Value
Carrier speed	80 m/s	Squint angle range	−10∼−12°
Signal bandwidth	920 MHz	Pitch angle	1.5°
Signal time width	25 us	Scene center range	12.6 km
3 dB beam width	1.4°	PRF	4000 Hz

Fig. 3 Google map of the experimental scene

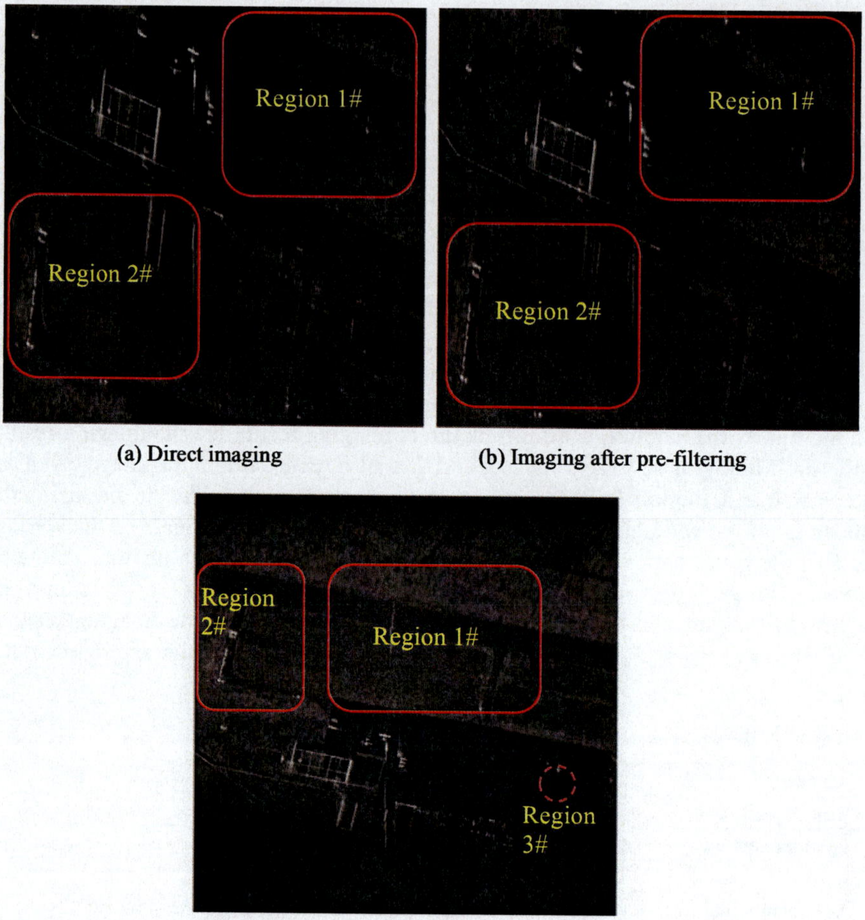

(a) Direct imaging

(b) Imaging after pre-filtering

(c) Imaging through the method in this article

Fig. 4 Imaging comparison in spotlight SAR mode

In Fig. 4c, the aircrafts, hangar, runway and surrounding objects on the apron are obvious, and the contours of the aircrafts on the apron are clearly visible, indicating that this method can better solve the spectrum aliasing problem in the spotlight SAR imaging process.

5 Conclusion

In order to solve the problem of azimuth spectrum aliasing which leads to the decline of spotlight SAR imaging effect, this paper proposes a high-resolution spotlight SAR imaging method based on "two-step processing". This method first compensates the error of the moving platform based on the inertial navigation echo signal data, and then uses Deramp to eliminate the azimuth spectrum aliasing. The method then uses the improved Range Doppler (RD) processing and phase compensation to achieve accurate range cell migration correction, and finally performs azimuth matched filtering to obtain the final spotlight SAR imaging results. The results of airborne spotlight SAR data processing have verified the effectiveness of this algorithm.

References

1. Zheng B, Mengdao X, Tong W (2005) Radar imaging technology. Publishing House of Electronics Industry, Beijing, pp 130–183
2. Carrara WG, Goodman RS, Majewski RM (1995) Spotlight synthetic aperture radar—signal processing and algorithms, Artech House, Boston, MA
3. Meta A, Mittermayer J, Prats P et al (2010) TOPS imaging with TerraSAR-X: mode design and performance analysis. IEEE Trans Geosci Remote Sens 48(2):759–s769
4. Mittermayer J, Moreira A, Loffeld O (1999) Spotlight SAR data processing using the frequency scaling algorithm. IEEE Trans Geosci Remote Sens 37(5):2198–2214
5. Lanari R, Tesauro M, Sansosti E et al (2001) Spotlight SAR data focusing based on a two-step processing approach. IEEE Trans Geosci Remote Sens 39(9):1993–2004
6. Jing W, Zhang L, Xing M et al (2009) Algorithm for wide-swath spotlight SAR data processing. Acta Electron Sin 37(3):470–475
7. Wei Y, Li C, Chen J et al (2012) Extended three-step focusing algorithm for spaceborne sliding spotlight SAR image formation. J Beijing Univ Aeronaut Astronaut 38(3):297–302
8. An D, Tao H-X, Tian J et al (2012) Extended two-step focusing approach for squinted spotlight SAR imaging. IEEE Trans Geosci Remote Sens 50(7):2889–2900
9. Reigber A, Alivizatos E, Potsis A et al (2006) Extended wavenumber-domain synthetic aperture radar focusing with integrated motion compensation. IEE Proc Radar Sonar Navig 153(3):301–310
10. Sun GC, Jiang XW, Xing M (2011) Focus improvement of highly squinted data based on azimuth nonlinear scaling. IEEE Trans Geosci Remote Sens 49(6):2308–2322
11. Wong H, Tat-soon Y (2001) New applications of nonlinear chirp scaling in SAR data processing. IEEE Trans Geosci Remote Sens 39(5):946–953

Research of Industrial Heat Sources Detection Method Based on Thermal Infrared Satellite Data

Jie Wang, Yu Liu, Hongzhong Lu, Jianwei Qi, and Ping Ai

Abstract Carbon based fuel burning in industrial plants accounts for a large proportion of greenhouse gas emissions, which has a significant impact on human health and climate change. Monitoring the operation status of these factories by means of thermal infrared (TIR) remote sensing technology is one of the important measures for environmental protection. This paper proposed an automatic detection method for industrial heat sources (IHSs) based on TIR data. Firstly, the land surface temperature data is retrieved, then the local abnormal high temperature pixels are extracted by sliding window smoothing, and finally the industrial heat sources are further screened by normalized vegetation index (NDVI) data. To verify the reliability of the proposed method, this paper used the Landsat8 data in Jiangsu and Hebei Province of China in 2019 to identify industrial heat sources. Through artificial judgment with optical satellite image, the detection accuracy was 83.78% in Jiangsu and 78.6% in Hebei respectively. Whether for large-scale plants or small workshop plants, the detection accuracy of proposed method is satisfactory. The validation results represent that this method could be used as a new technology for government or industry regulation.

Keywords Industrial heat sources · Land surface temperature · Local abnormal high temperature · Normalized vegetation index

J. Wang (✉) · Y. Liu · J. Qi · P. Ai
Land Satellite Remote Sensing Application Center of The Ministry of Natural Resources, Beijing, China
e-mail: wangjie0039@163.com

H. Lu
Qingdao Natural Resources Monitoring Center, Qingdao, China

L. Wang et al. (eds.), *Proceedings of the 7th China High Resolution Earth Observation Conference (CHREOC 2020)*, Lecture Notes in Electrical Engineering 757, https://doi.org/10.1007/978-981-16-5735-1_12

1 Introduction

With the acceleration of industrialization, the problem of greenhouse gases caused by industrial emissions has become increasingly prominent. Carbon based fuel burning in industrial facilities accounts for a large proportion of greenhouse gas emissions, which has a significant impact on human health and climate change. Obtaining the spatial distribution of industrial plants is conducive to understanding the greenhouse gas emission mode in industrial combustion, and to carrying out energy conservation and emission reduction [1, 2].

In addition, many small factories do not equip with environmental protection facilities, easy to become environmental pollution sources. Monitoring the operation status of these illegal factories is one of the important measures for environmental protection. As a result of extensive heat released by these factories in production, the land surface temperature of these plants will be significantly higher than that of the surrounding areas [3]. Therefore, it is possible to effectively identify industrial heat sources (IHSs) by means of thermal infrared (TIR) remote sensing technology.

Liu [4] and Zhang [5] used the VIIRS Nightfire product to detect the IHSs, but the Nightfire product can only identify the high temperature ignition points, and the resolution is too low to detect small IHSs. Xia [6] used the thermal anomaly index based on the ASTER data to detect the IHSs, however there is not enough TIR data available for aster, so it is difficult to ensure stability and continuity monitoring. In fact, Landsat8 would be a very good data source for industrial heat sources detection, with its high spatial resolution and stable orbit period.

2 Methodology

2.1 *The Land Surface Temperature Retrieving and Compositing*

In order to extract local abnormal high temperature pixels, Firstly, the land surface temperature (LST) would be retrieved based from satellite TIR data, and then the multiple coverage data of the same area are composed by maximum value composite algorithm to get the highest LST, thus effectively reducing the influence of atmospheric attenuation and sensor observation angle on IHSs detection. According to different satellite sensors, single channel algorithm [7, 8], split-window algorithm [9, 10] or other methods could be used for LST retrieval. This paper chose the radiative transfer equation method to retrieve the LST. The Radiance in the TIR band L_λ received by the satellite sensor consists of three parts: the surface emissive radiation passing through the atmosphere, the upward radiation $L\uparrow$ of the atmosphere and the downward radiation $L\downarrow$ reflected by the surface, which can be expressed as follows:

$$L_\lambda = [\varepsilon \cdot B(T_S) + (1 - \varepsilon) \cdot L \downarrow] \cdot \tau + L \uparrow \tag{1}$$

where, ε is land surface emissivity, τ is transmittance of the atmosphere in TIR band, Ts is the true temperature (k) of land surface, B(Ts) is brightness of the blackbody.

Therefore, the radiation brightness $B(T_S)$ with temperature T_S in TIR band can be written as:

$$B(T_S) = [L_\lambda - L \uparrow - \tau \cdot (1 - \varepsilon) \cdot L \downarrow]/(\tau \cdot \varepsilon). \tag{2}$$

Here, land surface emissivity can be retrieved by NDVI threshold method [11], L_λ can be obtained through the conversion of satellite DN values, the upward radiation $L\uparrow$, the downward radiation $L\downarrow$ and transmittance of the atmosphere τ can be obtained from NASA Atmospheric Correction Parameter Calculator according to satellite parameters. Consequently, T_S would be calculated by the function of Planck formula:

$$T_S = K_2/\ln(K_1/B(T_S) + 1). \tag{3}$$

For Landsat 8 Band 10: K_1 = 774.89 W/(m^2 μm sr), K_2 = 1321.08 K.

2.2 *Local Abnormal High Temperature Extracting*

The temperature of local abnormal high temperature pixel is usually much higher than the average temperature of some surrounding pixels. The higher the local temperature, the greater the difference between the smoothed temperature and the original temperature. Therefore, it is necessary to use a sliding window to calculate the average LST of each window as the temperature value of central pixel after smoothing. The size of the sliding window depends on the spatial resolution of the used satellite data. Usually the sliding window of 1 km is recommended to be selected, because the smooth effect of abnormal high temperature pixels in industrial plant under this window is the best, and the difference between the real LST and that after smoothing is the largest, which could extract the abnormal high temperature pixels well, and reduce the interference of the non high temperature abnormal points at the same time.

2.3 *Potential Industrial Heat Sources Detecting*

The original LST data need to be directly subtracted by smoothed LST, and the pixels with the difference values exceeding a certain threshold are considered as

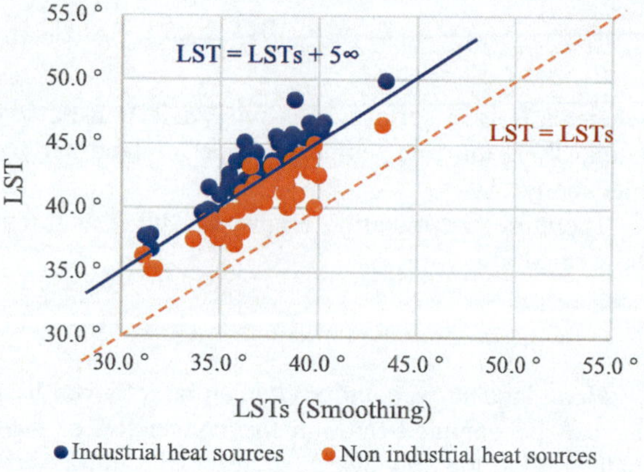

Fig. 1 Relationship of real LST and smoothed LST

the potential IHSs. If the threshold is too high, many real IHSs would be missed; on the contrary, if the threshold is too low, some non IHSs would be wrongly selected. In order to select a reasonable threshold value, this paper randomly selected 50 determined IHSs and 50 non IHSs through Google Earth map, and their real LST and smoothed LST are as shown in Fig. 1.

It can be seen from Fig. 1 that the temperature difference of 5° can be better divided into IHSs and non IHSs. All IHSs could be identified, with only about 10% of non IHSs detected wrongly. Although there are still some non IHSs identified in error, it can ensure almost no missed detection. Therefore, 5° would be used as a threshold to identify IHSs in this paper.

2.4 Industrial Heat Sources Further Screening

Among the potential ISHs in Sect. 2.3, there are some non IHSs that are wrongly detected, including mining points, bare land, residential buildings etc. For further screening out the right IHSs and eliminate the non IHSs wrongly detected, the normalized vegetation index (NDVI) can be used as a judgment factor.

This paper compared the NDVI values of the 50 determined IHSs and the 50 non IHSs selected, and the analysis results are shown in Fig. 2. The results show that if taking NDVI less than 0.55 as the additional criteria, nearly half of the non IHSs could be eliminated, which greatly reduces the false detection rate of this method.

Above all, the potential IHSs with NDVI less than 0.55 would be taken as the final identified IHSs in this method.

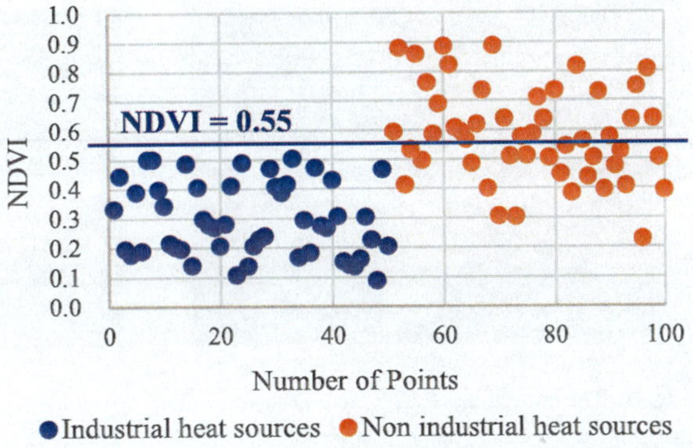

Fig. 2 Distribution of NDVI values of IHSs and non IHSs

3 Datasets and Study Area

3.1 Datasets

Landsat8 has a repeat cycle of 16 days, the spatial resolution of the thermal infrared band is 90 m, and that of the visible bands are 30 m. Through data fusion processing, the LST with 30 m resolution can be retrieved. Therefore, using Landsat8 TIR data, we can obtain high spatial resolution and continuous multi-phase LST data, so as to better detect IHSs. The NDVI with 30 m resolution can also be calculated by the near infrared band and red band.

Google Earth map was also used to visually discriminate correctness of the IHSs detected by the proposed method, and verify the accuracy of the algorithm.

Atmospheric parameters used in Eq. (2) could be calculated using the on-line Atmospheric Correction Parameter Calculator (ACPC, http://atmcorr.gsfc.nasa.gov).

3.2 Study Area

Jiangsu Province and Hebei Province of China were selected as the study areas of this paper, which have a large number of industrial plants, mainly including chemical plants, iron and steel plants, cement plants, metal smelting plants etc., the number of IHSs in Jiangsu and Hebei Province are in the top five in China.

4 Validation Results

4.1 Validation Results of Jiangsu Province

In this paper, we first used the data of Landsat8 in Jiangsu Province in 2019 to retrieve the LST and composed it according to the maximum value. Figure 3a

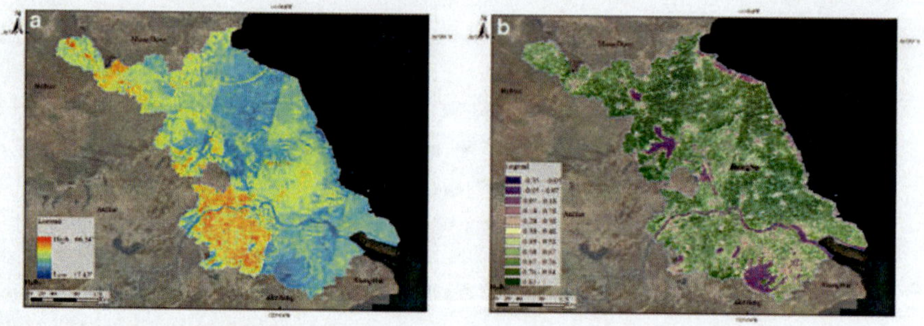

Fig. 3 The maximum LST and NDVI of Jiangsu Province in 2019: **a** LST; **b** NDVI

shows the maximum LST of Jiangsu Province with a resolution of 30 m in 2019. The highest temperature could reach 66°, which may be caused by the heat released by the IHSs.

According to the resolution of Landsat8 LST data, this paper took a rectangular box of 30 pixels rows and 30 pixels columns as the sliding window to smooth the LST data. Then, the NDVI data set of Jiangsu Province in 2019 was also calculated from Landsat8 near infrared band and red band that shown as Fig. 3b.

After data processing and analyzing, a total of 4611 IHSs in Jiangsu Province are identified, as shown in Fig. 4. Through visual interpretation based on Google Earth map, 748 non IHSs are eliminated, leaving 3863 correct IHSs.

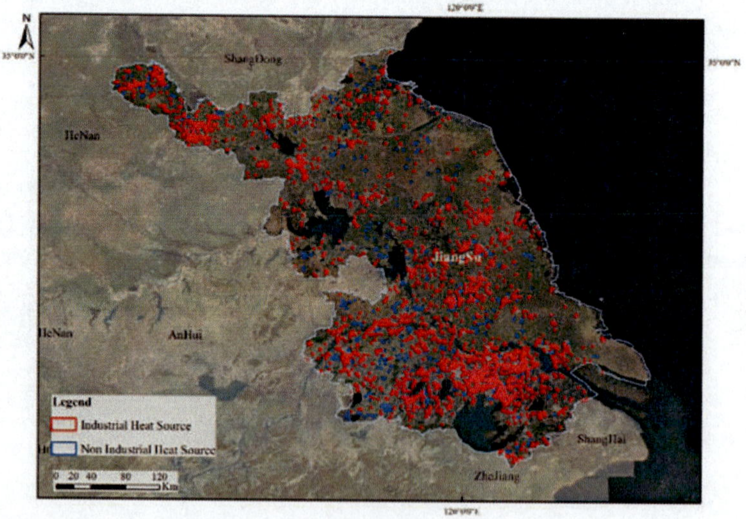

Fig. 4 Detection results of IHSs in Jiangsu Province

It can be seen from the results that the overall identification accuracy of this algorithm in Jiangsu Province reaches 83.78%, which has been able to meet the needs of remote sensing monitoring of heat source plants. It should be noted that these number does not mean that only these detected enterprises are in production, due to appropriate environmental protection measures, many plants would not release a lot of heat, so it could not be detected by this proposed method. In order to prove the detection effect of this method, Fig. 5 shows the high-resolution images of several different subcategories of IHSs plants detected by proposed method. The red solid line defining the boundary of the IHSs, is drawn manually.

4.2 Validation Results of Hebei Province

In this paper, we also detected the IHSs in Hebei Province for algorithm verification. Figure 6 shows the maximum LST and NDVI of Hebei Province with a resolution of 30 m in 2019. The highest temperature could reach 50°. Although

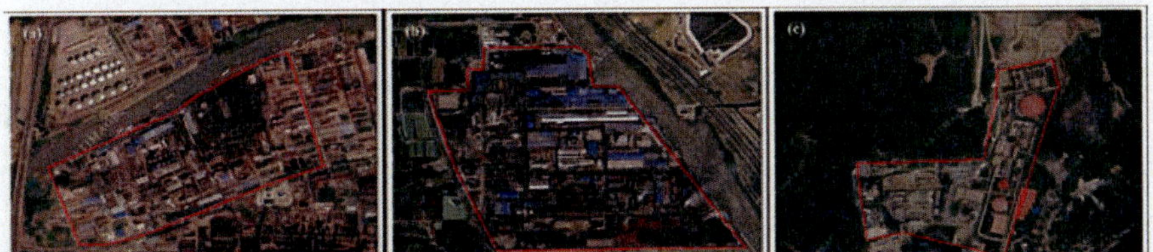

Fig. 5 High resolution images of several different subcategories of IHSs plants: **a** chemical plant; **b** steel plant; **c** cement plant; **d** metal material plant; **e** glass plant; **f** brick plant; **g** ceramic plant; **h** coking plant; **i** thermal power plant

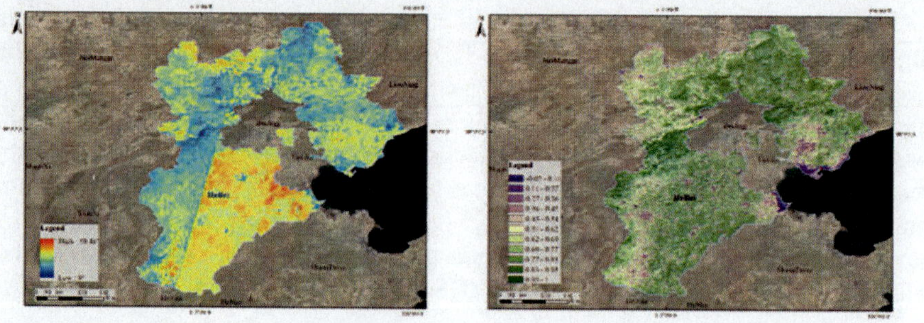

Fig. 6 The maximum LST and NDVI in 2019 of Hebei Province: **a** LST; **b** NDVI

there is obvious heterogeneity in the temperature data caused by different transit times of satellites, the accuracy of identification would not be affected. The spatial distribution of NDVI is consistent with that of IHSs in Hebei Province, and the NDVI of major industrial areas are very low relatively.

Finally, the detection results of Hebei Province are shown in Fig. 7. A total of 2501 IHSs are identified, with 535 non IHSs are eliminated, screening out 1966 correct IHSs.

The overall identification accuracy of this algorithm in Hebei Province is 78.6%. From the perspective of spatial distribution, the IHSs are mainly concentrated in Handan and Xingtai in the southwest and Tangshan in the Northeast of Hebei Province. Through comparative analysis, we find that this method identified a

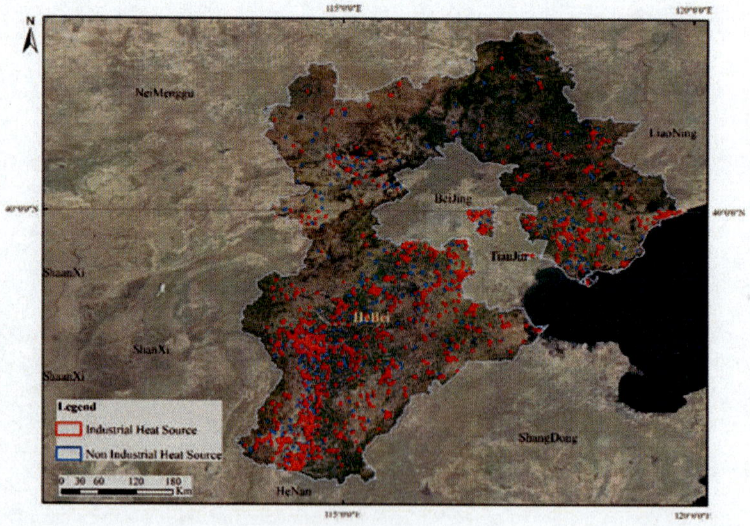

Fig. 7 Detection results of IHSs in Hebei Province

number of small factories which have not been registered with the Industry and Commerce Department. These factories usually do not have perfect environmental protection facilities, extremely easy to pollute the environment. The proposed method could effectively monitor these small illegal IHSs.

5 Conclusions

In this paper, an automatic detection method for industrial heat sources based on thermal infrared data was proposed, which can identify the industrial heat sources (IHSs) in production by extracting local abnormal high temperature pixels and combining with NDVI data. Jiangsu and Hebei Province of China, were chosen as the study areas, and Landsat8 data in 2019 was used to verify the proposed method. The results showed that the overall recognition accuracy could reach up to about 80%. In addition, it can identify the exothermic factories that are not listed in the government inventory, whether these are large-scale plants or small workshop plants, the detection accuracy is satisfactory. The results represent that the proposed method can be used as a new technology for government or industry regulation.

Acknowledgements This work was jointly supported by the 13th Five-Year Plan of Civil Aerospace Technology Advanced Research Projects under Grant D040401 and the National Natural Science Foundation of China (NO. 41801287).

References

1. Zhou K, Yang S, Shen C, Ding S, Sun C (2015) Energy conservation and emission reduction of China's electric power industry. Renew Sustain Energy Rev 45:10–19
2. Wu X, Zhao L, Zhang Y, Zhao L, Zheng C, Gao X, Cen K (2016) Cost and potential of energy conservation and collaborative pollutant reduction in the iron and steel industry in China. Appl Energy 184:171–183
3. Okwen R, Pu R, Cunningham J (2011) Remote sensing of temperature variations around major power plants as point sources of heat. Int J Remote Sens 32(13):3791–3805
4. Liu Y, Hu C, Zhan W, Sun C, Murch B, Ma L (2018) Identifying industrial heat sources using time-series of the VIIRS Nightfire product with an object-oriented approach. Remote Sens Environ 204:347–365
5. Zhang P, Yuan C, Sun Q, Liu A, You S, Li X, Liu M (2019) Satellite-based detection and characterization of industrial heat sources in China. Environ Sci Technol 53(18):11031–11042
6. Xia H, Chen Y, Quan J (2018) A simple method based on the thermal anomaly index to detect industrial heat sources. Int J Appl Earth Obs Geoinf 73:627–637
7. Qin Z, Karnieli A, Berliner P (2001) A mono-window algorithm for retrieving land surface temperature from Landsat TM data and its application to the Israel-Egypt border region. Int J Remote Sens 22(18):3719–3746
8. Jiménez-Muñoz JC, Cristóbal J, Sobrino JA, Sòria G, Ninyerola M, Pons X (2008) Revision of the single-channel algorithm for land surface temperature retrieval from Landsat thermal-infrared data. IEEE Trans Geosci Remote Sens 47(1):339–349

9. Becker F, Li ZL (1990) Towards a local split window method over land surfaces. Remote Sens 11(3):369–393
10. Wan Z, Dozier J (1996) A generalized split-window algorithm for retrieving land-surface temperature from space. IEEE Trans Geosci Remote Sens 34(4):892–905
11. Sobrino JA, Raissouni N, Li ZL (2001) A comparative study of land surface emissivity retrieval from NOAA data. Remote Sens Environ 75(2):256–266

Robust Detection and Tracking Algorithm for Pedestrian Targets with Posture Change

Tao Zhao, Jin-Sheng Xiao, Wen Wang, Hai-Gang Sui, Jing-Jin Ma, and Chuan Xu

Abstract In the intelligent monitoring system, pedestrian detection and tracking is the basis of behavioral analysis, pedestrian retrieval and other intelligent analysis technologies. In order to solve the problems caused by posture change and occlusion, this paper proposes a robust pedestrian target detection and tracking algorithm, which is based on YOLOv3 and DeepSORT. In order to reduce the impact of pedestrian posture change, we introduce a person re-identification feature that resists posture change on the tracking algorithm DeepSORT. For pedestrian occlusion, inspired by Spindle Net, we integrate seven local appearance features extracted from different regions of the human body together to form robust global appearance features, making the tracking more accurate. Experimental results prove that the algorithm proposed in this paper shows robust pedestrian detection and tracking performance on both subjective visual effects and objective metrics, which can meet the needs of actual applications.

Keywords Surveillance video · Anti-posture change · Pedestrian detection · Pedestrian tracking

T. Zhao · J.-S. Xiao (✉) · W. Wang
School of Electronic Information, Wuhan University, Wuhan 430072, China
e-mail: xiaojs@whu.edu.cn

H.-G. Sui
State Key Laboratory of Information Engineering in Surveying, Mapping and Remote Sensing, Wuhan University, Wuhan 430079, China

J.-J. Ma
Beijing Institute of Information Technology, Beijing 100000, China

C. Xu
School of Computer Science, Hubei University of Technology, Wuhan 430072, China

L. Wang et al. (eds.), *Proceedings of the 7th China High Resolution Earth Observation Conference (CHREOC 2020)*, Lecture Notes in Electrical Engineering 757, https://doi.org/10.1007/978-981-16-5735-1_13

1 Introduction

Pedestrian tracking, aiming to locate a pedestrian target in a video with an initial bounding box in the first frame, has been one of the most important problems in computer vision. With considerable progresses in the detection community, detection-based tracking has become the mainstream of visual tracking. Till now, there are two main camps for pedestrian detection algorithms: traditional machine learning based algorithms and deep learning based algorithms. Traditional machine learning based algorithms use manually designed features such as HOG, Haar, and LBP to describe images [1]. However, these methods usually result in poor performance in complex conditions. Deep learning based algorithms use convolutional neural networks to learn deep features of images automatically and usually obtain a better performance. RCNN (Regions with CNN features) [2] proposed by R. Girshick's is one of the first attempts to abstract the object detection problem into two stages. First, obtain a large number of proposed regions that may contain objects, and then classify each region. Since then, faster RCNN [3], mask RCNN [4] and other two-stage algorithms have been widely concerned because of high accuracy. However, the speed is slow. YOLO (You Only Look Once) [5] proposed by J. Redmon et al. is a pioneering work of single-stage object detection which has a much higher speed but lower accuracy than two-stage algorithms. To enhance performance of one-stage algorithms, recent state-of-the-art detectors do their own work. RefineDet [6] improves the detection performance by refining the SSD (Single Shot multibox Detector) algorithm. CenterNet [7] proposed by Duan et al. is almost the first attempt to transform the object detection problem into a key point estimation problem. For detection-based tracking, the performance of the object detection is very critical.

In the early days, visual tracking algorithms based on mean shift, particle filtering, and correlation filtering were widely used. But now, tracking algorithms based on deep learning have made considerable progress due to their robustness. The HDT (hedged deep tracking) algorithm proposed by Qi et al. [8] treats the features of different convolutional layers separately to estimate the position of the target more accurately. The Siamese-RPN network [9] proposed by Li et al. utilizes two types of sub-networks to integrate classification and regression. DeepSORT [10] proposed by Wojke, a multi-target tracking algorithm, adopts the data set in pedestrian re-identification to build an appearance model based on deep convolutional neural networks. Combining the model of multi-object segmentation and pedestrian re-identification [11], transforms the multi-pedestrian target tracking problem into a multi-object segmentation problem. Wang et al. [12], a multi-modal multi-target tracking framework, which can be trained in an end-to-end manner can maintain the reliability through independent multi-sensor features. Maksai and Fua [13] proposes a new training mode to improve the identity switching problem when the overlaps of the targets occur in the tracking phase. However, the above algorithms are less robust in the case of severe pedestrian occlusion and sharp posture changes.

Aiming at complex surveillance scenarios, combining YOLO3 [14] and DeepSORT, we propose a robust detection and tracking algorithm for pedestrian targets. Aiming at pedestrian occlusion and posture changes, a data association matching module and pedestrian feature extraction fusion network that resists posture changes are introduced. At the same time, the matching of target's appearance information and motion information is also taken into consideration, which improves the accuracy and robustness of pedestrian detection and tracking in surveillance scenarios.

2 Robust Detection and Tracking Algorithm

Due to the changes of pedestrians' posture, and the complexity of application scenarios, pedestrian detection and tracking are still facing challenges. The pedestrian detection algorithm YOLOv3 combined with the tracking algorithm DeepSORT is relatively inadequate under the complex monitoring situations with pedestrian occlusion and posture change. Therefore, this paper focuses on the robustness of the algorithm and proposes a robust pedestrian detection and tracking algorithm which is resistant to posture changes and occlusion.

2.1 *Association Matching Based on Motion and Appearance Information.*

Association matching of data is a crucial part of object tracking. Some simple tracking algorithms only take the association of motion information into consideration, which often results in the inaccurate matching. To match and track the object more accurately, the algorithm proposed in this paper combines motion information and appearance information as the criterion of data association assignment. We employ recursive Kalman filtering which performs data processing through optimized autoregression to predict the next position of the tracking target. The principle is shown in Fig. 1.

Kalman filtering first predicts the position of the tracked target at the next moment, and then updates the predicted position through the detector. For the position measurement, it is performed by calculating the standard deviation between the actual value of the currently detected target position and the tracking prediction value.

$$D^I(i,j) = (d_i - y_i)^T S_i^{-1}(d_i - y_i). \tag{1}$$

where $D^I(i,j)$ represents the degree of matching between the ith detection result and the jth tracking trajectory prediction result, S_i is the covariance matrix predicted by

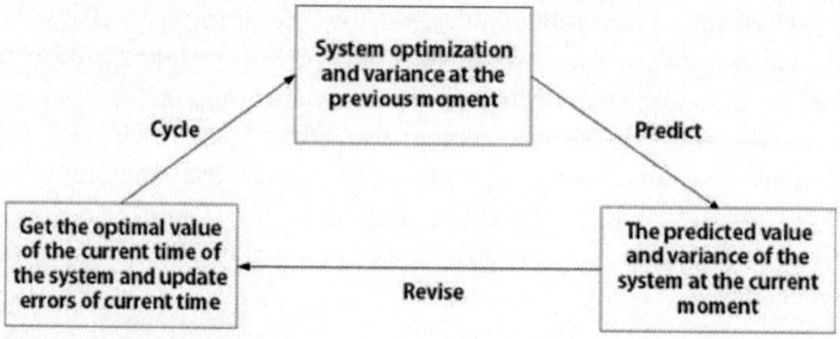

Fig. 1 Principles of Kalman filtering

Kalman, d_j represents the actual value of the jth target position, and y_i represents the predicted value of the ith tracker for the target position. Save the features of each tracking result, calculate the minimum cosine distance between the features and the feature vector of the detection result, and measure the appearance information. The calculation formula is:

$$D^{II}(i,j) = \min\left\{1 - v_j^T v_k^i \middle| v_k^i \in R_i\right\}. \tag{2}$$

where $D^{II}(i,j)$ represents the appearance matching degree between the jth detection result and the ith tracking target, and v_k^i represents the feature vector of the kth frame of the ith tracker.

2.2 *Appearance Feature Extraction Against Posture Change*

Aiming at the problem of tracking ID switches caused by pedestrian pose changes and inter-pedestrian occlusion in the video, we extract the features of different regions of the human body to form robust features to resist pose changes. The appearance features resisting pose changes are mainly extracted by multi-feature fusion. Input the image of the pedestrian and the response graph is generated by RPN. The bone joint points (x_i, y_i) represent the points with the largest response values in the response graph.

$$(x_i, y_i) = \underset{x\in[1,W],y\in[1,H]}{\arg\max} \; R_i(x, y). \tag{3}$$

According to the extracted 14 key points of the human body, human body is divided into several parts [15].

The 14 key points $P_A, P_B, P_{C1}, P_{C2}, P_{D1}, P_{D2}, P_{E1}, P_{E2}, P_{F1}, P_{G1}, P_{H1}, P_{F2}, P_{G2}, P_{H2}$ constitute 7 different point sets $J_1, J_2, J_3, J_4, J_5, J_6, J_7$. As shown in Fig. 2, they correspond to the pedestrian's head, limbs (left upper limb, right upper limb, left lower limb, right lower limb), upper body and lower body respectively. Use R_n, where $n(1,\ldots,7)$, to represent seven different regions:

$$R_n = [x_{lt}y_{lt}x_{rb}y_{rb}] = \min\left[\min_{x\in J_n}(x), \min_{y\in J_n}(y), \max_{x\in J_n}(x), \min_{y\in J_n}(y),\right]. \quad (4)$$

where (x_{lt}, y_{lt}) represents the coordinates of the upper left corner of the rectangle, (x_{rb}, y_{rb}) represents the coordinates of the lower right corner of the rectangle.

The pedestrian object is divided into 7 different human body regions, including three macro sub-regions, namely the head, upper body, and lower body, and four micro-sub regions, namely the limbs. We construct the multi-feature fusion network that resists posture changes through these regions. The network structure is shown in Fig. 3.

Make the images of seven different human body regions and the whole image of the pedestrian as the inputs. Through the division of human body regions and the fusion of regional features, we can obtain a more robust pedestrian feature vector for tracking. Fig. 4 shows the detection and tracking results in the case of pedestrian occlusion and posture changes.

Thanks to the feature extraction of anti-posture changes and the data association matching that combines motion information and appearance information, the algorithm proposed in this paper improves the problem of tracking ID switches caused by the change of pedestrian posture and occlusion while tracking.

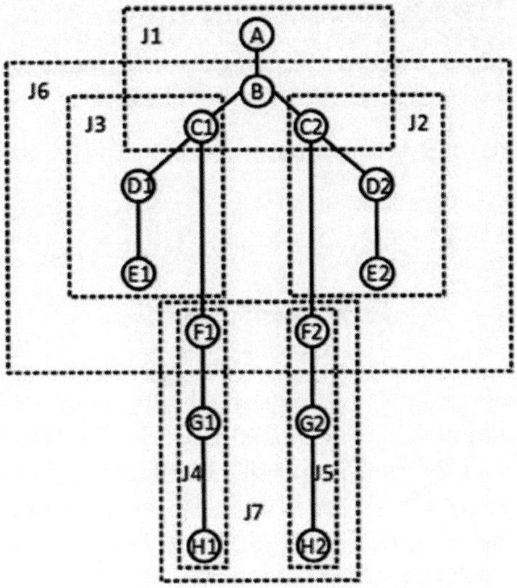

Fig. 2 Distribution of key points of human body

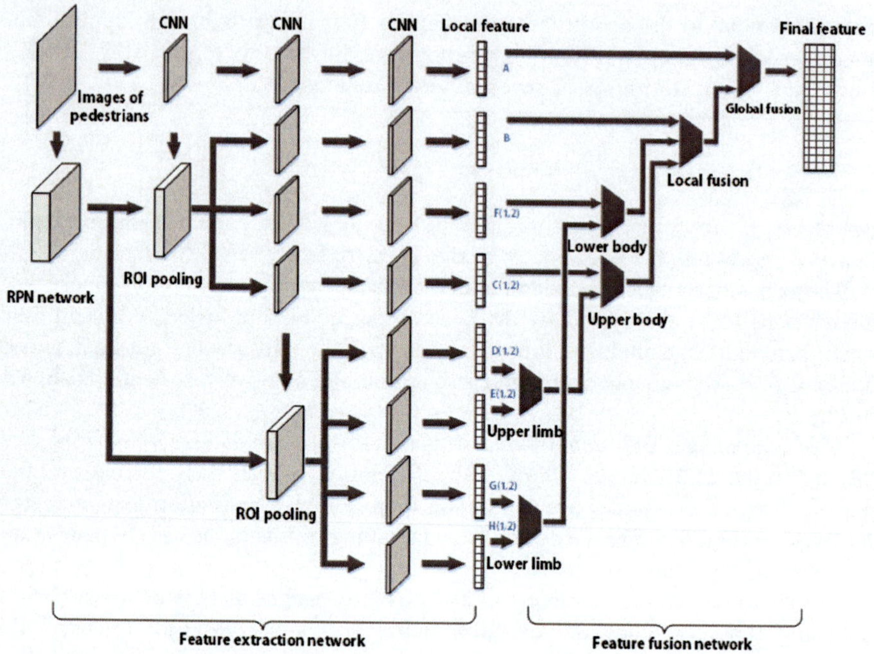

Fig. 3 Network structure of multi-feature fusion

Fig. 4 Results of pedestrian detection and tracking

3 Experiments and Results

3.1 Datasets and Settings

We conduct our experiments on both MOT17 dataset and actual application scenarios. The MOT17 dataset is a public dataset for multi-object detection and tracking, and is specifically used for pedestrian tracking. Also, In order to test the performance of the algorithm in practical applications, we conduce the experiments on actual application scenarios which is collected from campus monitoring. Considering that the images collected in real scenes may be interfered by fog and light, we preprocess the images based on [16, 17]. Our evaluation metrics follow most commonly used metrics include Precision, Recall, ID_Sw, MOTP and MOTA.

We perform our experiments on a computer with 16.0 GB of memory, Intel (R) Core i7-7700K@4.20GHz, and NVIDIA Tesla K40c graphics card. And the experimental framework is based on keras with tensorflow backend, python3, OpenCV, etc.

3.2 Comparison with Other Algorithms

Select the algorithm combining YOLOv3 and DeepSORT, the algorithm combining CenterNet and DeepSORT, and YOLO_Tracking as comparison algorithms. The experimental results on actual scenarios are shown below (Figs. 5, 6, 7 and 8).

It can be seen from the results that the algorithm combining CenterNet and DeepSORT has a poorer detection and tracking performance on remote small objects in the surveillance video, and there are more missed detection. The YOLO_Tracking algorithm can hardly handle the occlusion and posture change. The algorithm combining YOLOv3 and DeepSORT can detect remote smaller pedestrian objects in the video sequence, but the robustness is poor, and there still exists the problem of ID number switches. The algorithm proposed in this paper is more robust to pedestrian occlusion and posture changes, and it is better for detecting remote small objects. Test different videos, quantify the results, and analyze time complexity of the algorithm. The statistics are shown in Table 1.

It can be seen from Table 1 that compared with other comparison algorithms, the algorithm proposed in this paper has less ID switches and higher precision rate and recall rate. Although the algorithm in this paper costs more time, it still meets real-time requirements.

Fig. 5 Detection and tracking results of YOLOv3 + DeepSORT

Fig. 6 Detection and tracking results of CenterNet + DeepSORT

Fig. 7 Detection and tracking results of YOLO_Tracking

Fig. 8 Detection and tracking results of proposed method

Table 1 Comparison of tracking results of different algorithms

Algorithm	ID_sw	Precision (%)	Recall (%)	Time/Frame (ms)
YOLOv3 + DeepSORT	572	91.5	87.2	178.6
CenterNet + DeepSORT	503	87.4	77.8	116.3
YOLO_Tracking	659	90.6	84.6	111.2
Proposed	506	92.7	94.5	196.1

3.3 Posture Change and Occlusion Experiments

We conduct the experiments on the MOT17 dataset. The video sequence in the MOT17 dataset contains a large number of pedestrians, where exists a mass of pedestrian occlusion and pedestrian posture change. The experimental results are shown in Fig. 9.

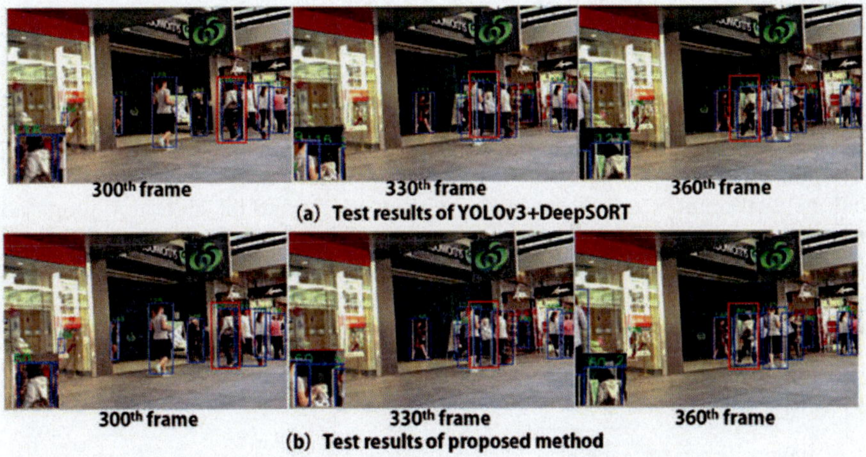

Fig. 9 Comparison of detection and tracking test results under pedestrian occlusion

Fig. 9 shows the comparison of the test results of YOLOv3 combined with the DeepSORT algorithm (for a) and that of the proposed algorithm in the MOT dataset (for b). Pay attention to the pedestrians marked with a red rectangular. As shown in Fig. 9, in the scene with occlusion, the pedestrian ID number in the red rectangular frame jumps from 116 to 123 in (a) due to posture change and occlusion, but in (b), the ID keeps the number 60 unchanged. From the figure, it can be intuitively shown that the proposed algorithm in this paper is robust to pedestrian posture changes and occlusion in surveillance video. The comparison of road scene detection and tracking results in the MOT dataset is shown in Fig. 10 and Table 2. As can be seen

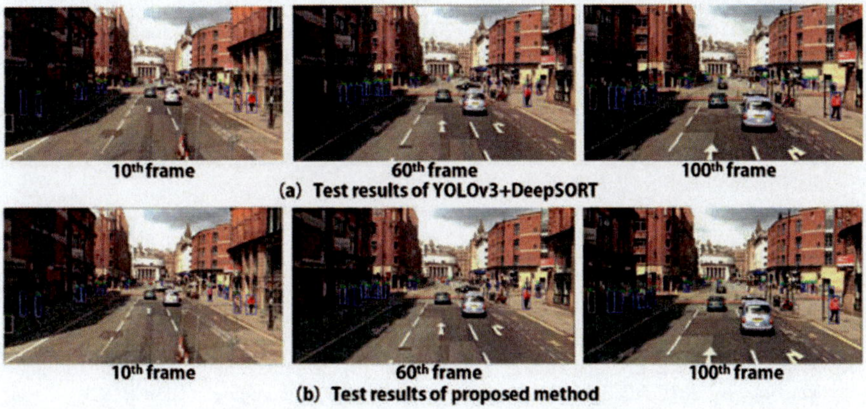

Fig. 10 Comparison of detection and tracking results in road scenes

Table 2 Comparison of MOT17 data set detection and tracking results

	ID_sw	MOTA (%)	MOTP (%)	Time cost/frame (ms)
YOLOv3 + DeepSORT	829	36.1	76.2	174.6
Proposed	651	36.5	76.3	190.4

that the total number of pedestrian ID switches is less and the algorithm proposed improves the robustness of resisting posture changes and occlusion. From the perspective of time complexity, the proposed algorithm improves the tracking performance to a certain extent without a significant increase in time cost.

4 Summary

In actual monitoring scenarios, frequent posture change and occlusion among pedestrians have all affected the accuracy of pedestrian detection and tracking to a certain extent. Thus, this paper proposes a robust pedestrian detection and tracking algorithm. It aims to reduce the effects of pedestrian posture change and pedestrian occlusion problems, improve the robustness of pedestrian detection and tracking, and reduce the switches of pedestrian ID numbers. The algorithm proposed in this paper has a good performance in both subjective visual experience and objective indicators, which can meet the needs of pedestrian detection and tracking in intelligent monitoring.

References

1. Su S, Li S, Chen S, Cai G, Wu Y (2012) A survey on pedestrian detection. Acta Electron Sin 40(04):814–820
2. Girshick R, Donahue J, Darrell T et al (2014) Rich feature hierarchies for accurate object detection and semantic segmentation. Proceedings of the IEEE conference on computer vision and pattern recognition. Columbus, pp 580–587
3. Ren S, He K, Girshick R et al (2015) Faster R-CNN: towards real-time object detection with region proposal networks. Adv Neural Inf Process Syst. Montreal, pp 91–99
4. He K, Gkioxari G, Dollar P, Girshick R (2020) Mask R-CNN. IEEE transactions on pattern analysis and machine intelligence 42(2):386–397
5. Redmon J, Divvala S, Girshick R et al (2016) You only look once: unified, real-time object detection. In: Proceedings of the IEEE conference on computer vision and pattern recognition. Las Vegas, pp 779–788
6. Zhang S, Wen L, Bian X et al (2018) Single-shot refinement neural network for object detection. In: Proceedings of the IEEE conference on computer vision and pattern recognition. Salt Lake City, pp 4203–4212
7. Duan K, Bai S, Xie L, Qi H, Huang Q, Tian Q (2019) CenterNet: keypoint triplets for object detection. In: 2019 IEEE/CVF international conference on computer vision (ICCV), Seoul, Korea (South), pp 6568–6577

8. Qi YK, Zhang SP, Qin L, Yao H, Huang Q, Lim J, Yang MH (2016) Hedged deep tracking. In: Proceedings of 2016 IEEE conference on computer vision and pattern recognition. Las Vegas, NV, USA, IEEE, pp 4303–4431
9. Li B, Yan JJ, Wu W, Zhu Z, Hu X (2018) High performance visual tracking with siamese region proposal network. In: Proceedings of 2018 IEEE/CVF conference on computer vision and pattern recognition. Salt Lake City, UT, USA, pp 8971–8980
10. Wojke N, Bewley A, Paulus D (2017) Simple online and realtime tracking with a deep association metric. In: 2017 IEEE international conference on image processing (ICIP), Beijing, pp 3645–3649
11. Tang S, Andriluka M, Andres B et al (2017) Multiple people tracking by lifted multicut and person re-identification. In: Proceedings of the IEEE conference on computer vision and pattern recognition, pp 3539–3548
12. Wang Q, Zhang L, Bertinetto L, Hu W, Torr PHS (2019) Fast online object tracking and segmentation: a unifying approach. In: 2019 IEEE/CVF conference on computer vision and pattern recognition (CVPR), Long Beach, CA, USA, pp 1328–1338
13. Maksai A, Fua P (2019) Eliminating exposure bias and metric mismatch in multiple object tracking. In: 2019 IEEE/CVF conference on computer vision and pattern recognition (CVPR), Long Beach, CA, USA, pp 4634–4643
14. Redmon J, Farhadi A (2018) Yolov3: an incremental improvement. arXiv preprint arXiv:1804.02767
15. Zhao H, Tian M et al (2017) Spindle net: person re-identification with human body region guided feature decomposition and fusion. In: 2017 IEEE conference on computer vision and pattern recognition (CVPR), Honolulu, HI, pp 907–915
16. Xiao J, Shen M, Lei J, Zhou J, Klette R, Sui H (2020) Single image dehazing based on learning of haze layers. Neurocomputing 389:108–122
17. Xiao J, Tian H, Zhang Y, Zhou Y, Lei J (2018) Blind video denoising via texture-aware noise estimation. Comput Vision Image Underst 169:1–13

Comparative Analysis of Chinese High-Resolution Satellite Data for Sugarcane Classification Based on U-Net Model

Chen Chen, Linjiang Lou, Xinyuan Gao, and Yu Liu

Abstract Monitoring sugarcane in China is important for the sugarcane industry which needs harvest progress information during the harvest season. The satellite image is one of the cost-effective and dynamic data sources for sugarcane classification recently. However, there are not many previous works for the classification of sugarcane with high-resolution satellite images especially sub-meter resolution data at present. Deep learning with a high performance of classification in agriculture was used in recent research. In this study, Chinese high-resolution satellite data based on the U-Net model was chosen to get a more precise segmentation of sugarcane in Laibin. GaoFen-1(GF-1) image with 2 m resolution and GaoFen-2 (GF-2) image with 0.8 m resolution were compared. GF-2 image has a good performance in the OA and Kappa coefficient compared with the GF-1 image which shows that a high-resolution image can get better segmentation results of sugarcane than the low resolution using the same data and method. Furthermore, to get more precise results of sugarcane classification, two different growth stages of sugarcane GF-2 image were chosen: tillering period data in May and a grand growth period in August. The result shows the grand growth period is suitable for sugarcane classification with a better improvement in the OA and Kappa coefficient.

Keywords Sugarcane · U-Net · Chinese high-resolution satellite

1 Introduction

Sugarcane is a major crop that plays an important role in China's socioeconomic. With the development of satellite technology, the satellite image is one of the effective ways of sugarcane classification with certain accuracy in recent years. The Moderate Resolution Imaging SpectroRadiometer (MODIS) sensor was chosen to classify sugarcane by Xavier et al. [1]. Rudorff et al. studied sugarcane which was converted using a series MODIS data from 2000 to 2008 in Paraná State, Brazil [2].

C. Chen · L. Lou (✉) · X. Gao · Y. Liu
Land Satellite Remote Sensing Application Center, MNR, Beijing, China

L. Wang et al. (eds.), *Proceedings of the 7th China High Resolution Earth Observation Conference (CHREOC 2020)*, Lecture Notes in Electrical Engineering 757, https://doi.org/10.1007/978-981-16-5735-1_14

Mulianga estimated sugarcane yeild from the MODIS data in Western Kenya [3]. Landsat Thematic Mapper (TM) and Landsat Enhanced TM Plus (ETM+) were used for identifying, classifying, calculating, and estimating sugarcane [4–8]. However, these sensors data are not suited for sugarcane samples collection in small scale fields with a resolution larger than 1 m per pixel [3, 9]. Moreover, it does not suit the area in china with small fields less than the size of the Landsat image 30 m per pixel.

In recent years, data from sub-meter spatial resolution satellite sensors such as QuickBird, GeoEye, WorldView, Resurs-P1, Pleiades, GF-2, BJ-2, KOMPSAT, DEIMOS, JL, are GJ available for agricultural identification and classification with higher precision than low-resolution satellite image. Sertel and Yay identified vineyard areas using object-based classification of Worldview 2 data in Turkey [10]. Upadhyay et al. indicated worldview2 was important for specific crop classification in an Indian study area [11]. But, there are a few previous research using high spectral resolution satellite data to classify sugarcane which could bring more accurate results [12].

A promising technique, Deeping Learning (DL), used in satellite image processing with a high performance of classification accuracy studied in many types of research. It was applied in land type classification, disease detection, fruit counting, and field estimation [13]. Kamilaris and Prenafeta-Boldú summarized 23 previous research about DL applied in agriculture and most of them reached better results compared with other techniques and only one case had worse performance [13].

In this paper, Chinese high-resolution satellite data: GF-1image with 2 m resolution and GF-2 image with 0.8 m resolution are used for sugarcane classification. One of the DL models, the U-Net model is chosen. Laibin, Guangxi of China is selected as a test area with 60% sugar coverage of china. Moreover, further analyses using two different growth stages of sugarcane are given.

2 Study Area and Data

2.1 Study Area

Guangxi is in the southern area of China with a sub-tropical climate which is suitable for sub-tropical crop growth such as rice, bananas, and other vegetables, especially sugarcane. Nearly 60% sugar of china is produced by the sugarcane of Guangxi.

The study area is in Laibin city which is one of the important sugarcane production bases in Guangxi. The geographical coordinates are 108°24′E–110°28′E, 23°16′N–24°29 N′ in the middle area of Guangxi. Sugarcane has six major growth periods in Laibin city.

1. sowing period in early March
2. germination period from mid-March to April
3. tillering period in early May
4. grand growth period in mid-June to September
5. mature period in November
6. harvest period from mid-November to mid-March in the following year

2.2 Data

Recently, the Chinese high-resolution satellite has developed such as Ziyuan3-01, Ziyuan3-02, Ziyuan3-03(ZY3-01, ZY3-02, ZY3-03), GF-1, GF1-B, GF1-C, GF1-D, Gaofen-6(GF-6), especially sub-meter satellite GF-2 and Gaofen-7(GF-7). In this paper, GF-1 with 2 meters spatial resolution and GF-2 satellite data with 0.8 meters spatial resolution are compared. Data description of GF-1 and GF-2 is shown in table 1.

Considering the growth stages with the obvious texturing in satellite images, the GF-1 images, and GF-2 images from June to November after the grand growth period were chosen for the study area. The GF-1 image captured on 10 June 2018 and the GF-2 image on 11 September and 9 August 2015 were acquired.

Ground truth data of sample data and training data were collected by filed works in June and October 2018 respectively. In total, 10,209 samples were separated for training points and validation points in a random pattern.

Table 1 GF-1 and GF-2 satellite image parameters

Parameter	GF-1		GF-2	
Spectral range	Panchromatic image	0.45–0.90 μm	Panchromatic image	0.45–0.90 μm
	Multispectral image	0.45–0.52 μm	Multispectral images	0.45–0.52 μm
		0.52–0.59 μm		0.52–0.59 μm
		0.63–0.69 μm		0.63–0.69 μm
		0.77–0.89 μm		0.77–0.89 μm
Resolution	Panchromatic image	2 m	Panchromatic image	0.8 m
	Multispectral image	8 m	Multispectral images	3.2 m
Swath width	60 km		45 km	
Re-visit period	41 days		69 days	

3 Methods

3.1 Data Processing

The original sensor corrected satellite images with panchromatic and multispectral sensors that were processed after radiation calibration and atmospheric correction. The imagery was ortho-rectified to the World Geodetic Survey 1984(WGS84) datum. Then, each multispectral image was matched to the panchromatic image respectively. At the last, the panchromatic data and the multispectral data were merged.

3.2 U-Net Model

The full convolutional network is one of the deep learning models with more precise segmentation compared with the classical CNN model. FCN is the end-to-end segmentation with more elegant architecture. However, it is not suited to limited training images [14].

In this paper, one of the FCN, the U-Net model is chosen which can get the desired output with location information. Moreover, the model is suited to a few training inputs with a more precise output [14]. The Targets mask U-Net model by Han et al. is used in this research and more details are shown in Fig. 1 [15].

The architecture of the U-Net in this research has 7 levels which are consisted of three encode layers (contracting path), a bridge layer, and three decode layers (expansive path). Two convolutions and a max pooling operation are repeated at each level which is a traditional convolutional network architecture.

The input images are transferred to 32-component feature vectors as compact representations and the feature channels are doubled in each contracting path with two 3×3 convolutions and a pooling operation as follows. In each expansive path layer, the feature channels are converted into half many features followed by two 3×3 convolutions. The last layer is a 1×1 convolution layer with 64 component features.

4 Results and Discussion

4.1 Comparation of Different Satellite Images

In this research, the U-Net model was used to compare the extraction accuracy of the GF-1 image with the extraction accuracy of the GF-2 image. The confusion matrix and Kappa statistics were analyzed the prediction results of the U-Net model quantitatively. The use's accuracy, the producer's accuracy, overall accuracy (OA),

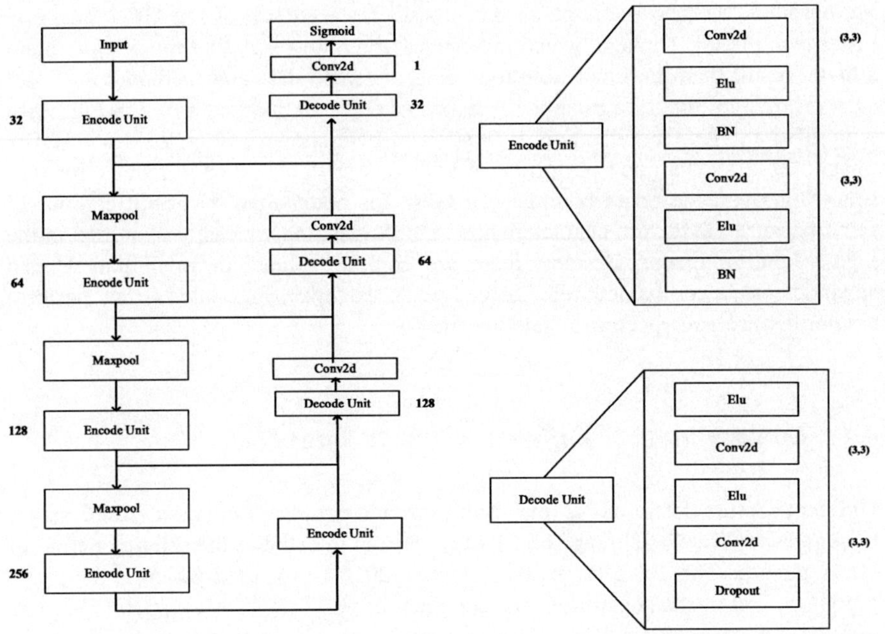

Fig. 1 The architecture of the target mask U-Net

and Kappa coefficients for the two type images were given in Tables 2 and 3. It can be indicated from Table 2, the OA of the GF-1 image is 86.32% and the Kappa coefficient is 0.40. The accuracy of the GF-2 image is 95.07% and the Kappa coefficient is 0.86 shown in Table 3, which gives a significant improvement in the OA and Kappa coefficient compared with the lower resolution image-GF1.

Table 2 Accuracy evaluation of GF-1 image

Land type	User's accuracy (%)	Producer's accuracy (%)
sugarcane	91.90	92.35
Other	49.14	47.02
Overall Accuracy	86.32	
Kappa coefficient	0.40	

Table 3 Accuracy evaluation of GF-2 image (capture date: August 9th, 2015)

Land type	User's accuracy (%)	Producer's accuracy (%)
sugarcane	84.70	93.97
Other	98.30	86.03
Overall accuracy	95.07	
Kappa coefficient	0.86	

Furthermore, the user's accuracy and producer's accuracy of the GF-2 image is better than the GF-1 image, which indicates that a high-resolution image can obtain a better result than the low resolution using the same data and method.

As shown in Fig. 2, segmentation results of GF-1 are marked as red in Fig. 2a–c with GF-1 image. Classification of GF-2 is in blue shown in Fig. 2d–f with GF-2 image. GF-2 shows good performance with few miss-classification and error parcels. Moreover, the result boundary of GF-2 fits better than the result contour of GF-1 because the texture characteristics of the sugarcane are easily identified in the high-resolution image. Besides, there are complex plants in small parcels and sugarcane is accompanied by maize, weeds, bananas, and others that perform similar texture and spectral in satellite images.

4.2 *Comparation of Different Growth Stages*

Different capture date of GF-2 images is chosen to compare two main growth stages of sugarcane. The GF-2 image on 11 May 2016 is selected as the tillering period of the sugarcane. The GF-2 image on 9 August 2015 is acquired which is the grand growth period of the sugarcane. The accuracy of GF-2 on 11 May 2016 is 92.03% and the Kappa coefficient is 0.73 shown in Table 4, which indicates data in the grand growth period gives a better improvement in the OA and Kappa coefficient compared to the tillering period image.

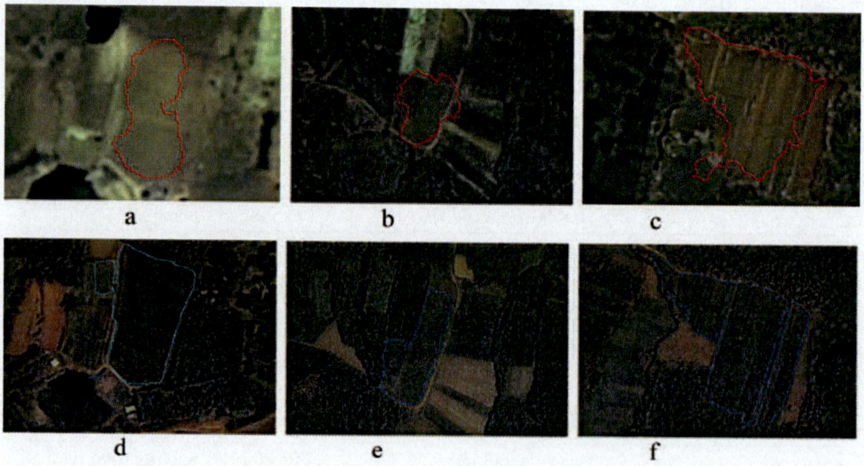

Fig. 2 Segmentation results **a**, **b**, **c** are GF-1 results, **d**, **e**, **f** are GF-2 results

Table 4 Accuracy evaluation of GF-2 image (capture date: May 11th, 2016)

Land type	User's accuracy (%)	Producer's accuracy (%)
sugarcane	97.82	92.52
Other	70.22	89.68
Overall accuracy	92.03	
Kappa coefficient	0.73	

5 Conclusion

This study indicates that the Chinese high-resolution satellite image is suitable for sugarcane classification in Laibin, Guangxi of China. GF-2 has a good performance in complex plant areas with more spectral and texture information compared to GF-1. Besides, the boundary of sugarcane plant areas is more clear extracting from the GF-2 image than the GF-1 image. By comparison of two sugarcane growth stages, the grand growth period is an ideal period of sugarcane segmentation.

References

1. Xavier AC, Rudorff BFT, Shimabukuro YE, Berka LMS, Moreira MA (2006) Multi-temporal analysis of MODIS data to classify sugarcane crop. Int J Remote Sens 27:755–768
2. Rudorff BFT, Adami M, Aguiar DAd, Gusso A, Silva WFd, Freitas RMd (2009) Temporal series of EVI/MODIS to identify land converted to sugarcane. In: 2009 IEEE international geoscience and remote sensing symposium, pp IV-252–IV-255
3. Mulianga B, Bégué A, Simoes M, Todoroff P (2013) Forecasting regional sugarcane yield based on time integral and spatial aggregation of MODIS NDVI. Remote Sens 5:2184
4. Aguiar DA, Rudorff BFT, Silva WF, Adami M, Mello MP (2011) Remote sensing images in support of environmental protocol: monitoring the sugarcane harvest in São Paulo State, Brazil. Remote Sens 3:2682
5. Gers CJ Relating remotely sensed multi-temporal LANDSAT 7 ETM+ imagery to sugarcane characteristics
6. Markley J, Raines A, Crossley R, Hogarth DM (2003) The development and integration of remote sensing, GIS and data processing tools for effective harvest management. Can J Cardiol 24:21–40
7. Rudorff BFT, Aguiar DA, Silva WF, Sugawara LM, Adami M, Moreira MA (2010) Studies on the rapid expansion of sugarcane for ethanol production in São Paulo State (Brazil) using landsat data. Remote Sens 2:1057
8. Vieira MA, Formaggio AR, Rennó CD, Atzberger C, Aguiar DA, Mello MP (2012) Object based image analysis and data mining applied to a remotely sensed Landsat time-series to map sugarcane over large areas. Remote Sens Environ 123:553–562
9. El Hajj M, Bégué A, Guillaume S, Martiné J-F (2009) Integrating SPOT-5 time series, crop growth modeling and expert knowledge for monitoring agricultural practices—the case of sugarcane harvest on Reunion Island. Remote Sens Environ 113:2052–2061
10. Sertel E, Yay I (2014) Vineyard parcel identification from Worldview-2 images using object-based classification model, SPIE, p 17

11. Upadhyay P, Ghosh SK, Kumar A, Roy PS, Gilbert I (2012) Effect on specific crop mapping using WorldView-2 multispectral add-on bands: soft classification approach, SPIE, p 14
12. Abdel-Rahman EM, Ahmed FB (2008) The application of remote sensing techniques to sugarcane (Saccharum spp. hybrid) production: a review of the literature. Int J Remote Sens 29:3753–3767
13. Kamilaris A, Prenafeta-Boldú FX (2018) A review of the use of convolutional neural networks in agriculture. J Agric Sci 156:312–322
14. Ronneberger O, Fischer P, Brox T (2015) U-net: convolutional networks for biomedical image segmentation
15. Liu MHHWGWY (2018) Targets mask U-Net for wind turbines detection in remote sensing images. Int Archives Photogrammetry Remote Sens Spatial Inf Sci 42:6

A Differential Flat Approach for Attitude Manoeuver Control of Agile Satellite

Deting Li, Yikang He, Shuyu Lin, Miao Li, and Kang Shi

Abstract This pater introduces a kind of control method for agile satellites which use SGCMG as actuators. Considering the SGCMG's singularities, an attitude trajectory planning method is designed based on the differential flatness theory firstly. This method considers the singularity features and the influence of gravity gradient torque. A sliding mode tracking controller is proposed based on the Modified Rodrigues parameters. Result of simulation shows that the SGCMG undergoing large maneuver without singularity and the tracking error is less than 0.03°, the attitude deviation in the steady state is less than 1×10^{-5}°.

Keywords Agile satellites · Single gimbal control moment gyroscope · Singularity · Differential flatness · Optimal control

1 Introduction

Along with the development of space technology, small satellites has become a hot spot in the development of new-generation satellites due to its high functional density, rapid development cycle and launch flexibility. As the core component unit, the attitude control system is the most important part of small satellite research. The typical attitude control system is composed of a jet system, a magnetic torque system, or a momentum wheel, etc., which are broadly employed in spacecraft with large rotational inertia, but are not appropriate for spacecraft with small rotational inertia. Single gimbal control moment gyroscope (SGCMG) is driven by electrical energy, which does not consume any fuel but has a large torque output capacity, making it an ideal device for attitude control of agile satellites.

Despite its large output torque, the SGCMG suffers from singularity problem, which complicates the attitude maneuver control process. To address the singular failure problem during maneuvering, numerous scholars have conducted related

D. Li (✉) · Y. He · S. Lin · M. Li · K. Shi
Shanghai Aerospace Control Technology Institute, Shanghai, China
e-mail: lideting1097@163.com

L. Wang et al. (eds.), *Proceedings of the 7th China High Resolution Earth Observation Conference (CHREOC 2020)*, Lecture Notes in Electrical Engineering 757, https://doi.org/10.1007/978-981-16-5735-1_15

studies. Zhao [1] employed a mixed solving strategy of pseudospectral method and direct targeting method to design a single-objective and multi-objective trajectory planning model, which is able to meet the SGCMG singularity saturation constraints. However, the mixed solving strategy adopts the targeting method to optimize the singularity metric value resulting in a computationally intensive and time-consuming algorithm.

The differential flatness theory was introduced by Fliess et al. [2, 3] in the twentieth century for nonlinear systems. In this article, a stellar online control method is designed by combining fast trajectory design and anti-interference tracking control algorithm. A rapid attitude path planning method is designed by referring to the research results of differential flatness theory in solving the optimal control problem. Incorporating the effects of environmental factors such as gravitational gradient moments and orbital angular velocity, the method considers the constraints on the beginning and end state of the star and the dynamics of the SGCMG, which enables the rapid derivation of an energy sub-optimal path that satisfies the constraints. An adaptive sliding-mode controller is obtained by combining sliding-mode control with self-adaptation. The controller is capable of overcoming the effects of external perturbations and gyroscopic execution errors on the trajectory of the tracked target.

2 Dynamic Model

2.1 Attitude Dynamics Model

Three coordinate systems are defined: an inertial reference system, an orbital coordinate system and a body coordinate system that is coupled to the spacecraft. The attitude is described by using modified Rodrigues parameters (MRP), which is the smallest implementation of the attitude description, and also has the characteristics of being less prone to singularity and high computational efficiency. The MRP-based attitude calculation is smaller in computation without loss of computational accuracy compared to quaternion [4], which helps to improve the efficiency of attitude calculation. The attitude transfer matrix represented by MRP is:

$$\mathbf{M}(\boldsymbol{\sigma}) = \mathbf{I}_3 + \frac{8[\boldsymbol{\sigma}\times][\boldsymbol{\sigma}\times] - 4(1-\boldsymbol{\sigma}^{\mathrm{T}}\boldsymbol{\sigma})[\boldsymbol{\sigma}\times]}{(1+\boldsymbol{\sigma}^{\mathrm{T}}\boldsymbol{\sigma})^2} \tag{1}$$

where $\mathbf{I}_3$ is a 3-D unit matrix and $[\boldsymbol{\sigma}\times]$ is vector cross product matrix:

$$[\boldsymbol{\sigma}\times] = \begin{bmatrix} 0 & -\sigma_3 & \sigma_2 \\ \sigma_3 & 0 & -\sigma_1 \\ -\sigma_2 & \sigma_1 & 0 \end{bmatrix} \tag{2}$$

The attitude kinematic differential equation describes of the MRP:

$$\dot{\sigma} = \mathbf{G}\omega \tag{3}$$

where $\boldsymbol{\omega}$ present attitude angular velocity, and $\boldsymbol{G}$ represent:

$$\mathbf{G} = \frac{1}{2}\left(\frac{1 - \boldsymbol{\sigma}^{\mathrm{T}}\boldsymbol{\sigma}}{2}\mathbf{I}_3 + \sigma\sigma^{\mathrm{T}} + [\boldsymbol{\sigma}\times]\right) \tag{4}$$

The rigid attitude dynamic equation in the satellite body coordinate:

$$\mathbf{J}\underline{\omega} = \boldsymbol{\tau}_{gg} - \boldsymbol{\omega} \times (\mathbf{J}\omega) + \mathbf{u} \tag{5}$$

where $\boldsymbol{\tau}_{gg}$ is gravity gradient moment, $\mathbf{u}$ is the attitude control torque, J is the moment of inertia.

2.2 SGCMG Dynamics Model and Singularity Analysis

The SGCMG is a momentum exchange device with three-dimensional output capability consisting of several (usually three or more) gyros. If each moment gyro has the same mode of angular momentum, the total angular momentum of the gyrogroups can be expressed as:

$$\mathbf{H} = h_0(\mathbf{A}\sin\mathbf{s} + \mathbf{B}\cos\mathbf{s})\mathbf{E}_3 \tag{6}$$

where: $\mathbf{A}$ and $\mathbf{B}$ is the mounting matrix associated with the configuration of SGCMG. $\mathbf{s}$ is he frame angle matrix of the SGCMG, $\sin \boldsymbol{s}$ and $\cos \boldsymbol{s}$ are the sine and cosine diagonal arrays of the frame angle. $\mathbf{E}_3$ is a 3-D unit vector whose elements are all 1.

The control torque applied to the spacecraft by the SGCMG is:

$$\mathbf{T}_{SGCMG} = -\dot{\mathbf{H}} - \boldsymbol{\omega} \times \mathbf{H} \tag{7}$$

the first part of the control moment $\dot{\mathbf{H}} = -(\mathbf{A}\cos\mathbf{s} - \mathbf{B}\sin\mathbf{s})\dot{\mathbf{s}}h_0$ is the torque generated by the rotation of the SGCMG's frame. The latter part $-\boldsymbol{\omega} \times \mathbf{H}$ is the gyroscopic torque produced by the change in the direction of the angular momentum of the satellite.

When SGCMG works, if systematic moments are continuously output in a direction, the accumulation of angular momentum values in a given direction may cause the angular momentum values to reach the maximum angular momentum of the SGCMG system which is known as angular momentum saturation is a form of failure of the SGCMG.

When SGCMG system frame angle is in some certain conditions where the target torque is orthogonal to the control torque, the SGCMG system cannot output the target torque and loses control ability, this phenomenon is called singular failure. All states of the gyro are singular when the target torque $\boldsymbol{u}_t$ and installation direction $\boldsymbol{g}_t$ are parallel. When the desired torque and the direction of installation are not parallel, then there are two directions of singular angular momentum [5] for each gyro:

$$\mathbf{n}_i = \pm(\mathbf{g}_i \times \mathbf{u}_t) \times \mathbf{g}_i / \|\mathbf{g}_i \times \mathbf{u}_t\| \tag{8}$$

When planning attitude trajectory, a determination of singularity based on the current frame angle is necessary, which is to define a singularity state measurement that determines the degree of singularity associated with different combinations of frame angles. According to the principle of frame angle singularity, where singularity happens when there is no $(\mathbf{A}\cos\mathbf{s} - \mathbf{B}\sin\mathbf{s})^{-1}$ existence of it, to measure the current state of frame angle singularity by using the singularity metric [6, 7], the singularity metric is defined as

$$D = \det((\mathbf{A}\cos\mathbf{s} - \mathbf{B}\sin\mathbf{s})(\mathbf{A}\cos\mathbf{s} - \mathbf{B}\sin\mathbf{s})^{\mathrm{T}}) \tag{9}$$

When $D = 0$, the system is in a singularity condition, and the larger the D, the further it is from the singularity point.

3 Attitude Trajectory Planning Based on Differential Flatness Theory

3.1 *Differential Flatness System Design*

It has been shown in the literature [8] that the satellite attitude system described by the attitude kinematics Eq. 3 and attitude dynamics equations (5) represents a differential flatness system with $(\sigma_1, \sigma_2, \sigma_3)$ as a flat output. In this paper, the study incorporates the SGCMG steering system, i.e., the state volume frame angle (s_1, s_2, s_3, s_4) is introduced into the system. The SGCMG is a redundant system, the frame angle velocity obtained by solving Eqs. (1) and (2) is non-unique, so the attitude control system considering the frame angle does not have the differential flatness property and cannot directly develop the trajectory design based on the differential flatness. To expand the flatness property of the system, the solution to the optimal pseudo-inverse manipulation law [9] is constrained to define the frame angle as:

$$\dot{\mathbf{s}} = -(\mathbf{A}\cos\mathbf{s} - \mathbf{B}\sin\mathbf{s})^{\mathrm{T}}\left((\mathbf{A}\cos\mathbf{s} - \mathbf{B}\sin\mathbf{s})(\mathbf{A}\cos\mathbf{s} - \mathbf{B}\sin\mathbf{s})^{\mathrm{T}}\right)^{-1}\frac{\dot{\mathbf{H}}}{h_0} \tag{10}$$

The frame angle can be uniquely represented by the flatness output, then it is a Differential Flatness system.

In combination with the attitude dynamics and kinematic equations, select $\mathbf{z} = \boldsymbol{\sigma}$ as flat output, then the system state variables $(\boldsymbol{\omega}, \mathbf{H}, \dot{\mathbf{s}})$ can be expressed by the flat output. And only the gravity gradient moment is considered for the system environmental disturbance moment, then the expression for the system of differential flatness is given by the formula:

$$\begin{cases} \boldsymbol{\omega} = \frac{1}{(1+|\boldsymbol{\sigma}|^2)^2} G(\boldsymbol{\sigma})^{\mathrm{T}}\dot{\sigma} + \boldsymbol{\omega}_o(\boldsymbol{\sigma}) \\ \dot{\mathbf{H}} = \mathbf{J}\underline{\omega} + \boldsymbol{\omega} \times (\mathbf{J}\omega) - \boldsymbol{\tau}_{gg} - \boldsymbol{\omega} \times \mathbf{H} \\ \dot{\mathbf{s}} = \mathbf{C}^{\mathrm{T}}(\mathbf{C}\mathbf{C}^{\mathrm{T}})^{-1}\frac{\mathbf{u} + \boldsymbol{\omega} \times \mathbf{H}}{h_0} \end{cases} \tag{11}$$

$\boldsymbol{\tau}_{gg}$ for the gravity gradient moment, $\boldsymbol{\omega}_o$ for the orbital angular velocity of the satellite.

3.2 Flat Output Base Function

Parameterization of the flat output is crucial for trajectory optimization, by which it transforms the optimal control problem that contains differentiation into a nonlinear planning problem [10]. When the virtual domain parameter γ is introduced, the virtual and time domains are transformed into each other. In other words, the same trajectory can be described by different speed histories. The virtual and time domains are transformed into each other:

$$t(\gamma_F) = \int_0^{\gamma_F} \frac{d\gamma}{\lambda(\gamma)} \tag{12}$$

where the $\lambda(\gamma)$ is a virtual speed, whose expression is $\lambda(\gamma) = \frac{d\gamma}{dt}$.

Bezier Curve has been widely employed in computer graphics and mechanical geometry shape design due to its capability to approximate complex curves and its high computational efficiency. Considering the velocity factor $\lambda(\gamma) = \frac{1}{t_f}$, where t_f is the attitude maneuver time, the attitude trajectory can be described by the Bezier polynomial in the virtual domain.

Given control points $p_0, p_1, p_3, \ldots, p_n$, then the Bezier polynomial $p(\gamma)$ of the virtual domain constrained by these control points is:

$$p(\gamma) = \sum_{i=0}^{n} p_i \beta_{i,n}(\gamma), \quad \gamma \in [0, 1] \tag{13}$$

where $\beta_{i,n}(\gamma)$ is the n-dimensions Bernstein polynomial:

$$\beta_{i,n}(\gamma) = C_n^i (1-\gamma)^{n-i} \gamma^i, \quad \gamma \in [0, 1] \tag{14}$$

The properties of Bezier polynomials [11] are:

$$\begin{array}{ll} p(0) = p_0, & p(1) = p_n \\ \frac{\mathrm{d}p(0)}{\mathrm{d}\gamma} = n(p_1 - p_0), & \frac{\mathrm{d}p(1)}{\mathrm{d}\gamma} = n(p_n - p_{n-1}) \end{array} \tag{15}$$

In the virtual domain each MRP can be represented by a Bezier polynomial:

$$\sigma_j(\gamma) = \sum_{i=0}^{n} a_{j,i} \beta_{i,n}(\gamma) \quad j = 1, 2, 3 \tag{16}$$

By imposing end-point pose and angular velocity constraints on Eq. 16, partial control points can be derived. If the endpoint constraints are known as $[\boldsymbol{\sigma}_0, \dot{\sigma}_0, \boldsymbol{\sigma}_f, \dot{\sigma}_f]$, where the $\dot{\sigma}_0, \dot{\sigma}_f$ angular velocity can be derived from Eq. 3, then control points:

$$\begin{cases} a_{j,0} = \sigma_0(j), a_{j,1} = \sigma_0(j) + t_f \frac{\dot{\sigma}_0(j)}{n} \\ a_{j,n} = \sigma_f(j), a_{j,n} = \sigma_n(j) - t_f \frac{\dot{\sigma}_f(j)}{n} \end{cases} \quad j = 1, 2, 3 \tag{17}$$

The number of control points requirement for a Bezier polynomial with N_B endpoints constraint is $n \geq N_B + 1$, at least one control point parameter that can be used to optimize the path as an optimization variable.

3.3 Non-Linear Planning Models

By using the flat output to describe the state variables, and the trajectory planning problem is transformed into a nonlinear planning problem.

The nonlinear planning problem has different performance metrics for different task requirements. In this paper, we define the performance metrics with maximum angular momentum as the optimization goal:

$$Q = \max(|\mathbf{H}|) \tag{18}$$

There are several constraints of the nonlinear planning model considered: the maximum speed of the SGCMG frame, the singularity of the SGCMG system and the upper limit of angular momentum saturation. The trajectory planning issue can be transformed into a nonlinear planning issue. Take the amount of state given by Eq. 11 the objective function can be expressed as:

$$\min \quad Q = |\mathbf{H}[p_0, p_1, p_3, \cdots, p_n]| \tag{19}$$

$$s.t.\begin{cases} \dot{s}_i[p_0, p_1, p_3, \cdots, p_n] \le \dot{s}_{\max} \\ D[p_0, p_1, p_3, \cdots, p_n] \ge D_{\min} \end{cases} \tag{20}$$

where: $p_0, p_1, p_3, \cdots, p_n$ are a set of control points of Bezier curve.

3.4 Attitude Tracking Controller Design

The attitude trajectory based on the differential flatness theory takes into account the influence of the orbital angular velocity and the gravity gradient moment, but in the actual maneuvering process, it is still have the influenced of the uncertain environmental moment and the unstable torque of actuators. A suitable controller is required to quickly converge the actual attitude trajectory to the planned open-loop optimal trajectory.

Defining the attitude tracking error, there is

$$\boldsymbol{\sigma}_{\mathrm{e}} = \boldsymbol{\sigma} - \boldsymbol{\sigma}_c \tag{21}$$

where the subscript c means target MRPs.

Likewise, using angular velocity deviation to represent the deviation between the actual angular velocity and the target angular velocity, there are

$$\boldsymbol{\omega}_{\mathrm{e}} = \boldsymbol{\omega} - \boldsymbol{\omega}_c \tag{22}$$

Substituting derivation of Eqs. (22) into 5, the kinematic differential equation for angular velocity deviation is obtained as follows

$$\mathbf{J}\dot{\omega}_{\mathrm{e}} = -[\boldsymbol{\omega}\times]\mathbf{J}\omega + \boldsymbol{\tau}_{gg} + \mathbf{u} - \mathbf{J}\dot{\omega}_{\mathrm{c}} \tag{23}$$

Derive for Eq. (3) and substituting it into Eq. 23, we can deduce the kinematic differential equations for attitude deviation and angular velocity:

$$\mathbf{u} = \mathbf{J}\mathbf{G}^{-1}\ddot{\sigma}_{\mathrm{e}} + \mathbf{J}\dot{\omega}_{\mathrm{c}} + [\boldsymbol{\omega}\times]\mathbf{J}\omega - \mathbf{J}\mathbf{G}^{-1}\dot{\mathbf{G}}\omega_e \tag{24}$$

Define the switch function of the sliding mode controller as

$$\mathbf{s} = \dot{\sigma}_e + \lambda \sigma_e \tag{25}$$

The reaching law adopts the exponential reaching law [12] in the form of

$$\dot{\mathbf{s}} = -\varepsilon \mathrm{sgn} s - \mathbf{k} s, \quad \varepsilon > 0, \mathbf{k} > 0 \tag{26}$$

Substituting Eq. 26 into Eq. 24, simplifying the controller is:

$$u = (\mathbf{J} - k\mathbf{E}_{3\times 3} - \lambda \mathbf{E}_{3\times 3} - \mathbf{J}\mathbf{G}^{-1}\mathbf{G})\boldsymbol{\omega}_e - k\lambda \mathbf{J}\mathbf{G}^{-1}\boldsymbol{\sigma}_e - \mathbf{J}\mathbf{G}^{-1}\varepsilon \mathrm{sgn} s + [\boldsymbol{\omega}\times]\mathbf{J}\omega \tag{27}$$

4 Simulation Examples

The satellite's inertia moment is $J = \mathrm{diag}(1552, 1045, 1174)$, orbital angular velocity is approximately 0.063025°/s. Attitude control actuator uses SGCMG with 50Nms angular momentum magnitude for each. The maneuver time is 45 s. The Satellite's Euler angle maceuver from [0.6141 1.0320 −0.8322]° to [0 0 61]°. With the SQP algorithm for nonlinear optimization, the number of control points of Bezier curve is 5. Add disturbance torque $T_d = [1.42 \ -0.9641 \ 4.01]^T$ sin (0.08) × 10^{-4} Nm into the simulation process.

The MRP parameter and attitude angular velocity of the satellite is present at Figs. 1 and 2. The variation curve of the error Euler angle is shown in Fig. 3 for better to illustrate the control result. It can be observed that: within the first 10 s, in order to complete the initial attitude capture, the control mode is variable structure control, with chattering phenomena. When the error becomes sufficiently small, the control mode will be changed to feedback control, so that the actual attitude can track the target attitude smoothly in a relatively small error. The tracking error in the stable tracking process is within 0.03°, when the maneuver complete the attitude error within 1 × 10^{-5}°. The variation of the control torque with time is shown in Fig. 4. The variation curve of the singularity measure is shown in Fig. 5, it can be observed that: the singularity measurement value is always greater than 1 during the maneuvering process, SGCMG is always away from the singularity state with a stable control ability.

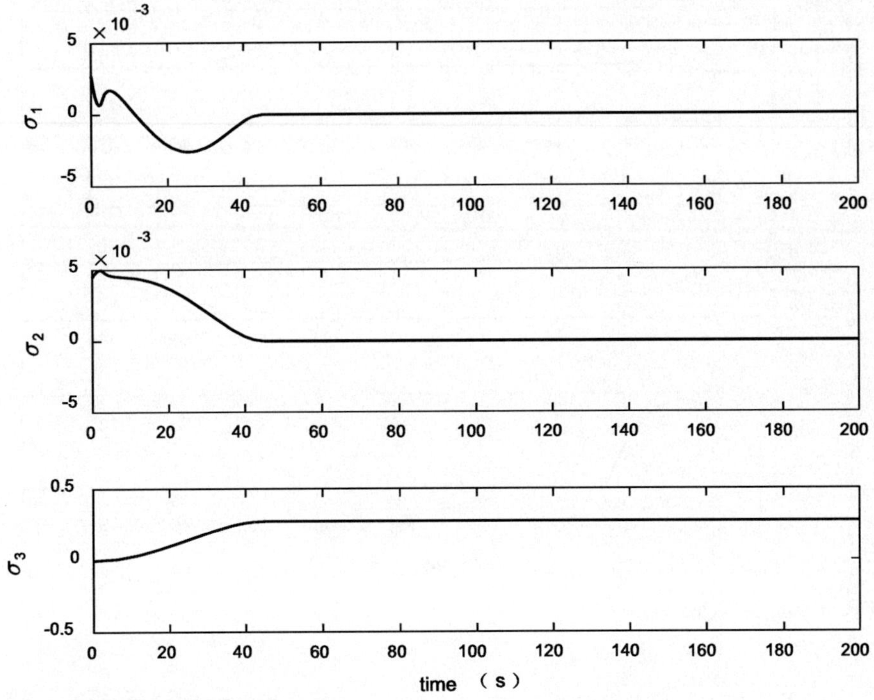

Fig. 1 Attitude by MRP parameter

5 Summary

In this article, it's focuses on the attitude maneuvering issue of small satellites utilizing SGCMG as the attitude actuator, considering the characteristic of singular saturation properties, a rapid reference trajectory planning method based on differential flatness theory combined with a sliding-mode controller for control is designed. This approach takes into account the effects of orbital angular velocity, gravitational gradient moments and spatial perturbation moments, providing good control accuracy within the constraint range and requiring less computation time, which satisfies the requirements of on-board control.

The method converts the optimal control issue into a nonlinear optimization issue based on differential flatness theory which simplifies the computational model and greatly improves planning efficiency.

The sliding mode control algorithm which is designed based on MRP is able to track the desired attitude under the influence of finite disturbance with little error to guarantee the tracking accuracy, so that the whole maneuver control torque, angular momentum and singularity measurement are all within the limited range.

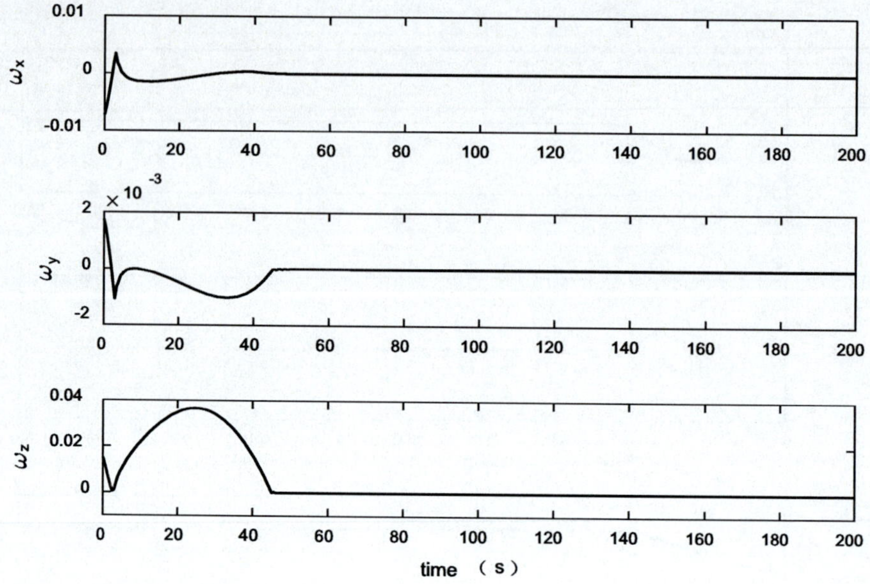

Fig. 2 Angular velocities

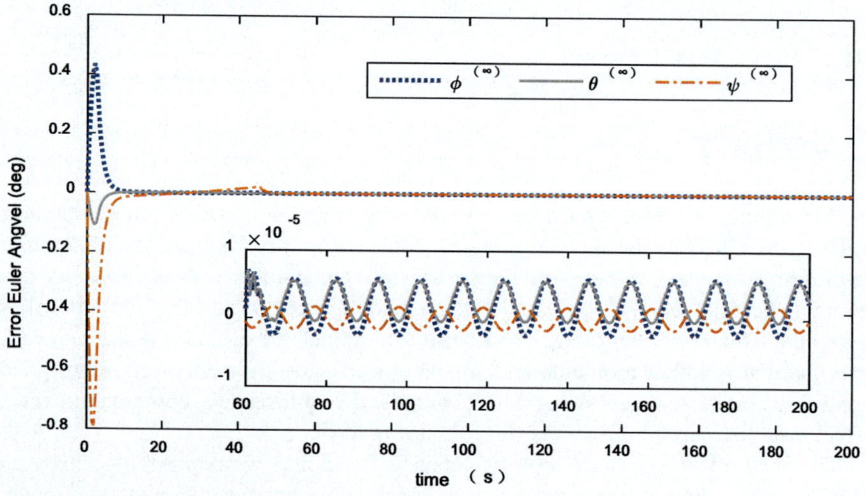

Fig. 3 Attitude tracking error with Euler angle

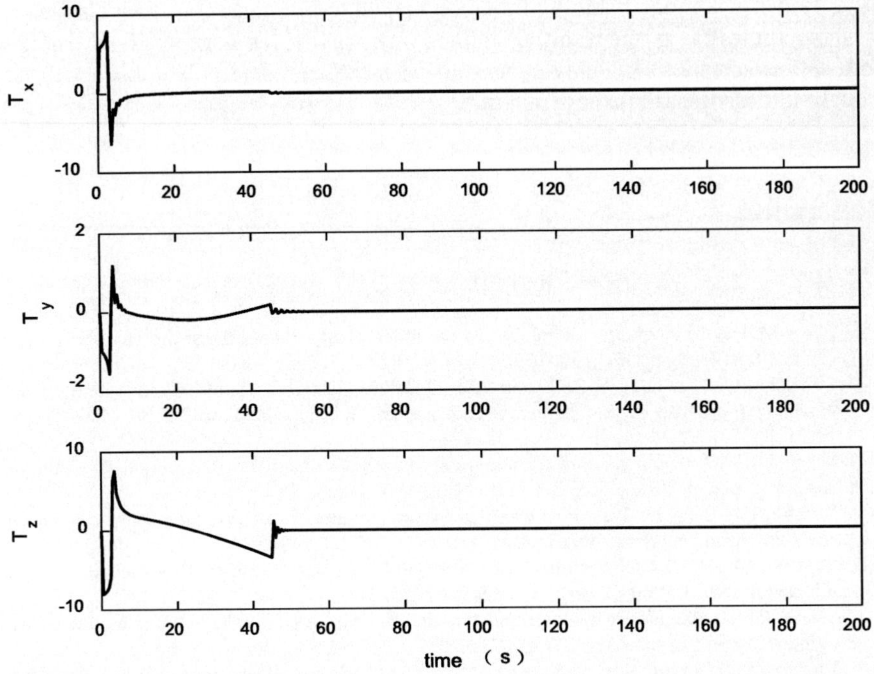

Fig. 4 Control torque

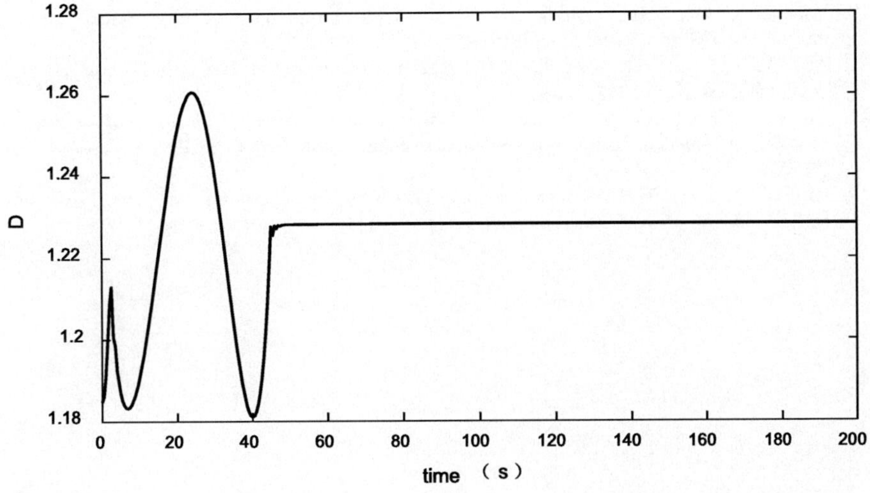

Fig. 5 Singular parameter

However, the limited ability of the polynomial to formulate the curve also leads to the optimization of the resulting attitude trajectory which is generally an energy sub-optimal solution. In subsequent studies, more complete polynomials will be sought to describe the flatness output.

References

1. 赵乾. 考虑力矩陀螺奇异与饱和的空间站姿态机动控制策略与路径规划[D]. 长沙: 国防科学技术大学 (2015)
2. Fliess M, Lévine J, Martin P et al (1993) On differentially flat nonlinear systems. Nonlinear Control Syst Design 25(3):159–163
3. Louembet C, Cazaurang F, Zolghadri A et al (2007) Design of algorithms for satellite slew manoeuver by flatness and collocation. American control conference. IEEE, New York, pp 3168–3173
4. Wang Y (2007) Comparison of quaternion and rodrigues parameters on attitude algorithm of secondary planet. Chin Space Sci Technol 20(4):858–863
5. Leeghim H, Bang H, Park JO (2009) Singularity avoidance of control moment gyros by one-step ahead singularity index. Acta Astronaut 64(9):935–945
6. Paradiso JA (1992) Global steering of single gimballed control moment gyroscopes using a directed search. J Guid Control Dyn 15(5):1236–1244
7. Wie B (2004) Singularity analysis and visualization for single-gimbal control moment gyro systems. J Guid Control Dyn 27(2):271–282
8. Louembet C, Cazaurang F, Zolghadri A et al (2007) Design of algorithms for satellite slew manoeuver by flatness and collocation. American control conference. IEEE, New York, pp 3168–3173
9. Xia Y, Pu F, Li S et al (2016) Lateral path tracking control of autonomous land vehicle based on ADRC and differential flatness. IEEE Trans Industr Electron 63(5):3091–3099
10. Yakimenko OA (2000) Direct method for rapid prototyping of near-optimal aircraft trajectories. J Guid Control Dyn 23(5):865–875
11. Acar T, Agrawal PN, Neer T (2017) Bezier variant of the Bernstein-Durrmeyer type operators. RM 72(3):1341–1358
12. Cui HT, Cheng XJ (2012) Anti-unwinding attitude maneuver control of spacecraft considering bounded disturbance and input saturation. SCIENCE CHINA Technol Sci 55 (9):2518–2529
13. Lappas VJ, Steyn WH, Underwood CI (2002) Attitude control for small satellites using control moment gyros. Acta Astronaut 51(1):101–111

A Voxel-Based Fusing Method for Aerial Laser Scanning and Oblique Image Point Cloud Via Noise-and-Occupancy-Aware

Shiming Li, Qing Zhu, Han Hu, Xuming Ge, and Chuncheng Zhu

Abstract High resolution remote sensing is the main data source for urban 3D modeling, among which airborne laser scanning and oblique photogrammetry have become the two most important data acquisition methods. In the complex urban environment with tall buildings, these two methods have their own Pros and Cons. The airborne LiDAR point cloud with high precision is limited by the low spatial resolution and single top viewing angle, which fails to accurately describe the facade information of the building. The oblique image-based point cloud, which has the advantages of multiple views and rich texture, is limited by the noise, the sharp features cannot be accurately expressed either. The fusion of the two kinds of data brings new potential for breaking through the bottleneck of single data source. The completeness and accuracy of city description by point cloud data are expected to be improved, so as to provide accurate data support for target extraction, 3D reconstruction, spatial analysis and disaster protection. In view of the significant differences between the two kinds of point cloud data in spatial resolution, coverage and noise level, this paper proposes a cross-source point cloud fusion method that takes both mass data processing efficiency and local detail fidelity into account. First, the method is based on a robust registration algorithm, which extracts accurate point cloud subsets from image-based point clouds affected by noise, and corrects the initial registration results provided by inertial navigation system and image control points. Second, an adaptive voxel segmentation method based on plane structure perception is proposed to establish correspondence of cross-source point clouds. Finally, cross-source point clouds are filtered for mutual exclusion. Experiments show that the proposed method integrates the two groups of point clouds, and the obtained results complete the vacancy of the LiDAR point cloud and improve the accuracy of the image-based point cloud.

S. Li (✉) · Q. Zhu · H. Hu · X. Ge
Faculty of Geosciences and Environmental Engineering, Southwest Jiaotong University, Chengdu 611756, Sichuan, China
e-mail: lishiming@my.swjtu.edu.cn

S. Li · C. Zhu
Sichuan Shihuizhitu Spatial Information Technology Co. Ltd., Chengdu 610036, Sichuan, China

L. Wang et al. (eds.), *Proceedings of the 7th China High Resolution Earth Observation Conference (CHREOC 2020)*, Lecture Notes in Electrical Engineering 757, https://doi.org/10.1007/978-981-16-5735-1_16

Keywords Voxel-based · Image-based point cloud · LiDAR point cloud · Point cloud fusion · The completeness and accuracy

1 Introduction

Starting in the 1990s, laser scanning has been used in urban 3D data collection. The data density can be obtained at dozens of points per square meter with accuracy up to centimeter level, but data acquisition is still relatively expensive [1]. Thanks to the upgrade with calculate capacity and image dense matching algorithm optimization, we can from a large amount of unstructured image data set for 3D reconstruction [2], this makes the oblique photogrammetry can quickly obtain city data, together with the laser scanning became the present two methods of standard for remote sensing [3]. Compared with triangular meshes, point cloud does not need to store or maintain a polygon mesh connectivity or topological consistency, directly based on point cloud processing is more effective directly. Therefore, using airborne laser scanning and oblique photogrammetry dense matching to get point cloud data is the important basis of smart-city application data and real 3D reconstruction [4].

The data acquisition of single method has obvious defects, so it is limited in data application. Airborne LiDAR point cloud has high data accuracy and low noise level, but the extracted building plane is usually incomplete, especially in the facade area of the building, and the small structure cannot perfect the sampling [5]. Oblique photographic point cloud has extremely high data density and abundant texture information. However, due to the influence of occlusion, data source quality and matching algorithm, there is a large amount of noise, which enhances the difficulty of data mining, affects the extraction and analysis of detailed information, and results in point feature degradation [6]. According to the analysis in Table 1, the two methods are highly complementary. The comprehensive utilization of the two means can break through the limitations of scanning range, time, collection perspective and collection accuracy, and obtain three-dimensional data in a multi-dimensional and comprehensive manner [7], providing accurate data support for target extraction, 3D reconstruction, spatial analysis and disaster protection.

For the synthesis of two or more sets of point cloud data, the usual approach is to perform accurate registration to make its spatial reference consistent. There is a residual error in the transformation of global coordinate system by using the inertial navigation system or image control points in the acquisition system itself. Further accurate registration is still necessary [8]. In addition, the collection mechanism of cross-source point cloud is very different, with different data density and noise level, the point cloud after registration only achieves the consistency of spatial datum, but it will produce new contradictions, such as local stratification and conflict [9]. Reasonable mutex selection and fusion operations are required to obtain a unified point cloud model for the whole scene [10]. Massive point cloud data need to occupy a large amount of computing resources, so how to efficiently

Table 1 Comparison of airborne laser scanning and oblique photogrammetry

	Oblique photogrammetry	Airborne laser scanning
Density	>300 p/m^2	3–30 p/m^2
Common platforms	Mainly light and small drones with an altitude of less than 300 m	Mainly with large-load manned aircraft with 500–3000 m altitude
Main source of error	Lower image resolution, matching algorithm error, and underlying occlusion caused by viewing angle	Navigation system transmission error, geometric correction residual
Data accuracy	Affected by image quality and matching accuracy, it can usually reach decimeter level	$\leq$ 15 cm And the elevation accuracy is better than the plane accuracy
Advantage	Complete coverage, higher density, with texture information	High efficiency, high precision, penetration
Demand	Need to reduce the noise level, need to penetrate the top cover	Need filling holes, need to increase structural details, need to increase density

process large scene data is an important problem that needs to be overcome. At the same time, the fused data should be valid for fidelity local details. Multi-scale processing requirements must be considered in the fusion phase.

Therefore, this paper proposes a systematic cross-source point cloud fusion strategy for noise and overlay perception, aiming at the significant differences and fusion requirements existing in airborne laser point clouds and image-based point clouds obtained through dense matching of oblique images. Firstly, the method is based on robust registration algorithm to extract accurate point cloud subsets from image point clouds affected by severe noise, and the initial registration results provided by inertial navigation system and image control points are corrected. Secondly, an adaptive voxel segment method based on plane structure perception is proposed to establish spatial mapping relationships across source point clouds. Finally, cross-source point clouds are selected for mutual exclusion effectively. Experiments show that this method effectively fuses the two groups of point clouds into a whole data set, and the obtained results fill in the vacancy of laser point cloud and improve the precision of tilted point cloud.

2 Review

Point cloud registration is to solve the spatial alignment of point clouds from different sources. It usually solves the alignment problem of two sets of point clouds or multi-station continuous point clouds. The simple method is to set up reasonable targets before point cloud acquisition and identify them in the data [11]. The laser scanning system usually has an inertial navigation system which can be used to solve the global coordinates of the point cloud [12]. For image point cloud, using image control point to obtain global coordinates is an effective method.

In fact, when the scanning area is very large, especially in complex cities, it is difficult to set up sufficient targets to meet the requirements of identification, and the accuracy of the inertial navigation system also makes the aligned data still have residual error, which still needs to be accurately registered. At present, a large number of point cloud registration studies, from coarse to fine strategy has become the paradigm [13]. If the data acquisition system itself cannot provide the initial global coordinates, then the method of feature extraction can also be combined for coarse registration [14], such as exhaustive matching strategy like RanSAC, 4PCS, etc., as the initial value of further accurate registration [15].

ICP [16] algorithm is the most widely used precision registration algorithm, through iterative calculation of the minimum distance between points, to solve the point cloud rigid body transformation of the rotation component R and translational T. Many studies have been carried out to optimize the ICP algorithm, so as to improve the algorithm to better resistance to noise and overlapping areas, and extend the point element to solve the corresponding relationship between line and surface elements [17]. However, image point clouds have too much random noise and are usually clustered at the boundary of structural changes, which makes it difficult to obtain accurate registration results even with a good initial registration value. Noise suppression becomes a necessary step for image point cloud itself.

Point cloud registration after the principle contradiction is the root cause of the differences caused by the error categories and the size is different, adjusted in the pretreatment process for the formation of the constraint error is a kind of effective means, through calculation [18] or identification [19] multi-source data of mutual connection points to optimize image connection relations can reduce the error produced by the dense matching. However, some data that need to be fused have already completed the preprocessing steps and are irreversible. In addition, the adjustment of flight strips based on navigation is a relatively effective means [20]. The cross-source data takes the strip as a unit or segments the strip, and the constrained optimization results are based on the perspective relationship or scanning angle. Such methods require the support of additional information, such as flight trajectory, which is often missing during fusion. For the point cloud data without additional auxiliary information, reasonable space segmentation and independent processing with different strategies can meet the complex requirements of fusion.

Bódis-Szomorú et al. [21] used tetrahedral structure for spatial segmentation and combines line of sight collinear relation to deal with stratification error, but it is time-consuming and low precision. Li et al. [9] realized the mutual exclusion screening of open space data based on voxel over-segmentation [22] and graph cutting [23], but the over-segmentation was inefficient, so it could only process the small scene data obtained by the ground acquisition system. Therefore, it is necessary to explore a method suitable for large scene data and able to take into account local detail fidelity.

3 Methodology

Due to the existence of multiple differences such as density and noise, point clouds of different sources do not correspond one-to-one. Establishing the correct correspondence between the cross-source point clouds is a necessary step for fusion processing. The whole space is replaced by several local blocks, which have consistent processing strategies and retain accurate information. For each local block, cross-source point cloud exists both, which includes the correspondence of geometric positions and logical correspondence of structures, so as to support adaptive methods and parameter selection and realize highly automated point cloud data fusion.

Therefore, data fusion of multiple types of point clouds requires a three-step operation. First, to achieve accurate registration of the geometric positions of multiple types of point clouds. Second, to extract structures from point clouds as a basis, to perform the same spatial segmentation on multiple types of point clouds, and to establish corresponding relationships. Third, the fusion processing method in the segmented area.

3.1 Multi-type Point Cloud Registration with Significant Detail Differences

Due to the bottleneck of the dense image matching method, the image-based point cloud cannot accurately and continuously sample small structures, especially the building edges and hollow railings. As shown in Fig. 1, due to the effects of occlusion and parallax during the dense image matching process, the edge structure of the roof contour is lost, which leads to the error of the top contour tightening inward. Therefore, traditional registration methods based on geometric feature points and lines are limited. On the other hand, in order to pursue the visual art effect, high-transmission materials such as glass have been widely used in urban architecture, which has led to a large amount of internal noise in laser scan data.

The plane is the most stable structural information that can be referenced in point cloud registration. The difficulty in ensuring the extraction of more than three corresponding surfaces restricts its use in traditional registration. When facing the fusion of a large amount of data in cities, the restrictions are broken because there are more buildings and road planes in the scene. ICP algorithm based on a sufficient number of accurate points screened out based on accurate planar structures can accurately register multiple types of point clouds with significant differences. The pipeline of registration is shown in Fig. 2.

Accurate point cloud normal vector computation can better support plane extraction, and it is also an important feature of subsequent multiple steps [24]. We adopt a denoisy normal direction estimation method to identify and eliminate outliers based on local neighborhood structures [25] to overcome the influence of noise prevalent in image point clouds (Fig. 3a) and optimize the normal vector estimation results at sharp features (Fig. 3b). The specific algorithm process is as follows.

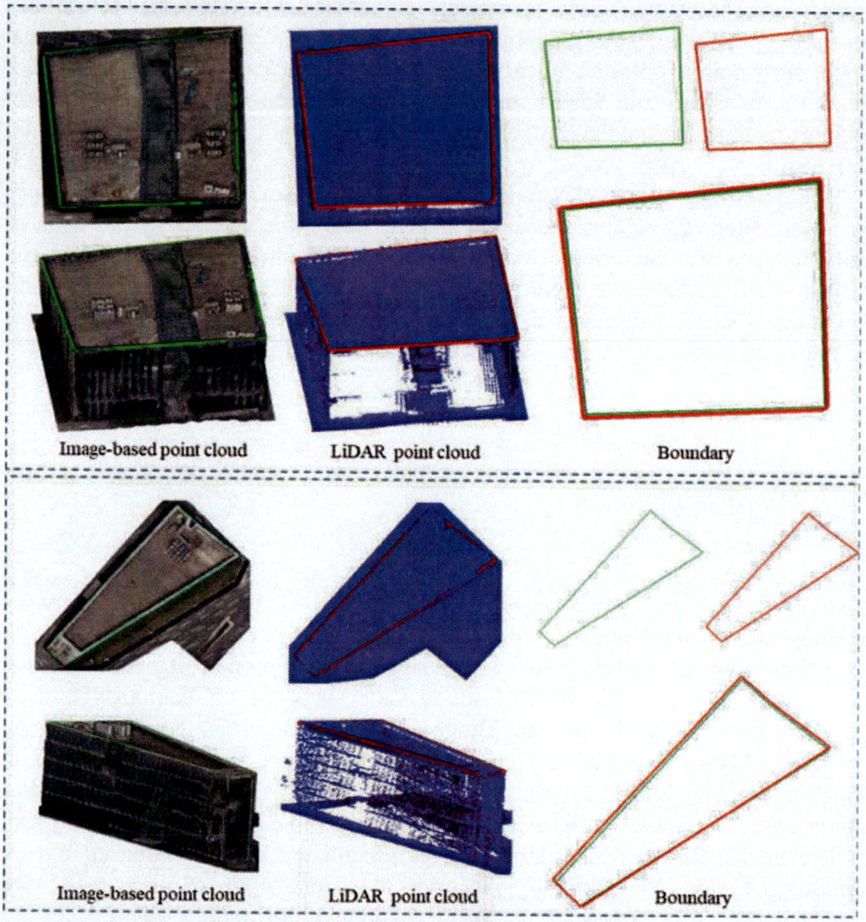

Fig. 1 The roof of the image-based point cloud tightens inward. Image-based point cloud has real color and the extracted boundary line is green. LiDAR point cloud is shown in blue and the extracted boundary line is shown in red

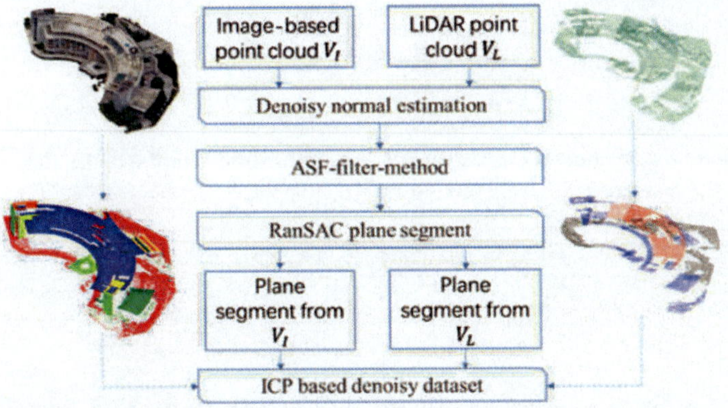

Fig. 2 The pipeline of registration

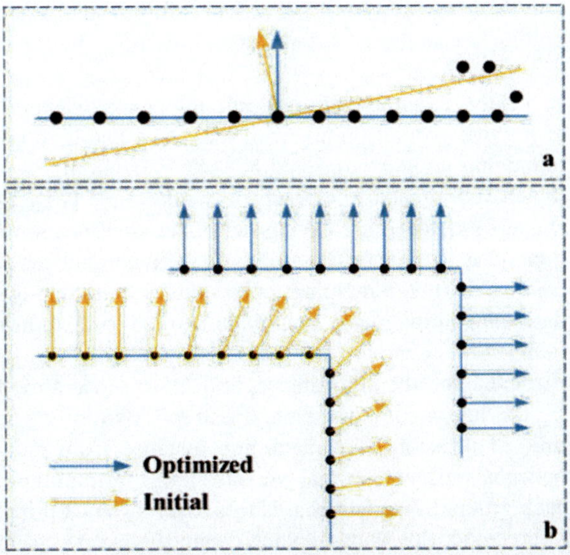

Fig. 3 Comparison of before and after normal estimation optimization. Among them, yellow indicates the result of normal estimation based on neighborhood points, and blue indicates the result of normal estimation based on outlier detection. **a** Eliminate local noise; **b** optimize sharp features

Any point $p(x_p, y_p, z_p)$ in the point cloud and its k-neighboring points form the point set V_k, calculate the covariance matrix $C_{3\times3}$ as Eq. 1.

$$C_{3\times3} = \frac{1}{k}\sum_{i=1}^{k}(p_i - p)(p_i - p)^T \tag{1}$$

where p_i is a neighboring point of p. Let $\lambda_1 > \lambda_2 > \lambda_3$ and e_1, e_2, e_3 be eigenvalue and eigenvector of $C_{3\times3}$. Then we can get the tangent plane at point p as Eq. (2). Because noise may exist in the point set V_k, outlier detection is required.

$$\begin{cases} e_3 = n_p(\mathrm{a}_p, \mathrm{b}_p, \mathrm{c}_p) \\ d_p = -(\mathrm{a}_p x_p + \mathrm{b}_p y_p + \mathrm{c}_p z_p) \\ f_p = \frac{\lambda_3}{\lambda_1 + \lambda_2 + \lambda_3} \end{cases} \tag{2}$$

Calculate the distance D_{Pi} from all neighborhood points p_i to the plane and calculate the variance Std_p according to Eq. (3).

$$\begin{cases} D_{pi} = \mathrm{a_p} x_{pi} + \mathrm{b_p} y_{pi} + \mathrm{c_p} z_{pi} + d_p \\ Std_p = \sqrt{\frac{1}{k-1} \sum_{i=1}^{k} \left(D_{pi} - \frac{1}{k} \sum_{i=1}^{k} D_{pi} \right)^2} \end{cases} \tag{3}$$

Set δ_{ne} as the noise detection threshold. When $D_{pi} > \delta_{ne} \Delta Std_\mathrm{p}$, p_i will be recognized as an outlier and will be removed from V_k, otherwise it will be retained. When all k points have been judged and identified, the remaining point cloud set V' is used to recalculate the covariance matrix and the eigenvectors. Outlier detection will perform again until the variance Std_p is small enough, then the robust normal of point p is calculated. After normal estimation by the above method, plane detection can be performed.

After the denoisy normal estimation, we first used ASF filtering algorithm [26] to remove ground points from the scene data and conduct plane detection for the remaining non-ground point sets. Random sampling consensus algorithm (RanSAC) is used to extract the stable plane [27]. This method is widely used in shape detection [28]. Taking the plane identification as an example, the RANSAC method is to select three points in a given data set and estimate the plane parameters. Test the remaining point clouds with the generated candidate planes to determine how many points the model can approximate well. After a certain number of iterations, the plane that possesses the largest percentage of inliers is extracted and the algorithm continues to process the remaining data.

The important parameters that the RANSAC algorithm needs to set include the time of iterations, the maximum distance from the point to the plane, the angle between adjacent normal vectors, and the minimum number of point clouds for each group. Since our goal is to extract a robust principal plane in the image-based point cloud, the point-to-plane distance constraint should be as small as possible to eliminate noisy points. The angle between adjacent normal vectors can be appropriately increased to overcome noise and extract backbone information. The minimum number of point clouds in each group should be large to accommodate the density of image-based point clouds.

A reasonable threshold can extract the main planar structure of the building as accurately as possible. Unfortunately, due to increasingly complex building structures and also the noise influence, the segmentation results obtained by RanSAC still have many redundant errors (as shown in Fig. 4b). We still need to filter the extracted planes again. The planes iteratively segmented by RanSAC algorithm need to be compared with the previous planes, which can realize the recognition of adjacent planes and merge or eliminate the smaller planes. The center point p_R of the plane segment point clusters, the normal angle $\Delta\theta_R$ between the planes, and the distance d_R between the plane center points need to be calculated according to Eq. (4).

$$\begin{cases} p_R = \frac{1}{n}\sum_{i=1}^{n} p_i \\ \Delta\theta_R = \theta(n_j, n_R) = \arccos\left(\frac{n_j \Delta n_R}{|n_j||n_R|}\right) \\ \Delta D_R = D(p_R - p_j) \end{cases} \tag{4}$$

In the above formula,p_i refers to the point cloud contained in the current plane,p_R is the calculated point cloud center point, n_R is the normal vector of the current plane,p_j and n_j are point cloud center points and normal of other planes obtained by RANSAC segmentation. ΔD_R is the distance between the center points of the plane.

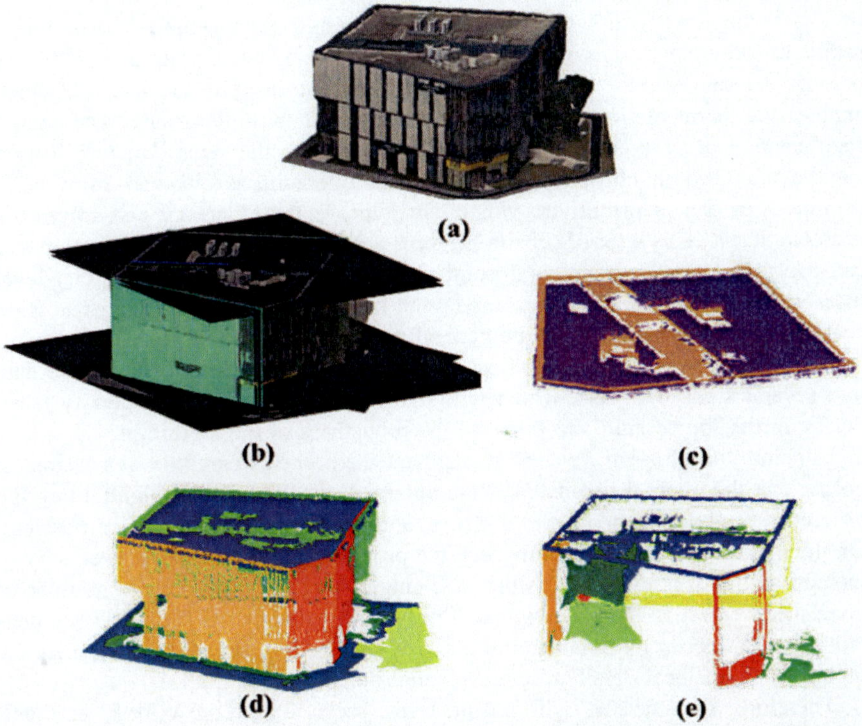

Fig. 4 Extracting principal plane of image-based point cloud based on RANSAC method. **a** Raw image-based point cloud data; **b** the roof is extracted into two planes; **c** plane extraction results still need further filtering; **d** the principle plane obtained after filtering; **e** the plane extracted removed after filtering

By setting the threshold to judge $\Delta\theta_R$ and ΔD_R, the recognition of adjacent planes can be realized and the smaller planes can be eliminated (as shown in Fig. 4c). Sometimes the necessary interactive action filtering is unavoidable. In the end we can get the results shown in Fig. 4d.

After the above process, we can segment a point set $V_{plane}^{(o)}$ from the image-based point cloud. This point set is segmented from the main planes such as buildings and roads contained in the scene. Because these planes have obvious topological relations, effective geometric constraints are formed. Moreover, our way of extracting the plane resulted in the final segmented point cloud being as accurate as possible. The next operation is to calculate the rotation matrix of the LiDAR point cloud and point set $V_{plane}^{(o)}$ by using the ICP algorithm to obtain accurate registration results of the two types of point clouds.

3.2 Voxel-Based Segmentation and Feature Description

In this section, we will introduce an effective multi-scale adaptive voxel-based segmentation method to establish the correspondence between point clouds. In this process, we perceive the plane structure of the point cloud in local voxel, which enables the segmentation process to be adaptive at multiple scales. The spatial segmentation process is necessary because points with the same structure feature can be managed in clusters. The subsequent operation is changed from point primitives to cluster primitives, which can improve the efficiency and solves the problem that it takes a lot of time to perform geometric structure segmentation and category extraction from original point clouds, especially when facing city-level massive data. In addition, cross-source point clouds have different data expression scales and precisions, and there are generally some differences in the description of the same location of the same object. Voxel-based segmentation divides the data into several small data sets, which will significantly reduce the complexity faced during further processing and improve the robustness of the algorithm.

In point cloud fusion, we need to segment the scene as carefully as possible to ensure that the internal geometry of the obtained voxel is simple, highlighting the differences caused by point cloud errors, and thus making error repair possible. On the other hand, in order to improve the processing efficiency, it is necessary to include as many points as possible in a single voxel and reduce the number of voxels in the scene. This shows that if only one scale is used for voxel segmentation, the efficiency requirements of massive data in large scenes and the accuracy requirements of local detail structures cannot be considered both.

Therefore, our method will segment the scene based on voxels, and will implement voxel segmentation at different scales based on valid structural information detection. Specifically, since buildings and roads with a large number of regular geometric structures are the most important feature categories of urban scenes, segmenting a point cloud into several point cloud clusters based on a planar structure will provide benefits for subsequent work such as point cloud classification.

First, a large initial step size d_{ini} set to perform voxel segmentation on point cloud data, and performs feature calculation and structure detection on the segmented voxels as a basis for further segmentation. The covariance matrix $M_{3\times3}$ is constructed using points set P belonging to voxel S according to Eq. 5.

$$\begin{cases} C_{3\times3} = \frac{1}{n}\sum_{i=1}^{n}(p_i - p_s)(p_i - p_s)^T \\ p_s = \frac{1}{n}\sum_{i=1}^{n} p_i \end{cases} \tag{5}$$

where p_i is the point belonging to voxel S,p_s is the center point of point set P. Let $\lambda_1 > \lambda_2 > \lambda_3$ and e_1, e_2, e_3 be eigenvalue and eigenvector of $M_{3\times3}$. Then the saliency features g_1, g_2, g_3 [29] of voxel S are calculated as Eq. 6.

$$\begin{cases} g_1 = \frac{\sqrt{\lambda_1}-\sqrt{\lambda_2}}{\sqrt{\lambda_1}} \\ g_2 = \frac{\sqrt{\lambda_2}-\sqrt{\lambda_3}}{\sqrt{\lambda_1}} \\ g_3 = \frac{\sqrt{\lambda_3}}{\sqrt{\lambda_1}} \end{cases} \tag{6}$$

Finally, the plane detection of voxel S can be performed as Eq. 7.

$$h_s = \begin{cases} planar, if (g_2 > g_1) \cap (g_2 > g_3) \\ nonplanar, otherwise \end{cases} \tag{7}$$

And the geometric features of each voxel S is calculated as Eq. 8.

$$\begin{cases} n_s(a_s, b_s, c_s) = e_3 \\ d_s = -(a_s x_s + b_s y_s + c_s z_s) \\ f_s = \frac{\lambda_3}{\lambda_1 + \lambda_2 + \lambda_3} \end{cases} \tag{8}$$

where n_s and d_s are the parameters of the tangent plane of voxel S, and is the curvature.

In addition, we need to measure the planarity from the point to the fitting plane in the voxel, which is expressed by $F_{planarity}$ calculated as Eq. 9 f_s

$$F_{planarity} = \frac{1}{n}\sum_{i=1}^{n} D(p_i, plane(S)) = \frac{1}{n}\sum_{i=1}^{n} \frac{(a_s x_i + b_s y_i + c_s z_i + d_s)}{a_s^2 + b_s^2 + c_s^2} \tag{9}$$

where n is the number of points contained in P,$p_i(x_i, y_i, z_i) \in P$. The plane parameters (a_s, b_s, c_s, d_s) of voxel S are calculated from the previous steps. The $F_{planarity}$ of voxels is an important feature to measure the dispersion and error of interior points.

In addition to the above feature calculations, the similarity of adjacent voxels also needs to be considered in our algorithm. We define the parameter $F_{consistency}$. The hypothesis that "points located on the same ground structure have similar properties" [30] is extended to consider the continuity of normal vectors of adjacent voxels.

In addition to the above feature calculations, the similarity of adjacent voxels also needs to be considered in our algorithm. We define the parameter $F_{consistency}$ to calculate as Eq. 10.

$$F_{consistency} = Num\left\{ S_j | S_j \in N, \theta\left(\underset{s}{\vec{n}} \cdot \underset{j}{\vec{n}}\right) < \delta \right\} \tag{10}$$

where N is the set of voxels adjacent to the current voxel S, $\underset{s}{\vec{n}}$ is the normal vector of the current voxel S and $\underset{j}{\vec{n}}$ is for the normal vector of the neighboring voxels S_j, $\theta\left(\underset{s}{\vec{n}} \cdot \underset{j}{\vec{n}}\right)$ indicates the angle between them, and $Num\{\}$ means the number of elements in the collection.

The voxel S and its features are described by one 10-tuple.

$$N_s = \{p_s(x_s, y_s, z_s), n_s(a_s, b_s, c_s), d_s, f_s, F_{planarity}, F_{consistency}\} \tag{11}$$

3.3 Adaptive Multi-Scale Segmentation and Fusion Processing

The two sets of point clouds that need to be fused should have the same voxel segmentation results, which are used to interoperate with each other in the fusion stage. Due to the influence of errors and other aspects, there are a large number of divergent areas. The two sets of point clouds in the same voxel space usually have different voxel feature vectors. When judging whether voxels are further segmented, the results of voxel features and structures calculated from image point clouds with higher noise levels shall prevail.

The main process of voxel segmentation is as follows:

- Step 1: Calculate the common bounding box of the two types of point clouds to ensure that the voxel segmentation can include the complete data set.
- Step 2: Set the initial step size s of the voxels, and divide the two types of point clouds into three-dimensional voxels respectively according to the relationship between points and voxels.
- Step 3: Label the voxel. There are four main types of labels: voxels with two types of point clouds, voxels with only LiDAR points, voxels with only image-based points, and empty voxels.

- Step 4: Perform structural detection and feature calculation on voxels with image-based points. The method is based on 3.2 and recorded as $N_s^{(o)}$.
- Step 5: For non-planar voxels, adjust the voxel step size and perform further segmentation.
- Step 6: Cycle steps 3 to 5 until all voxels are planar voxels or the voxel step reaches the minimum threshold. The voxel segmentation result is obtained at this time.
- Step 7: Perform structural detection and feature calculation on voxels with laser point clouds. The method is based on 3.2 and recorded as $N_s^{(l)}$.

$$\begin{cases} N_s^{(l)} = \left\{ \left(x_s^{(l)}, y_s^{(l)}, z_s^{(l)}\right), \left(a_s^{(l)}, b_s^{(l)}, c_s^{(l)}\right), d_s^{(l)}, f_s^{(l)} \right\} \\ N_s^{(o)} = \left\{ \left(x_s^{(o)}, y_s^{(o)}, z_s^{(o)}\right), \left(a_s^{(o)}, b_s^{(o)}, c_s^{(o)}\right), d_s^{(o)}, f_s^{(o)} \right\} \end{cases} \tag{12}$$

At this time, we have obtained multi-scale voxel segmentation results, each voxel has only one label of four categories in total. Among them, the voxels of the first type have voxel feature vectors $N_s^{(l)}$ and $N_s^{(o)}$ respectively calculated from the LiDAR and image-based points. The voxels of the second type only have voxel feature vectors $N_s^{(l)}$. The voxels of the third type only have voxel feature vectors $N_s^{(o)}$. The voxels of the fourth type lacks data or points are too sparse, and it is no longer necessary for further structure and error detection, so it will not participate in any further calculations. This will be the basis for our further fusion operation.

Each type of voxel has a different cause and requires a different treatment. The first type of voxel is the focus of our attention. The internal cross-source cloud provides a good spatial mapping where the LiDAR point cloud provides accurate planar structure information and therefore will be used as an accurate value within the voxel. Under the influence of various differences, the description of the same location by LiDAR point cloud and image-based point cloud will definitely be different. There are not only one form of this difference, such as a sudden appearance of a bump in the plane (as shown in Fig. 3a, mostly due to local noise matching in the image), obvious stratification can be seen (as shown in Fig. 3b, mostly due to inaccurate registration results) Obvious intersection of two planes (as shown in Fig. 3c, structural inaccuracy), etc. By comparing the voxel feature vectors obtained from the two types of point cloud calculations, we can identify the accuracy of voxels and guide the fusion operation.

The main indicators for measuring the difference between the two types of point clouds are calculated according to $N_s^{(l)}$ and $N_s^{(o)}$ is calculated as Eq. 13.

$$\begin{cases} F_{Plangle} = \theta\left(n_s^{(l)}, n_s^{(o)}\right) = \arccos\left(\frac{n_s^{(l)} \cdot n_s^{(o)}}{\left|n_s^{(l)}\right|\left|n_s^{(o)}\right|}\right) \\ F_{Plandis} = \left|d_s^{(l)} - d_s^{(o)}\right| \end{cases} \tag{13}$$

Among them, $n_s^{(l)}$ and $n_s^{(o)}$ are plane normal vectors calculated in the voxel according to the LiDAR point cloud and the image-based point cloud, respectively.$d_s^{(l)}$ and $d_s^{(o)}$ represent the distance from the center point of the voxel to the plane. The difference $F_{Plandis}$ is used to reflect the layering of the point cloud.$\theta\left(n_s^{(l)}, n_s^{(o)}\right)$ represents the angle between the plane, so the existence of $F_{Plangle}$ is to measure whether the angle between the two planes exceeds the limit value. If the above two thresholds do not exceed our limit, we can use LiDAR to determine the similarity of the normal vector of the image-based points within a certain range based on the premise that the LiDAR point cloud usually has a lower noise level, so that the image-based points The cloud is screened and entered into the final set of fusion points. If the limit value is exceeded, we consider that there is a large error in image-based point cloud in voxel and eliminate it.

The second type of voxel contains only LiDAR point cloud. Generally, image-based point cloud has complete scene coverage capabilities. It can be understood that the object area scanned by the LiDAR point cloud should have corresponding image-based point cloud data. However, due to the limited step size of the voxel segmentation, when there is a large divergence between the image-based point cloud and the LiDAR point cloud, a second type of voxel will appear (Fig. 5a, roof, image matching error is large; Fig. 5b, lawn, Lack of penetrating images) (Fig. 5).

Therefore, we can understand that when there is only LiDAR point cloud in voxel V_i, the image-based point cloud in the corresponding region will appear in voxel V_j due to noise influence and voxel step limitation. Generally, V_j and V_i are adjacent, and only image-based point clouds exist in V_j. Based on the constraints of this knowledge rule, when processing the second type of voxels, we will retrieve its

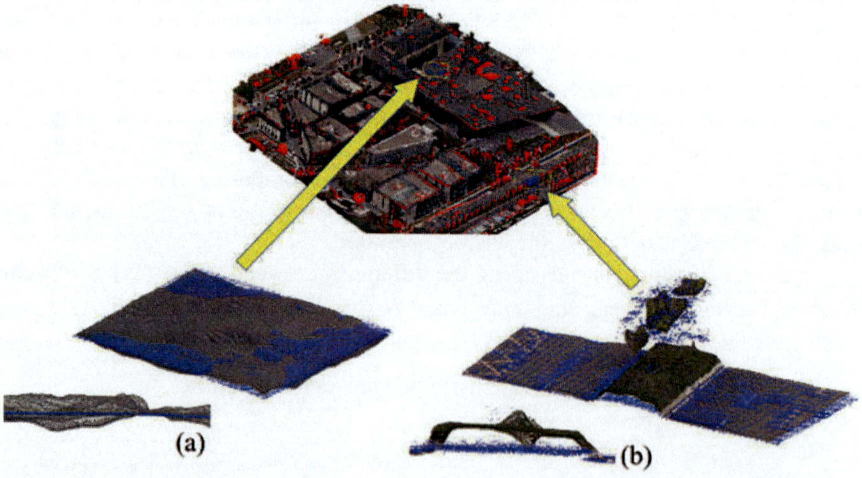

Fig. 5 The area where have a large divergence between the image-based point cloud and the LiDAR point cloud. **a** The roof; **b** Green belt of road

neighboring third type of voxels and merge the retrieved third type of voxels with the current voxel. At this point, the two types of point clouds in the current voxel will co-exist, and the subsequent processing strategy will be the same as the first type of voxel.

The third type of voxel contains only image-based point clouds inside. The absence of LiDAR point clouds here means that there are holes in the laser scan. Usually, this occurs in the building facade area. LiDAR data is also sparse and it is difficult to avoid the existence of hollow areas (such as the building facade area in the aerial perspective and the close-up occlusion in the ground perspective). In order to obtain a complete object, the usual method is to interpolate the interior to a plane or fit a surface based on a hole boundary. Due to the lack of internal real structure information, the data obtained by this method is often missing or even wrong. Although the image-based point cloud with full perspective coverage is doped with more noise, it still contains correct data, which can provide information missing from the corresponding vacancy area of the LiDAR point cloud.

Therefore, when processing the third type of voxel, the internal noise situation is first measured according to the parameter $F_{planarity}$ in the voxel vector. If the noise is small, the image-based point cloud will be retained directly to fill the hole. Otherwise, the moving least squares method [31] will be used to smooth the point cloud.

4 Experiments and Analysis

4.1 Dataset description

This experiment mainly conducts data fusion processing of image-based point cloud and LiDAR point cloud. The experimental data covers an area of about 200 m * 200 m in an industrial park of Shenzhen. The interior contains ground objects such as buildings, vegetation, roads and steps, among which the building is a steel frame structure. The image-based point cloud of this group of data was used to collect image data by the oblique photogrammetric camera equipped with five lenses on the UAV platform, and ContextCapture[1] software was used to process the image data. The obtained point data density was 400 points per square meter. The LiDAR point cloud of this set of data is collected by the UAV platform carrying small airborne laser scanning system, and the obtained point data density is 20 points per square meter.

[1] https://www.bentley.com/zh/products/brands/contextcapture.

4.2 Experiment Results

4.2.1 Cross-Source Point Cloud Registration

We use the method described in Sect. 3.1 to achieve image-based point cloud and airborne LiDAR cloud registration. It is different from the traditional method of feature recognition and registration based on multiple primitives. This method overcomes the effects of noise and structural inconsistency. At the same time, this method ensures the applicability of complex data, and the process is short and the parameter adjustment is simple, which is suitable for the efficient processing of large scene data.

Figures 6 and 7 shows the result of registration. First, the LiDAR point cloud uses the integrated navigation data to solve the point cloud into the global coordinates. Similarly, the image-based point cloud uses the image control point. There is an initial alignment between the two, and it can be seen from the section that there is an obvious inconsistency (a). The average distance between the two groups of point clouds is 3.256 m, and the RMSE is 1.682. On the basis of the initial alignment, the ICP algorithm registration results are shown in (b), where the average point distance is reduced to 0.949 m and RMSE is 1.309. We can see better consistency between the two sets of point clouds, but there is still a significant overall deviation, especially in the horizontal direction. (c) shows the registration results obtained by using the proposed method, with average point distance of

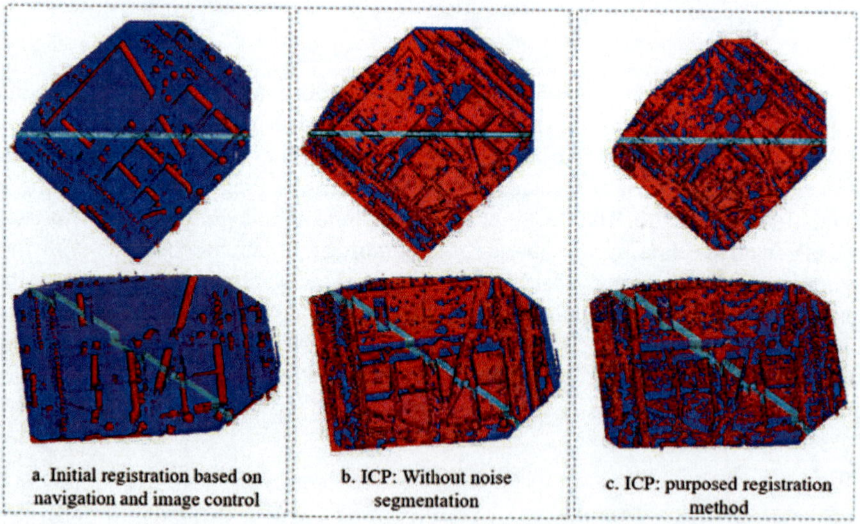

Fig. 6 The registration results are shown from the top view and side view. Red: LiDAR point cloud; Blue: image-based point cloud; Cyan: represents the position of the section in the point cloud

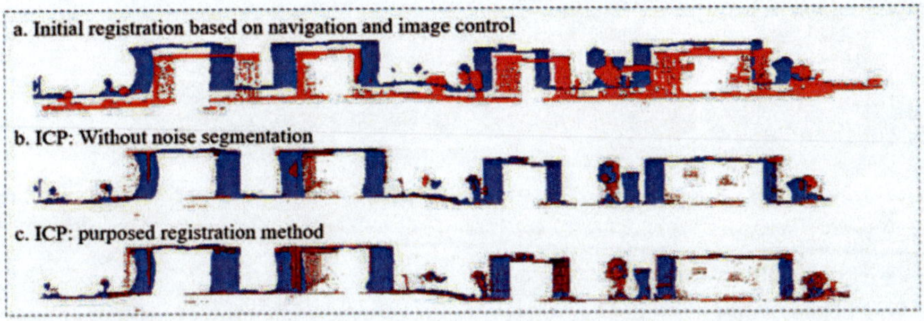

Fig. 7 The registration results are shown from the sectional view respectively

Table 2 The statistical results of registration

	The average distance between adjacent points	MSE	RMSE
Initial registration based on navigation and image control	3.25631	2.82787	1.68163
ICP: without noise segmentation	0.94886	1.71269	1.3087
ICP: purposed registration method	0.302516	0.613933	0.783539

0.302 m and RMSE of 0.784. It can be seen that the registration accuracy has been significantly improved. At the same time, we can also feel the optimization intuitively from the contrast of sections. The statistical results of registration are shown in Table 2.

4.2.2 Segmentation Results and Types of Voxels

Voxel segmentation is to establish local correspondence of cross-source point clouds. First, we preliminarily divided the scene data according to the step length interval of 1 m, and replaced the voxel with the geometric center point. We classified voxels according to the method in Sect. 3.3 and added different colors to the voxel center point. The first type of voxel was labeled as gray, the second type of voxels as red, and the third type of voxels as blue. The results were shown in Fig. 8 and Table 3.

From Fig. 8, we can see some intuitive and special phenomena. Firstly, from the perspectives of top (a) and bottom (b), it can be seen that there are a lot of second type voxels in the interior of the building. This is because the projection nature of the glass on the building surface makes the laser beam penetrate the surface of the building, and a large number of laser noise points appear inside the building. Efficient and accurate elimination of internal noise points is another topic. In our method, these internal noise points and image point clouds do not adjacency, so there is no need to process, but also can be manually interactive deletion. From the

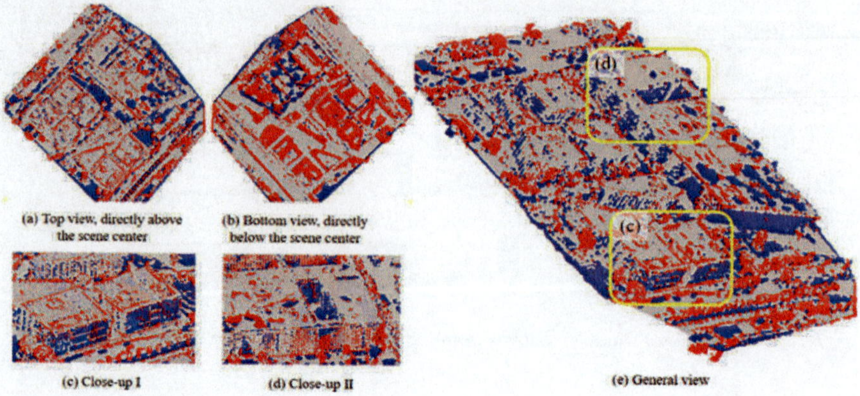

Fig. 8 The type of voxels obtained by initial segmentation. The center of each voxel replaces the original point cloud. Gray: voxels of the first type, which exists both point clouds. Red: the second type of voxel that only exists the LiDAR point cloud. Blue: the third type of voxel that exists only image-based point clouds

Table 3 The statistical results of voxel type categories

	First type voxel	Second type voxel	Third type voxel
Number	70,345	36,830	25,480
Percent (%)	53.03	27.76	19.21

close-up I(c),it can be seen that the top boundary of the building obviously exists second type voxels. This is due to the restriction of the principle of dense matching, the image-based point cloud will lose the sharp features of the structure, resulting in the indenting of the boundary. At the same time, there are a large number of the third type of voxels on the building facade, and the image-based point cloud can be a good supplement to the LiDAR point cloud to obtain complete data. A similar situation also occurs in close-up II(d). In the internal courtyard of the building, the difficulty of LiDAR point cloud coverage is increased due to the narrow space, which can be compensated by oblique photogrammetry with multiple perspectives. We can also see that there is significant noise in the image point cloud, which leads to the irregular appearance of second type or third type voxels at the top of the building.

For the initial segmentation of voxels, we used the method mentioned in Sect. 3.3 to conduct planar detection of the image-based point cloud within each voxel, and to conduct adaptive segmentation of non-planar voxels. As shown in Fig. 9, we adopted a building to show the adaptive voxel division results more clearly. It can be seen that voxel segmentation occurs again near the edge of the building and also in the place where distortion occurs.

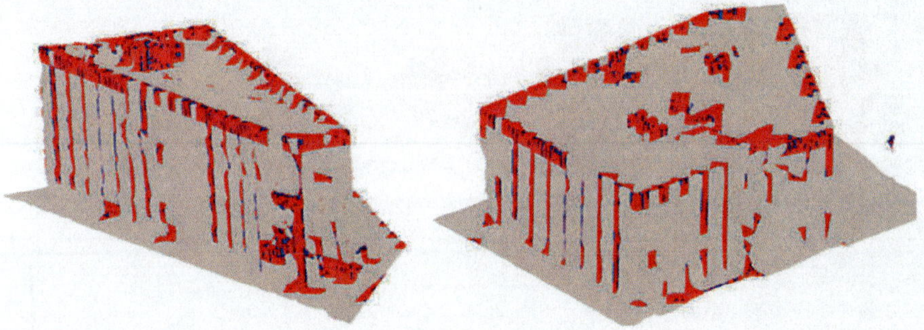

Fig. 9 Adaptive voxel segmentation. Gray: initial segmentation, red: re-segmentation, and blue: further re-segmentation

4.2.3 Cross-Source Point Cloud Fusion

Based on the above adaptive voxel segmentation results and the recognition of voxel types, two kinds of point cloud fusion operations are carried out. The fusion point cloud is shown in Fig. 10. As can be seen, the fused data realizes the overall coverage of the scene.

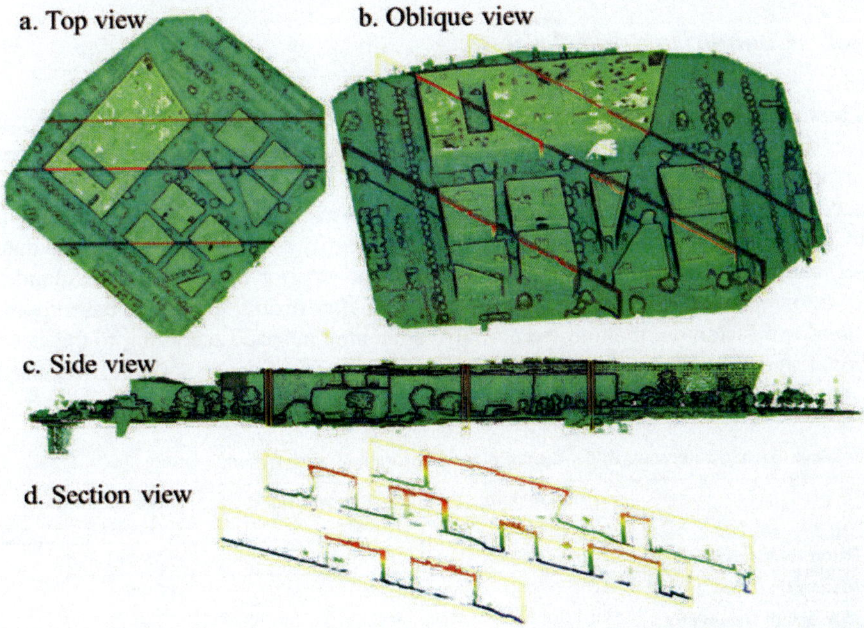

Fig. 10 Fusion point cloud

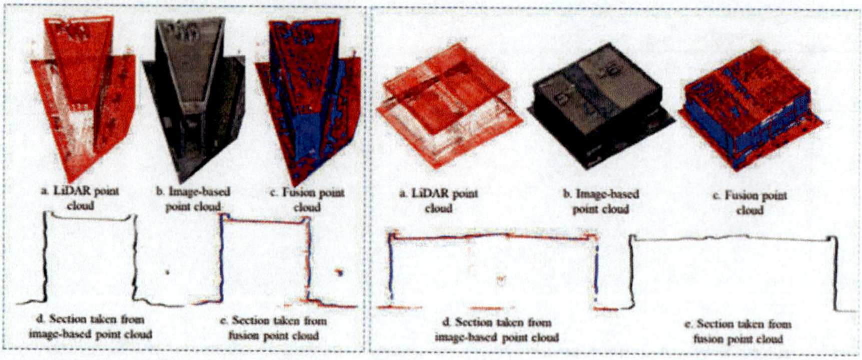

Fig. 11 The close-up for fusion point cloud. Red: LiDAR point cloud; blue: the image-based points obtained by mutex filtering

As shown in Fig. 11, we adopted two close-ups to show the fusion details. First, the three groups of point clouds (a, b and c) are compared, and it can be intuitively seen that the fusion point cloud effectively complements the vacancy of the LiDAR point cloud. Secondly, the section demonstrates effective quality improvement. The deformed part in image-based point cloud section (d) can be seen deleted in fusion point cloud section (e).

4.3 *Comparison and Analysis*

Through fusion processing, we have three data sets for comparison: LiDAR point cloud, image-based point cloud and fusion point cloud. We compared the number of points and average point density, and compared the number of effective voxels according to the same voxel segmentation. The corresponding statistics are shown in Table 4. These comparisons can reflect the integrity and point density of the data set, and it can be seen that fusion point cloud is superior to LiDAR point cloud.

In order to compare the accuracy before and after fusion, the image-based point cloud and fusion point cloud were firstly segmented in space according to the same voxel step size. For each voxel, the principal component analysis (PCA) was used

Table 4 Statistical comparison of point clouds before and after fusion

	LiDAR point cloud	Image-based point cloud	Fusion point cloud
Number of points	10,291,627	92,969,624	55,781,774
Mean point spacing (m)	0.1	0.01	0.05
Number of available voxels	107,175	95,825	132,655

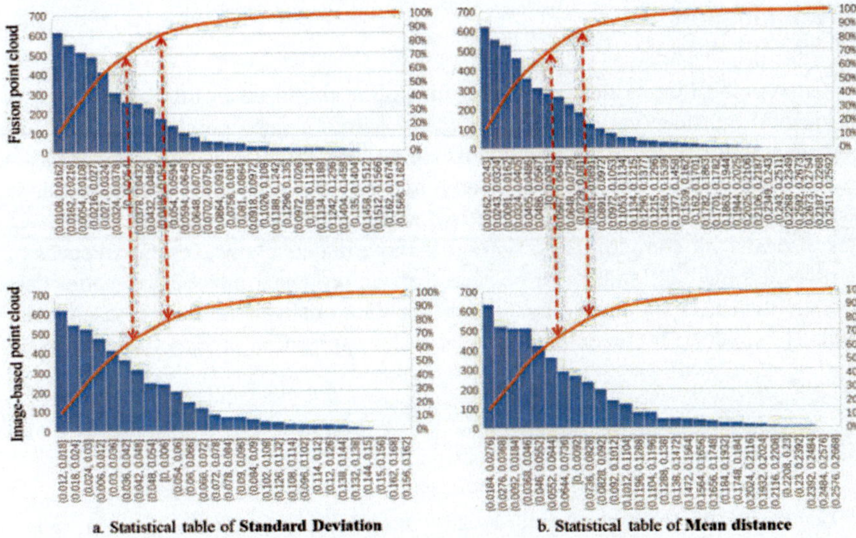

Fig. 12 The distribution of the mean distance and standard deviation

to calculate the main plane of the point cloud, and the distance from each point to the plane was calculated and then get the mean value and standard deviation of the distance. The distribution of the mean distance and standard deviation is then calculated and shown in Fig. 12. It can be seen that the distribution of mean value and standard deviation of the fused data is obviously better than that of image-based point cloud. This proves that the accuracy of the fused point cloud is optimized.

Through the above comparison, it can be seen that the method proposed in this paper can effectively overcome the inconsistency of the cross-source point cloud and realize the fusion of the two kinds of data. The fused data has higher point density and better data coverage than the LiDAR point cloud, and higher precision than the image-based point cloud. However, our approach still has some problems.

First, due to the lack of public data sets and ground-truth, we did not have more comparison data to compare the results of the fusion horizontally. Only through the reasonable design of the experimental, the existing comparative results are obtained to verify the effectiveness of the method.

Second, our method can only deal with the error of image-based point cloud within a reasonable range. Excessive deviation value is considered as "errors" and should be eliminated. The quality of image-based point cloud is related to the quality of image data collection. Although the resolution of experimental data we currently use is generally adopted in industry applications, there will also be image-based point clouds with higher noise level generated by images with lower resolution. We have not considered how to optimize "errors" at this time.

5 Conclusions

A cross-source point cloud data fusion method is proposed in this paper. Firstly, this method consider various inconsistencies across source point clouds and has good adaptability to noise. Secondly, this method proposes an adaptive voxel-based segmentation method and the local spatial mapping of cross-source point clouds is established. At the same time, this method establishes a set of effective processing rules to deal with complex fusion process. Experiments show the effectiveness of the proposed method, and it can be seen that the proposed method has better data coverage than LiDAR point cloud and better data quality than image-based point cloud. Thus, better 3D reconstruction effect and application prospect are expected.

References

1. Lei Y, Bo W (2018) Integrated image matching and segmentation for 3D surface reconstruction in urban areas. Photogram Eng Remote Sens 84(3):135–148
2. Mattei E (2017) Point cloud denoising via moving RPCA. Comput Graph Forum 36(8):123–137
3. Bo W, Linfu X, Hu H et al (2018) Integration of aerial oblique imagery and terrestrial imagery for optimized 3D modeling in urban areas. Isprs J Photogram Remote Sens 139:119–132
4. Zhu Q, Wang F (2018) Intact planar abstraction of buildings via global normal refinement from noisy oblique photogrammetric point clouds
5. Bastian A, Martin K, Andreas W (2016) Automatic extraction and regularization of building outlines from airborne Lidar point clouds. ISPRS Int Arch Photogram Remote Sens Spat Inf Sci XLI-B3:555–560
6. Xie L, Hu H, Zhu Q et al (2017) Hierarchical regularization of polygons for photogrammetric point clouds of oblique images. ISPRS Int Archives Photogram Remote Sens Spat Inf Sci XLII-1/W1:35–40
7. Zhu Q, Li S, Hu H et al (2018) Multiple point clouds data fusion method for 3D city modeling. Geomatics Inf Sci Wuhan Univ 43(12):1962–1971
8. Ravi P, Costas A, Chris H et al (2017) Automatic integration of 3-D point clouds from UAS and airborne LiDAR platforms
9. Wei L, Cheng W, Dawei Z et al (2018) A volumetric fusing method for TLS and SFM point clouds. IEEE J Sel Topics Appl Earth Obs Remote Sens 1–9
10. Shahram I, Richarda N, David K et al (2011) KinectFusion: real-time dynamic 3D surface reconstruction and interaction. Acm Siggraph
11. Tilen U, Žiga R, Mojcakosmatin F et al (2019) New target for accurate terrestrial laser scanning and unmanned aerial vehicle point cloud registration. SENSORS 19(14):3179
12. Itzik K, Sagi F (2012) Lidar and ins fusion in periods of GPS outages for mobile laser scanning mapping systems. Isprs Int Archives Photogram Remote Sens Spat Inf Sci XXXVIII-5/W12(5):231–236
13. Feng W, Han H, Xuming G et al (2020) Multientity registration of point clouds for dynamic objects on complex floating platform using object silhouettes. IEEE transactions on geoscience and remote sensing
14. Cheng L, Wu Y, Chen S et al (2018) A symmetry-based method for LiDAR point registration. IEEE J Sel Topics Appl Earth Obs Remote Sens 11(1):285–299
15. Nicolas M, Dror A, Niloyj M (2015) Super4PCS: fast global pointcloud registration via smart indexing. Comput Graph Forum 33(5):205–215

16. Paulj B, Neild M (1992) A method for registration of 3-D shapes. Proc SPIE Int Soc Opt Eng 14(3):239–256
17. Yue F, Zuo W, Zhang D (2010) ICP registration using principal line and orientation features for palmprint alignment
18. Zhu Q, Wang Z, Hu H et al (2020) Leveraging photogrammetric mesh models for aerial-ground feature point matching toward integrated 3D reconstruction. ISPRS J Photogram Remote Sens 166:26–40
19. Yongjun Z, Xiaodong X, Mengqiu W et al (2014) A new aerial image matching method using airborne LiDAR point cloud and POS data. Acta Geodaetica et Cartographica Sinica 43 (4):380–388
20. Toschi I, Remondino F, Rothe R et al (2018) Combining airborne oblique camera and Lidar sensors: investigation and new perspectives. Int Archives Photogram Remote Sens Spat Inform Sci
21. Bódis-Szomorú A, Riemenschneider H, Van Gool L (2016) Efficient volumetric fusion of airborne and street-side data for urban reconstruction. In: 23rd international conference on pattern recognition
22. Fiocco M, Bostrom G, Goncalves JGM et al (2005) Multisensor fusion for volumetric reconstruction of large outdoor areas. International conference on 3-d digital imaging & modeling
23. Kolmogorov V, Zabin R (2004) What energy functions can be minimized via graph cuts? IEEE Trans Pattern Anal Mach Intell 26(2):147–159
24. Alexandre B, Renaud M (2012) Fast and robust normal estimation for point clouds with sharp features. Comput Graphics Forum 31(5):1765–1774
25. Nurunnabi A, Nasser M, Imon A (2016) Identification and classification of multiple outliers, high leverage points and influential observations in linear regression. J Appl Stats 43(1–4):509–525
26. Hu H, Ding Y, Zhu Q et al (2014) An adaptive surface filter for airborne laser scanning point clouds by means of regularization and bending energy. Isprs J Photogram Remote Sens 92 (2):98–111
27. Fischler M (1987) Random sample consensus: a paradigm for model fitting with applications to image analysis and automated cartography. Read Comput Vision 726–740
28. Lin L, Fan Y, Zhu H et al (2017) An improved RANSAC for 3D point cloud plane segmentation based on normal distribution transformation cells. Remote Sens 9(5)
29. Bisheng Y, Zhen D, Gang Z et al (2015) Hierarchical extraction of urban objects from mobile laser scanning data. ISPRS J Photogramm Remote Sens 99:45–57
30. Charalambos P (2013) A framework for automatic modeling from point cloud data. IEEE Trans Pattern Anal Mach Intell 35(11):2563–2575
31. Mederos B, Velho L, Figueiredo L (2003) Moving least squares multiresolution surface approximation. In: 16th Brazilian symposium on computer graphics and image processing (SIBGRAPI 2003), 12–15 Oct 2003, Sao Carlos, Brazil

A Fast and Robust UAV Images Mosaic Method

Guangrui Yu, Changliang Ha, Chunlin Shi, Lianbing Gong, and Lili Yu

Abstract Unmanned Aerial Vehicle (UAV) can respond quickly and provide image data at the first time in mapping support tasks, such as battlefield environment monitoring and natural disaster damage assessment. However, UAV is prone to fast flight speed and poor attitude stability, which brings a new challenge to image mosaic relying on feature matching. In this paper, an efficient and robust UAV image mosaic method is proposed. The main idea is to take graph theory and POS data to image mosaic process. Firstly, we estimate the range of aerial photography based on POS data, and construct the UAV images relationship diagram. Then, according to the attitude information, the images in relation set are divided into stable group and disturbance group, and the image classification connection graph is generated. By defining the weighted topological graph and constructing the minimum spanning tree, we search key frames and reference image. Finally, we derive the homography matrix between adjacent images according to imaging model. The mosaic image is generated after global registration and images fusion. Experimental results show that our proposed approach can solve the mosaic problems of narrow overlap and texture deficient images, and produce superior results in reducing projection distortion and shortening registration time consumption.

Keywords UAV images · POS data · Projection plane · Homography matrix · Mosaic

---

G. Yu
College of Mechanical Engineering, Dalian University of Technology, Dalian 116024, P. R. China

G. Yu · C. Ha
32023 Troops, Dalian 116023, P. R. China

C. Shi (✉)
61206 Troops, Beijing 100043, P. R. China

L. Gong · L. Yu
China Nuclear Industry Fifth Construction Co., Ltd., Shanghai 201512, P. R. China

L. Wang et al. (eds.), *Proceedings of the 7th China High Resolution Earth Observation Conference (CHREOC 2020)*, Lecture Notes in Electrical Engineering 757, https://doi.org/10.1007/978-981-16-5735-1_17

1 Introduction

With the increasing demand of geographic information, remote sensing technology has become a necessary high-tech means [1]. As a powerful supplement to satellites and manned aircraft, UAV has the advantages of flexible mobility, no casualties, little influence by weather, and high cost performance [2]. The remote sensing images are acquired by airborne aerial camera, which can directly reflect the most comprehensive surface information. However, due to the limitation of camera focal length, high resolution and flight height, the range of single image acquired by UAV is small. In order to monitor large target area and control the overall situation, we need to merge multiple images with overlapping range into a single scene image [3–7].

For a long time, as a main product of geographic information, image mosaic map is very important for monitoring large-area area. It can be used as a working map of task planning, strategic deployment and even map production. The key to image stitching is geometric transformation, which is to project different pose images onto the same reference plane [8]. At present, there are mainly three transformation methods, including digital surface model (DSM) based, ground control point (GCP) based, and image tiepoint based schemes [9]. Among them, DSM-based scheme requires aerial triangulation and point cloud network construction in advance, so that images can be corrected by three-dimensional terrain information [10]. Due to splicing process is rigorous and high-accuracy. DSM-based scheme is applied to produce digital orthophoto map (DOM) by current mainstream commercial software (Pix4d, Agisoft photoscan, Context capture). GCP-based scheme refers to the relationship between ground control points and corresponding image points, so as to realize global image transformation [11, 12]. When GCPs reach a certain number, high-precision geometric transformation can also be realized. Tiepoint-based scheme is a geometric transformation method between images, which uses matching point pairs to derive the transformation matrix [13]. It can be seen that although the DSM-based and GCP-based schemes can obtain accurate mosaic results, but they need accurate 3D surface information and sufficient number GCPs. Therefore, tiepoint-based scheme is more suitable for UAV rapid detection. However, all these image mosaic methods rely on feature similarity to match one image with another. In recent years, a series of classical algorithms have been developed, such as SIFT, SURF, CensurE, OGB, LDB. Subsequently, FAST, ORB and BRISK fast detection algorithm have emerged [14–16].

By combining with the random sample consistency (RANSAC) algorithm, the transformation estimation can be reliable [9]. After applying the global registration strategy, the error accumulation phenomenon can be effectively eliminated and the transformation matrix can be further optimized to achieve a wide range of image mosaic [17]. It can be seen that image registration cannot be achieved without the extraction of a large number of feature points, which is time-consuming, and it is difficult to achieve rapid response to natural disasters and emergency rescue; when the image overlap is narrow or contains large areas of water, forest, desert and other texture deficient areas, it is easy to make matching errors or even unable to splice.

Kim et al. [9] proposed an image mosaic method using resampling grid to solve the mosaic problem of narrow overlap UAV images. In the process of registration, using FAST and BRISK algorithms to extract image connection points, and distinguish them into model points and checkpoints. By setting corresponding points that vary with the overlap degree, the reliability of narrow overlap image registration is improved. Wu [18] proposed a texture adaptive extraction method to solve the problems of few connection points and difficult matching in the process of image mosaic in desert and other areas. Firstly, rough matching was completed by establishing image pyramid, then the original image was partitioned, and the entropy of gray level co-occurrence matrix is used to evaluate the image texture index. The research shows that this method can not only extract more connection points, but also ensure the uniform distribution of points.

Although the above research has solved the problem of image mosaic in special cases to a certain extent, it still does not escape the difficulties of feature point extraction and matching. Yahyanejad et al. [19] proposed a hybrid mosaic method based on metadata and image. Metadata method uses GPS, IMU and other auxiliary sensors to obtain the position and direction of camera, and then uses them to calculate conversion relationship. This method overcomes new problems of low altitude and small-scale UAV images, but the splicing process is not completely independent of the image features. Chen et al. [20–22] studied the image mosaic method based on pose and deduced the transformation matrix of two or more images registration process in order to solve the problem of slow speed of traditional image mosaic. Since there is no need to extract and match image features, the efficiency is greatly improved. However, due to the accuracy of POS data measured by GPS and IMU, the ground object dislocation is strict. And the selection principle of reference image is not considered, which results in the error accumulation phenomenon is not well eliminated.

In order to eliminate the dependence of image mosaic on gray and texture features, and reduce the error accumulation caused by image deformation, attitude disturbance and matrix operation, we study a fast and robust UAV image mosaic method which focus on two aspects: (1) Projection plane selection. In the image relation set, the image with large attitude angle is selected as the disturbance group, and the other images are taken as the stable group, and the image classification connection graph is generated. By defining the weighted topology map, the minimum spanning tree is constructed, and the key frame and reference image are found to ensure the minimum number of global image registration. (2) Images global alignment. Firstly, the high-precision POS data is used to restore the imaging model and deduce the homography matrix between adjacent images. Secondly, the global matrix of all images transform to the reference plane is derived from the transformation relationship between adjacent images; Finally, the panoramic mosaic image is generated by image fusion. The specific stitching process is shown in Fig. 1.

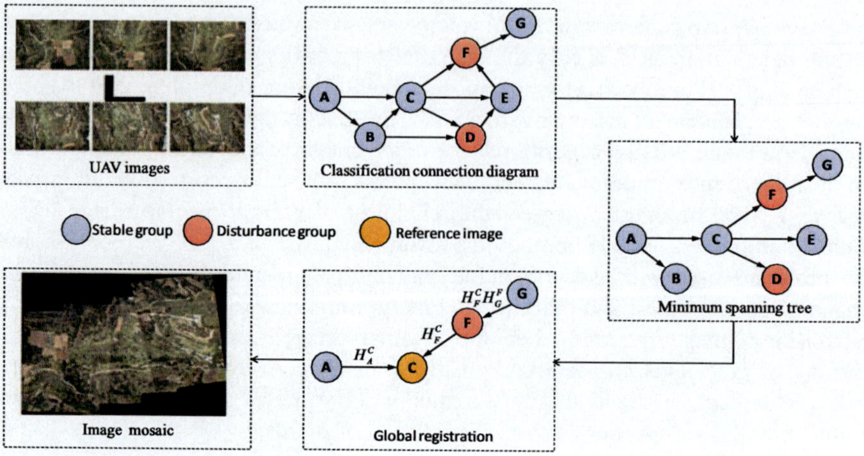

Fig. 1 Splicing process of UAV images

2 Key Frame and Reference Selection

Due to the light weight of UAV, it is prone to appear unstable flight state. When the image inclination angle is pretty large, this will cause the error that cannot be eliminated by the perspective transformation, and even cause the image perspective distortion in a large area. Especially for low altitude flight, the selection of reference image will directly affect the visual effect of mosaics. Therefore, in the process of image mosaic, it is crucial to select the appropriate reference image to be stitched.

2.1 The Influence of Image Attitude and Quantity on Mosaics

As shown in Fig. 2, two images are taken with the building as the object, one of which is shooting directly at the building, and the other is shooting at an angle relative to the building. From the results of two stitching, we can find that reference image will produce a great visual difference for the image mosaic. For UAV aerial photography, aerial camera is generally required to observe the earth vertically and obtain mosaic map based on the horizontal image.

In the registration process of large number images, each image is projected to the reference image plane according to certain rules, which will cause errors to accumulate gradually. That is, the farther image is from reference, the more times matrix operation is and the greater image deformation will be. Therefore, it is necessary to find the reference image which can minimize the total propagation error of the model.

(a) Take oblique image as reference (b) Take horizontal image as reference

Fig. 2 Visual effect of datum selection on mosaic

On the other hand, for the method of image registration by feature matching, the transformation matrix can be adjusted uniformly by adjustment optimization before splicing. In this paper, POS data is used to realize image mosaicking. Considering the high degree of image overlap in general, we try to reduce the number of images involved in stitching, so as to reduce registration times, and further achieve the purpose of reducing error accumulation. As shown in Fig. 3, eight aerial images of mission area were stitched. For the first time, the first image is used as reference (Fig. 3a). The second time, central image is used as reference to complete eight images mosaic (Fig. 3b). The third time, also taking central image as reference, four images were extracted to complete the mosaic (Fig. 3c).

It can be seen that when the first image is used as the reference for splicing, the error accumulation of the last image is obvious. By switching the reference, the local deformation of a single image caused by error accumulation can be effectively

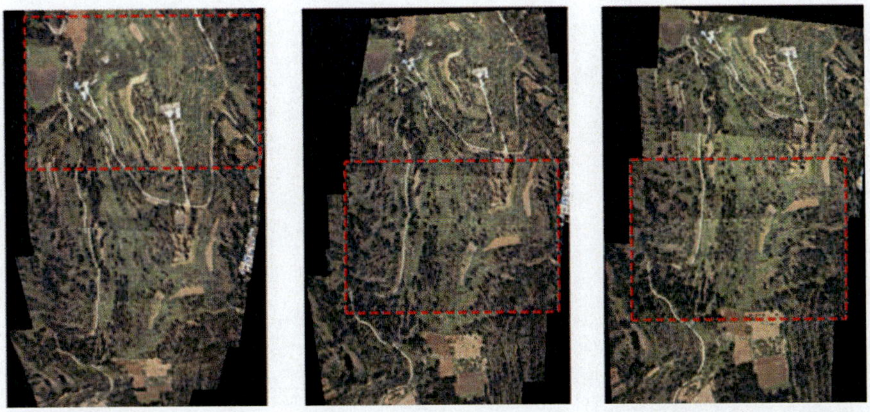

(a) (b) (c)

Fig. 3 Influence of image datum and quantity on mosaic error reduced. On the premise of satisfying the image coverage, reasonable reduction of the number of images to be spliced can further improve the image mosaic effect.

2.2 Image Relationship Construction

For a large number of UAV images, completing the construction of image relationship in advance is the foundation of image network construction. The image relationship includes the relationship map of all images and the relative relationship between two images that make up the stereopairs.

In this paper, image relationship is constructed based on POS data, the camera imaging model can be expressed as:

$$\lambda m = KR[I \quad -C]M \tag{1}$$

where m and M represent the coordinates of the target point in the image coordinate system and the world coordinate system respectively, λ is the scale factor, K is the parameter matrix constructed by the internal orientation elements (main distance and main point coordinate), R is the rotation matrix composed of attitude angle, C is the station coordinate, and I is the unit matrix. As shown in Fig. 4, we can calculate the four corner coordinates of two adjacent images and get the external rectangular area S according to Formula (1).

In order to eliminate the image pairs with narrow overlapping or small area, we use Formula (2) to filter.

Fig. 4 Image overlap range

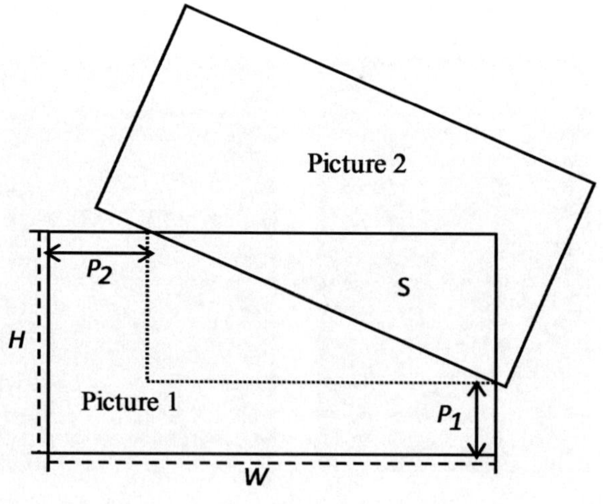

$$\begin{aligned} W - P_2 &\geq r_0 W \\ H - P_1 &\geq r_0 H \end{aligned} \quad (2)$$

r_0 is the overlap ratio. We can estimate image overlap according to formula (3).

$$S = \frac{(W - P_2) \times (H - P_1)}{W \times H} \quad (3)$$

After using POS to estimate the overlapping range of images, UAV image relationship diagram can be constructed. For a large number of images, only image pairs need to be selected from edge set for registration. Compared with exhaustive method, the computational efficiency of this method is improved from $O(n^2)$ to $O(n)$.

2.3 Image Classification

In order to avoid the large angle image becoming reference for mosaic process and reduce the influence of perspective error on global registration, UAV images are divided into disturbance group and stability group before splicing. As shown in Fig. 5, the light green rectangles represent stable group images, and the red rectangles represent disturbance group images.

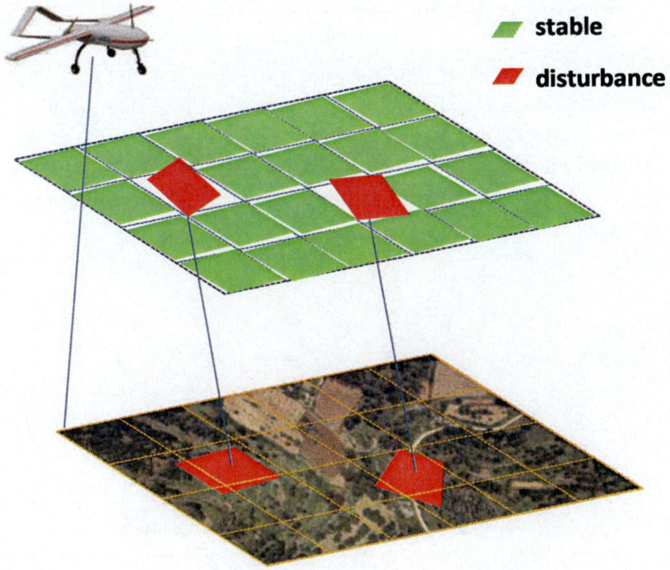

Fig. 5 UAV images grouping

Disturbance group means that the angle of a certain image exceeds the set threshold due to adverse factors such as wind speed and wind direction in flight process. Stable group refers to UAV in normal cruise state, image photographed is approximately horizontal, and inclination angle is less than threshold value. There is a large registration error between disturbance group images and surrounding stable images [23]. When images are spliced, we should try our best to avoid using disturbance group images, especially can't use it as reference.

In the field of reduce image registration error, this paper considers two points: One is to reduce the number of images involved in stitching, and the other is to avoid disturbance images participating in splicing. In the previous section, POS data has been used to determine the image relationship set $G = (V, E)$, so as to label two types of images, construct image classification connection diagram is shown in Fig. 6.

2.4 *Reference image selection*

In image relation set $G = (V, E)$, V is the vertex set, E is the edge set, and the weight of edges is set to w_{ij}, which indicates the importance of the connection between images. As shown in Eq. 4, we consider using the image overlap and the inclination angle of the newly added image to measure the weight of matching pair.

$$\begin{cases} w_{ij} = r_w w_p + (1 - r_w) w_\alpha, r_w \in (0, 1) \\ w_p = \frac{p_{ij}}{p_{\max}} \\ w_\alpha = \sin\alpha \end{cases} \tag{4}$$

where r_w represents the ratio of two weights w_p and w_α,p_{ij} and $p_{\max}$ represents the overlap degree between images and the maximum overlap degree of all images, α represents the maximum angle from elevation and roll. Therefore, the smaller w_{ij} is, the larger span of image connection is and the smaller inclination angle is. Thus, as shown in Fig. 7, we can use prim algorithm to construct the minimum spanning tree which traverses all the vertices.

Fig. 6 Classification connection diagram of images

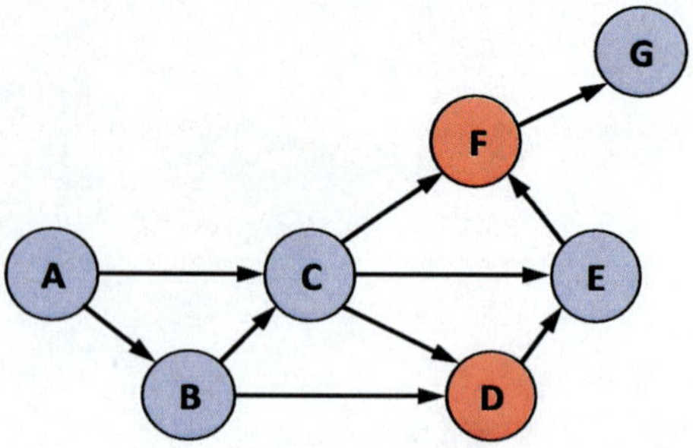

Fig. 7 Minimum spanning tree of images

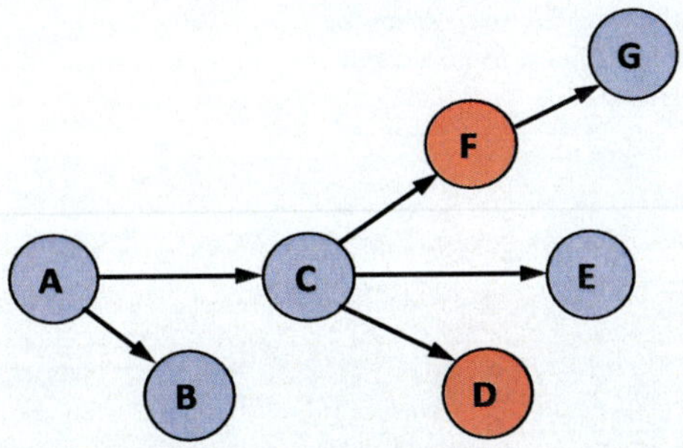

The minimum spanning tree determines the best image pair we need. After the first and last frames of the route are given, the shortest path of two images can be determined by breadth first search algorithm. For the minimum spanning tree of the above image, $A \rightarrow C \rightarrow F \rightarrow G$ is the minimum path and the only path, and A, C, F, G is the keyframe of image mosaic.

On reference plane selection, only the key frame images need to be stitched, so minimum spanning tree root node or geometric center cannot be used as the benchmark. In this paper, we use POS data for image keyframe stitching. The factors that cause error accumulation mainly include projection transformation times and disturbing group images. Set the weight of each edge as shown in Formula (5).

$$p_{ij} = \begin{cases} 1, & I_i \text{ and } I_j \in S \\ 1+\delta, & I_i \text{ or } I_j \in D \\ 1+2\delta, & I_i \text{ and } I_j \in D \end{cases} \tag{5}$$

S refers to stable group, D refers to disturbance group, and δ is the error adjusting factor. Using Floyd-Warshall shortest path algorithm, we obtain the optimal reference of stable group images in weighted graph, which is node C of above path. Figure 8 shows the minimum depth global registration with the reference image as root node.

For multi route stitching, the key frames can be acquired one by one to ensure that the range of surveyed route is covered by a small number of high-quality images. Then, the classification connection map is reconstructed for key frame

Fig. 8 Global registration of image keyframes

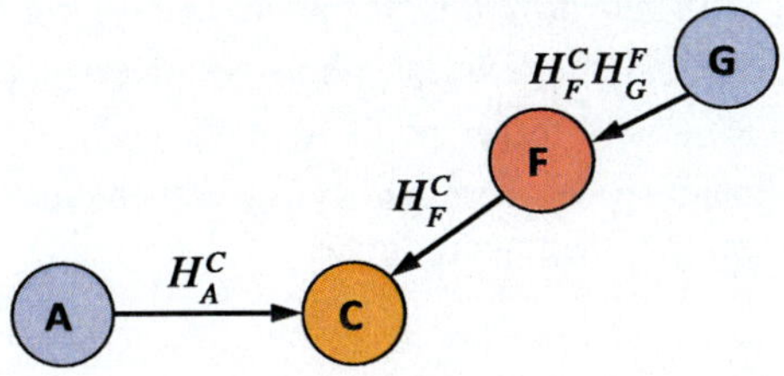

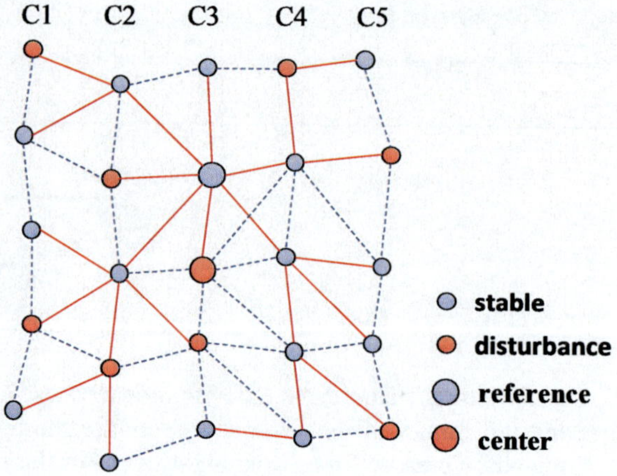

Fig. 9 Minimum spanning tree of multi route weighted graph

images of the whole survey area. The shortest path algorithm is used to set the weights in Formula (5) to construct the minimum spanning tree. Figure 9 shows reference images of five routes. Finally, global transformation of all nodes to the reference image is completed.

Therefore, we construct the minimum spanning tree for breadth first search, which can eliminate the large disturbance image and obtain the most appropriate image key frame. In the selection of reference image, it is different from finding the root node and geometric center point of all nodes, but finding the node with the shortest weighted path in the key frame image to ensure the minimum total error accumulation and make a good preparation for the next step of stitching.

3 Image Registration

POS data refers to six exterior orientation elements including camera station coordinates and rotation angles, which are generally obtained by GPS and IMU. The accuracy of POS data affects the processing process of UAV images. Differential UAV can obtain accurate station coordinates, but attitude acquisition is relatively low. When image feature points cannot be extracted and used, we can use POS to solve the image mosaic problem in special areas. Next, we discuss the derivation of homography matrix from POS data of images.

As shown in Fig. 10, aerial camera continuously exposed at space point O_1, O_2 to obtain image I_1, I_2 with certain overlap. $(X_W, Y_W, Z_W)^{\mathrm{T}}$ is the world coordinate of the ground target point A, $A_1(X_{c_1}, Y_{c_1}, Z_{c_1})^{\mathrm{T}}$ and $A_2(X_{c_2}, Y_{c_2}, Z_{c_2})^{\mathrm{T}}$ represent two image space coordinates of A respectively, $a_1(u_1, v_1, 1)^{\mathrm{T}}$ and $a_2(u_2, v_2, 1)^{\mathrm{T}}$ represent the image coordinates of A on image I_1, I_2 respectively.

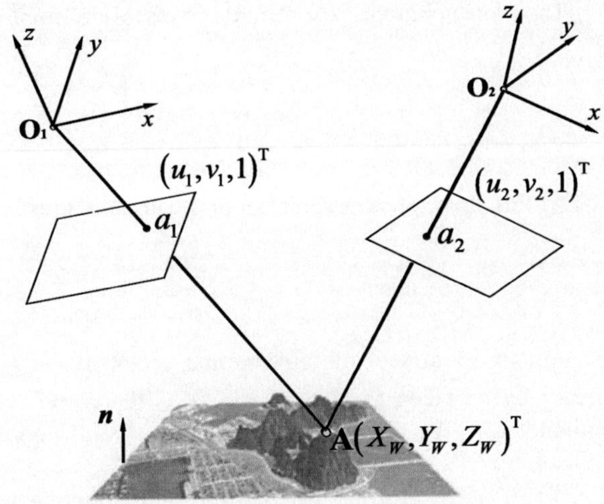

Fig. 10 Stereo relative imaging model

If we set image I_1 in Fig. 10 as world coordinate system, Formula (6) can be obtained according to the principle of photographic imaging.

$$Z_{c_1}\begin{bmatrix} u_1 \\ v_1 \\ 1 \end{bmatrix} = K\begin{bmatrix} X_w \\ Y_w \\ Z_w \end{bmatrix} \tag{6}$$

K is a parameter matrix calibrated by the camera, which can be expressed by the focal length f of camera, pixel coordinates (u_0, v_0) of main point, and pixel sizes dx and dy in the horizontal and vertical coordinates.

$$K = \begin{bmatrix} f/dx & 0 & u_0 \\ 0 & f/dy & v_0 \\ 0 & 0 & 1 \end{bmatrix} \tag{7}$$

UAV moves from O_1 to O_2, the position change is represented by t, which can be determined by coordinates of two stations (x_1, y_1, z_1) and (x_2, y_2, z_2). Rotation matrix is represented by R, which can be determined by the change angle (φ, θ, ξ) of pitch, roll and yaw.

$$t = \begin{bmatrix} x_2 - x_1 & y_2 - y_1 & z_2 - z_1 \end{bmatrix}^{\mathrm{T}} \tag{8}$$

$$R = R_\varphi R_\theta R_\xi = \begin{bmatrix} 1 & 0 & 0 \\ 0 & \cos\varphi & \sin\varphi \\ 0 & -\sin\varphi & \cos\varphi \end{bmatrix}\begin{bmatrix} \cos\theta & 0 & \sin\theta \\ 0 & 1 & 0 \\ -\sin\theta & 0 & \cos\theta \end{bmatrix}\begin{bmatrix} \cos\xi & \sin\xi & 0 \\ -\sin\xi & \cos\xi & 0 \\ 0 & 0 & 1 \end{bmatrix} \tag{9}$$

Therefore, photography formula at exposure point O_2 can be expressed as:

$$Z_{c_2}\begin{bmatrix}u_2\\v_2\\1\end{bmatrix}=KR\begin{bmatrix}X_w\\Y_w\\Z_w\end{bmatrix}-KRt \tag{10}$$

At this time, photographic ground can be expressed as:

$$\frac{n^{\mathrm{T}}}{d}X=-1 \tag{11}$$

Among them, n is the unit normal vector of the ground plane, flat ground can generally be taken as $\begin{bmatrix}0 & 0 & 1\end{bmatrix}^T$, X is the world coordinates of target point on ground, d is the distance from origin of world coordinate system to the ground plane.

By substituting Eqs. (11) into (10) and adopting ground point coordinates, we can obtain Eq. (12).

$$Z_{c_2}\begin{bmatrix}u_2\\v_2\\1\end{bmatrix}=KR\begin{bmatrix}X_w\\Y_w\\Z_w\end{bmatrix}+\frac{KRtn^{\mathrm{T}}}{d}\begin{bmatrix}X_w\\Y_w\\Z_w\end{bmatrix} \tag{12}$$

By combining Eq. (6) with (12), the relationship between image I_1,I_2 can be deduced.

$$\begin{bmatrix}u_2\\v_2\\1\end{bmatrix}=\frac{Z_{c_1}}{Z_{c_2}}KR\left(I+\frac{t}{d}n^{\mathrm{T}}\right)K^{-1}\begin{bmatrix}u_1\\v_1\\1\end{bmatrix} \tag{13}$$

$\frac{Z_{c_1}}{Z_{c_2}}$ is a non-zero scale constant, so the homography transformation matrix between two images is expressed as follows:

$$H=KR\left(I+\frac{t}{d}n^{\mathrm{T}}\right)K^{-1}=KRK^{-1}+\frac{1}{d}KRtn^{\mathrm{T}}K^{-1} \tag{14}$$

Global registration of multiple images is to calculate the transformation matrix of all images to reference. In image $I_1, I_2, \ldots, I_n$, if homography transformation matrixes between adjacent images are $H_{2,1}, H_{3,2}, \ldots, H_{n,n-1}$, reference image is I_m, the projection transformation matrix H_i of each image can be expressed as follows:

$$H_i=\begin{cases}(H_{i+1,i}\cdots H_{m,m-1})^{-1} & i<m\\ I & i=m \quad i,m=1,2\ldots n\\ H_{m+1,m}\cdots H_{i,i-1} & i>m\end{cases} \tag{15}$$

4 Experiment and Conclusion

4.1 Experimental Data

We choose FW100 UAV with Sony RX1R II camera. UAV flying altitude is 250 m, and image pixel is 6000 × 4000. Figure 11 and Table 1 show the image and POS data of a route.

4.2 Image Mosaic

We stitched images of above single route. The processing steps are as follows:

(1) Key frame selection and datum determination. The image relation set is constructed from POS data. At the same time, minimum spanning tree is constructed by taking 5° as boundary value between image stabilization group and

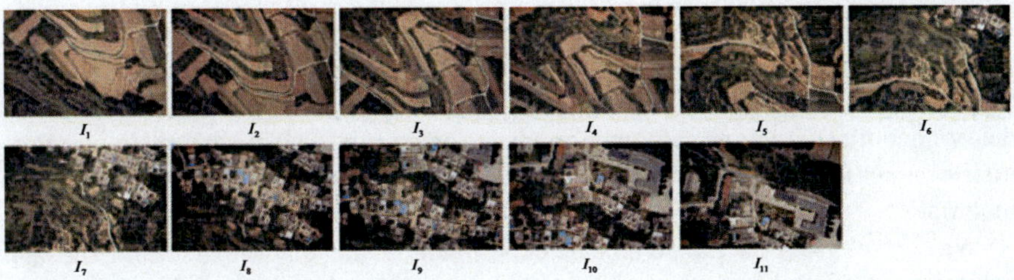

Fig. 11 Images of single route

Table 1 Images POS data

Image (ID)	Position (m)			Attitude (°)		
	X	Y	Z	α	γ	β
DSC08717.JPG	417,802.06	3,821,154.90	667.72	352.5149	−1.2157	−1.6986
DSC08718.JPG	417,802.85	3,821,211.70	666.31	353.7801	2.1644	0.2930
DSC08719.JPG	417,804.37	3,821,272.90	666.31	353.7995	2.5348	−2.1829
DSC08720.JPG	417,807.44	3,821,332.10	667.08	355.3409	−1.1544	−2.2079
DSC08721.JPG	417,809.64	3,821,391.22	664.81	351.4400	4.3015	−1.2022
DSC08722.JPG	417,811.29	3,821,450.04	665.88	351.0462	5.0474	1.8022
DSC08723.JPG	417,813.16	3,821,511.40	666.85	348.7402	2.5965	−0.2943
DSC08724.JPG	417,814.46	3,821,572.60	667.27	350.4567	5.7461	0.5010
DSC08725.JPG	417,816.47	3,821,631.37	668.03	353.1908	5.0337	−1.1509
DSC08726.JPG	417,818.06	3,821,693.41	668.04	352.8059	3.7119	−1.0458
DSC08727.JPG	417,819.75	3,821,754.82	668.46	351.9266	6.0306	−0.5097

disturbance group. Through longitude breadth first search, we determined that key frames of images to be assembled were DSC08717.jpg, DSC08720.jpg, DSC08722.jpg, DSC08723.jpg, DSC08725.jpg and DSC08727.jpg, with the name of $P_1, P_2, P_3, P_4, P_5, P_6$. According to the shortest path method, reference image is DSC08723.jpg, namely P_4.

(2) Image preprocessing. Firstly, Images need to be corrected. Then, in order to meet computer memory requirements, we process images to be assembled to 1/10 of the original size, and focal length is also scaled in the same proportion.
(3) Image registration. According to Eq. (14), we can calculate homography transformation matrix between adjacent images, and further realize the projection transformation from each image to datum by Eq. (15). The global registration results of the above data are shown in Table 2.
(4) Image fusion. After projection transformation of all images to reference, in order to eliminate "ghost image" and fuzzy phenomenon in the overlapping area, we use fusion method to achieve smooth transition between images, and then achieve good visual effect. In this paper, we use average fusion method.

After above steps, we completed key frame image mosaic based on POS, and obtained global mosaic image, which is called M-POS for short. Compared with the following splicing methods: (1) POS mosaic with all images is referred to as A-POS; (2) feature mosaicking involving all images is referred to as A-SIFT; (3) feature mosaicking involving only key frame images is referred to as M-SIFT. M-POS method determines reference image by the shortest path, A-POS method determines first image as reference, and the other two splicing methods use geometric center to determine reference. Figure 12 shows image mosaic results of each method.

We can see from the image mosaic experiment results that the first three mosaics achieved the overall coverage of the task area, and the last method failed to access the network. It can be seen that image mosaic can be realized by using POS information and feature matching. Without correction of GCPs, DSM and DEM, stitching effect of them depends on the accuracy of POS data and feature matching

Table 2 Image global registration

Global matrix	Result			Global matrix	Result		
H_1	0.8946	−0.0999	−6.6660	H_4	1.0000	0.0000	0.0000
	0.0354	0.8281	700.6113		0.0000	1.0000	0.0000
	0.0000	−0.0001	0.9511		0.0000	0.0000	1.0000
H_2	0.8791	−0.1423	52.8834	H_5	1.0277	−0.0547	26.9623
	0.0728	0.8209	383.3662		0.0765	1.0509	−315.3825
	0.0000	−0.0001	0.9810		0.0000	0.0001	1.0000
H_3	1.0623	−0.0180	−16.8938	H_6	1.0598	−0.0226	10.4021
	0.0694	1.0702	86.1175		0.0593	1.0857	−604.3944
	0.0001	0.0001	1.0069		−0.0000	0.0001	0.9727

(a) M-POS

(b) A-POS

(c) A-SIFT

(d) M-SIFT

Fig. 12 Comparison of UAV images mosaic results

respectively. Moreover, the error accumulation caused by POS information splicing method is obviously less than that of feature matching method. Compared with A-POS mosaic method, M-POS method can better control errors generation and propagation after improving the number and projection benchmark. A-SIFT and M-SIFT stitching methods have better convergence effect on image features with obvious adjacent texture features. When texture features are less or overlap is low, it will seriously affect the matching accuracy, and even cannot match.

M-POS mosaic method in this paper not only discards the interference of disturbing image in A-POS and A-SIFT methods, but also avoids the complex process of feature matching in A-SIFT and M-SIFT methods, effectively solves the limitation of image overlap and texture, and improves image mosaic efficiency. However, the dislocation of objects is still obvious in image mosaic, which shows that the mosaic method in this paper relies heavily on POS data accuracy and surface elevation. Therefore, in face of emergency support needs, or meet the image overlapping range is small, large area of water, desert, forest and other texture deficient areas, M-POS mosaic method can effectively solve image mosaic problem, and obtain a better panoramic mosaic image.

5 Conclusion

In this paper, considering the problem that UAV image quantity is large and pose is easily disturbed, we apply graph theory knowledge to image mosaic process reasonably. By establishing edge weights based on image inclination and overlap, we generate minimum spanning tree, determine image key frames, and get reference. Finally, through M-POS method in this paper, we can reduce cumulative error impact caused by the number and disturbance of images while ensuring the whole area coverage. In the process of image registration, M-POS mosaic method avoids complex process of feature matching, but uses precise pose parameters obtained by differential UAV system to derive homography transformation matrix between adjacent image pairs, and further calculates projection transformation relationship from the global image to reference. After image fusion, a large field of view images are obtained. Experimental results show that the proposed method completely avoids the complex process of feature point extraction and matching, and image registration efficiency is significantly improved. Compared with feature-based mosaic method and all images involved strategy, it can better reduce cumulative error. However, due to the influence of POS data acquisition accuracy and undulating terrain, there are inevitably errors in homography transformation matrix. As a result, some ground object dislocations will appear in the joint between adjacent images. Therefore, the mosaic method in this paper is more suitable for emergency needs, and can solve splicing problem of special aerial photography conditions. Higher precision POS system and necessary correction model will become the key to improve image mosaic effect in the future.

References

1. Li D (2008) Development prospect of photogrammetry and remote sensing. Geomatics Inf Sci Wuhan Univ 12:1211–1215
2. Li D, Li M (2014) Research progress and application prospect of UAV remote sensing system. Geomatics Inf Sci Wuhan Univ 39(05):505–513
3. Xu Y, Ou J, He H et al (2016) Mosaicking of unmanned aerial vehicle imagery in the absence of camera poses. Remote Sens 8(3):204
4. Forlani G, Dall Asta E, Diotri F et al (2018) Quality assessment of DSMs produced from UAV flights georeferenced with on-board RTK positioning. Remote Sens 10(2):311
5. Xiang R, Sun M, Jiang C et al (2014) A method of fast mosaic for massive UAV images. SPIE, Beijing, PEOPLES R CHINA
6. Ge Y, Wen G, Yang X (2016) A fast mosaicking method for small UAV image sequence using a small number of ground control points. IEEE, pp 90–94
7. Jia Y, Su Z, Zhang Q et al (2015) Research on UAV remote sensing image mosaic method based on SIFT[J]. Int J Signal Process Image Process Pattern Recogn 8(11):365–374
8. Su MS, Hwang WL, Cheng KY (2004) Analysis on multiresolution mosaic images. IEEE Trans Image Process 13(7):952–959
9. Kim J, Kim T, Shin D et al (2017) Fast and robust geometric correction for mosaicking UAV images with narrow overlaps. Int J Remote Sens 2557–2576
10. Jhan JP, Rau J, Huang C (2016) Band-to-band registration and ortho-rectification of multilens/multispectral imagery: a case study of MiniMCA-12 acquired by a fixed-wing UAS. ISPRS J Photogramm Remote Sens 114:66–77
11. Masek J (2009) Automated registration and orthorectification package for Landsat and Landsat-like data processing. J Appl Remote Sens 3(1):1–20
12. Toutin T (2004) Review article: geometric processing of remote sensing images: models, algorithms and methods. Int J Remote Sens 25(10):1893–1924
13. Brown M, Lowe DG (2007) Automatic panoramic image stitching using invariant features. Int J Comput Vision 74(1):59–73
14. Rosten E (2006) Machine learning for high-speed corner detection. ECCV, Graz, Austria
15. Rublee E, Rabaud V, Konolige K et al (2011) ORB: an efficient alternative to SIFT or SURF. ICCV, Barcelona, Spain
16. Leutenegger S, Chli M, Siegwart RY (2011) BRISK: binary robust invariant scalable keypoints. ICCV, Barcelona, Spain
17. Yaoyao Y (2012) Research on key technology of UAV image fast mosaic. PLA Information Engineering University
18. Wu X (2017) Research on UAV image positioning optimization technology. PLA Information Engineering University
19. Yahyanejad S, Wischounigstrucl D, Quaritsch M et al (2010) Incremental Mosaicking of images from autonomous, small-scale UAVs[C]. MA, USA, Boston
20. Gao X, Zhu S, Li R (2018) Image mosaic method of small UAV Based on pose information. Sci Surveying Mapp 43(2):70–76, 82
21. Cheng Z, Zhang L (2016) An aerial image mosaic method based on UAV pose information. Acta Geodaetica et Cartographica Sinica. 45(06):698–705
22. Fan J, Zuo X, Li Z (2018) UAV remote sensing image mosaic method based on homography matrix. Sci Surveying Mapp 43(06):99–105
23. Xie R, Tu J, Yao J et al (2019) A robust projection plane selection strategy for UAV image stitching. Int J Remote Sens 40(8):3118–3138

Comparative Research on Water Body Extraction Methods Based on SPOT Data

Linjiang Lou, Chen Chen, Xinyuan Gao, Kun Liu, Minmin Li, and Yajie Fu

Abstract With the rapid development of science and technology, it has become an inevitable trend to use remote sensing technology to monitor, investigate and analyze urban waters. The rapid and accurate extraction of water bodies from remote sensing images has become an important water extraction method. Based on SPOT4 images, this paper applied Ostu Threshold method and Maximum Likelihood Classification method to conduct experiments on water body extraction and compared the advantages and disadvantages of supervised classification and unsupervised classification in urban waters area extraction. Experimental results show that the Maximum Likelihood Classification method has the highest accuracy, and the Threshold Method has the fastest efficiency, but small rivers cannot be distinguished.

Keywords SPOT-4 · MNDWI · Water extraction method · Image segmentation

1 Introduction

Water resources play a pivotal role in national economic development. In recent years, global climate changes have become increasingly severe, ecological environment has continued to deteriorate, and disasters have occurred frequently. With the rapid development of remote sensing technology, it has been an inevitable trend to apply remote sensing data on water resources protection and monitoring, disaster analysis and early warning. The rapid and accurate extraction of water body information from remote sensing images is an important method and approach for water diversion monitoring and flood disaster assessment.

L. Lou · C. Chen (✉) · X. Gao · K. Liu · M. Li
Land Satellite Remote Sensing Application Center, MNR, Beijing, China

Y. Fu
Natural Resources Survey and Planning Institute of Shanxi Province, 3 Wangjing Road, Taiyuan, China

L. Wang et al. (eds.), *Proceedings of the 7th China High Resolution Earth Observation Conference (CHREOC 2020)*, Lecture Notes in Electrical Engineering 757, https://doi.org/10.1007/978-981-16-5735-1_18

In water extraction research based on satellite remote sensing, optical images are usually preferred because of their easier visual interpretation. At present, there are many techniques for water extraction, with two main types, i.e. supervised classification and unsupervised classification. McFeeter [1] proposed the Normalized Difference Water Index (NDWI) based on green and near-infrared bands by Landsat satellite images. The NDWI index presents good performance in extracting water regions. Xu [2] modified the NDWI proposed by McFeeter, and proposed the modified NDWI (MNDWI) by introducing the middle-infrared band. Shen et al. [3] proposed the use of Gaussian Normalized Difference Water Index (GNDWI) to extract river water models, so that the index can ensure the continuity of river extraction to a greater extent. Yuan et al. [4] proposed the method of adaptively determining the local range threshold based on the Otsu method, and applied the HJ-1A/1B satellite image NDWI to conduct water extraction experiments in the Minjiang River Basin. By applying the NOAA.AVHRR data and comparing single-band threshold segmentation, Tasseled Cap Transformation and the NDWI index, Jain et al. [5] upheld that the NDWI extraction result was the best. Cao et al. [6] proposed an object-oriented segmentation method which combined the spectral, morphology and textual features and could extract the water area excellently. Wang et al. [7] used U-Net Model and Random Forest Model to extract water information from GF-1 high-resolution remote sensing images based on which the water extraction results and efficiency of the two models were compared and analyzed. The water extraction results generated from the U-Net Model are more consistent with the manual visual interpretation results.

Therefore, the paper conducted a comparative study based on SPOT images using the common methods of supervised classification and unsupervised classification to analyze the effect on water body identification of the SPOT satellite remote sensing data as well as the advantages and disadvantages of the two methods mentioned above, which has provided an optimal strategy for SPOT images water body extraction.

2 Study Area and Data

Cagayan de Oro is a large port city of the Philippines located in the southern part of the Philippines. The central coordinates of the image of the test area are N8° 30′ 58‵‵N, E124° 40′ 51″E. The remote sensing data selected in this paper is SPOT4 optical image with the resolution of 20 m. The experimental data were acquired on July 14th 2011with band 2 (B), band 3 (G), band 4 (G) and band 11 (SWIR) (Fig. 1).

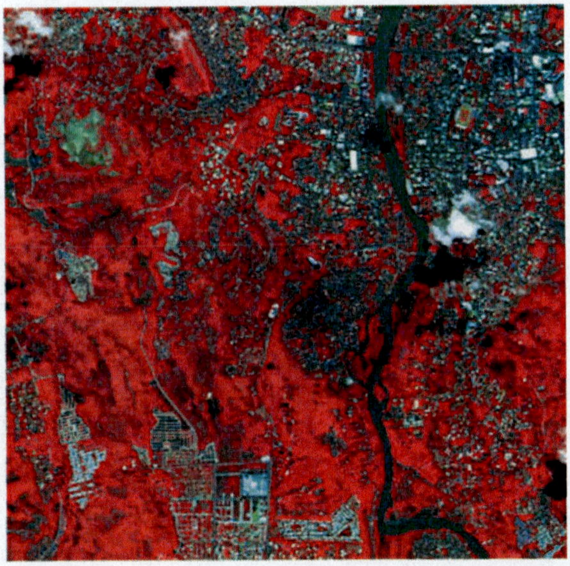

Fig. 1 Test area of image data

3 Material and Methods

To ensure the accuracy and speed of information extraction of the lake water, the images were preprocessed by radiometric calibration, atmospheric correction, orthorectification and cropping. The purpose of atmospheric correction is to eliminate the influence of such factors as atmosphere and illumination on ground objects reflection, invert the true reflectivity of ground objects, and improve the extraction accuracy.

(a) Improved normalized index

In this paper, the improved normalized index MNDWI could not only extract water bodies within the range of cities and towns, but also easily distinguish shadows and water bodies. Moreover it solved the difficult problem of removing shadows in water extraction. The calculation formula is as follows (Fig. 2):

$$\mathrm{MNDWI} = \frac{(\rho(\mathrm{Green}) - \rho(\mathrm{SWIR}))}{(\rho(\mathrm{Green}) + \rho(\mathrm{SWIR}))} \tag{1}$$

(b) Otsu thresholding algorithm

The Otsu thresholding algorithm is also called the maximum between-class variance algorithm. This standpoint is motivated by a conjecture that well threshold classes would be separated in gray levels, and conversely, a threshold giving the best separation of classes in gray levels would be the best threshold [8].

Fig. 2 MNDWI image

Suppose the pixels of a figure be represented in L gray levels, 1, 2, …, L. The number of pixels at level i is denoted by n_i and the total number of pixels by $N = n_1 + n_2 + \ldots + n_L$. The gray-level histogram is normalized and regarded as a probability distribution:

$$p_i = n_i/N \; p_i \geq 0 \quad \sum_{i=1}^{L} p_i = 1 \tag{2}$$

Dichotomizing all the pixels into two classes C0 and C1 by a threshold at level k, the probabilities of class occurrence and class mean value are given by:

$$\omega_0 = \sum_{i=1}^{k} p_i = \omega(k) \tag{3}$$

$$\omega_1 = \sum_{i=k+1}^{L} p_i = 1 - \omega(k) \tag{4}$$

and

$$\mu_0 = \sum_{i=1}^{k} ip_i/\omega_0 = \mu(k)/\omega(k) \tag{5}$$

$$\mu_1 = \sum_{i=k+1}^{L} ip_i/\omega_1 = \frac{\mu_T - \mu(k)}{1 - \omega(k)} \tag{6}$$

are the zeroth and the first-order cumulative moments of the histogram up to the kth level, respectively, and

$$\mu_t = \mu(L) = \sum_{i=1}^{L} ip_i \tag{7}$$

is the total mean level of the original figure.

The object function (the criterion measures) in the Otsu thresholding algorithm is:

$$\eta = \sigma_B^2/\sigma_T^2 \tag{8}$$

where

$$\sigma_B^2 = \omega_0(\mu_0 - \mu_T)^2 + \omega_1(\mu_1 - \mu_T)^2 = \omega_0\omega_1(\mu_1 - \mu_0)^2 \tag{9}$$

$$\sigma_T^2 = \sum_{i=1}^{L}(i - \mu_T)^2 p_i \tag{10}$$

The optimal threshold k^* that maximizes η is selected in the following sequential search by using the simple cumulative quantities:

$$\eta(k) = \sigma_B^2(k)/\sigma_T^2 \tag{11}$$

$$\sigma_B^2(k) = \frac{[\mu_T\omega(k) - \mu(k)]^2}{\omega(k)[1 - \omega(k)]} \tag{12}$$

and the optimal threshold k^* is

$$\sigma_B^2(k^*) = \max_{1 \le k < L} \sigma_B^2(k) \tag{13}$$

(c) Maximum Likelihood Classification

Maximum Likelihood Classification is an image classification method of establishing a set of nonlinear discriminant functions based on Bayesian decision criteria, assuming that various distribution functions are normally distributed, and selecting training areas for calculating the attribution probability of each sample area to be classified [9]. The discriminant formula is:

$$P(\omega_i/X) = \frac{P(\omega_i) \times P(X/\omega_i)}{P(X)} \quad (14)$$

$P(\omega_i)$ is the prior probability of the ω_i category, which indicates the probability of all pixels of the ω_i category in the classifier, i.e. the proportion of distribution area of each category in the image. The prior probability is of greater help in determining the accuracy of pixel category attribution at the category junction. In the formula, $P(X)$ represents the probability of pixel appearance, and its influence can be ignored. Therefore, the category of pixel X is jointly determined by $P(\omega_i)$ and $P(\omega_i/X)$. $P(\omega_i)$ and $P(\omega_i/X)$ change proportionally. Large prior probability will increase the pixel classification error, while the small prior probability will increase the pixel misclassification.

$$P(\omega_i) = \frac{P(L_i) \times U_{ai}}{P(a_i)} \quad (15)$$

By using the user accuracy U_{ai} obtained from the confusion matrix and the relation equations between the producer accuracy result and the analysis $P(a_i)$, the $P(L_i)$ is the probability of this category in the classification result, and $P(\omega_i)$ is the probability of appearance of ω_i category in the image.

4 Result Analysis

This paper applied Otsu method. It can be found in the histogram that in general the image has a multi-peak distribution and the peaks of the water body are relatively small. Yet, considering that the MNDWI water body value was greater than 0, the experiment used the domain values greater than 0 to perform segmentation experiments, with 0.23 segmentation threshold. The experimental results showed that Ostu method could not extract small rivers and could not distinguish the shadows caused by mountains and clouds, with miss-detections (Shown in Fig. 3a).

Furthermore, in order to verify the feasibility of the Maximum Likelihood Classification method, the paper firstly collected samples of water bodies, buildings and mountains, and conducted experiments using Maximum Likelihood Classification. The experimental results showed that the Maximum Likelihood method could effectively avoid miss-detections (Shown in Fig. 3b).

In terms of target detection, the detection results cover the following four situations: situation A means that there are no pixels in the reference image and no pixels are detected in the detection image; B means that there are no pixels in the reference image but there are pixels detected in detection image, which is generally

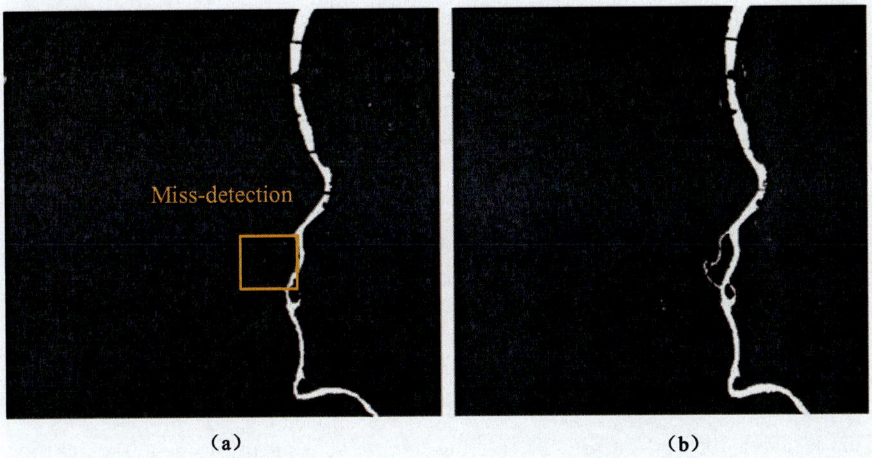

Fig. 3 **a** Segmentation image of Ostu threshold method, **b** segmentation image of maximum likelihood method

called false alarm; C indicates that there are targets in the reference image but they are not found in the detection image, which is called missed detection; D indicates that targets are found both in the reference image and in the detection image, which means the targets are successfully detected [10]. The detection rate P_D and the false alarm rate P_{FA} are defined based on the above four situations.

$$P_D = \frac{D}{C+D} \tag{16}$$

$$P_{FA} = \frac{B}{A+B} \tag{17}$$

To compare the detection performance among different thresholding methods quantitively, we introduced the index of Figure of Merit [11], which is expressed:

$$FoM = \frac{N_{td}}{N_{fa} + N_{gt}} \tag{18}$$

where N_{td} is the number of detected pixels, while N_{fa} is the number of false-alarm pixels. The larger the value of FOM, the better the detection ability.

It can be found in Table 1 that the overall detection rate can reach more than 75%, and the FoM performance are all above 99%. Among them, the Maximum Likelihood Classification method has the best performance, with 99.5% of FoM, and given the pre-production process of samples, it has longest running time. The Otsu Threshold method has the fastest extraction result and high accuracy, and the K-means Clustering method has the highest misclassification and the lowest performance.

Table 1 Comparison among different methods

		Ostu		Maximum likelihood	
		Others	Water	Others	Water
True	Others	244,345	111	244,311	145
	Water	1158	4386	977	4567
Detection rate		0.7911		0.8238	
False-alarm rate		0.00045		0.00059	
FoM		0.9945		0.9949	
Running time		1.09 s		30 s	

5 Conclusion

This research focused on SPOT4 satellite remote sensing image data, the solution on how to choose supervised classification and unsupervised classification to extract water bodies, and also applied Otsu method and Maximum Likelihood Classification method in method selection and extraction efficiency and effects test. The following conclusions are obtained:

Based on SPOT4 satellite images, the extraction performance of the Maximum Likelihood Classification method and the Threshold method are similar. When an emergency flood occurs, the Threshold method has the best effect and fast running time, and can quickly extract and analyze the disaster water areas. When extracting general water bodies, the Maximum Likelihood Classification method can effectively circumvent the problem of the indiscernibility of small rivers and the misclassification of shadow areas by the Threshold method, and it has a more accurate extraction of water bodies. Moreover, regarding the extraction efficiency, Otsu method performs better than Maximum Likelihood Classification method.

In summary, both of the two extraction methods have sound effects on water body extraction, with the unsupervised classification and supervised classification having different advantages and disadvantages. At the same time, how to reduce the classification omission rate and false alarm rate and better improve the extraction accuracy will be the key research direction in the future.

References

1. McFeeters SK (1996) The use of the normalized difference water index (NDWI) in the delineation of open water features. Int J Remote Sens 17(7):1425–1432
2. Xu HQ (2005) A study on information extraction of water body with the modified normalized difference water index (MNDWI). J Remote Sens 5:589–595
3. Shen ZF et al (2013) Automatic and high-precision extraction of rivers from remotely sensed images with Gaussian normalized water index. J Image Graph 18(4):421–428
4. Yuan Z et al (2016) Extraction of water body information using adaptive threshold value and OTSU algorithm. Remote Sens Inf 31(005):36–42

5. Jain SK et al (2006) Flood inundation mapping using NOAA AVHRR data. Water Resour Manage 20(6):949–959
6. Cao K, Jiang N, Li X-G (2007) The extraction of water information in urban areas based on SPOT 5 image using object-oriented method. Remote Sens Land Resour (2):27–30 [In Chinese]
7. Wang N et al (2020) Application of U-net model to water extraction with high resolution remote sensing data. Remote Sens Land Resour 32(1)
8. Otsu N (1979) A threshold selection method from gray-level histograms. IEEE Trans Syst Man Cybern 9(1):62–66
9. Tan KC, Lim HS, Matjafri MZ (2011) Comparison of neural network and maximum likelihood classifiers for land cover classification using landsat multispectral data. Open systems. IEEE
10. Beck JR, Edward KS (1986) The use of relative operating characteristic (ROC) curves in test performance evaluation. Arch Pathol Lab Med 110(1):13–20
11. Foulkes SB, Dooth DM (2000) Ship detection in ERS and Radarsat imagery using a selforganising Kohonen neural network. In: Proceesings of nova scotia conference on ship detection in coastal waters. Digby, NS, Canada

Research on National Land Resource Accounting Method

Xinyan Zheng, Yu Zhou, Tao Cheng, Jin Liu, Chen Chen, and Lu Yi

Abstract The accounting on national natural resources is the basis of acknowledgement of the asset value of our national resources, which is also the important criterion on the supervision, examination and management of national resource assets owned by the whole people of China. Land resource is not only the significant component of the natural resources, but also the spatial carriers of the natural resources such as mineral, forest, grass, wet land and water. In recent thirty years, the implementation of the system of paid use of land has made the value of the assets of the land resources more and more explicitly, which has laid more and more economic and social significance on the state governance, macro economy, national wealth, and the property owners. The content, method of accounting on land resources is the key point of the accounting on natural resources. This essay brings up the accounting method based on the physical quantity and value of the land resource assets. The physical quantity accounting could be relied on the basic data of the accounting area, and be calculated based on the forward operator, backward operator, and forward and backward operator method. The value accounting, based on the benchmark land price of the accounting area, could be measured on the comprehensive value on the whole accounting area weighed by the area at different levels. Though this accounting method is practicable, the problems in the management process need to be taken seriously. The conclusion is that the quality of the national land resource asset accounting relies on the solid data fundamentals and the relatively matching managing system.

X. Zheng · T. Cheng · J. Liu
National Geomatics Center of China, 28 Lianhuachi West Road, Haidian District, Beijing 100830, China

Y. Zhou (✉) · L. Yi
Guangdong Provincial Institute of Land and Resource Surveying and Planning, Unit 6, Building 1, No. 468 Huangshi East Road, Guangzhou 510030, Guangdong, China
e-mail: gdghy_pjs@gd.gov.cn

C. Chen
Land Satellite Remote Sensing Application Center, MNR, 1 Baishengcun, Haidian District, Beijing 100048, China

L. Wang et al. (eds.), *Proceedings of the 7th China High Resolution Earth Observation Conference (CHREOC 2020)*, Lecture Notes in Electrical Engineering 757, https://doi.org/10.1007/978-981-16-5735-1_19

Keywords Land resource assets · Asset accounting

1 Introduction

Ever since 1950s, many countries, such as Canada, Norway, Finland, France, Indonesia and Mexico etc., have gradually started their individual natural resources accounting practices. Before 2000, the research on natural resource accounting was mainly discussed in academic circles in China. Ever since the release of the Overall Program of the Ecological Civilization Reform by the Chinese government in 2015, the natural resources balance sheet was brought up explicitly, the guide book of natural resources balance was established, the accounting on natural resources such as water, land, forest etc. was specified, the account of the quantity of physical objects was created, and the classification and statistical specification was set up, in order to evaluate the changing state of the natural resources. This has signified that China has officially launched the work on the accounting on natural resources and establishment of the balance sheet of the natural resources [1].

The newly Ministry of Natural Resources of the People's Republic of China has just established in order to carry out all types of natural resources asset owner responsibilities for all people around China, unified survey and registration, establish a system of paid use of natural resources, solve the problem of the owner of natural resources, overlapping of spatial planning, and further strengthen the properties of natural assets.

Land is not only the significant component of natural resources, but also the spatial carrier of the natural resources such as mineral, forest, grassland, wet land, and water etc. The natural and economic nature of land has already become the critical factor of related economic system that can't be neglected, which has significant economic and social meaning not only for the macro economy, common wealth, but also for the owners. Therefore, the land resources could be defined of property rights, commercialized, based on the resource properties. The accounting on land resources has laid solid basis on the following natural resources accounting for all the people and balance sheet establishment. This essay focused on the accounting theory and method of the land resources, brought up the potential problems with the management view, and gave some suggestions for the better system of the land resource management system to improve the accounting quality.

2 National Land Resource Accounting in Canada

In 1956, the Canadian economist Anthony Scott thought that natural resources should be given more significance in the gross national welfare, and their annual change should be reflected in the national accounts. Entering 1990s, with the symbol of publishing the 1993 version of SNA, environmental and resource accounts are internationally officially taken into the national account framework. Nowadays, of all the mature developed economy, Canada Statistical Bureau has established dedicated National Environmental and Resource Account System, and was released official guidebook in 2006 which had introduced the accounting principle and method of the accounting on the stock and value of the natural resources.

The Canadian System of Environmental and Resource Accounts (short for CSERA) mainly include three accounts: the first one is Natural Resource Stock Accounts, which mainly records all different types of physical and monetary accounts of natural resources and also the change accounts because of natural factors or productivity. This is the basis of national resource balance sheet of Canada; the second one is Material and Energy Flow Accounts (short for MEFA), which mainly record the material flow between environment and economy and the material and energy cross the entire economy; the third one is Environmental Protection Expenditure Accounts (short for EPEA), which record the expenditure in the environmental protection in all the departments in national economy [2].

Canada natural resource stock account records the annual stock of natural resource, which could be divided into three accounts, as underground resource, forest and land accounts. According to the difference of the unit of the measurement, the accounts could be divided into physical and monetary accounts. Land resource is compiled as a special national resource stock account, using more detailed spatial framework based on the significance of geographical location in the land resource characteristics, which is updated mainly in 5 years via the population and agriculture census, and some indicators might be updated two to three years. Nowadays, Canada NRSA only evaluates on the agriculture, residential and commercial lands by the market price approach. A comparison is made between China and Canada on the natural resource accounting as the following Table 1 [3]:

3 Objects and Content of the Accounting

The clear concept of land resource assets is a significant prerequisite of the accounting of the land resource assets. Land resource assets are those that have already been used by human beings and could be used in the future and could bring forward the economic benefits for the owners of the property rights [4].

Table 1 Comparison between China and Canada on the natural resource accounting

	China	Canada
Accounting subject	Central and local government	Canadian Statistical Bureau
Accounting area	The natural resources which could have economic value	Natural Resource Stock which could have economic value
Accounting classification	Forest, grass, mineral, land, energy, water resources	Natural Resource Stock, land and ecological system
Measuring method	Present earning value method	Price and discount methods (Net Present Value); market value method

The accounting of national land resource assets indicates the work of statistically checking and measuring the total amounts and structure change and reflecting the balance situation of land resource owned by all the people according to constitution and law at certain time and space, in terms of physical, value and quality, based on the reasonable evaluation [5]. National land resource asset accounting includes economic value accounting, ecological value accounting and social value accounting. Considering the ecological and social value is hard to measure through market, the land resource accounting only includes the physical and asset value accounting. The physical accounting is the basis of the asset value accounting, and could directly indicate the amount, quality, location and change according to the unit of physical quantity in certain amount of time. The physical account, which is based on the land resource investigation, is the survey, statistics, surveillance and evaluation of the objective entity. The value account, which is given price based on the physical account, could reflect the total indicate of the land resource asset, and give a comprehensive evaluation to the land resource change process.

The natural resources owned by all the people include land, mineral, forest, grassland, wetland, water and sea. The object of national land resource asset account is the national land, which includes agriculture land, construction land and unused land. Considering the overlap relationship between the land resource and forest, grassland and water, the account objectives don't include those who directly overlap.

The land classification is significant because it will influence the beginning and ending account of the year of the physical account, which is the core of the physical account, and ultimately influence the value account. In the past, the natural resources were only mainly classified as mineral, water, land resources etc. by the scholars in China, which lack of a relatively more detailed classification. With the reorganization of the Ministry of Natural Resource of the People's Republic of China and the promotion of the Third National Land Survey of China, the more detailed classification more detailed land survey data will greatly benefit the physical account by providing more precise and detailed land data.

4 Accounting Method

4.1 Physical Accounting Method

In order to account physical national land, the area of the national land should firstly be settled down by the data of registration, supply and reservation, then mark the area of the national land accounting, and make statistics on all types of physical volume according to the land use.

The area of national land could be set by adding up the property data or deducting the collective land or combining these two methods:

(1) Forward solution to statistically set the national land area: extract the all the registration data of national and collective land, national land-use right registration data, cadastral survey data, real estate data, national land supply data, land reservation red-line data etc., spatially combine them together to become the original area of national land accounting. Besides, the controversial land area (between nation and community) should be filtered in the accounting area, and be marked as the controversial land. The national-owned but not registered land, which include unregistered town residential land, unregistered public service infrastructure land, unregistered special land, unregistered railway etc., should be supplemented by referring to the reliable survey data.
(2) Backward solution to statistically set the national land area: firstly extract the community-owned land from the spatial data, such as community-owned registration data and village industrial park etc., and spatially merge together to become the area of community-owned land. Besides, unregistered common public service infrastructure lands, unregistered village roads, unregistered town roads will be merged into the area of community-owned land. The remaining area that deducted the community-owned land from the administrative area would be considered the national land accounting area. At last, add the state-owned but not registered land from the survey data.
(3) Forward and backward combined method to statistically set the national land area: the overlapping area based on overlaying the area that are conducted by the forward and backward method will be considered the national land area, the un-overlapped parts will be investigated on site or refer to the database to make sure the land ownership. The community-owned land will be deleted, and the remaining area will be the land of uncertain ownership and marked the controversial land or uncertain ownership land.

The patches of the physical volume accounting analyzed by overlaying of the national land area calculated by forward, backward, forward–backward combined methods separately with the Third National Land Survey data will be accounted respectively by agriculture land, building area, grassland, water body, ditch, village residential land etc., and be collected together to become the national land resource physical volumes for all types of land, which will be the basis of value accounting.

4.2 Value Accounting Method

Nowadays, different countries have controversial understanding on the concept of value, and no common consensus has been accomplished. Thus, the value of natural resource assets that are calculated by different countries varies from each other. The value accounting gives important guidance to the physical accounting, while the reliable physical accounting is the basis for the value accounting.

Prof. Li [5] thought that the value of natural resource asset should include two parts: one is that the potential social value is contained within the natural resource asset itself; the other is the current social value within the labor of human society of the natural resource asset.

Currently, in practice, the value of national land volume is measured on limited-year land price, whereas the characteristics of the rights of the year for all types of land mainly rely on the value based on statutory highest year of the complete land use rights. National land value accounting includes state-owned agriculture land value account and state-owned construction land value account, which adopt the area-weighted method at different levels to measure the basic comprehensive land price in the accounting units.

5 Problems of the State-Of-Art Accounting

Nowadays, as all provinces in China are initiating the national land resource asset accounting, they all have accumulated various experiences on the accounting. As common standards have not come into consensus, and accounting methods vary in different provinces, there are no comparison in different provinces, and many problems also reveal as follows:

(1) **Problems in fundamental data**

As fundamental data at national, provincial, city and county level are the basis for all the accounting, the precise, common-standard, common format fundamental data are significant. However, the status-in-qua spatial data in different provinces might not match with each other; the statistical data may vary from each other because of the different meaning of the same content; some places are still lack of key data.

Fortunately, more and more precise and detailed fundamental data are now being produced by the Chinese government, such as the Third National Land Survey of China data, China's National Geographic Conditions Census data etc. These data will greatly benefit the accounting work.

(2) **Problems in physical account**

Currently, the most obstacle is the different agreement on the accounting principles, as the old land survey and land registration could not provide the precise property boundary, which makes difficult for the connection of results with physical volume and annual land use change survey. Therefore, the physical account method should emphasis on getting close to the reality, and convergence with other results.

(3) **Problems in value account**

According to the current practical experience in some province, the land price is the most significant problem. If the price information management is not so standard, for example, no unified baseline land price database standard, the comparison of all the different cities is of no meaning. Currently, the land price is influenced at a certain level by administrative management, thus the objectivity and rationality of the land price is also influenced and damaged at some point.

(4) **Problems in land resource management**

With the practices in the process of land resource asset accounting, some flaws are discovered in the work of land resource management: firstly, various amount of key attributes information are missing, the red-line areas of the parcels are in discord with practical land use area, the current time phase of the parcel is not updated etc.; secondly, data is missing or none or not complete, such us the lack of information on the registration data, lack of spatial data, lack of history data etc.

6 Suggestions on the Accounting

As the population and the people's needs grow, the contradictions between man and nature are getting increasingly sharp. Protecting natural resources and building ecological civilization have become the common topic of human being's sustainable development. Chinese government has attached highly importance to build the ecological civilization by initiating the natural resource asset accounting, hoping to realize the goal of sustainable development and protection of natural resources. The fundamental goal of implementing national resource asset accounting is to realize the sustainable utilization and protection of natural resources. By knowing the different experiences in different provinces on the accounting of land resource assets, the following suggestions and conclusions are brought out:

(1) **Initiating the national resource stock account is the necessity of our national economical account system development**

The lack of environmental resource account, especially natural resource stock account, will directly result in lack of complete control of the current status and future of our national natural resource. Under the new background of obvious energy resource constraint, the decision making and the mid and long term implementation will be greatly influenced. At the same time, the lack of natural resource stock accounts will not be beneficial to the current assets liability of our nation and government. Overall, by the related conduct of those developed countries, such as Canada, it is a necessity to compiling our national resource environmental account as soon as possible.

(2) **Physical account is the important prerequisite of the natural resource stock account system**

According to the introduction of Canadian natural resource stock account, we get to know that physical account is the prerequisite of evaluating the value of the stock on one hand, and the basis of conducting related prediction research on the other hand. As the property data management is in disorder, historical remained problems are severe.

(3) **More detailed and precise natural resource classified data is important for the physical account**

As more detailed and precise natural resource classified data will make the data of the initiating and ending of the year of the physical account more precisely. The data of the Third National Land Survey data will be of great help to the physical and value accounting. With these data, the survey and calculation will be of the same standard and same classification at national, provincial, city and county level.

(4) **The land resource management needs to be promoted**

Ever since the establishment of China, with the important tasks in different era, the natural resources, such as land, mineral, water, forest, grassland etc., are managed by different administrations, which led to the missing registration information, the different data standards and formats on the same features, spatially overlaid features etc. The newly organized Ministry of Natural Resource of China is the department for all the natural resources, which solve the managerial problem from multiple management to the single management. This will greatly benefit the solving of the problems such as information island problem, historical property owner missing problem, spatial planning overlapping problem etc.

(5) **Research on establishing the national land resource asset accounting system for China**

As different countries vary in the national land resource asset accounting systems, China needs to establish its own system to cater the situation in China, which will have the common technique and accounting rule framework. Only under this common framework does the physical and value account could be of real true value for the sustainable development and ecological protection.

References

1. Zhu DL, Zhang H, Duan WJ, Du T (2019) Discussion on the logical rules of natural resources asset accounting and the methods of land resource asset accounting. J China Land Sci 33 (11):1–7
2. Lei JX, Wang Y (2016) Accounting method and enlightenment on Canada's natural resource asset account. J Finance Econ 01:34–40
3. Liu L (2019) Comparison and enlightenment of Chinese and foreign natural resource assets accounting. J New Probe Theory 3:9–12
4. Zhu DL, Du T (2017) Accounting method and result analysis of cultivated land resource asset in China. J China Land Sci 10:23–31
5. Li JC (1992) Environmental value and economical accounting. J Environ Prot (7)

A Configuration Approach and Effectiveness Evaluation of Modular Reconfigurable Satellite

Fu Xin, Mai Xiamei, Zhang Haolong, and Xue Fengtong

Abstract Modular reconfigurable satellites are characterized by low cost and short cycle of development, launch, operation and maintenance, which can maximize the interests of developers and operators while meeting the needs of users. However, this field is still in its infancy and has huge market potential. On the basis of the domestic and international research on modular reconfigurable satellite structure and technology, this paper proposes a new configuration of modular reconfigurable satellite. The key technologies, operation and maintenance modes, and commercial application prospects are studied according to the configuration. In this paper, a modeling of this modular satellite configuration is established in a simulation system, and its performance in earth observation application scenarios is evaluated by means of simulation data and evaluation index system. The evaluation results show that this configuration can improve the effectiveness of modular reconfigurable satellites.

Keywords Modular · Reconfigurable satellite · Configuration · Effectiveness evaluation

1 Introduction

With the development of modular reconfigurable technology, the traditional model of satellite development, launch, operation and maintenance is gradually being overturned. In terms of research and development, modular reconfigurable satellites can be mass-produced and standardized. In terms of launch, they can be lifted off through "free ride" launches or "space bus" and then assembled and placed into orbit by robots or service systems. In terms of operation and maintenance, the

F. Xin (✉) · Z. Haolong · X. Fengtong
China Academy of Aerospace Systems Science and Engineering, Beijing 100048, China
e-mail: fuxinyymy@163.com

M. Xiamei
The Earth Observation Systems and Data Center of CNSA, Beijing 100101, China

L. Wang et al. (eds.), *Proceedings of the 7th China High Resolution Earth Observation Conference (CHREOC 2020)*, Lecture Notes in Electrical Engineering 757, https://doi.org/10.1007/978-981-16-5735-1_20

modular reconfigurable satellites can extend their life continuously through on-orbit module maintenance, reconfiguration and upgrading operations. Therefore, modular reconfigurable satellites have the advantages of low cost and short cycle of development, launch, operation and maintenance, which can maximize the interests of developers and operators while meeting the needs of users. Overall, the modular satellite technology is still in the research and validation stage and the business model is still in its infancy with great market potential.

A modular spacecraft is a fast-response spacecraft that is functionally independent, physically independent and integrated through standard interfaces to achieve the functions of the entire spacecraft system [1]. The configuration of a modular reconfigurable satellite is a complex system program due to its wide ranges of professional fields and high technical difficulties and directly related to the target mission. The configuration of modular reconfigurable satellite needs to take users' requirements as input, select corresponding modules from the standard modules, and make calculations on attitude control, power balance, interface coordination and other aspects in order to achieve the optimal structure and functions of the satellite.

2 Development of Modular Reconfigurable Satellite

2.1 Foreign Status and Development

The development of modular satellites can be roughly divided into four stages: origin, initial development, rapid development and industry refinement [2].

(1) Origins stage

Research on modular satellites originated in 1980, when the United States began to develop the Hubble Space Telescope (HST). Due to the limited launch conditions and the harsh environment in orbit at that time, the optical, electrical, thermal and configuration design of HST need to meet harsh requirements. Some countries, mainly the United States, began a series of explorations on modular satellites [3].

(2) Initial development

From the 1980s to around 2000, the rudiment of modular satellites emerged and underwent initial development with NASA Goddard Space Center, Mosher L, Edward and other scholars conducting a series of studies on multi-mission modular spacecraft (MMS) [4]. As research progresses, DARPA and some European scholars carried out researches on the definition of modular satellites.

(3) Rapid development

After 2003, research on modular satellites has entered a rapid development stage, gradually moving from concept and subsystem research to whole-satellite development and series planning. During this period, Japan's New Energy Development

Organization (NEDO) launched a five-year panel extension satellite (PETSAT) program and Tokyo Institute of Technology proposed a reconfigurable robotic satellite constellation [5]. The United States Air-Force Laboratory has developed the modular satellite as PnPSat with plug-and-play capabilities and auto-configuration system. DARPA successfully demonstrates the validation of the "Orbital Express" program with the Future Fast, Flexible, Fractionated, Free-flying Satellite as Program F6 program. The MIT-led SWARM program investigates modularity, wireless communications and standard interfaces [6–10].

(4) Industry refinement

In the last three years, the development of modular reconfigurable satellite technology has been gradually refined to meet industry needs and adapt to different application scenarios.

In military, the United States issued the "Black Jack" low earth orbit constellation system program in 2018. The American Aerospace Corporation announced in 2018 the "Hive" on-orbit autonomous assembly modular spacecraft platform concept composed by mass production, small, intelligent unit module on-orbit which could self-assembly and reconfiguration, in the event of threats which can be disintegrated to achieve self-protection. In 2020, DARPA "geosynchronous orbit satellite robotic services" program completed payload design review.

In scientific research, the experimental device of NASA's "Robotic On-orbit Refueling Mission 3" program has completed the technical validation of on-orbit refueling on the International Space Station. ESA has developed the "e.Deorbit" robot to provide on-orbit upgrading, maintenance and refueling services.

In commercial, the French Airbus Group's "O.Cubed", the United States Altius Space Company's "Bulldog", Japan's Astroscale's "ELSA-d" space clean-up program, the United States Chandahar Space Company's "InsureSat" remote sensing detection satellite, the United Kingdom's "ELSA-d" space shuttle, and the United States Chandahar Space Company's "InsureSat" remote sensing satellite, Effective Space's "SpaceDrone" on-orbit life extension service, Germany's "iBoss" on-orbit plate extension assembly service and so on are currently undergoing research on modular satellites and related on-orbit services. In addition, Made In Space and Northrop Grumman Innovative Systems have launched a number of on-orbit support satellites/robots, and the former orbital ATK company announced the development of a "mission robotics vehicle" [11, 12].

2.2 Domestic Status and Development

Research related to modular reconfigurable satellite has been carried out in China since 2008, and relevant research results have been published by China Aerospace Science and Technology Corporation, Chinese Academy of Sciences, Space Engineering University, Information Engineering University, Harbin Institute of Technology, the Northwestern Polytechnical University and some others.

(1) Theoretical Conceptual Studies

Beijing Institute of Spacecraft System Engineering suggested that modular design needs to consider the system optimization of the whole satellite, provide multiple installation interfaces, solve the strong coupling problems between modules, change the design process of modules and reasonably design of the space unlocking [13]. Some scholars from Harbin Institute of Technology believe that the existing variable configuration satellites are mainly designed to meet the needs of satellite modularity or heat dissipation, and only the on-orbit deployment is considered, not the on-orbit convergence [14]. Shanghai Academy of Spaceflight Technology divides modular satellites into three independent modules: power module, platform center module and payload module, then proposes structural, protocol, and application layer construction plans for the architecture design [15]. Shenyang Institute of Automation of the Chinese Academy of Sciences proposed a concept of modular satellites combined with reconfigurable robots [16].

(2) Technology Configurations Studies

Northwestern Polytechnical University has conducted a series of studies on modular satellite technology, including the proposal of transformation satellites such as hinge-spread, rack-and-pinion and screw-nut types and kinematic and dynamic characterization associated. They also proposed key techniques for reconfigurable cellular stars, including configuration optimization, information fusion, control of multicellular structures, and spatial teleoperation for fine assemblies [17–19]. National University of Defense Technology proposed a scheme for the construction of a modular small satellite automatic measurement and control system [20]. Yewei Yang of the Equipment Academy studied a space-based laser energy transmission technology based on the design concept of separating modular satellites [21]. Based on the reference configuration proposed by Changchun Institute of Optics, Fine Mechanics and Physics of Chinese Academy of Sciences, it is believed that high stiffness and easy-to-process carbon fiber composites should be chosen as the main material [22]. He Guangsong of the Aerospace Engineering University proposed a modular satellite application system based on the storage module library and studied the simulation of modular satellite power system [23]. A research team from Harbin Institute of Technology proposed a hierarchical planning method based on connection matrix for satellite architecture [24]. Fuling Guan of Zhejiang University carried out design research on extensible structures of modular in space [25].

2.3 *Typical Cases Studies*

Currently, internationally developed examples of modular satellites include the Phoenix and Hive Programs in the United States, the Panel Extension Satellite Program in Japan and the iBOSS Program in German. From these, the Phoenix and iBOSS were selected for focused study.

(1) The Phoenix Program

DARPA's Phoenix program, based on the "Orbital Express" and "FREND" programs, believes that the relationship between cost, quality, performance and time of existing satellites can be deconstructed. Inexpensive miniature "semi-finished satellites", thrusters, tool belts, etc. with common shapes, different functions and only core functions could be delivered by commercial off-the-shelf (COTS).

The system of the Phoenix program consists mainly of three subsystems: the cellular satellites named "Cells", the on-orbit delivery system of payload named "PODS" and the Space Server. Cells are inexpensive miniature "half-built satellites" with only core components, common shapes but different functions, and placed as payloads within the PODS. The PODS is a structural container for cells, robotic end-execution tools, and other electronic equipment. The Servicer is equipped with a robotic arm and a hand-eye vision system, and is capable of on-orbit capture, maintenance and other Manipulation capabilities. It can reconstruct new satellites using antennas on decommissioned satellites in orbit and carry new satellites for orbital transfer, as well as serve as a relay communication station and wireless energy transmission station for new satellites. Its key technologies are shown in the Fig. 1.

DARPA split the Phoenix program into the New Phoenix and the Robotic Service for Geosynchronous Orbiting Satellites (RSGS) programs in 2015 to develop an on-orbit service spacecraft and a modular reconfigurable spacecraft respectively. And then, they spawned the "Dragonfly" program in 2018 to validate an autonomously assembled antenna in orbit.

(2) iBOSS program

The iBOSS programme (Intelligent Block for On-orbit Services System), funded by the German Aerospace Center (DLR) and carried out by several research institutions, aims to develop, manufacture and applicate a new modular spacecraft "iSAT" to support on-orbit reconfiguration and service expansion. In terms of configuration, a "building block" model based on electronic payloads has been proposed to convert traditional satellite subsystems into specific standardized modules, which not only enables on-orbit maintenance, but also enhances and replaces common satellite infrastructure and helps other satellites to achieve on-orbit life extension.

The results of the iBOSS program include not only the iSAT intelligent modular satellite, but also the Intelligent Computer-Aided Satellite Design (iCASD) software and the Virtual Test Bed (VTi) for spacecraft simulation. iSAT consists of a building block (iBLOCK), a central structure and an intelligent space system interface "iSSI". ICASD is capable of generating configuration-specific iSATs, taking into account users' and mission's requirements and the constraints of mission-specific components and modules. Any iSAT can be tested and verified in detail in a spacecraft simulation system provided by the Virtual Test Bed (VTi). ICASD and VTi help to reduce the time and costs of development and ensure the quality and reliability of satellite. The key technologies are shown in the Fig. 2.

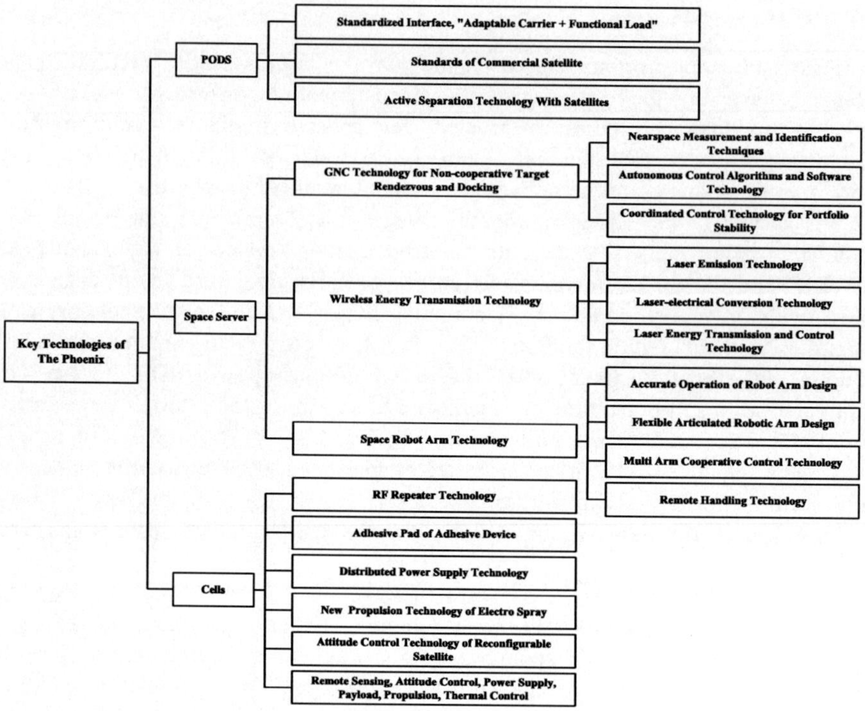

Fig. 1 Key technologies of the phoenix

2.4 Summary

From a technical point of view, at present, modular reconfigurable satellites internationally use building block modules as the main configuration which differences are whether there is a central structure, the modules are in the form of blocks or cubes, whether flexible connections are used, and on-orbit robots are required to assist in reconfiguration or intelligent self-reconfiguration. Research on modular reconfigurable satellite configurations has focused on the optimal configuration of the mission, power consumption and cost, but few studies on the performance evaluation of modular satellites have been conducted [26].

From a management point of view, the vast majority of modular satellite programs are currently led by the military and governments, while commercial space agencies are more focused on on-orbit reconfiguration and the development of on-orbit services. Commercial on-orbit service systems, processes and capabilities are gradually evolving from start-up to maturity period, and the maintenance risks and costs are gradually decreasing, so civil and commercial companies will invest more in this area [27].

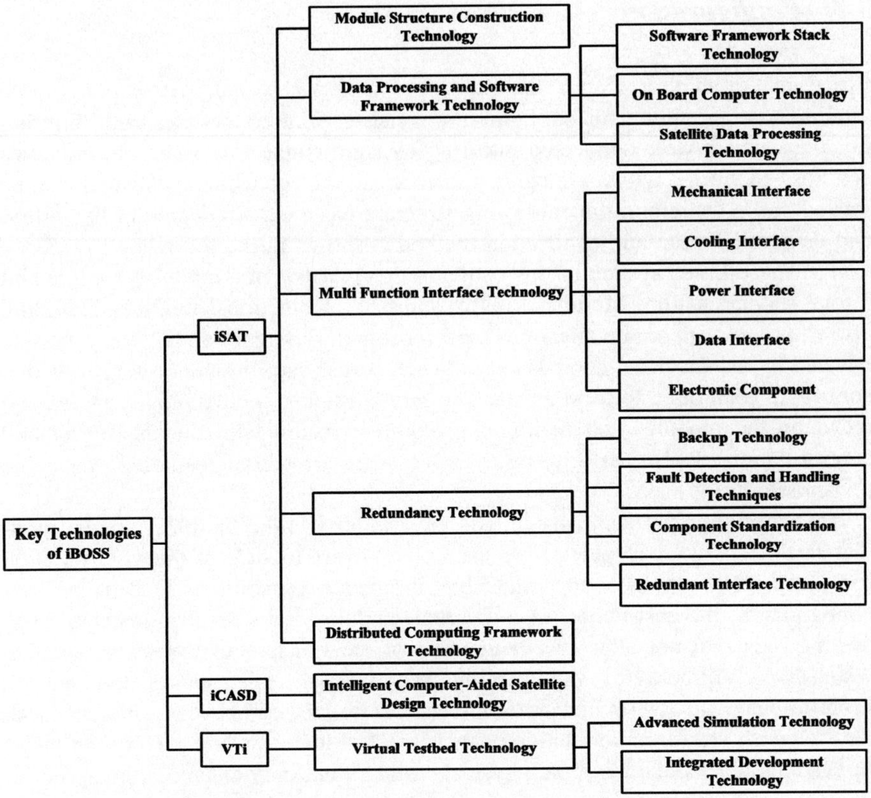

Fig. 2 Key technologies of iBOSS

3 A New Configuration of Modular Reconfigurable Satellite

The modular reconfigurable satellite configuration has high requirements for the stability, flexibility, versatility, destructibility, and scalability of the overall structure. In addition, for Earth observation application scenarios, the modular reconfigurable satellite configuration also needs to meet the requirements of flexible assembly of sensors in different bands, stable and high attitude orbit accuracy, agile imaging, large and freely controllable amplitude, stereo imaging capability, fast response as possible and shortened revisit cycle [28, 29].

3.1 Configuration

Based on the above research, this study proposes a new modular reconfigurable satellite configuration. This configuration consists of cubic modules with differing functions which were connected into ring by rigid connecting rods. The modules are cross-connected with an integrated interface to the connecting rod which can be rotated 360°. Therefore, the entire ring structure has a certain degree of flexibility, and its configuration and connection method is shown in the following Fig. 3.

The space-based system of this configuration consists of a modular satellite and a total service station. Modular satellite consists of functional modules including power, propulsion, communications, and sensors in different bands. Every module covered by a solar panel outer layer and each corner carrying connecting rods that connect to each other to form a ring. The service station could storage, release and recycling the modules and maintaining communication with the measuring and control stations. It also performs tasks such as auxiliary orbital positioning and relay transmission.

This configuration has many unique advantages. Unlike the block model modular satellites that are shielded from each other, every module of this configuration has independent power supply and heat dissipation capabilities. Due to the ring configuration, the destruction of individual modules and a small number of connecting rods will not affect the realization of the function of the whole satellite which greatly improving the stability and damage resistance of the system. Front-to-back and side-to-side view effects can be created between different modules to create stereoscopic relative or multispectral images when performing tasks. In evading attacks, the isotropic physical structure not only enhances the agility of

Fig. 3 Concept design

the satellite, but also provides the possibility for missiles to pass through the middle of the satellite, in addition to propulsion the satellite to move and dispersal to evasion. When adding or subtracting modules, the configuration reduces the difficulty of autonomous recognition calculations of rotational inertia and center of mass, which reduces the probability of rolling and side tumbling.

3.2 Key Technologies

The key technologies involved in this modular reconfigurable satellite configuration can be categorized as both hardware technology and software technology. In terms of hardware technology, the enclosure needs to be simple and stable, the interface needs to be standardized and integrated design, and the power supply and structural reconstruction design should comply with standards. In terms of software technology, system refactoring capabilities, post-module reorganization coordination, exception handling capabilities, and data standardization are required.

3.2.1 Structure of Single Module

The single module structure needs to withstand the force, housing, support and heat at launch, as well as the environmental effects of radiation, temperature and electromagnetism after entry into orbit. The following requirements also need to be met [30, 31].

(1) Symmetry: axial symmetry and 90° rotational symmetry need to be ensured so that other modules can be connected at will without restricting their relative orientation.
(2) Low complexity: minimize the complexity of structures and designs in order to avoid failures during on-orbit mission execution and then reducing failure rates.
(3) Low weight: the total mass of the module should be kept as low as possible in relation to the costs of material and launch.
(4) Low capacity: for interfaces, the interface should take up as little of the module's available capacity as possible.
(5) Sensor: the module has an interface detection sensor that detects if another module is connected and measures the rotational offset between interfaces.

3.2.2 Segmentation and Integration Techniques for Functional Modules

Segmentation and integration techniques for functional modules mainly include segmentation, design, manufacturing of the modules and the connection design of mechanical, electrical, thermal, etc. [32]. Define which functions are the common features of this type of module for analysis, and then carry out a reasonable division of functional modules to ensure that the module design should be independent and complete. And all types of interfaces should be standardized to enable rapid assembly, integration and testing.

3.2.3 Spatial Capture and Identification Techniques

When tracking approach and capture targets such as modules, the target needs to be characterized to enable the estimation of relative position and attitude. Accurate identification, measurement and capture call for an extremely high precision, accuracy, real-time information acquisition and processing system, which directly affects rendezvous and docking and the effects of mandate implementation [33, 34].

3.2.4 Multi-Function Interface Standardization Techniques

This configuration requires a unified development standard for the interface on the one hand, and a multifunctional design of the interface on the other hand in order to realize the plug-and-play satellite. The mechanical, electrical, thermal and data interfaces need to be integrated, and redundant interfaces need to be designed [35]. Mechanical interfaces are used to create rigid structures between modules for physical connections. Since it is difficult to achieve the power required for each module locally generated by the module's solar cells, an electrical interface is required to coordinate the energy transfer on each module. This configuration has different locations for the distribution of heat sources and heat sinks, and an architecture with heat sink interfaces will greatly reduce the power requirements of the thermal control system. The data interface is used as the primary data bus for critical system data transfer as well as non-critical payload data transfer. The limited surface space of the module makes it impossible to set up multiple interfaces at the same time, and too many interfaces can lead to structural instability and difficulties in on-orbit operation, thus requiring the development of multifunctional interfaces [36]. This configuration greatly improves the balance efficiency of heat and data transmission.

3.2.5 Coordinated Control Techniques for Satellite Stability

The weight and momentum characteristics of the reconstituted, assembled or repaired satellites will change abruptly, which will not only fail to meet the requirements of control performance, but will also lead to the destabilization of the whole system and may cause the combination to tumble in space [37, 38]. The ring configuration structure is more stable and relatively effective in avoiding the occurrence of space tumbling. In addition, the rotational inertia and the position of the center of mass of the reconstituted satellite are identified in orbit autonomously, and the corresponding robust control or adaptive control algorithm is designed to ensure the stability of the whole system. In addition, the rotational inertia and the center position of mass of the reconstituted satellite are identified in orbit autonomously, and the corresponding robust control or adaptive control algorithm is designed to ensure the stability of the whole system.

3.2.6 Satellite-Based Data Distribution and Processing System

The technology is primarily used to enable functions such as the distribution of data and execution of procedures between multiple modules, task state management and results evaluation. The distribution needs to balance the task distribution, heat dissipation, power consumption and process load of multiple modules and dynamically select the most appropriate module to perform the task [39].

3.2.7 Simulation and Effectiveness Evaluation Techniques

The assemble of modular structures in 3D space generates a large number of solutions, and the search for the best combination of solutions is tedious and laborious. The technique is mainly used to achieve qualitative and quantitative calculations of satellites that with a less quantity, excellent function and stable structure [40]. In addition, the cost, lifetime and power consumption of the mission will also have an impact on the composition and function of the satellite, which requiring the technology to make a comprehensive measurement. For different satellite designs, it is necessary to put them into the simulation of space environment and mission scenarios for effectiveness evaluation, and assist selection and optimize the program based on the simulation results finally.

3.3 Commercial Operation and Maintenance

In the commercial operation and maintenance model, commercial components of the module are orbited through low-cost commercial launches, such as "free ride" launches or "space bus" and stored in the server station. The operator calculates and simulates the optimal architecture of the satellite by the intelligent construction software and injects the instructions into the server when receives the mission requirements. The server assembles the satellite in orbit and releases it to the target orbit, and then provides guarantee services such as energy transmission, attitude orbit maintenance, parts replacement and data relay. When the mission is completed, the server recovers, dismantles and stores the modules for use in other missions. The modules can also be installed on a partially disabled satellite by transaction through the on-orbit service market.

3.4 Prospects for Commercial Applications

Some unique business applications can be derived from the unique advantages of this configuration, such as on-orbit module leasing rather than satellite leasing. Rental of one or more modules for Earth observation, such as two modules forming stereoscopic relative or different-band sensor imaging, which can be achieved by their own rotation rather than reconstructed by the server, which could save much time and propulsion costs. Thus, the service life of the satellite can be extended by module replacement, and the utilization efficiency can be increased during the service life, thus extending the profitability time of the satellite. As the module can be independently powered and cooled, which can be combined with other satellites that have lost some of their original functions by the on-orbit trading of secondary market and recycled at mission completion.

4 Effectiveness Evaluation of the Configuration

This study is based on the earth observation scenario, uses the simulation model as the data source and the ADC model method to evaluate the effectiveness of the satellite of this configuration.

4.1 Performance Evaluation Based on EO Scenarios

For modular remote sensing imaging satellites, the mission effectiveness is defined as the extent to which the EO mission requirements are met under specified conditions [41]. Using the ADC (Availability, Dependability, Capability) model analysis method for effectiveness evaluation, it can be decomposed into maintainability, reliability, imaging capability, image transmission capability, image processing and application capability. The specific structure is subdivided into maintenance degree, repair time, reliability, failure rate, average lifetime, image acquisition range, quantity of acquisition, image accuracy, time effectiveness of imaging, and completion time of the task, width, resolution, revisit period, and attitude maneuver capability are selected for simulation evaluation. The orbit-related metrics are not considered because the block modular satellite model was chosen as a control with the same orbit as that configuration of satellite. The basic structure for assessing specific performance indicators is shown in the Fig. 4.

The objective of the scenario is to observe the surface area covering 1000 km × 300 km at three randomly selected positions A1 (116°20′E, 39°56′N), A2 (120°34′W, 34°43′N) and A3 (13°2′E, 52°31′N) around the world. Both the block modular satellite and the satellite of this configuration mainly focus on observation in the visible band with the time set for the sunshine conditions at the

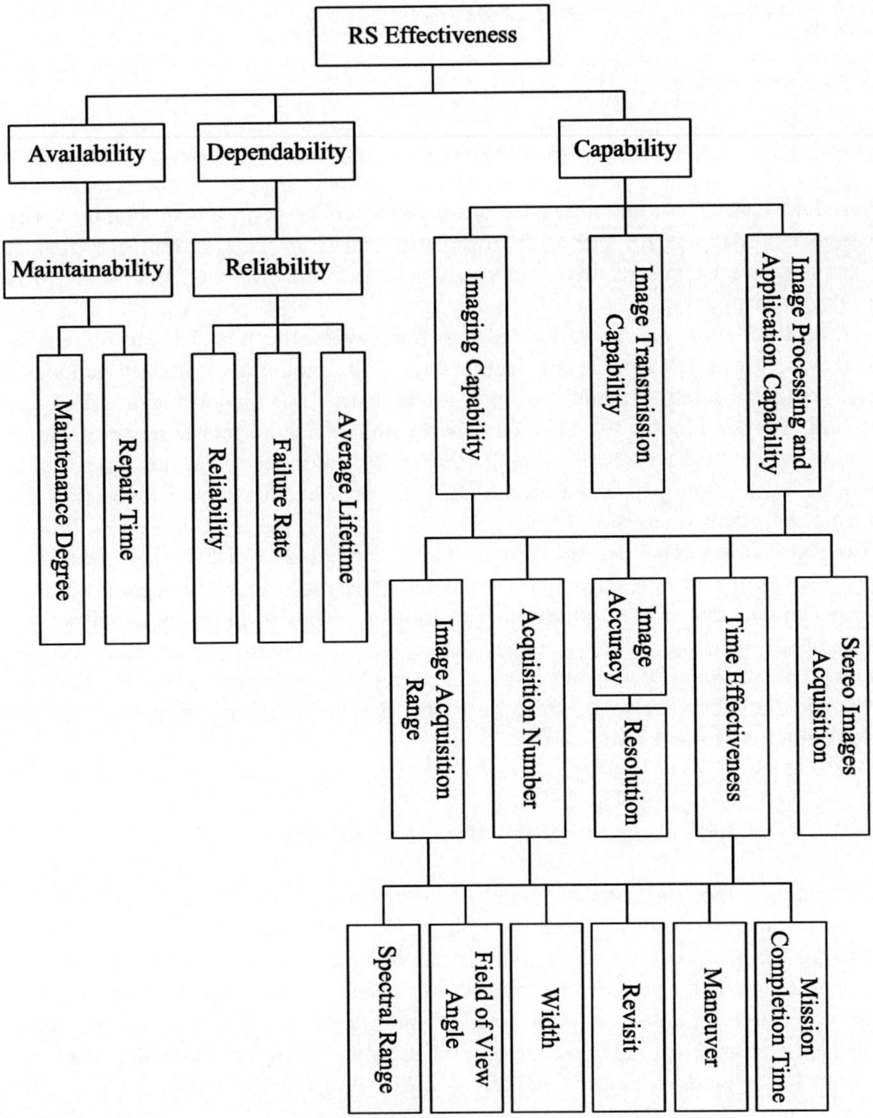

Fig. 4 The basic structure of evaluation index

autumn equinox, the orbital altitude set for 500 km, and the resolution of not less than 2 m to meet the demand. The scenario designed does not involve ground measurement and control, data relay transmission and reception, data processing and application; it does not involve targets of specific terrain or meteorological influences such as rains, clouds and others.

4.2 Simulation System and Models

4.2.1 Overview of the Simulation System

In order to objectively carry out the effectiveness evaluation of the above-mentioned modular satellites, a simulation system for effectiveness evaluation based on Data Distribution Service protocol is built to form the simulation verification environment for the earth observation mission. This system can provide simulation data input for the effectiveness evaluation and its architecture is shown in the following Fig. 5.

The simulation system is divided into four layers: the model layer, the engine layer, the tool layer and the application layer. The model layer mainly includes a new modular satellite model, a block modular satellite model and a target area model. The simulation system establishes a unified model interface specification, and the engine calls are realized under the unified interface of various models. The engine layer developed based on the DDS protocol enables model invocation and data distribution interaction by encapsulating models into a dynamic link library. The simulation engine mainly includes the service of data distribution, simulation process control, time management, conceptual analysis, model management, entity management and engine interface. The tool layer is the functional tools of the simulation system, including model management, scenario design and development, test design, 2D/3D state display. The application layer realizes the situation display and data analysis, which lays the data foundation for the subsequent performance evaluation [42, 43].

4.2.2 Model Construction of the Modular Satellites

A functional modular satellite model is constructed for the EO scenario, with the main parameters including satellite orbit, satellite attitude and imaging capability. Among them, the orbital model parameters include semi-major axis, eccentricity, orbital inclination, perigee magnitude, ascending intersection equator and true perigee angle. Satellite attitude model parameters include manoeuver speed, manoeuver range, pointing accuracy and stability. Imaging capability modeling parameters include resolution, width, side heave, spectral band and sub-satellite track (Fig. 6).

4.3 Analysis of Evaluation Results

The research combines qualitative and quantitative data to analyze the evaluation results. The maintainability and credibility results are mainly of qualitative analysis and the capability results are mainly quantified analysis based on simulation data.

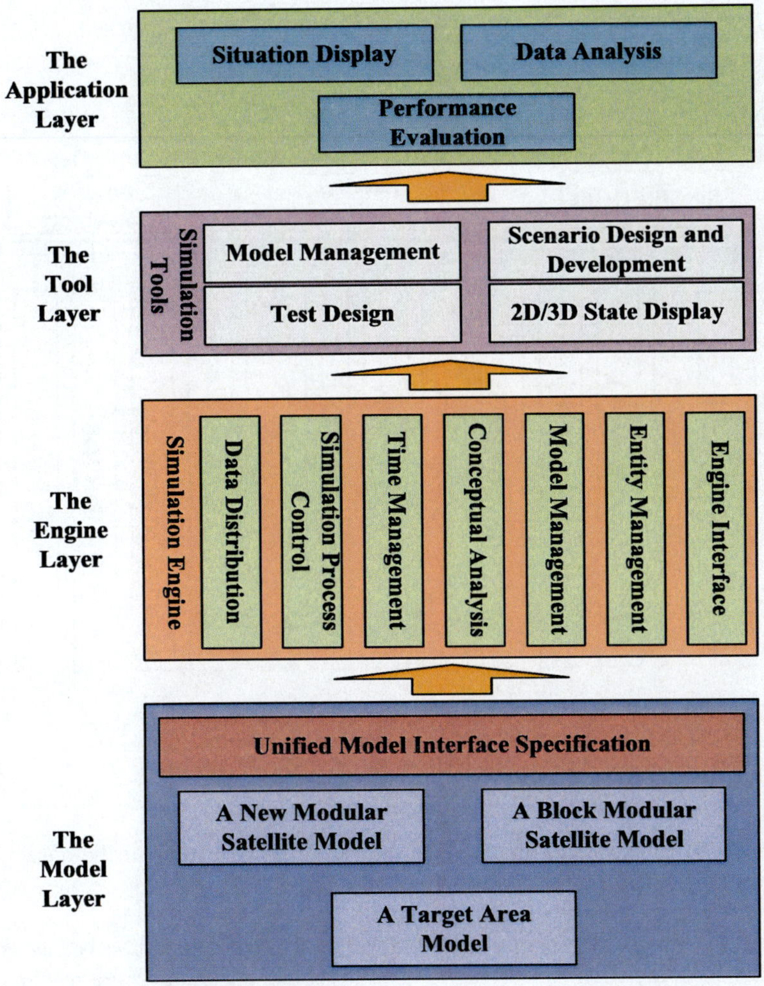

Fig. 5 The architecture of the simulation system

4.3.1 Analysis of Maintainability and Credibility

In terms of the maintainability, there is no blocking of each other and no large deviation of the center of mass during reorganization, which greatly reduces the difficulty of repair and thus shortens the repair time. In terms of reliability, the multiple modules of this configuration are connected in parallel, so the damage to a single module or a small number of connecting links does not affect the realization of the whole satellite function, so that the average life of the satellite is significantly extended with a failure rate comparable to that satellites of other configurations.

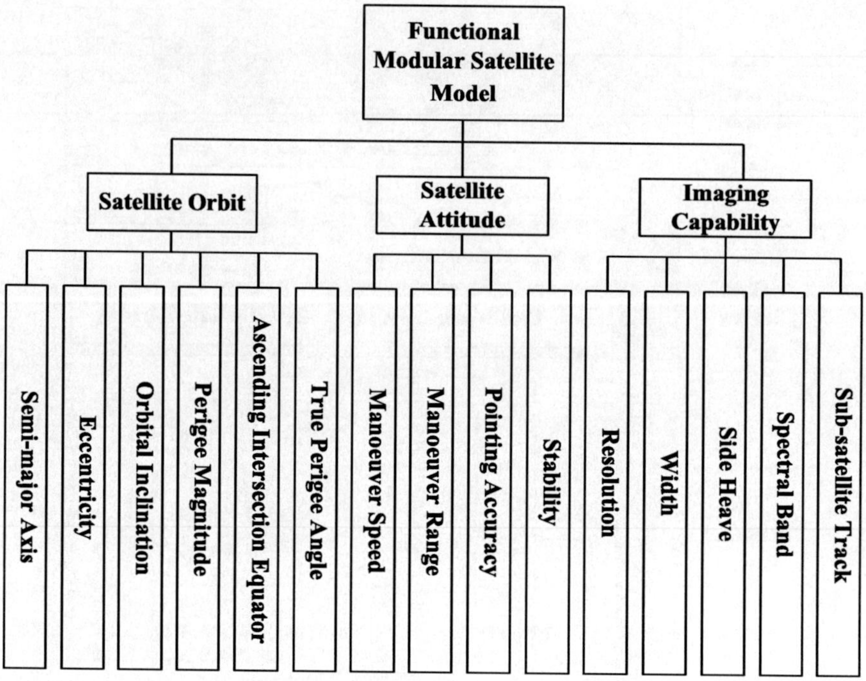

Fig. 6 The framework of the satellite model

4.3.2 Analysis of Imaging Capability

Based on the EO scenario, the average values of the indicators of the observation capacity for the three regions of satellites in different configurations are calculated, and the results of the assessment are shown in the following Table 1.

From the simulation results, the revisit time of this configuration is shorter and the mission completion time is significantly shorter, but the average resolution is lower than that of the block satellite under the same half field angle, due to the larger side pendulum angle. The reason is that when the side pendulum angle reaches a certain limit, the spatial resolution will be less than 2 m which is not sufficient for mission accomplishment. In order to achieve the specified resolution, the side pendulum angle of this configuration is limited to a certain extent, and thus the side pendulum capability is not inversely correlated with the mission completion time. However, the imaging capability assessment results of this configuration are significantly better than those of the block modular satellites. In general, the evaluation results show that this configuration can improve the effectiveness of modular reconfigurable satellites.

Table 1 The assessment results of The EO scenario

	The satellite of the block	The satellite of this configuration
Half field angle (°)	±8°	±8°
Width (km)	10	12
Revisit time (h)	8.46	6.31
Side pendulum angle (°)	±10°	±45°
Mission completion time (h)	658.21	189.58
Average resolution (m)	1.93	1.99
Stereo imaging capability	Not	Do

5 Conclusion

Compared with traditional block modular satellites, this configuration has higher availability and reliability. Combining the EO scenario and simulation data, this configuration also has obvious advantages in revisit period, imaging efficiency, acquisition range and stereo images acquisition. The evaluation results show that this configuration can improve the performance of modular reconfigurable satellites, but there is still space for refinement due to the limitations of the scenario settings and simulation system.

Based on the domestic and international research on configuration and technology of modular reconfigurable satellites, this study proposes a new configuration of modular reconfigurable satellite and carries out modeling and simulation on it, and uses ADC method to evaluate the effectiveness with the simulation data and evaluation index system, which provides the direction for the development of market and improvement of related technology.

References

1. Xiaoqian C, Jianping Y, Wen Y et al (2009) On-orbit service technology of spacecraft. China Astronautic Publishing House, Beijing, pp 238–241
2. Li D, Zhu Q, Zhu X (2010) Task oriented remote sensing information focusing service. Science Press, Beijing, pp 11–13
3. Schmidt J, Phillips MR (2013) A distributed, redundant navigation and fault detection system for DARPA system F6. Aiaa Guidance, Navigation, & Control
4. Bartlett RO (1978) NASA standard multi mission modular spacecraft for future space exploration. In: 16th American Astronautical Society and Deutsche Gesellschaft fuer Luftund Raumfahrt, God-dard Memorial Symposium, Washington, D.C., AAS
5. Nakasukas S, Sugawaray Y et al (2007) A novel satellite concept “panel extension satellite (PET-SAT)”consisting of plug-in, modular, functional panels. In: Proceedings of 21st annual conference on small satellites. AIAA, Washington D.C.
6. Martin M, Lyke J (2013) Modular open network architecture (MONARCH): transitioning plug-and-play to aerospace. In: Aerospace conference, 2013 IEEE, IEEE, pp 1–10

7. Fronter House DC, Center K, Strunce B et al (2010) Pnpsat-2 SPA technology testbed initial results and development status. In: IEEE aerospace conference
8. Darpa TTO (2020) System F6 on-orbit demonstration testbed broad agency announcement [2020–10]. https://www.fbo.gov/spg/ODA/DARPA/.CMO/DARPA-BAA-1232/listing.html
9. Mathieu C, Weigel AL (2005) Assessing the flexibility provided by fractionated spacecraft. In: Proceedings of the AIAA space 2005 conference and exposition. AIAA, Long Beach, California. 10. 2514/6.2005-6700
10. Brown O, Eremenko P (2006) The value proposition for fractionated space architectures. In: Proceedings of the AIAA space 2006 conference and exposition. AIAA, San Jose, California. https://doi.org/10.2514/6.2006-7506
11. Davis JP, Mayberry JP, Penn JP (2020) ON-Orbit servicing: inspection, repair, refuel, upgrade, and assembly of satellites in space, [2020-10]. https://aerospace.org/sites/default/files/2019-05/DavisMayberry-Penn_OOS_04242019.pdf
12. Buchen E (2014) SpaceWorks'2014 nano/microsatellite market assessent. In: Proceedings of the AIAA/USU conference on small satellites. North Logan, UT, US
13. Xu Q, Fan L, Gao H, Lu Q, Shi H (2014) Integrated configuration of remote sensing satellite platform and payload. Space Return Remote Sensi 35(04):9–16
14. Chong Z, Hongwei G, Zongquan D, Rongqiang L, Shoubin L (2018) Structural design and performance evaluation of variable configuration truss type satellite platform. J Harbin Inst Technol 50(01):11–17
15. Wang Z, Xia C, Kang Z et al (2018) Modular nuclear-powered spacecraft design and its key technologies. Shanghai Aerospace
16. Ding J, Liu J, Zhang X, Gao Q (2018) Research on key technologies of reconfigurable modular spacecraft. In: 2018 Software defined satellite summit conference abstracts collection. Software defined satellite technology alliance, p 49
17. Hao L, Geng L, Ma S et al (2012) Spacecraft design and dynamics simulation with rack-and-pinion configuration. J Syst Simul 024(008):1606–1611
18. Ma S, Liu G, Luo H (2010) Spacecraft design and dynamics simulation with hinge deployment configuration. China Space Sci Technol 06:49–56
19. Huang P, Chang H, Lu Z, Wang M (2016) Key technologies and perspectives of reconfigurable cell satellites for on-orbit services. Astronautics 37(01):1–10
20. Fu W, Zhang S, Zhang R et al (2013) Modular automatic test system construction for small satellite measurement and control. Spacecraft Eng 022(002):104–107
21. Yang Y, Li X, Zhang D (2014) Research on laser energy delivery technology based on separated module spacecraft. Laser Infrared 02:126–130
22. Yuan J (2020) Research on the structure of carbon fiber honeycomb interlayer optical machine for space camera. University of Chinese Academy of Sciences (Changchun Institute of Optical Precision Machinery and Physics, Chinese Academy of Sciences
23. He G, Li X, Zhang H, Wang Q (2018) Research on modular spacecraft application system based on storage module library. Comput Measur Control 26(06):203–206
24. Wang W, Ye D, Sun Z, Tang S, Chen X (2019) Hierarchical planning for on-orbit self-reconstruction of modular reconfigurable satellites. J Aeronaut 40(09):276–288
25. Shen X, Zeng L, Jin Y, Zhang Q (2017) Research status and development trend of on-orbit assembly technology. Manned Space 23(02):228–235+244
26. Andreev AS, Leonova AV, Popov AM (2020) Demand for satellite reconfigurability. In: IOP Conf Series Mater Sci Eng 822:012036
27. Zhou Z, Wu Z, Ji Y (2008) Development and application of space-to-earth observation technology. China Eng Sci 06:28–32
28. Chen Q, Zhang J, Cao X (2017) Current status and key technologies of cluster spacecraft network development. J Harbin Inst Technol (4)
29. Zheng Y, Shuai S, Kaichun Z et al (2019) Key technologies for nanosatellite development. Spacecraft Eng 28(01):1–9
30. Li C, Yu Z, Chen J (2019) Review of high resolution starborne SAR imaging and image quality improvement methods. Radar J 8(06):717–731

31. Zhang K, Zhu Z, Xia L (2015) Technical analysis of small satellite modular design. Spacecraft Eng 24(06):107–115
32. Zhao C, Guo H, Deng Z et al (2018) Structure design and performance evaluation of variable configuration truss-type satellite platform. Harbin Gongye Daxue Xuebao/J Harbin Inst Technol 50(1):11–17
33. Zheng T, Libin X, Lv Q, Sun J, Li P, Gao S, Yin Z (2019) Super-resolution imaging of CX-6 (02) micro-nanosatellite. J Remote Sens 23(02):196–204
34. Chen N, Liu X, Gong J (2018) Design and implementation of satellite sensor spatio-temporal coverage capability evaluation and optimal combination system. Surveying Mapping Bull (09):69–73
35. Pang Y, Li Z, Chen X, Zhang Z, Huang J (2016) Research on modular reconfigurable space system. Spacecraft Eng 25(03):101–108
36. Guo D, Liu L, Chen Y, Li W, Cheng S (2020) A conceptual design method for reconfigurable spacecraft based on visualization models. J Beijing University Aeronaut Astronaut 46 (07):1405–1411
37. Wen F, Cheng Y, Jiang B, Liu W (2014) Reconfigurability analysis of satellite attitude control system. Astronautics 35(02):185–191
38. Wang R, Wang J, Hu Xi (2017) Analysis of uncontrolled positioning accuracy of optical satellite photography. J Surveying Mapping (03):69–74
39. Guo D, Liu L, Chen Y, Li W, Cheng S (2020) Reconfigurable spacecraft conceptual design method based on visualization model. J Beijing University Aeronaut Astronaut 46(07): 1405–1411
40. Li S, Ren D (2018) Research on the development of simulation and performance assessment techniques for remote sensing satellite system system. In: Proceedings of the 19th CCSSTA annual conference on china system simulation technology and its applications (19th CCSSTA 2018). Chinese Society of Automation System Simulation Professional Committee pp 238–241
41. Liu F (2017) Research on mission effectiveness assessment method for remote sensing satellite system. University of Chinese Academy of Sciences (National Space Science Center, Chinese Academy of Sciences)
42. Zhang X, He N, Wang R, Wang W (2018) A modular micro-small satellite docking mechanism modeling simulation. Astronaut 39(04):368–375
43. Zhang Y, Jiang T, Jiang G, Yu K, Yu Y (2019) Three-dimensional reconstruction of visual SLAM inverse depth filtering by feature method. J Surveying Mapping 48(06):708–717

FD-LinkNet: A Encoder-decoder Structure Network for High Resolution Satellite Imagery Rural Road Extraction

Lin Yi, Miao Yang, Liang Shuang, Peng Xiangyang, and Song Wentao

Abstract Due to the backward construction of rural road infrastructure and lack of supervision, unreasonable use of land and frequent safety accidents have occurred. At present, road extraction is a hot task in the area of high resolution satellite imagery target sensing. It is the basis of rural road management to extract rural road through the high resolution satellite imagery. Based on this, the state of art road extraction semantic segmentation neural network named D-LinkNet is applied and improved in this paper. The improved semantic segmentation neural network is named FD-LinkNet. The FD-LinkNet network uses encoder-decoder architecture and adds dilated convolution layers in center part. The encoder part takes advantages of ResNet34 pretrained on ImageNet dataset. Dilated convolution layers with skip connections is applied to full scale to enlarge the receptive filed of feature points without reducing the feature maps. Finally, the decoder part output the binary images as the binary semantic segmentation results. Compared with UNet, and D-LinkNet, The best IoU scores on the validation set and test set of FD-LinkNet are 0.6819, which has increased by 8.3%, and 5.5% respectively. The trained network in this paper can effectively extract rural road, which is benefit for road management departments to detect and superbise rural road.

Keywords High resolution satellite imagery · Semantic segmentation neural network · Rural road extraction · D-LinkNet

1 Introduction

Road extraction from satellite images has been a hot research field in past few years, because of the development of artificial intelligence and big data technology. The introduction of deep convolutional neural networks (DCNN) is greatly improved

L. Yi (✉) · M. Yang · P. Xiangyang · S. Wentao
CETC Les Information System Co.Ltd, Nanjing, Jiangsu, China

L. Shuang
China Design Group Co.Ltd, Nanjing, Jiangsu, China

L. Wang et al. (eds.), *Proceedings of the 7th China High Resolution Earth Observation Conference (CHREOC 2020)*, Lecture Notes in Electrical Engineering 757, https://doi.org/10.1007/978-981-16-5735-1_21

the accuracy of target extraction. Road extraction has been widely used in geographic information updating, city planning, road monitoring, car navigations, etc.

However, with the continuous innovation of remote sensing technology, the measurement accuracy has reached the sub-meter level, and the resolution has also broken through the decimeter (dm) level, and there is a trend for further development. Accompanied by the details of the background information of the ground features are more abundant. Non-road information such as vegetation shadows, vehicle flow, high-rise building occlusion, pedestrian flow, etc. is very complicated as interference noise, which makes it more difficult to extract target information, that is, road information. The identification becomes more difficult. In addition, in the road network identification process, roads of different textures will show different spectral characteristics [1], and the phenomenon of same-spectrum foreign matter or same-spectrum different spectra may also appear; most of the current road extraction uses semi-automatic methods, plus A series of problems such as the poor robustness of the above algorithm itself, the low recognition accuracy of traditional methods, and the cumbersome process have brought great challenges to road network extraction.

At present, the deep learning theory represented by DCNN has shown great development potential in image classification [2, 3] and target detection [4, 5]. In the field of image semantic segmentation, fully-convolutional network (FCN) [6] architecture is the most advanced and common. The FCN can produce a segmentation map for an entire input image through single forward pass. The latest excellent semantic segmentation networks [7–10] are all derived from FCN. Li et al. [11] proposed a linear integrated convolution algorithm for CNN training large sample data sets. This algorithm can predict the probability that a pixel area is a road, and at the same time assign a label to each pixel to determine whether it is a road area. The applicable scene is road Images with rough edges and contours; Xia et al. [12] proposes a weakly-supervised extraction algorithm based on the classification of different ground materials. The algorithm uses Deep Lab to build a learning network, and the ResNet network structure is responsible for testing the stability. Conditional Random Field (CRF) repairs the boundary, and uses the spectral angle distance to connect adjacent road sections in a sliding window; Dong et al. [13] proposes an improved graph cut road detection algorithm, which can use Orchard-Boumand clustering algorithm to cluster Road and non-road pixels are used to construct the area items in the Gibbs energy penalty function, and then use the max flow algorithm to segment the generated weight map to extract the road information.

In the field of satellite image road extraction, the most used is generating pixel-level labeling of roads [14, 15], which handling the road extraction task as a binary semantic segmentation task. D-LinkNet [16] is the best performing semantic segmentation neural network in road extraction. The D-LinkNet model architecture adds dilated convolution layers in the central area, which can maximize the range of the receptive field and promote multi-scale feature fusion, without causing loss of feature map resolution, and as much as possible to preserve the spatial detail information of the road.

In this paper, D-LinkNet is applied to the research of road extraction, and a series of improvement measures are made to the model based on the characteristics of the road in the image to improve the segmentation accuracy of the network model in the high-resolution remote sensing image. The main research work and improvement measures of this paper are as follows:

(a) Full dilated convolution layers: full dilated convolution layers is added to the traditional encoder-decoder structure to obtain multi-scale information.
(b) Center part: In the central part between encoder and decoder, an end-to-end model architecture is proposed based on the Encoder-Attention-Decoder structure, which can automatically learn multi-level and multi-scale road feature information.

2 Method

2.1 Network Architecture

The D-LinkNet network model was proposed by the Pattern Recognition Laboratory of Beijing University of Posts and Telecommunications. The model that achieved the highest score in the 2018 CVPR (Computer Vision and Pattern Recognition) Global Satellite Imagery Road Extraction Challenge (DeepGlobe Road Extraction Challenge) project. The model in this paper is based on the D-LinkNet network architecture, which consists of three parts: encoder constructed by the residual network, center part and decoder.

The model uses LinkNet as the basic framework and adopts an Encoder-Decoder architecture. The main idea of the model is that the coding area encodes road information onto feature information, and the decoding area maps the coded road feature information to space for segmentation. The most important thing is that on the basis of adding the convolution layer to the central area, the network itself does not increase the learning parameters, which greatly reduces the difficulty of training. For road extraction tasks, D-LinkNet adds a dilated convolution layer in the center area to further identify the receptive field of road feature point information. The advantage of the hole convolutional layer over the pooling layer in LinkNet is that it does not lose spatial information while ensuring the resolution of the feature map (Fig. 1).

In D-LinkNet, the coding area is composed of an initial convolution module with a size of 7×7 and stride = 2 and 4 residual modules. The residual module uses ResNet34 as a pre-training network structure, and uses jump connections to enhance the generalization and characterization capabilities of regional blocks. The central part is the dilated convolution layer module, which is a connection mode that coexists in series and parallel. In addition, due to the different shapes and sizes of rural roads, in order to improve the extraction accuracy of rural roads, this paper

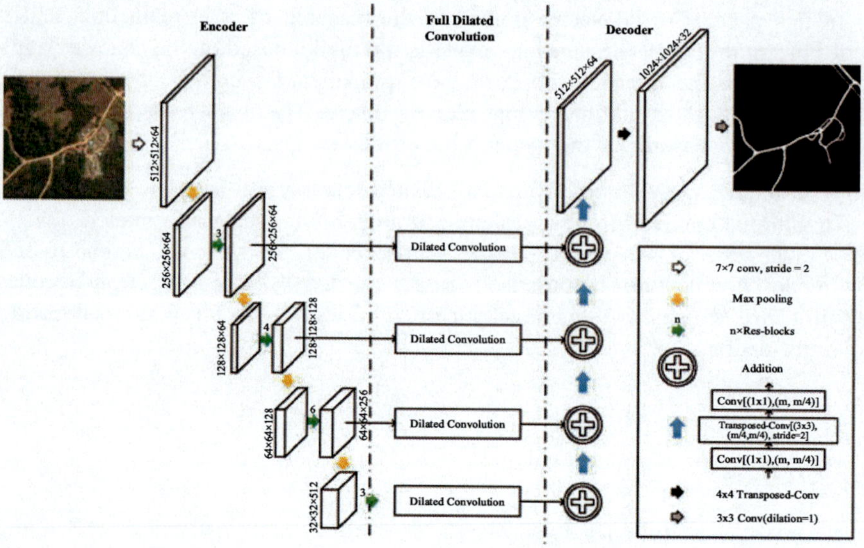

Fig. 1 FD-Linknet neural network for rural road extraction

modifies the central part to a full dilated convolution layer to further process and perceive features of different scales encoded by the encoder part.

2.2 *Full Dilated Convolution*

FD-LinkNet uses full dilated convolution layers with skip connections in the center part.

The bottom dilation convolution structure shows (the other three hole convolution structures are similar to it), as shown in Fig. 2. The five branches in the central part have joined the cavity convolution operation module. The so-called hollow convolution is essentially a special "pooling" operation. Compared with the ordinary pooling process, the advantage of hole convolution is that it can increase the range of the receptive field without loss of characteristic information, thereby outputting more information. The gray arrows in the figure indicate the depth of the neural network. The depths of the five branches are 4, 3, 2, 1, 0 (0 stands for identity mapping), and the size of the receptive field is 15, 7, 3, 1, 0. The structure can complete the fusion of features with multiple depths and sizes without loss of resolution. It adopts the feature fusion method of feature layer "stacking" without post-processing calculation process. After that, a 1 * 1 convolutional layer is added for feature fusion. Finally, a road prediction probability map with the same pixel

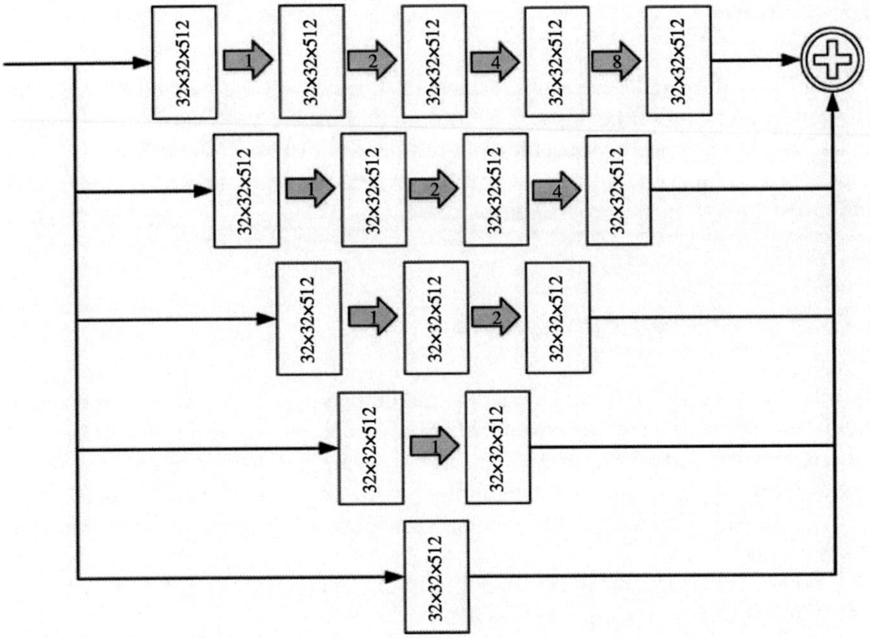

Fig. 2 The center dilation convolution of bottom part in FD-LinkNet

level as the input image resolution is obtained through the normalized Sigmoid function and Relu activation function. According to the set threshold, the binary image of remote sensing image road segmentation prediction is finally obtained.

3 Experiments

In order to demonstrate the accuracy and effectiveness of the road extraction method from high-resolution remote sensing images proposed in this paper, a series of horizontal comparison experiments were carried out for the setting of the weight ratio of different hyperparameters within the FD-LinkNet model. In addition, for the exterior of the improved model, a longitudinal comparison experiment with the current mainstream road segmentation method is carried out to visually show the comparison between the improved network in this paper and the current mainstream classic network road prediction effect.

3.1 Dataset

On the DeepGlobe road extraction dataset [17], we tested the method proposed in this paper. The data set consists of 6226 training images, 1243 verification images and 1101 test images. The resolution of each image is 1024 × 1024. The data set is expressed as a binary segmentation problem, where roads are marked as foreground and other objects are marked as background.

3.2 Experimental Environment

In order to make an objective evaluation and comparison of the road segmentation prediction effect of the improved network model proposed in this paper, all experiments are carried out under the same software and hardware environment.

Software environment: the operating system is ubantu 18.04 with 32 g of memory; the deep learning framework is Pytorch1.6; the programming environment is Python 3.6.

Hardware environment: CPU model is Intel (R) Xeon (R) Gold 5115 CPU @ 2.40 GHz, 8 GeForce RTX 2080Ti GPUs.

3.3 Image Preprocessing

In this study, the following methods are used to amplify data, increase the amount of data injected into the D-LinkNet network model, and improve the model's segmentation accuracy.

(a) Random zoom: randomly zoom the image up to 10%.
(b) Random offset: Randomly offset the image up, down, left, and right by up to 10%.
(c) Random pull up: Randomly pull the image up to 10% vertically or horizontally.

After the above four transformations, the 1024 * 1024 part of the center of the image is intercepted, and the insufficient part is filled with 0.

This research uses OpenCV to perform data enhancement on the image by color transformation in the HSV space. After data enhancement, the original image and the transformed image are shown in Fig. 3.

In addition, when testing the model, this study used time-to-test time augmentation (TTA) to create multiple different versions of the original image, including horizontal flipping of the image, vertical flipping of the image, and diagonal flipping of the image (predicted each image is 2 × 2 × 2 = 8 times), and then restore the output to match the original image. Then, the probability of each prediction is averaged using 0.5 as the prediction threshold to generate a binary output.

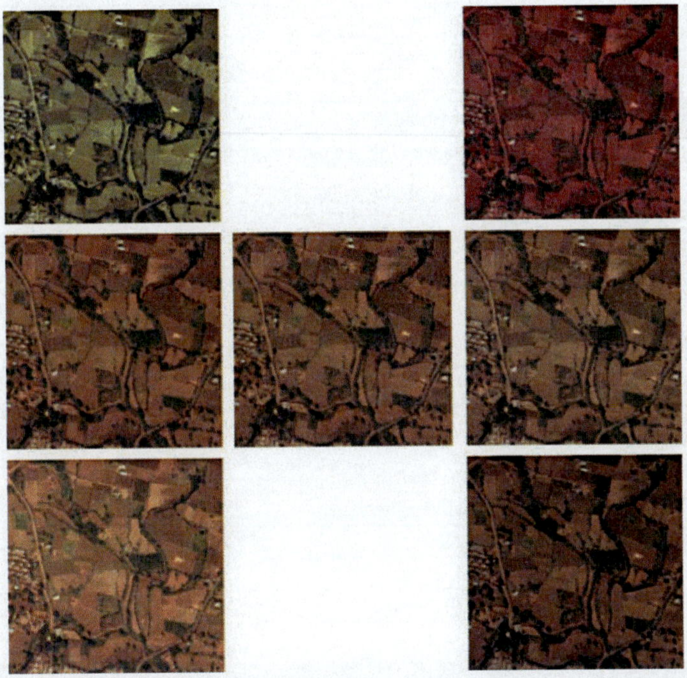

Fig. 3 High-resolution remote sensing image with enhanced data (center is the original image)

3.4 Model Training and Evaluation Criteria

In order to prevent the model from over fitting, this research carried out data enhancement, including horizontal flip, vertical flip, diagonal flip, color jitter, image shift, and zoom. For FD-LinkNet, this study uses the sum of BCE (binary cross entropy) and Dice Loss as the loss function. The calculation formula is as follows:

$$L = 1 - \frac{\sum_{i=1}^{N} |p_i \cap GT_i|}{\sum_{i=1}^{N} (|p_i| + |GT_i|)} + \sum_{i=1}^{N} BCELoss(p_i, GT_i) \tag{1}$$

This research chooses Adam as the optimizer. Initially set the learning rate to 0.0002, and when the training loss is observed to decrease slowly, the learning rate is increased by 2 times, and when there are 5 loss reductions, the training is exited. The batch size in the training phase is fixed at 4, and the network convergence in this study requires about 200 epochs (Table 1).

Table 1 Training parameter settings

Parameters	Values
Batch size	4
Learning rate	0.0002
Threshold	5
Epoch	200

4 Results

In this study, we trained our model on the DeepGlobe road extraction dataset. In order to compare with the FD-LinkNet network model, we also trained two network models, including a deep UNet with 7 pooling layers as The benchmark model, which can cover images with a size of 1024 × 1024; a D-Linknet that uses a bottom-level cavity convolution. The results of the three models are shown in the Table 2 and Fig. 4.

Table 2 Rural road extraction results

Neural network	IOU
U-Net	0.6294
D-LinkNet	0.6412
FD-LinkNet	0.6819

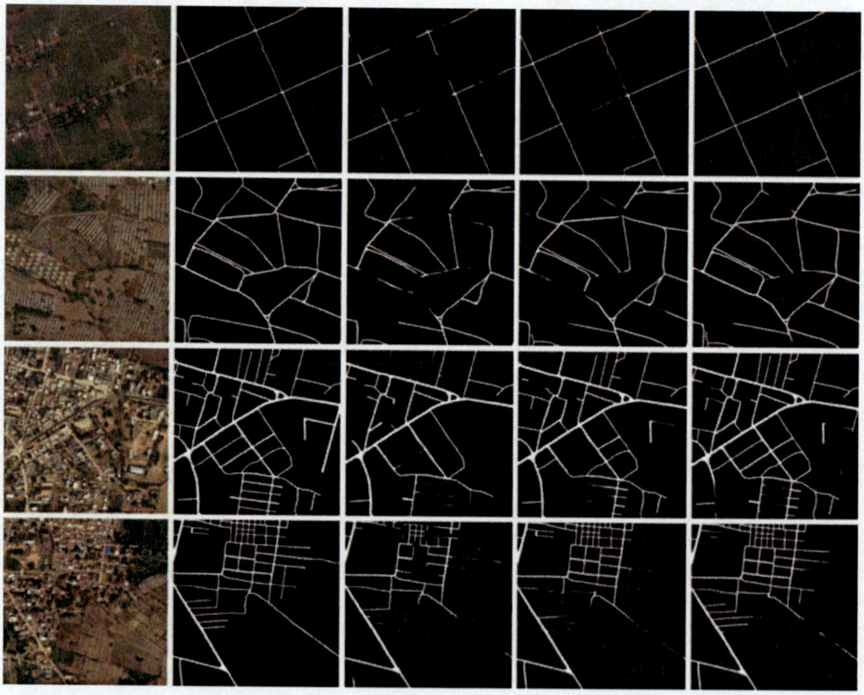

Fig. 4 Rural road extraction results

It can be found from the Table 2 and Fig. 4 that our benchmark model Unet has a larger receptive field, and the resolution of the central feature map is 8 × 8, which is too small to save detailed spatial information, and the connectivity of the agricultural road extraction results is poor. D-LinkNet adds hole convolution to make the network have better presentation capabilities, but it only uses dilation convolution in the fifth down sampling layer, which cannot cover 1024 × 1024 images and extracts multi-scale agricultural road information. FD-LinkNet adds dilation convolution to all down-sampling layers, which enables the network to extract multi-scale information, especially some small rural roads. Compared with D-LinkNet, FD-LinkNet obtains a larger receptive field and retains detailed information at the same time. It can also be seen through the IOU indicators that the FD-LinkNet network has the best results, and the extraction effect of roads is more accurate.

5 Conclusion

Based on the D-LinkNet segmentation network, this paper proposes an improved D-LinkNet network model to solve the problems of "false detection", "missing detection", and "misdetection" in high-resolution image road extraction. The model is named FD-LinkNet. The model adds dilation convolution in all down-sampling layers to expand the receptive field of the model to perceive multi-scale information and improve the extraction accuracy of small farm roads. Experiments show that the performance of the FD-LinkNet network model proposed in this article on the DeepGlobe road extraction dataset is better than the original D-LinkNet segmentation network

Acknowledgement This work was supported by the National Key Research and Development Program of China under Grant 2019YFB2102002.

References

1. Xiong S, Lei-Min LI, Yu-Qing H (2014) Unstructured road detection based on wavelet transform and K-means. J Comput Eng 40(2):158–161
2. Xu R et al (2018) Attention-mechanism-containing neural networks for high-resolution remote sensing image classification. J Remote Sens 10(10):1602
3. Ma W et al (2019) Double-branch multi-attention mechanism network for hyperspectral image classification. J Remote Sens 11(11):1307
4. Zaimbashi A, Li J (2020) Tunable adaptive target detection with Kernels in colocated MIMO radar. J IEEE Trans Signal Process 68:1500–1514
5. Corona E et al (2018) Active garment recognition and target grasping point detection using deep learning. J Pattern Recogn 74:629–641

6. Long J, Shelhamer E, Darrell T (2015) Fully convolutional networks for semantic segmentation. In: Proceedings of the IEEE conference on computer vision and pattern recognition
7. Ronneberger O, Fischer P, Brox T (2015) U-net: convolutional networks for biomedical image segmentation. In: International conference on medical image computing and computer-assisted intervention. Springer, Cham
8. Badrinarayanan V, Kendall A, Cipolla R (2017) Segnet: a deep convolutional encoder-decoder architecture for image segmentation. J. IEEE Trans Pattern Anal Mach Intell 39(12):2481–2495
9. Chaurasia A, Culurciello E (2017) Linknet: exploiting encoder representations for efficient semantic segmentation. 2017 IEEE visual commun image process (VCIP) IEEE
10. Chen L-C et al (2018) Encoder-decoder with atrous separable convolution for semantic image segmentation. In: Proceedings of the European conference on computer vision (ECCV)
11. Li P et al (2016) Road network extraction via deep learning and line integral convolution. In: 2016 IEEE international geoscience and remote sensing symposium (IGARSS), IEEE
12. Xia W et al (2017) A weakly supervised road extraction approach via deep convolutional nets based image segmentation. In: 2017 International workshop on remote sensing with intelligent processing (RSIP), IEEE
13. Pei D, Fanhuai S (2015) Road detection method based on small unmanned aerial vehicle image. J Comput Eng (2015)
14. Zhang Z, Liu Q, Wang Y (2017) Road extraction by deep residual u-Net. J IEEE Geosci Remote Sens Lett 99:1–5
15. Alshehhi et al (2017) Hierarchical graph-based segmentation for extracting road networks from high-resolution satellite images. J ISPRS J Photogrammetry Remote Sens
16. Zhou L, Zhang C, Wu M (2018) D-LinkNet: LinkNet with pretrained encoder and dilated convolution for high resolution satellite imagery road extraction. In: 2018 IEEE/CVF conference on computer vision and pattern recognition workshops (CVPRW) IEEE
17. Demir I et al (2018) DeepGlobe 2018: a challenge to parse the earth through satellite images

Geolocation Accuracy of Bistatic InSAR Configuration with Geostationary Transmitter and LEO Receivers

Yuekun Wang, Zhonghao Wei, Feng Tian, and Long Zhuang

Abstract Spaceborne GEO-LEO bistatic InSAR system with broad application prospects is an effective approach to obtain high-precision digital elevation model (DEM). Nevertheless, the geometry of bistatic interferometry is complex compared with traditional InSAR system, meanwhile the range equation is in the form of round-trip delay. In this paper, the InSAR geolocation equations, based on the GEO-LEO bistatic geometry, are established by combining SAR Doppler equation, phase equation and range equation. Furthermore, the geolocation errors caused by all error sources are derived by calculating the partial derivatives. Simulation experiments are carried out to analyze the geolocation errors, which can obtain meter-level precision and mainly caused by the absolute interferometric phase error, the range error and the baseline measurement error.

Keywords GEO-LEO · Bistatic · InSAR · Geolocation

1 Introduction

Spaceborne bistatic InSAR configuration is characterized by the flexible baseline adjustment ability for different terrains [1]. Therefore, it has become the main InSAR mode for high accuracy and efficiency. The LEO formation InSAR configuration can provide the desired spatial resolution and signal-to-noise (SNR) performance with less system complexity, but the revisit cycle is too long to satisfy the real-time demand [2]. The GEO-LEO InSAR configuration based on the geostationary transmitter and LEO receiver has the advantages of flexibility, lightweight receiving platform and low cost [3, 4]. It is conducive to the industrialization of remote sensing satellites and standardization of data application.

Y. Wang (✉) · Z. Wei · L. Zhuang
Nanjing Research Institute of Electronics Technology, Nanjing, China
e-mail: wangyk92@163.com

F. Tian
National Key Laboratory of Radar Signal Processing, Xidian University, Xi'an, China

L. Wang et al. (eds.), *Proceedings of the 7th China High Resolution Earth Observation Conference (CHREOC 2020)*, Lecture Notes in Electrical Engineering 757, https://doi.org/10.1007/978-981-16-5735-1_22

The solution of GEO-LEO bistatic InSAR geolocation is an important method to obtain the DEM data, and the positioning result directly determines the accuracy of DEM products. The Newton iteration method is a general approach for bistatic InSAR geolocation [5]. However, the iterative way is time-consuming and not conducive to accuracy analysis. Meanwhile, the geometry of GEO-LEO bistatic InSAR system is significantly different from the traditional LEO formation one [6, 7]. Therefore, based on the geometry of GEO-LEO bistatic InSAR configuration, the InSAR geolocation equations are established and the geolocation accuracies related to each error source are derived and analyzed.

In this paper, we derived and analyzed the geolocation accuracy of GEO-LEO bistatic InSAR configuration. In Sect. 2, the GEO-LEO InSAR geolocation equations are established. Then, the partial derivative of each error parameter based on the geoloaction equations is calculated, and the geoloaction accuracies related all error sources are obtained. In Sect. 3, the performance of the InSAR geoloaction is analyzed by simulation experiments. Conclusions close the paper in Sect. 4.

2 Geolocation Accuracy of GEO-LEO Bistatic InSAR Configuration

2.1 Bistatic InSAR Geolocation Equation

Here, we choose the Doppler equation and the range equation corresponding to the master image and the range equation corresponding to the slave image as the target location equations of the one-station fixed GEO-LEO bistatic InSAR system, which can be expressed as follows:

$$\begin{cases} \left\|\vec{S}_G - \vec{S}_T\right\| + \left\|\vec{S}_{L_m} - \vec{S}_T\right\| = R_G + R_{L_m} = R_m \\ \frac{\left(\vec{S}_T - \vec{S}_{Lm}\right)\cdot \vec{V}_{Lm}}{\lambda \cdot \left\|\vec{S}_T - \vec{S}_{Lm}\right\|} = f_{dc,m} \\ \left\|\vec{S}_G - \vec{S}_T\right\| + \left\|\vec{S}_{L_s} - \vec{S}_T\right\| = R_G + R_{L_s} = R_s \end{cases} \tag{1}$$

where $\|\cdot\|$ represents vector modulus. R, $\vec{S}$ and $\vec{V}$ are the slant range, the position vector, and the velocity vector, respectively. The subscript G, T, L_m and L_s stand for GEO transmitter, ground target, LEO #1 receiver and LEO #2 receiver, respectively. λ is the radar wavelength. $f_{dc,m}$ is Doppler centroid frequency corresponding to the master image. R_m can be obtained by Eq. (2). R_s can be calculated by interferometric measurement, given by Eq. (3).

$$R_m = r_{m0} + (m-1)\cdot \Delta r_m \tag{2}$$

$$R_s = R_m + \frac{\lambda}{2\pi} \cdot \Delta\varphi \tag{3}$$

where r_{m0} is the nearest two-way slant range corresponding to the master image. f_s is the range sampling frequency and $\Delta r_m = c/f_s$ is the sampling interval of echoes in the range direction. m represents pixel positions in the range direction. $\Delta\varphi$ is the absolute interferometric phase. Assuming that the 'stop-and-go' hypothesis error has been corrected in SAR imaging processing. Thus, the master image can be considered as a reference for the InSAR processing then the phase center position of the LEO SAR slave antenna corresponding to each pixel (m, n), through the interpolation of the initial azimuth time and its registration bias, can be obtained.

2.2 Geolocation Accuracy Analysis

The parameters $\vec{S}_G$, $\vec{S}_{L_m}$, $\vec{V}_{L_m}$, $\vec{S}_{L_s}$, R_m and $\Delta\varphi$ are utilized to obtain the InSAR geolocation result. However, these parameters is measured or processed with errors, which cause the final InSAR positioning errors. To analyze the relation between the parameter errors and positioning errors, we establish the sensitivity equations by calculating the partial derivative of the parameter equations. Then the sensitivity equations are derived by means of Cramer Rule.

The vectors in formula (1) is expressed by three-axis coordinates, namely $\vec{S}_G = (S_{Gx}, S_{Gy}, S_{Gz})$,$\vec{S}_T = (x_T, y_T, z_T)$, $\vec{S}_{L_m} = (S_{L_mx}, S_{L_my}, S_{L_mz})$, $\vec{S}_{L_s} = (S_{L_sx}, S_{L_sy}, S_{L_sz})$, $\vec{V}_{L_m} = (V_{L_{mx}}, V_{L_{my}}, V_{L_{mz}})$, the relationship between each error source and positioning error is as follows.

Assuming the range measurement error is ΔR_m, the following Eq. (4) can be obtained from Eq. (1):

$$\begin{cases} \mathrm{M}(x_T, y_T, z_T) = \sqrt{(x_G - x_T)^2 + (y_G - y_T)^2 + (z_G - z_T)^2} \\ \qquad + \sqrt{(x_{L_m} - x_T)^2 + (y_{L_m} - y_T)^2 + (z_{L_m} - z_T)^2} - R_m = 0 \\ \mathrm{D}(x_T, y_T, z_T) = (x_T - x_{L_m}) \cdot V_{L_{mx}} + (y_T - y_{L_m}) \cdot V_{L_{my}} + (z_T - z_{L_m}) \cdot V_{L_{mz}} \\ \qquad - \lambda \cdot f_{dc,m} \cdot \sqrt{(x_T - x_{L_m})^2 + (y_T - y_{L_m})^2 + (z_T - z_{L_m})^2} = 0 \\ \mathrm{S}(x_T, y_T, z_T) = \sqrt{(x_G - x_T)^2 + (y_G - y_T)^2 + (z_G - z_T)^2} \\ \qquad + \sqrt{(x_{L_s} - x_T)^2 + (y_{L_s} - y_T)^2 + (z_{L_s} - z_T)^2} - R_m - \frac{\lambda}{2\pi} \cdot \Delta\varphi = 0 \end{cases} \tag{4}$$

Based on (4), the sensitivity equations can be expressed as:

$$\frac{\partial x_T}{\partial R_m} = -\frac{|J_{Rmx}|}{|J|}\ \frac{\partial y_T}{\partial R_m} = -\frac{|J_{Rmy}|}{|J|}\ \frac{\partial z_T}{\partial R_m} = -\frac{|J_{Rmz}|}{|J|} \tag{5}$$

The symbol $|\cdot|$ stands for the determinant of the matrix. The Jacobian matrix in the formula (5) is:

$$J = \begin{bmatrix} \frac{\partial M}{\partial x_T} & \frac{\partial M}{\partial y_T} & \frac{\partial M}{\partial z_T} \\ \frac{\partial D}{\partial x_T} & \frac{\partial D}{\partial y_T} & \frac{\partial D}{\partial z_T} \\ \frac{\partial S}{\partial x_T} & \frac{\partial S}{\partial y_T} & \frac{\partial S}{\partial z_T} \end{bmatrix} \tag{6}$$

$$J_{Rmx} = \begin{bmatrix} \frac{\partial M}{\partial R_m} & \frac{\partial M}{\partial y_T} & \frac{\partial M}{\partial z_T} \\ \frac{\partial D}{\partial R_m} & \frac{\partial D}{\partial y_T} & \frac{\partial D}{\partial z_T} \\ \frac{\partial S}{\partial R_m} & \frac{\partial S}{\partial y_T} & \frac{\partial S}{\partial z_T} \end{bmatrix} \quad J_{Rmy} = \begin{bmatrix} \frac{\partial M}{\partial x_T} & \frac{\partial M}{\partial R_m} & \frac{\partial M}{\partial z_T} \\ \frac{\partial D}{\partial x_T} & \frac{\partial D}{\partial R_m} & \frac{\partial D}{\partial z_T} \\ \frac{\partial S}{\partial x_T} & \frac{\partial S}{\partial R_m} & \frac{\partial S}{\partial z_T} \end{bmatrix}$$
$$J_{Rmz} = \begin{bmatrix} \frac{\partial M}{\partial x_T} & \frac{\partial M}{\partial y_T} & \frac{\partial M}{\partial R_m} \\ \frac{\partial D}{\partial x_T} & \frac{\partial D}{\partial y_T} & \frac{\partial D}{\partial R_m} \\ \frac{\partial S}{\partial x_T} & \frac{\partial S}{\partial y_T} & \frac{\partial S}{\partial R_m} \end{bmatrix} \tag{7}$$

Thus, the positioning error $\Delta\vec{S}_T^{\Delta R_m}$ can be calculated as:

$$\Delta\vec{S}_T^{\Delta R_m} = \left[\frac{\partial x_T}{\partial R_m}, \frac{\partial y_T}{\partial R_m}, \frac{\partial z_T}{\partial R_m}\right] \cdot \Delta R_m \tag{8}$$

Similarly, the positioning error $\Delta\vec{S}_T^{\Delta\Phi}$ can be expressed as

$$\Delta\vec{S}_T^{\Delta\varphi} = \left[\frac{\partial x_T}{\partial \varphi}, \frac{\partial y_T}{\partial \varphi}, \frac{\partial z_T}{\partial \varphi}\right] \cdot \Delta\varphi \tag{9}$$

Assuming the three-axis satellite position errors of GEO transmitter and LEO master receiver are ΔS_{Gx}, ΔS_{Gy}, ΔS_{Gz} and ΔS_{Lmx}, ΔS_{Lmy}, ΔS_{Lmz}, respectively, there is

$$\begin{cases} \Delta\vec{S}_T^{\Delta S_{Gx}} = \left[\frac{\partial x_T}{\partial S_{Gx}}, \frac{\partial y_T}{\partial S_{Gx}}, \frac{\partial z_T}{\partial S_{Gx}}\right] \cdot \Delta S_{Gx} \\ \Delta\vec{S}_T^{\Delta S_{Gy}} = \left[\frac{\partial x_T}{\partial S_{Gy}}, \frac{\partial y_T}{\partial S_{Gy}}, \frac{\partial z_T}{\partial S_{Gy}}\right] \cdot \Delta S_{Gy} \\ \Delta\vec{S}_T^{\Delta S_{Gz}} = \left[\frac{\partial x_T}{\partial S_{Gz}}, \frac{\partial y_T}{\partial S_{Gz}}, \frac{\partial z_T}{\partial S_{Gz}}\right] \cdot \Delta S_{Gz} \end{cases}$$
$$\Delta\vec{S}_T^{\Delta S_G} = \Delta\vec{S}_T^{\Delta S_{Gx}} + \Delta\vec{S}_T^{\Delta S_{Gy}} + \Delta\vec{S}_T^{\Delta S_{Gz}} \tag{10}$$

$$\begin{cases} \Delta\vec{S}_T^{\Delta S_{Lmx}} = \left[\frac{\partial x_T}{\partial S_{Lmx}}, \frac{\partial y_T}{\partial S_{Lmx}}, \frac{\partial z_T}{\partial S_{Lmx}}\right] \cdot \Delta S_{Lmx} \\ \Delta\vec{S}_T^{\Delta S_{Lmy}} = \left[\frac{\partial x_T}{\partial S_{Lmy}}, \frac{\partial y_T}{\partial S_{Lmy}}, \frac{\partial z_T}{\partial S_{Lmy}}\right] \cdot \Delta S_{Lmy} \\ \Delta\vec{S}_T^{\Delta S_{Lmz}} = \left[\frac{\partial x_T}{\partial S_{Lmz}}, \frac{\partial y_T}{\partial S_{Lmz}}, \frac{\partial z_T}{\partial S_{Lmz}}\right] \cdot \Delta S_{Lmz} \end{cases} \tag{11}$$
$$\Delta\vec{S}_T^{\Delta S_{Lm}} = \Delta\vec{S}_T^{\Delta S_{Lmx}} + \Delta\vec{S}_T^{\Delta S_{Lmy}} + \Delta\vec{S}_T^{\Delta S_{Lmz}}$$

Assuming the three-axis velocity errors of LEO master receiver are ΔV_{Lmx}, ΔV_{Lmy} and ΔV_{Lmz}, there is

$$\begin{cases} \Delta\vec{S}_T^{\Delta V_{Lmx}} = \left[\frac{\partial x_T}{\partial V_{Lmx}}, \frac{\partial y_T}{\partial V_{Lmx}}, \frac{\partial z_T}{\partial V_{Lmx}}\right] \cdot \Delta V_{Lmx} \\ \Delta\vec{S}_T^{\Delta V_{Lmy}} = \left[\frac{\partial x_T}{\partial V_{Lmy}}, \frac{\partial y_T}{\partial V_{Lmy}}, \frac{\partial z_T}{\partial V_{Lmy}}\right] \cdot \Delta V_{Lmy} \\ \Delta\vec{S}_T^{\Delta V_{Lmz}} = \left[\frac{\partial x_T}{\partial V_{Lmz}}, \frac{\partial y_T}{\partial V_{Lmz}}, \frac{\partial z_T}{\partial V_{Lmz}}\right] \cdot \Delta V_{Lmz} \end{cases} \tag{12}$$
$$\Delta\vec{S}_T^{\Delta V_{Lm}} = \Delta\vec{S}_T^{\Delta V_{Lmx}} + \Delta\vec{S}_T^{\Delta V_{Lmy}} + \Delta\vec{S}_T^{\Delta V_{Lmz}}$$

Assuming the three-axis baseline errors between LEO receivers are ΔB_{Lx}, ΔB_{Ly} and ΔB_{Lz}, there is

$$\begin{cases} \Delta\vec{S}_T^{\Delta B_{Lx}} = \left[\frac{\partial x_T}{\partial B_{Lx}}, \frac{\partial y_T}{\partial B_{Lx}}, \frac{\partial z_T}{\partial B_{Lx}}\right] \cdot \Delta B_{Lx} \\ \Delta\vec{S}_T^{\Delta B_{Ly}} = \left[\frac{\partial x_T}{\partial B_{Ly}}, \frac{\partial y_T}{\partial B_{Ly}}, \frac{\partial z_T}{\partial B_{Ly}}\right] \cdot \Delta B_{Ly} \\ \Delta\vec{S}_T^{\Delta B_{Lz}} = \left[\frac{\partial x_T}{\partial B_{Lz}}, \frac{\partial y_T}{\partial B_{Lz}}, \frac{\partial z_T}{\partial B_{Lz}}\right] \cdot \Delta B_{Lz} \end{cases} \tag{13}$$
$$\Delta\vec{S}_T^{\Delta B_L} = \Delta\vec{S}_T^{\Delta B_{Lx}} + \Delta\vec{S}_T^{\Delta B_{Ly}} + \Delta\vec{S}_T^{\Delta B_{Lz}}$$

Considering these error sources are independent from each other, the total positioning error ΔS_T can be calculated as:

$$\Delta S_T = \sqrt{(\Delta S_T^{\Delta R_m})^2 + (\Delta S_T^{\Delta \varphi})^2 + (\Delta S_T^{\Delta S_G})^2 + (\Delta S_T^{\Delta S_{Lm}})^2 + (\Delta S_T^{\Delta V_{Lm}})^2 + (\Delta S_T^{\Delta B_L})^2} \tag{14}$$

Thus, we have established the relationship between all error sources and positioning errors in the GEO-LEO bistatic InSAR configuration.

Table 1 Simulation parameters

Parameters	Symbol	Value	Units
Position error of GEO transmitter	$\Delta\vec{S}_G$	(3, 3,3)	m
Position error of LEO master receiver	$\Delta\vec{S}_{Lm}$	(0.1, 0.1,0.1)	m
Velocity error of LEO master receiver	$\Delta\vec{V}_{Lm}$	(5, 5, 5)	mm/s
Range error	ΔR_m	3	m
Absolute interferometric phase error	$\Delta\varphi$	27	deg
Baseline error between two LEO satellites	$\Delta\vec{B}_L$	5	mm

3 Simulation Experiments

To analysis the geolocation accuracy of bistatic InSAR configuration with geostationary transmitter and two LEO receivers, the measurement errors of each error source are adopted in Table 1.

According to the simulation parameters and the relationship of the geolocation accuracy and error sources in Sect. 2, $\Delta\vec{S}_T^{\Delta R_m}$, $\Delta\vec{S}_T^{\Delta\Phi}$, $\Delta\vec{S}_T^{\Delta S_G}$, $\Delta\vec{S}_T^{\Delta S_{Lm}}$, $\Delta\vec{S}_T^{\Delta V_{Lm}}$ and $\Delta\vec{S}_T^{\Delta B_L}$ are 1.6, 4.0, 0.8, 0.2, 0.9 and 1.6 m, respectively. Therefore, the absolute interferometric phase error $\Delta\varphi$ has the great impact on the geolocation accuracy. While the GEO and LEO satellite position and velocity errors $\Delta\vec{S}_G$, $\Delta\vec{S}_{Lm}$, $\Delta\vec{V}_{Lm}$ have less impact on the positioning accuracy. The positioning error caused by range error ΔR_m and baseline measurement error $\Delta\vec{B}_L$ is moderate. According to Eq. (14), the total positioning error is about 4.76 m.

The previous simulation experiment is carried out with the lookangle of 40°. Here, the additional experiments are carried out to obtain the positioning errors caused by each error source with the lookangles of LEO satellite from 25° to 60°. The experimental results are shown in Fig. 1. It can be seen from Fig. 1 that the positioning errors caused by the error sources increase with the change of lookangle, except the parameter $\Delta\vec{B}_L$. The red solid curve represents the total positioning error caused by all error sources. Obviously, the total positioning error reaches the smallest when the lookangle is 40° under the simulation parameters in Table 1.

4 Summary

In this paper, we derived and analyzed the geolocation accuracy of GEO-LEO bistatic InSAR configuration. Based on the geometry, the InSAR geolocation equations are established, then the geolocation errors caused by all error sources are derived by calculating the partial derivatives. The simulation results validate the

Fig. 1 The positioning error varies with the lookangle of the LEO satellite

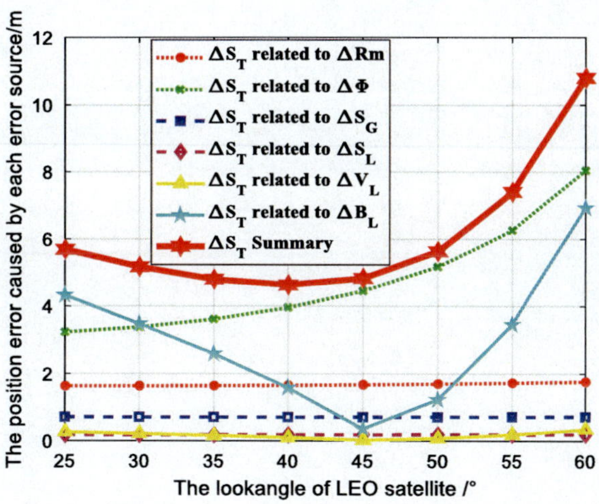

effectiveness of GEO-LEO bistatic InSAR mode, which can obtain meter-level geolocation precision and mainly caused by the absolute interferometric phase error, the range error and the baseline measurement error.

References

1. Krieger G, Moreira A (2006) Spaceborne bi- and multistatic SAR: potential and challenges. IEE Proc Radar Sonar Navig 153(3):184–198
2. Krieger G, Moreira A, Fiedler H et al (2007) TanDEM-X: a satellite formation for high-resolution SAR interferometry. IEEE Trans Geosci Remote Sens 45(11):3317–3341
3. Sarabandi K, Kellndorfer J, Pierce L (2003) GLORIA: geostationary/low-earth orbiting radar image acquisition system: a multi-static GEO/LEO synthetic aperture radar satellite constellation for earth observation. IEEE Int Geosc Remote Sens Symp Toulouse (2):773–775
4. Wang Y, Liu Y, Li Z, Suo Z, Fang C, Chen J (2017) High-resolution wide-swath imaging of spaceborne multichannel bistatic SAR with inclined geosynchronous illuminator. IEEE Geosci Remote Sens Lett 14(12):2380–2384
5. Zeng T, Zhu M, Hu C, Tian W, Long T (2015) Experimental results and algorithm analysis of DEM generation using bistatic SAR interferometry with stationary receiver. IEEE Trans Geosci Remote Sens 53(11):5835–5852
6. Sun Z, Wu J, Yang J, Huang Y, Li C, Li D (2016) Path planning for GEO-UAV bistatic SAR using constrained adaptive multiobjective differential evolution. IEEE Trans Geosci Remote Sens 54(11):6444–6457
7. Sun ZC, Wu JJ, Pei JF, Li ZY, Huang YL, Yang JY (2016) Inclined geosynchronous spaceborne-airborne BiSAR: performance analysis and mission design. IEEE Trans Geosci Remote Sens 54(1):343–357

Preliminary Evaluation of the Stereo Mapping Accuracy of the Gaofen-7

Liping Zhao, Zhi Fang, and Qifeng Chu

Abstract The Gaofen-7 (GF-7) is China's new-generation sub-meter stereo mapping satellite and an important part of the CHEOS (China High Resolution Earth Observation System). The stereo mapping camera carried on GF-7 allows acquisition of along-track stereoscopic pairs with a swath width greater than 20 km, a resolution better than 0.8 m, and a base-to-height ratio of approximately 0.6. GF-7 stereo data over Sichuan and Heilongjiang test sites are used for the evaluation and analysis of 1:10,000 scale stereo mapping accuracy. The test results show that, according to China's mapping specifications, products such as digital line graphic (DLG), digital terrain model (DTM), digital surface model (DSM) and digital orthophoto map (DOM) can meet the accuracy requirements for hilly, mountainous and alpine terrain. For the flat terrain type with the highest accuracy indicators, products such as DTM, DSM and DOM can also achieve the required accuracy; but for the DLG products, the elevation accuracy is not always able to meet the specifications, which need to be further studied.

Keywords Gaofen-7 · Stereo · Accuracy · 1:10,000 · DLG · DTM · DSM · DOM

L. Zhao (✉)
Land Satellite Remote Sensing Application Center, Ministry of Nature Resource, Beijing, China
e-mail: zhaolpwww@163.com

Z. Fang
The Third Geoinformation Mapping Institute of Ministry of Nature Resource, Chengdu, China

Q. Chu
Heilongjiang Institute of Geomatics Engineering, Harbin, China

L. Wang et al. (eds.), *Proceedings of the 7th China High Resolution Earth Observation Conference (CHREOC 2020)*, Lecture Notes in Electrical Engineering 757, https://doi.org/10.1007/978-981-16-5735-1_23

1 Introduction

The Gaofen-7 (GF-7) is the first sub-meter resolution optical stereo mapping satellite in China, and reached orbit in November 3, 2019. GF-7 is part of the CHEOS (China High Resolution Earth Observation System) family with the primary objective to provide high-resolution stereo images and high precision laser altimetry data for 1:10,000 scale stereo mapping, natural resources monitoring and management, urban and rural construction, statistical investigation, etc.

As a new generation of mapping satellites, the instruments onboard the spacecraft are a double-line array stereo mapping camera (SMC) and a laser altimeter. The SMC for obtaining the stereo images consists of two telescopic cameras, one of which points to the backward at 5°, and the other is offset by 26° forward in the flight direction, called the BWD camera and FWD camera, respectively. The BWD camera can provides panchromatic (PAN) imagery of 0.65 m resolution and four-band multispectral (MS) imagery with blue, green, red and near-infrared (NIR) at 2.6 m resolution, and the FWD camera is able to acquire 0.8 m panchromatic data. Swath width of stereo images is more than 20 km (see Tables 1 and 2).

Table 1 Main characteristics of the GF-7

Launch date	November 3, 2019
Design life	8 years
Altitude	506 km
Repeat cycle	59 days
Revisit time	5 days
Pointing capability in cross-track	±32°

Table 2 Specifications of the GF-7 stereo camera

Imager type	Push-broom with TDI capability
Swath width (stereo)	>20 km
GSD at nadir	FWD PAN: 0.8 m BWD PAN: 0.65 m BWD MS: 2.60 m
Wavelength	Panchromatic: 0.45–0.90 μm Multispectral: Blue: 0.45–0.52 μm Green: 0.52–0.59 μm Red: 0.63–0.69 μm NIR: 0.77–0.89 μm
Off-nadir angle along track	FWD: +26° BWD: −5°
Base-to-height ratio	0.6

GF-7 is also the follow-up satellite of the Ziyuan-3 (ZY-3). The resolution of stereo image has been optimized from 2.5 m of ZY-3 to 0.8 m, and the mapping accuracy is increased from 1:50,000 scale to 1:10,000 scale. At present, there are three ZY-3 satellites operating simultaneously in orbit. They were launched in January 2012, May 2016 and July 2020 respectively. ZY-3 and GF-7 can form a stereoscopic constellation, which can greatly improve the observation frequency and data acquisition efficiency.

1:10,000 scale stereo mapping is the primary task of the GF-7 mission. In this paper, two experimental areas in Sichuan and Heilongjiang, China are selected to carry out trial production and accuracy analysis of digital line graphic (DLG), digital terrain model (DTM), digital surface model (DSM) and digital orthophoto map (DOM).

2 Geometric Model

2.1 *Overview*

The geometric stability of the spacecraft bus and the stereo camera is a prerequisite for high-precision stereo mapping. The stability of the structure is solved by the installation of the integrated bracket of the main payloads, such as the stereo camera, laser altimeter, star sensor and gyro. The three-axis attitude control with high stability is realized by suppressing the disturbance driven by the satellite sailboard and data transmission antenna. Based on the multi-field star map fusion technology, the accuracy of the three-axis attitude measurement is further improved.

2.2 *Camera Characteristics*

In order to ensure the imaging quality and geometric stability of the camera during in-orbit operation, the stereo mapping camera integrates such space camera design technologies as high geometric stability optical system, precise temperature control measures, flexible support and deployable hood.

2.3 *RPC Model*

In general, the geometric accuracy of the RPC (Rational Polynomial Coefficients) model constructed by the ratio of cubic polynomials is equivalent to that of the corresponding strict physical model, which will not affect the geometric analysis.

In this paper, the RPC model attached with the image data is used for the bundle adjustment and analysis of stereo mapping accuracy.

3 Dataset

3.1 Study Area

The two research areas are located in Heilongjiang Province in northeast China and Sichuan Province in southwest China, respectively, as shown in Figs. 1 and 2. The Sichuan test area contains all four terrain categories required by China's mapping specifications, namely flat, hilly, mountainous and alpine terrain. The experimental area of Heilongjiang has small topographic relief, including flat and hilly terrain.

3.2 GF-7 Data

The basic information of GF-7 data used in this research is shown in Table 3. Figures 3 and 4 show the thumbnail of each scene in the Sichuan test area. The data acquisition time is after April 4, 2020, that is, more than 5 months after the launch

Fig. 1 The schematic diagram of the study area and image coverage (Sichuan)

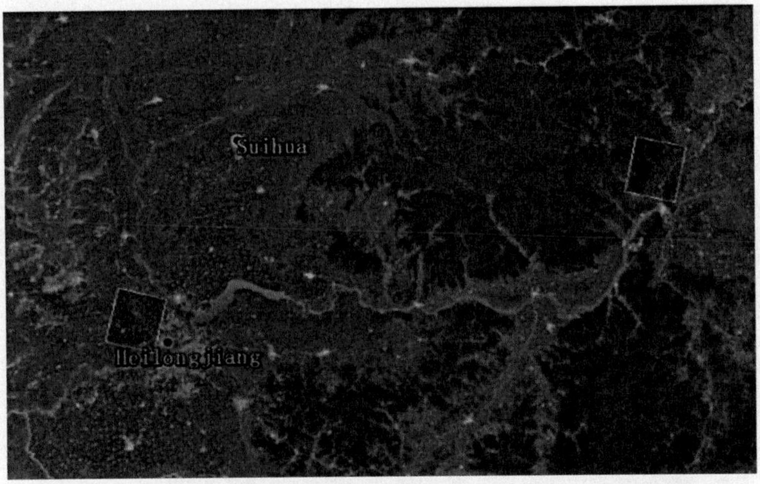

Fig. 2 The schematic diagram of the study area and image coverage (Heilongjiang)

Table 3 The GF-7 image data of the study area

Area	Data name	Date	GSD (meter)	Roll angle (degree)
Sichuan	GF7_SC_0430	2020–04–30	FWD PAN: 0.79 BWD PAN: 0.66 BWD MS: 2.63	−7.78
	GF7_SC_0529	2020–05–29	FWD PAN: 0.77 BWD PAN: 0.65 BWD MS: 2.60	5.63
Heilongjiang	GF7_HLJ_0404	2020–04-04	FWD PAN: 0.77 BWD PAN: 0.65 BWD MS: 2.58	6.14
	GF7_HLJ_0429	2020–04–29	FWD PAN: 0.80 BWD PAN: 0.68 BWD MS: 2.71	12.29

of the satellite. At this time, the status of the spacecraft bus and cameras has been solidified, and the overall geometric performance is stable, which can be used for geometric analysis. In addition, the roll angles of these data are all less than 13°, regardless of its influence on the geometric accuracy evaluation.

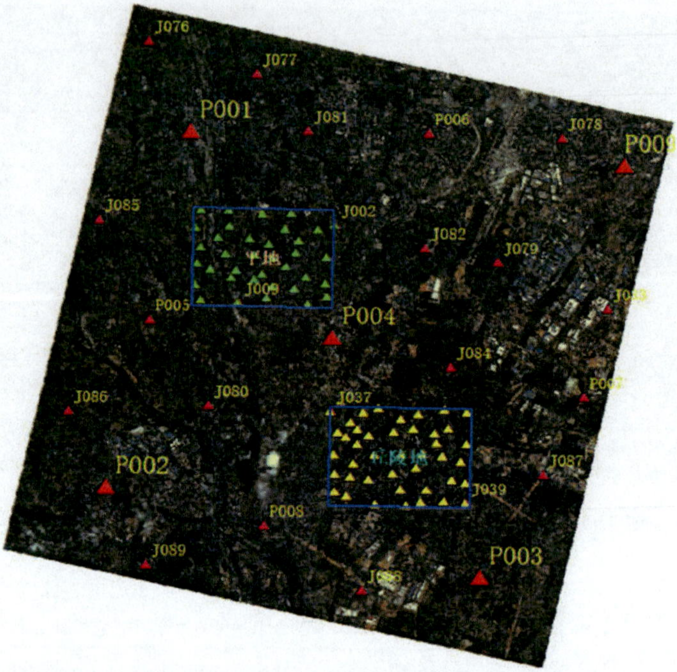

Fig. 3 The schematic diagram of the stereo mapping test area (flat and hilly terrain). The red triangles are the control points of the GF-7 stereo pair, and the pink triangles are the check points. The upper rectangle is the flat map area, the lower rectangle is the hilly map area, and the green and yellow triangles are the check points of the two maps respectively

3.3 GCP and ICP

The ground control points (GCP) and independent check points (ICP) used for the block adjustment of the stereo pairs, as well as the check points used for product accuracy evaluation of thc DEM, DSM, DOM, DLG, etc., are all measured by GPS in the field, with the horizontal and vertical accuracy better than 0.2 m. The corresponding image coordinates are obtained through manual measurement, which can reach sub-pixel accuracy. The point distribution is shown in Figs. 3 and 4.

4 Adjustment

For the GF-7 stereo pairs, a bundle block adjustment with GCP, ICP and tiepoints is performed, and the bias and affine correction in image space is estimated according to the previous experience [1–5]. Generally speaking, the root mean square error

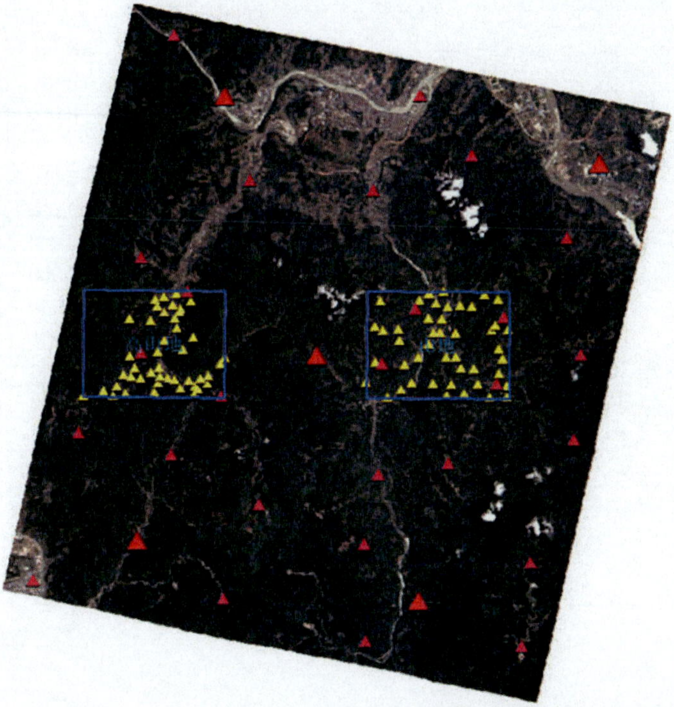

Fig. 4 The schematic diagram of the stereo mapping test area (mountainous and alpine terrain). The red triangles are the control points of the GF-7 stereo pair, and the pink triangles are the checkpoints. The rectangle on the right is the mountain map range, the rectangle on the left is the alpine map range, and the yellow triangles are the checkpoints of the two maps respectively

(RMSE) of the tie-points between the BWD and FWD panchromatic images is about 0.2 pixels, which is mainly caused by the tiny geometric distortion residual and slight attitude vibration.

Tables 4 and 5 summarize the adjustment accuracy in image space and object space in the Sichuan test area, and the following rules can be found:

(1) There are still slight systematic errors in the bias model;
(2) Based on the affine correction, the image accuracy in sample and line direction are both about 0.5 pixels;
(3) Although the affine model does not require many GCP, the accuracy and distribution of the control points will affect the accuracy of the adjustment.

Table 4 The accuracy of the GF-7 Sichuan test image (unit: pixel)

No	Image	Model	GCP/ICP	ΔSample		ΔLine	
				Mean	Std. Dev	Mean	Std. Dev
1	GF7_SC_0430_BWD	Bias	GCP: 25	0.000	0.859	0.000	0.616
2		Affine	GCP: 25	0.000	0.383	0.000	0.412
3		Affine	GCP: 5	0.000	0.216	0.000	0.326
			ICP: 20	−0.136	0.446	−0.160	0.448
4	GF7_SC_0430_FWD	Bias	GCP: 25	0.000	0.890	0.000	0.405
5		Affine	GCP: 25	0.000	0.596	0.000	0.356
6		Affine	GCP: 5	0.000	0.450	0.000	0.216
			ICP: 20	−0.248	0.634	−0.056	0.415
7	GF7_SC_0529_BWD	Bias	GCP: 27	0.000	0.964	0.000	0.529
8		Affine	GCP: 27	0.000	0.313	0.000	0.393
9		Affine	GCP: 5	0.000	0.088	0.000	0.065
			ICP: 22	−0.122	0.345	0.336	0.430
10	GF7_SC_0529_FWD	Bias	GCP: 27	0.000	1.043	0.000	0.715
11		Affine	GCP: 27	0.000	0.564	0.000	0.627
12		Affine	GCP: 5	0.000	0.448	0.000	0.347
			ICP: 22	−0.292	0.665	−0.099	0.754

5 Mapping Product

According to China's mapping specifications, after the adjustment of GF-7 stereo image pairs, the DLG is collected, the DSM and DTM are extracted, and the DOM is also produced. Using the GF-7 stereo image pairs, the DLG is collected and edited by experienced operators in stereo measurement environment. In the process of the DSM and DTM extraction, the DLG data is used as initial values and constraints, and the gross errors are removed through manual interactive editing. The DOM is made using PAN and MS image data of GF-7 BWD camera with the support of DTM. Finally, the accuracy of these products is evaluated using the checkpoints measured in the field.

5.1 DLG

Figures 5, 6, 7 and 8 respectively show the partial schematic diagrams of DLG products of different terrain types such as flat, hilly, mountainous and alpine terrain in Sichuan experimental area. Tables 6 and 7 list the horizontal and vertical

Table 5 The accuracy of the GF-7 Sichuan stereo pairs (unit: meter)

No	Stereo Pairs	Model	GCP/ICP	ΔNorth		ΔEast		ΔHeight	
				Mean	Std. Dev	Mean	Std. Dev	Mean	Std. Dev
1	GF7_SC_0430_BWD + FWD	Bias	GCP: 25	0.000	0.402	0.000	0.378	0.000	0.530
2		Affine	GCP: 25	0.000	0.255	0.000	0.301	0.000	0.413
3		Affine	GCP: 5	0.000	0.131	0.000	0.250	0.000	0.590
			ICP: 20	0.118	0.293	-0.095	0.324	0.109	0.381
4	GF7_SC_0529_BWD + FWD	Bias	GCP: 27	0.000	0.307	0.000	0.264	0.000	0.989
5		Affine	GCP: 27	0.000	0.257	0.000	0.239	0.000	0.727
6		Affine	GCP: 5	0.000	0.083	0.000	0.183	0.000	0.420
			ICP: 22	−0.146	0.301	−0.117	0.250	−0.482	0.822

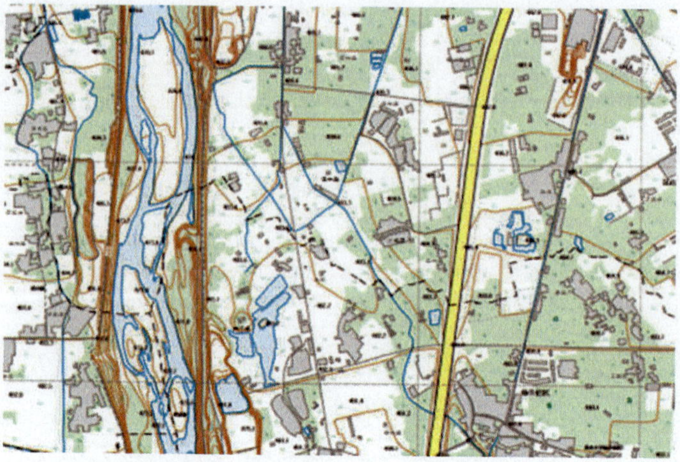

Fig. 5 The schematic diagram of the partial DLG (flat terrain)

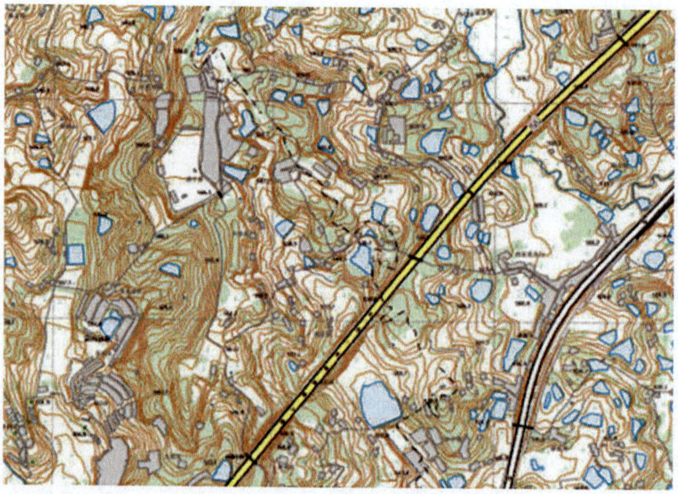

Fig. 6 The schematic diagram of the partial DLG (hilly terrain)

accuracy in the Sichuan and Heilongjiang test area. It should be noted here that the error is the accuracy of the stereo pair, not the accuracy of the DLG product in Sichuan area.

It can be found from Tables 6 and 7 that according to the current mapping specifications in China, the DLG accuracy can meet the requirements under the conditions of hilly, mountain and alpine terrain. However, the accuracy

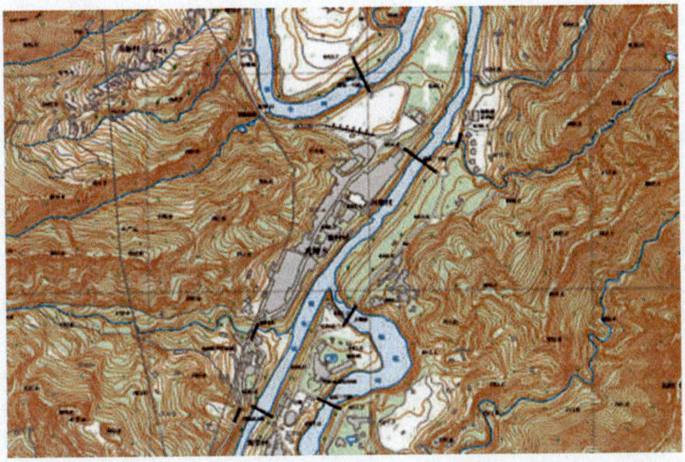

Fig. 7 The schematic diagram of the partial DLG (mountainous terrain)

Fig. 8 The schematic diagram of the partial DLG (alpine terrain)

Table 6 The accuracy of the DLG in the Sichuan test area (unit: meter)

Terrain		ICP	Max err	RMSE
Flat	Horizontal	36	1.496	0.860
	Vertical	36	0.483	0.218
Hilly	Horizontal	42	3.239	0.939
	Vertical	42	1.565	0.601
Mountainous	Horizontal	49	1.871	0.682
	Vertical	49	1.176	0.574
Alpine	Horizontal	47	1.741	0.741
	Vertical	47	0.871	0.379

Table 7 The accuracy of the DLG in the Heilongjiang test area (unit: meter)

Terrain		ICP	Max err	RMSE
Flat	Horizontal	42	1.207	0.734
	Vertical	42	−0.816	0.521
Hilly	Horizontal	28	0.998	0.542
	Vertical	28	−0.782	0.426

Table 8 The accuracy requirements for the 1:10,000 scale DLG product (unit: meter)

Terrain	Horizontal Accuracy (RMSE)	Vertical Accuracy (RMSE)	
		Elevation point	contour
Flat	5.0	0.35	0.5
Hilly	5.0	1.2	1.5
Mountainous	7.5	2.5	3.0
Alpine	7.5	4.0	6.0

requirements of flat terrain are very challenging for the GF-7 stereo image pair, and the accuracy of DLG products cannot always meet the 1:10,000 scale specification requirements (see Table 8). This issue needs to be discussed later.

5.2 *DSM and DTM*

Figures 9, 10, 11 and 12 respectively show the partial schematic diagrams of the DSM and DTM products for the 4 terrain categories in Sichuan test area. Tables 9, 10, 11 and 12 summarize the vertical accuracy in the Sichuan and Heilongjiang test area. Both the DSM and DTM products are accurate enough to meet the 1:10,000 scale specification requirements (see Tables 13 and 14). In addition, the extraction of DSM and DTM is the main goal of the GF-7 stereo mapping satellite, which requires further in-depth analysis in the future.

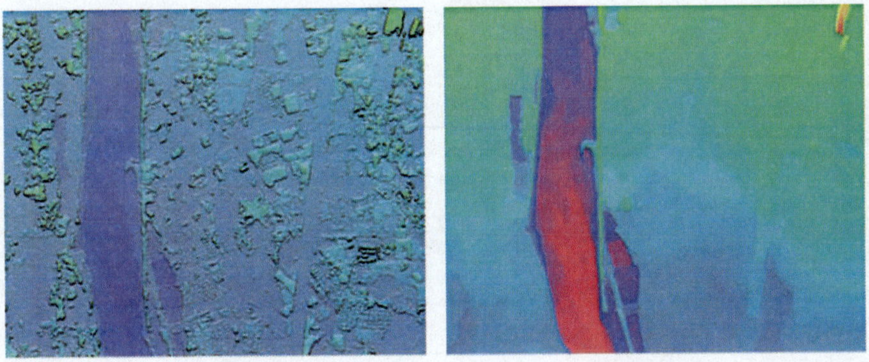

(a) DSM (b) DTM

Fig. 9 The schematic diagram of the partial DSM and DTM (flat terrain)

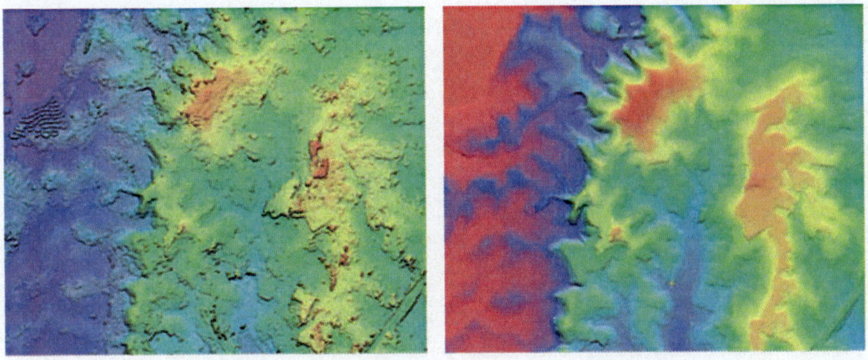

(a) DSM (b) DTM

Fig. 10 The schematic diagram of the partial DSM and DTM (hilly terrain)

5.3 DOM

Figures 13, 14, 15 and 16 respectively show the partial schematic diagrams of the DOM products for the 4 terrain categories in Sichuan experimental area. Tables 15 and 16 list the horizontal accuracy in the Sichuan and Heilongjiang test area. All DOM products are far more accurate than the 1:10,000 scale specification requires (see Table 17).

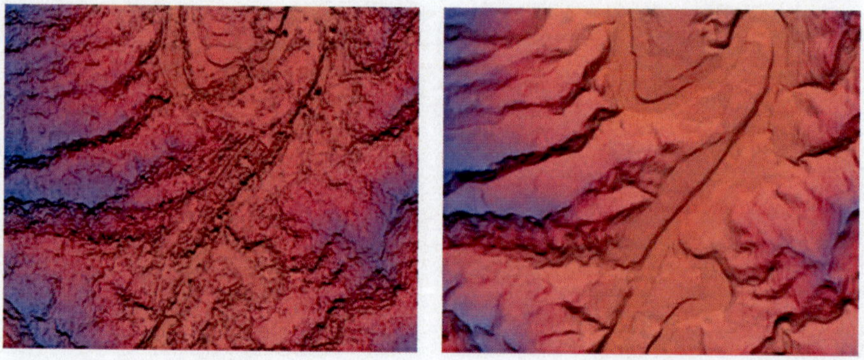

(a) DSM (b) DTM

Fig. 11 The schematic diagram of the partial DSM and DTM (mountainous terrain)

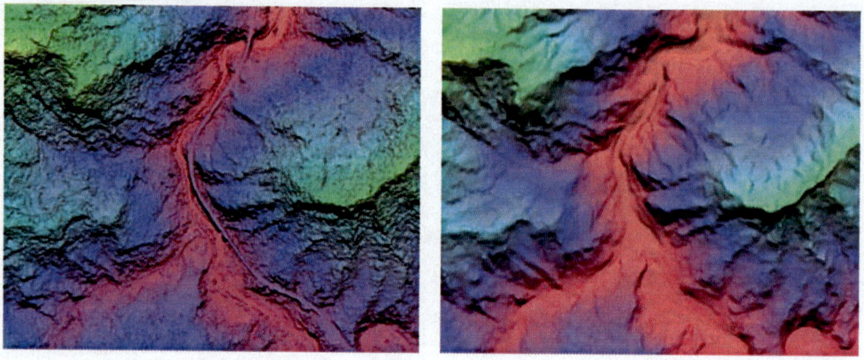

(a) DSM (b) DTM

Fig. 12 The schematic diagram of the partial DSM and DTM (alpine terrain)

Table 9 The vertical accuracy of the DSM in the Sichuan test area (unit: meter)

Terrain	ICP	Max err	RMSE
Flat	33	−1.428	0.690
Hilly	41	−2.921	1.106
Mountainous	39	−2.542	0.807
Alpine	35	−2.980	0.939

Table 10 The vertical accuracy of the DTM in the Sichuan test area (unit: meter)

Terrain	ICP	Max err	RMSE
Flat	33	−1.274	0.601
Hilly	41	−2.693	1.203
Mountainous	39	−3.322	1.473
Alpine	35	3.573	1.269

Table 11 The vertical accuracy of the DSM in the Heilongjiang test area (unit: meter)

Terrain	ICP	Max err	RMSE
Flat	36	−1.017	0.366
Hilly	34	1.486	0.632

Table 12 The vertical accuracy of the DTM in the Heilongjiang test area (unit: meter)

Terrain	ICP	Max err	RMSE
Flat	24	1.733	0.813
Hilly	22	1.335	0.616

Table 13 The vertical accuracy requirements for the 1:10,000 scale DSM product (unit: meter)

Terrain	RMSE		
	Level 1	Level 2	Level 3
Flat	0.35	0.50	0.70
Hilly	0.85	1.20	1.70
Mountainous	1.75	2.50	3.30
Alpine	3.50	5.00	6.70

Table 14 The vertical accuracy requirements for the 1:10,000 scale DTM product (unit: meter)

Terrain	RMSE		
	Level 1	Level 2	Level 3
Flat	0.5	0.7	1.0
Hilly	1.2	1.7	2.5
Mountainous	2.5	3.3	5.0
Alpine	5.0	6.7	10.0

6 Conclusion

Through the preliminary evaluation and analysis of the test data in the two experimental areas in Sichuan and Heilongjiang, it is found that GF-7 offers a good stereo geometric performance, and in most cases it can meet the accuracy requirements of 1:10,000 scale stereo mapping. The accuracy of DSM, DTM and DOM products can meet the mapping specification. For DLG products, except for

Fig. 13 The schematic diagram of the partial DOM (flat terrain)

Fig. 14 The schematic diagram of the partial DOM (hilly terrain)

Fig. 15 The schematic diagram of the partial DOM (mountainous terrain)

Fig. 16 The schematic diagram of the partial DOM (alpine terrain)

Table 15 The horizontal accuracy of the DOM in the Sichuan test area (unit: meter)

Terrain	ICP	Max err	RMSE
Flat	36	1.958	1.018
Hilly	41	4.175	1.415
Mountainous	48	2.574	0.911
Alpine	46	1.862	0.743

Table 16 The horizontal accuracy of the DOM in the Heilongjiang test area (unit: meter)

Terrain	ICP	Max err	RMSE
Flat	46	1.929	0.760
Hilly	41	1.287	0.619

Table 17 The horizontal accuracy requirements for the 1:10,000 scale DOM product (unit: meter)

Terrain	RMSE
Flat and hilly	5.0
Mountainous and alpine	7.5

the elevation accuracy challenges in the flat terrain, the accuracy specifications are met under other conditions. In general, the GF-7 is very suitable for large-area stereo mapping. In the follow-up work, the collection accuracy of DLG and the extraction quality of DSM and DTM need to be further discussed.

Acknowledgements This work was funded by the Special Fund for *China High Resolution Remote Sensing Application in Surveying and Mapping (Phase II)* (Grant No.: 42-Y30B04-9001-19/21)

References

1. Grodecki J (2001) IKONOS stereo feature extraction—RPC approach, ASPRS annual conference St. Louis
2. Grodecki J, Dial G (2003) Block adjustment of high-resolution satellite images described by rational functions. Photogramm Eng Remote Sens 69:59–70
3. Jacobsen K (2017) Problems and limitations of satellite image orientation for determination of height models. Int Arch Photogramm Remote Sens Spatial Inf Sci XLII-1/W1:257–264
4. Jacobsen K (2018) Systematic geometric image errors of very high resolution optical satellites. Int Arch Photogramm Remote Sens Spatial Inf Sci XLII-1:233–238
5. Zhao L, Fu X, Dou X, Liu H, Fang Z (2018) Comparison and analysis of accuracy of elevation extraction based on the ZY-3 01 and 02 satellites stereoscopic images. Int Arch Photogramm Remote Sens Spatial Inf Sci XLII-3:2379–2384. https://doi.org/10.5194/isprs-archives-XLII-3-2379-2018

Research on Cloud Detection Method of GaoFen-6 Wide Camera Data

Shiyun Ke, Mi Wang, Jinshan Cao, and En Long

Abstract Cloud detection is one of the research hotpots in high-resolution remote sensing images, and is an important part of remote sensing image preprocessing. The traditional physical threshold method based on pixel is inefficient when the amount of data is large. It is also easily affected by the noise points in the image, and the appropriate spectral threshold is difficult to determine. In view of the above problems, this paper starts from the GF-6 wide camera data, studies the spectral characteristics of GF-6 wide camera data, and selects the appropriate spectral detection band. Then, in order to improve the universality, using the maximum inter class variance method (OTSU) to adaptively obtain the appropriate cloud detection threshold. Finally, the image is segmented by simple linear iterative clustering algorithm (SLIC). Super pixel is used as the processing unit of subsequent image processing to replace single pixel for cloud detection. It both improves validity and accuracy. In this paper, by doing experiments and quantitative evaluation, it can be proved that the cloud detection algorithm can effectively and accurately detect the cloud region in the image. The accuracy was all higher than 90%, and the average accuracy of the experiment is 94.21%.

Keywords Cloud detection · Super-pixel segmentation · Spectral features · GaoFen-6 · Dynamic threshold

---

S. Ke · M. Wang (✉)
State Key Laboratory of Information Engineering in Surveying, Mapping and Remote Sensing, Wuhan University, Wuhan 430079, China
e-mail: wangmi@whu.edu.cn

J. Cao
School of Computer Science, Hubei University of Technology, 28 Nanli Road, Wuhan 430068, China

E. Long
Beijing Institute of Remote Sensing Information, 2 Xiaoyingdong Road, Beijing 100192, China

L. Wang et al. (eds.), *Proceedings of the 7th China High Resolution Earth Observation Conference (CHREOC 2020)*, Lecture Notes in Electrical Engineering 757, https://doi.org/10.1007/978-981-16-5735-1_24

1 Introduction

The launch of GaoFen series of satellites indicates that China's satellite technology is in the forefront of the world, and various technical indicators are rising step by step, also, the amount of satellite image data is increasing exponentially. However, cloud cover hinders the process of information and intelligent processing of satellite remote sensing images. The global cloud cover data show that more than 60% of the area above the earth's surface is covered by clouds [1], which makes a large amount of ground object information covered by clouds, and a great part of the value of satellite data is lost. Consequently, cloud detection is one of the research hotpots in high-resolution remote sensing images, and is an important part of remote sensing image preprocessing.

In this regard, domestic and foreign scholars has done a lot of researches on this topic. At present, cloud detection algorithms can be divided into physical threshold method, texture analysis method and pattern recognition method. Texture analysis method makes use of the texture features and spatial relations of the image. For example, Christidoulos analyzed and counted 55 kinds of texture features on meteorological satellite data, and automatically detected clouds based on multi texture feature space [2]. However, the accuracy of cloud detection is not high because of the variety and complexity of cloud types and the lack of uniform features in texture and shape. The cloud detection methods of pattern recognition mainly use artificial neural network and support vector machine and so on. Latry proposed a cloud detection technology, which uses the geometric parallax generated by multispectral reflectance and panchromatic/multispectral stereo angle as the input of SVM algorithm for cloud detection [3]. Although it has high precision and good effect, it is not suitable for large-scale engineering applications for the complex algorithm principle. In practical production, the most commonly used method is physical threshold method, because it is simple and easy to operate. The classical multi band fixed threshold method has the multi group threshold cloud detection method which uses 14 MODIS bands proposed by Ackerman. In this method, a fixed threshold is set for each band, which is used to judge a single pixel in 14 bands, and the detection results of the pixel are considered comprehensively [4]. However, when the underlying surface is desert, coastline and river, the detection effect is not ideal. Afterwards, due to the diversity of remote sensing image data and the improvement of cloud detection accuracy, fixed threshold gradually developed into dynamic threshold, multi band combination threshold and so on. The commonly used dual channel dynamic threshold method often selects visible light and near-infrared channels, but the threshold calculated by this method is easily affected by solar altitude angle and season, so it is not suitable for medium and high latitude regions [5]. Izak van ZYL Marais proposed a multi-channel fusion method, which combines the blue, green, red and near-infrared channels into a gray image, and sets the threshold value for cloud detection [6]. This method has high detection accuracy and efficiency, therefore, it shows advantages in solving the problem of real-time processing of satellites in orbit. Although, with the continuous progress of satellite technology, the band that can be detected by the sensors on the satellite is also increasing. The original four channel combination is not the most suitable band combination for each satellite image.

Most of the above methods are based on specific satellite sensors or underlying surface conditions. But, in fact, the parameters of sensors and underlying surface are complex and diverse, so a single cloud detection method is difficult to meet the requirements. In practical applications, physical spectral features, texture features and pattern recognition methods are often combined. Such as automatic multi feature combination (MFC) method, it combines physical threshold method and texture analysis method. Firstly, they using the physical threshold method to detect the cloud in the image by using the spectral characteristics of the cloud under a certain threshold value. Then, according to the texture and spatial characteristics of the cloud, the preliminary cloud mask is precisely detected by using the geometric and texture features of the cloud. The experimental results show that the accuracy of this method is higher than that of single spectral feature or texture feature [7].

In this paper, for GF-6 wide camera image, we propose a cloud detection algorithm based on super pixel segmentation and dynamic threshold. First of all, previous studies have shown that the multi-channel combination method is effective for cloud detection. Therefore, according to the spectral characteristics of GF-6 data, we select the appropriate band combination for spectral threshold detection according to its band characteristics. Then, by using simple linear iterative clustering algorithm, adjacent pixels with similar texture, color and brightness are clustered into image blocks, which is called super pixel. On the one hand, it introduces the spatial relationship and texture features of the image, on the other hand, it reduces the image dimension. Finally, doing histogram equalization on the image, and then using OTSU algorithm to dynamically select the threshold to further improve the universality of the algorithm. We compare the experimental results with those of multispectral fixed threshold method by visual interpretation and quantitative evaluation. The experimental results show that the quality index of this method is better than that of multi spectral threshold method, the average accuracy is 94.21%, and the detection accuracy is all higher than 90%.

2 Methods

Although many algorithms have emerged in cloud detection, physical threshold method is still popular in practical application because of its simple principle and easy implementation. In the physical threshold method, the most important things are selecting the threshold and improving the efficiency. On the premise of rapid development of satellite technology and computer technology, the traditional fixed threshold shows its disadvantage in accuracy. Because the differences in sensors, underlying surface conditions, shooting time and other conditions result in a big gap between satellite images. Only relying on fixed threshold is not enough, so the dynamic threshold method is proposed. It can dynamically select the appropriate threshold according to each image to adapt to a variety of situations. In this paper, we will use the maximum inter class variance method (OTSU) to select the dynamic threshold. In image segmentation, OTSU is one of the best algorithms when searching for threshold. It is easy to calculate and is not affected by image brightness and contrast. The algorithm principle will be shown in detail in Sect. 2.2.

The traditional physical threshold method is based on pixels for cloud detection. On the one hand, some abnormal pixels with high brightness may be mistakenly detected as clouds. On the other hand, when the amount of data is very large, the detection efficiency based on pixels will be reduced. In this case, we use the super pixel. A super pixel is an image block that integrates a single adjacent pixel. The pixels in the image block have similar texture, color, brightness and other attribute characteristics [8]. It divides the pixels into different sub regions according to the similarity degree of attributes, which not only makes the image block contain more image content information, especially spatial information, but also greatly reduces the dimension of the original image, which significantly improves the efficiency of subsequent image processing algorithms. We use super pixel instead of a single pixel as the basic unit of subsequent cloud detection. Calculating the spectral mean value of super pixel, and detecting the super pixel which is greater than the threshold value as cloud. The algorithm will be shown in detail in Sect. 2.1.

2.1 Super Pixel Segmentation Algorithm Based on SLIC

Researchers have proposed a variety of hyperpixel segmentation algorithms. Achanta made an empirical comparison on the adhesion ability, speed, storage efficiency and the impact on segmentation performance of image boundary based on five latest super-pixel algorithms. It is proved that the simple linear iterative clustering (SLIC) segmentation algorithm is better if comprehensive considering the shape and compactness of super pixels, running speed and boundary preservation [9].

In this algorithm, CIELab color space is used to replace the original three primary color mode of color image, and use three color values (L, a, b) of each pixel and its spatial coordinates (x, y) to form a 5-Dimensional feature vector. The distance between the 5-Dimensional feature vectors of two pixels measures the similarity between them. Closer distance represents a greater degree of similarity. The distance of vector is used as the standard to realize local clustering. The super pixels generated by SLIC algorithm are arranged orderly and compact. Not only the shape is good, but also the efficiency of the algorithm is high, and the segmentation result of the object is much better. The specific steps are as follows:

① Transforming the original color image represented by three primary colors into the CIELab color space composed of brightness and two color channels.
② Setting **K** seed points randomly, which are the initial cluster centers.
③ In the 3 × 3 window (or rectangular window of other sizes) around the seed point, calculating the gradient value for each pixel, and the formula is as follows:

$$\begin{aligned} Gradient(x,y) &= d_x(i,j) + d_y(i,j) \\ d_x(i,j) &= I(i+1,j) - I(i,j) \\ d_y(i,j) &= I(i,j+1) - I(i,j) \end{aligned} \tag{1}$$

The point with the minimum gradient value in the window is used to replace the original randomly set seed point. This step is to prevent the random seed point from being at the unreasonable position,such as the edge of the image or the noise pixels.

④ In the 2S × 2S window around the new seed point, measuring the similarity between all pixels and seed points according to Formula (2). Each pixel may be calculated more than once. Assigning the pixel to the super pixel of the seed point which has the maximum distance from the pixel. The similarity measurement relationship is as follows:

$$\begin{aligned} d_{lab} &= \sqrt{(l_k - l_i)^2 + (a_k - a_i)^2 + (b_k - b_i)^2} \\ d_{xy} &= \sqrt{(x_k - x_i)^2 + (y_k - y_i)^2} \\ D_i &= d_{lab} + \frac{m}{S} d_{xy} \end{aligned} \tag{2}$$

where, the meaning of parameters in the formula are as the Table 1:

Table 1 The meaning of parameters in the formula

Parameter	Meanings
$[l, a, b]$	The color value of the pixel
$[x, y]$	Coordinates of pixels
d_{lab}	Color difference between pixel K and pixel I in CIELab color space
d_{xy}	Difference in spatial distance between two pixels
S	Side length of superpixels
m	Scale parameters to balance spatial position and color
D_i	The degree of similarity between two pixels

⑤ Iterative optimization. In the fourth step, each pixel is clustered into the corresponding image block. Calculating the average value of the pixels in the same image block, and the average pixel is taken as the new clustering center, and then repeating the previous steps until the distance between the two clustering centers is within the threshold range, that is, when the seed points no longer change, the iteration ends.

⑥ Merge outliers to enhance connectivity. After the iteration, the size of the super-pixel may be too small, or a single small area may not be connected with the super-pixel to which it belongs. In this case, it is necessary to reclassify the problematic pixel or the super-pixel area that is too small to the adjacent super-pixel with similar attributes, so as to ensure that each super-pixel is completely connected.

Only one parameter needs to be set for super-pixel segmentation, that is, the number of initial superpixel **K**. This value not only determines the operation results of superpixel segmentation, but also affects the efficiency of the algorithm. Therefore, when using SLIC super-pixel segmentation experiment, the key point is to determine the number of initial superpixels. The setting of parameters should be selected according to different data sets.

2.2 OTSU Algorithm

OTSU is an algorithm to determine the threshold value of binary image segmentation, also known as the maximum inter class variance method. It thinks that if the image can be divided into two categories according to a certain attribute, there is a **T** value, which makes the difference between the two types of pixels divided by the **T** value as the segmentation standard is the largest (the maximum variance), and the **T** value is the best threshold when the image is divided into two categories.

Assuming that the image can be divided into c_1 (less than threshold **T**) and c_2 (greater than threshold **T**), the average spectral mean of these two types of pixels is m_1, m_2, and the global mean of the image is m_G. The probability that pixels are classified into c_1 and c_2 is p_1 and p_2 respectively. There are:

$$p_1 * m_1 + p_2 * m_2 = m_G \quad (3)$$

$$p_1 + p_2 = 1 \quad (4)$$

According to the concept of variance, the expression of variance between classes is:

$$\delta^2 = p_1(m_1 - m_G)^2 + p_2(m_2 - m_G)^2 \quad (5)$$

where,

$$p_1 = \sum_{i=0}^{k} p_i$$
$$m_1 = \frac{1}{p_1} * \sum_{i=0}^{k} ip_i \quad (6)$$
$$m_2 = \frac{1}{p_2} * \sum_{i=0}^{L-1} ip_i$$

In this experiment, based on multi-channel fusion image, we will use the spectral characteristics of different ground objects and clouds to distinguish cloud and non cloud. It is mainly utilize the high reflectivity of clouds in the visible and near-infrared bands. In the image, it appears as a "highlight" area.

2.3 *Optimization and Combination of SLIC and OTSU*

In this paper, we use the SLIC super pixel segmentation algorithm to generate the super-pixel object, and then the spectral mean value of the super-pixel is calculated according to formula (7). Then using OTSU algorithm to adaptively get the threshold. After,We judge the super pixels whose spectral mean is greater than or equal to the threshold as clouds, and the rest are non clouds. In this way, super pixel is used to replace the original single pixel for subsequent image processing, which introduces spatial relationship and reduces image dimension.

$$gray_i = \sum_{k=1}^{n_i} gray_k \quad (7)$$

where, $gray_i$ is the gray value of the super pixel **i**, n_i is the total number of pixels contained in the super-pixel **i**, and $gray_k$ is the gray value of the pixel **k** in the super-pixel **i**.

In the process of generating superpixels, the most important thing is to determine the number of initial superpixels **K**. According to individual image resolution and feature types, different optimal **K** values should be selected for different data sets and different purposes. Based on the experimental statistical analysis of GF-6 wide camera data, it is found that the algorithm has the best cloud detection effect when the number of super pixels **K** is **height** × **width/100**, that is, the initial size of super pixels is 100. Where, **height** and **width** are the length and width of the image respectively. The detailed analysis will be shown in Sect. 3.2.

OTSU algorithm can select the threshold **T** with the largest difference between the two types of pixels according to each image, which enhances the credibility and universality of threshold selection.This method is very suitable for cloud detection, because the difference of spectral characteristics between cloud and other ground objects is obvious, and it is easy to be used in practical production. However, the maximum inter class variance method is not good enough when there is a big difference in the foreground background size ratio or there are no obvious two types of pixels in the image, so we make corresponding adjustments according to the actual situation:

(1) Before the threshold is calculated, do the histogram equalization on the image. The essence of histogram equalization is to redistribute the gray level, expand the gray range occupied by a large number of pixels, so that the number of pixels in each gray value is more uniform. Histogram equalization can not only increase the contrast of the image, but also can highlight the texture and details of the image. In this way, when the threshold is selected according to OTSU algorithm, the threshold **T** can be obtained more accurately according to the maximum inter class variance.
(2) Define a range for the threshold: between 160 and 230. If the threshold is not in this range, taking **T** = 200 as the threshold value of cloud detection. In order to ensure that the two kinds of pixels segmented by the threshold are cloud or cloud like with high reflection and other ground objects, so as to prevent the accurate threshold **T** not obtained when there are no clouds or all clouds in the image.

3 Results and Discussion

3.1 Experimental Datasets

This experiment is based on GF-6 satellite to study the cloud detection method of GF-6 wide camera image. GF-6 satellite is the first optical remote sensing satellite with red edge band data in China. It also has a unique purple band (0.40~0.45 μm). It has high spatial and spectral resolution. The coverage of the wide camera is up to 800 km, and the image quality is high, which perfectly reflects the high performance index of China's satellites.

GF-6 satellite is equipped with two cameras, one is panchromatic and multispectral high-resolution camera, with an imaging width of 90 km, and the other is a multispectral medium resolution camera (wide camera), with an imaging width of 800 km. We will study the cloud detection method for the image data of the wide camera. The main parameters of the wide camera [10] are shown in the Table 2:

The image data used in this experiment mainly include: cloud, vegetation, residential areas, water, mountains and so on. The spectral curve of typical features in the image is shown in the following figure:

Table 2 Main performance parameters of GF-6 satellite wide range camera

Camera type	Linear CMOS push broom
Spectral band range	Blue band (B1): 0.45 ~0.52 μm Green band (B2): 0.52~0.59 μm Red band (B3): 0.63~0.69 μm Near infrared band (B4): 0.77~0.89 μm Red edge band 1(B5): 0.69~0.73 μm Red edge band 2(B6): 0.73~0.77 μm Purple band (B7): 0.40~0.45 μm Yellow band (B8): 0.59~0.63 μm
Spatial resolution	16 m
Imaging width	800 km
Quantification level	12 bit
Number of overlapping pixels	120 pixel
Focal length	548 mm
Ground sampling distance (GSD)	16 m

In the Fig. 1a, The four curves are the spectral curves of cloud, residential area, vegetation and water. We can find out that the spectral values of clouds in each band are much higher than those of other ground objects. In the Fig. 1b, the three curves are the spectral curves of cloud, highlight mountains and other mountains. Because the mountain area is mainly rock and soil, its spectral reflectance characteristics are more complex, the size of reflectivity has no fixed change characteristics, so there may be some areas with high reflectivity. As shown in the Fig. 1b, the spectral curves of some mountains (purple) have great difference with clouds within the eight spectral bands of GF-6. However, some bright mountain areas (yellow) also have small spectral curves differences with clouds, which is difficult to distinguish.

Through reading literature and research, we can know that, near 0.65 μm, the reflectance of vegetation and land is relatively low. Meanwhile, the reflectivity of clouds in this band is generally much higher than that of other typical ground

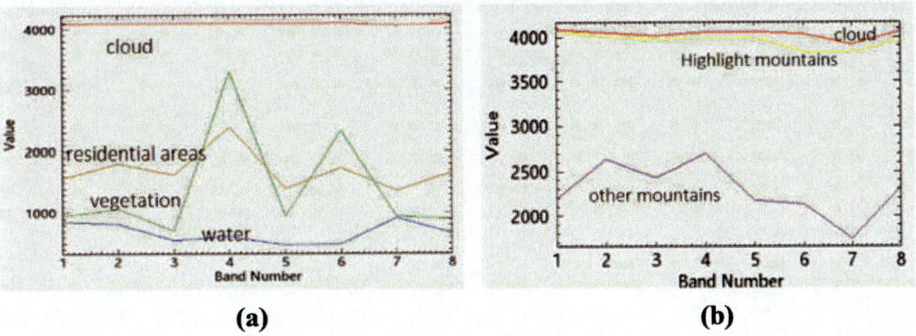

(a) (b)

Fig. 1 Spectral curves of typical objects in GF-6 image

objects, so visible light of 0.65 μm is often used as the preferred band for cloud detection [11]. In addition, the spectral characteristics of cloud in the near-infrared band are mainly affected by water content, showing the characteristics of absorption valley in the spectral curve. Therefore, the near-infrared band is often selected as the band for cloud detection [12]. The difference of spectral reflectance curves between snow and cloud is small. However, near the 1.64 μm band, the absorption of solar radiation by snow is strong, while the absorption of cloud is relatively weak, which results in the reflectivity of cloud higher than that of snow in this band. Therefore, the normalized indices of 0.65 μm and 1.64 μm can be used to distinguish cloud and snow [13]. But the index is sensitive to coastline, we can be modified by adding purple band [14].

To sum up, combined with the spectral characteristics of clouds and the band information of GF-6 wide range camera, we can see that, In the eight spectral bands of GF-6, the spectral curves of cloud and other ground objects such as water and vegetation are quite different, which is much higher than that of other typical ground objects. Especially in band B3, B5 and B7, the spectral curve difference between other ground features and clouds is the largest, which is shown in the image that the gray difference between cloud and other ground objects is large. However, it is difficult to distinguish it from snow and some mountains. After that, we will use the high reflection of cloud, that is, the gray difference between cloud and other ground objects in some band to detect cloud. According to the spectral band range of GF-6 wide range camera, the bands of cloud detection are red band (B3): 0.63 ~ 0.69μm, red edge band 1 (B5): 0.69 ~ 0.73μm and purple band (B7): 0.40 ~ 0.45μm. In order to make the experimental results clearer, we use ERDAS to cut all the images into 2791 × 2225 pixels, which is convenient for observation and display.

3.2 Comparative Analysis of Different Parameter Settings

Set the initial super pixel size to 50, 100, 500. Some experimental results and detailed analysis are shown below, where **S** is the size of the initial superpixel. The relationship between the size **S** and the number of super pixels **K** is as follows:

$$K = \frac{N}{S^2} \tag{8}$$

where, **N** is the total number of image pixels.

Figure 2 shows five images and their cloud detection results under different parameter settings, Among them, **(a)**, **(b)**, **(c)**, **(d)**, **(e)** are the original image, **(a1)**, **(b1)**, **(c1)**, **(d1)**, **(e1)** are the superposition display of the original image and the cloud detection result image when **S** = 100; **(a2)**, **(b2)**, **(c2)**, **(d2)**, **(e2)** are the superposition display of the original image and the cloud detection result image

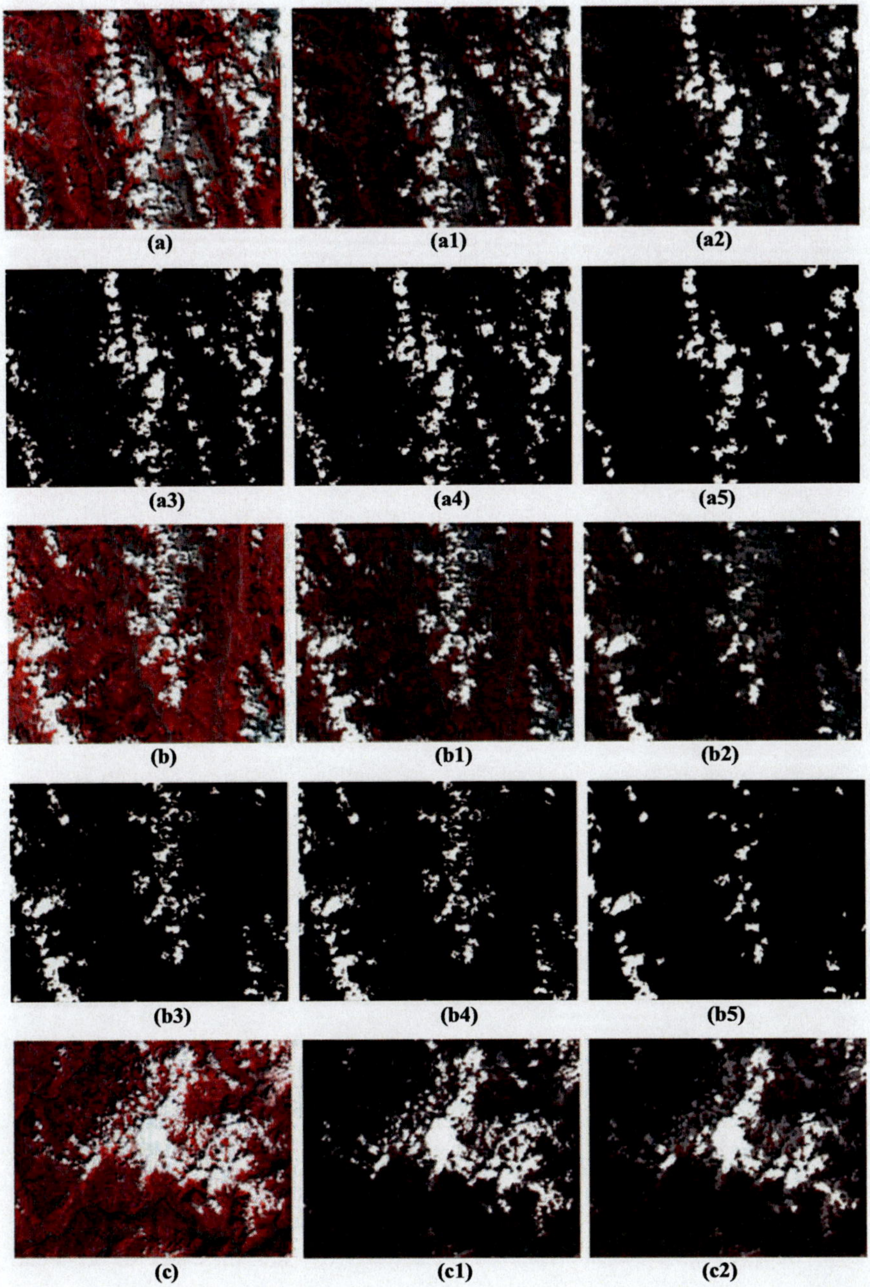

Fig. 2 Cloud detection results of some images

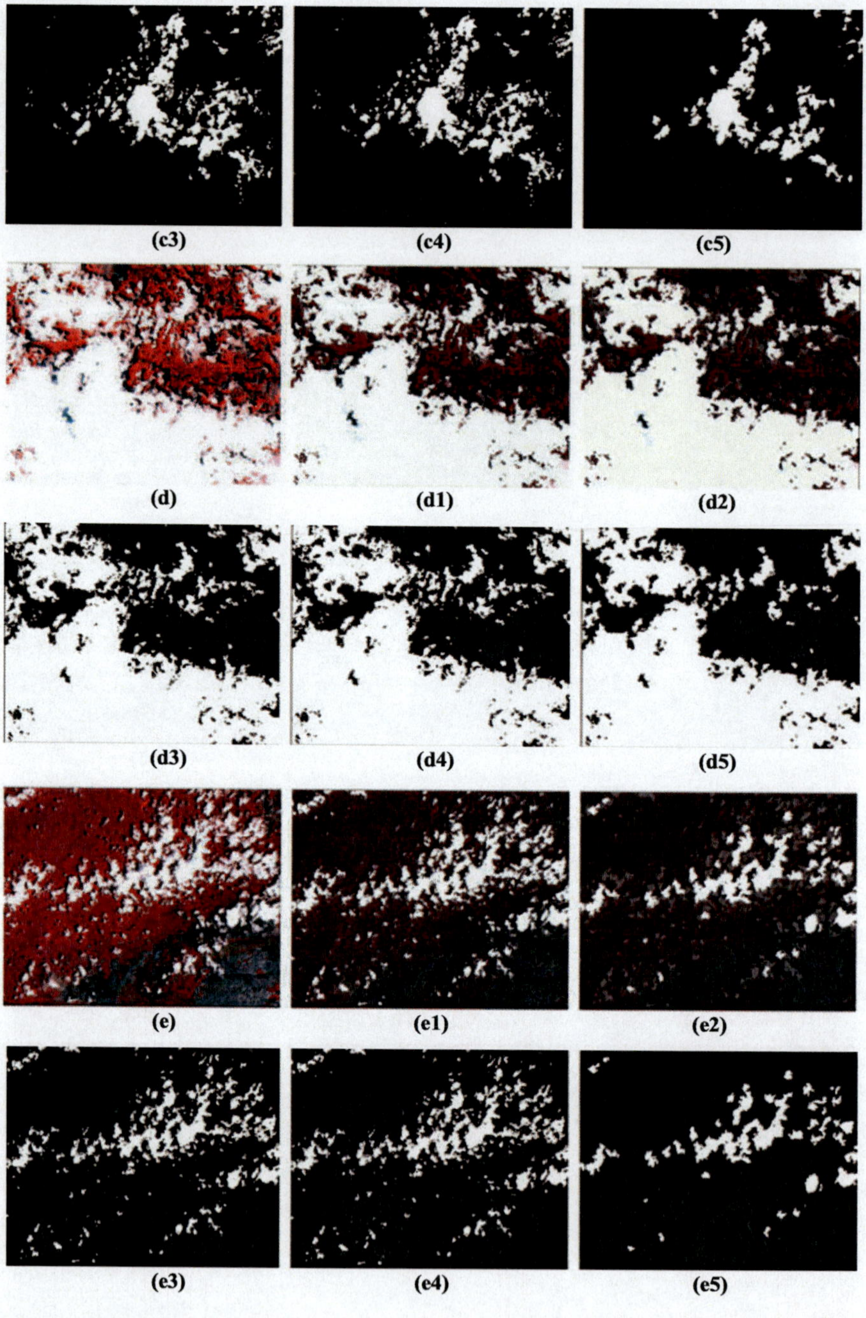

Fig. 2 (continued)

Table 3 The number of pixels in cloud detection under different super pixel sizes

	S = 50 (%)	S = 100 (%)	S = 500 (%)
Image a	11.74	11.74	8.69
Image b	8.26	8.26	5.32
Image c	12.44	12.44	8.85
Image d	58.81	58.81	56.60
Image e	12.97	12.97	8.41

when **S** = 500; **(a3)**, **(b3)**, **(c3)**, **(d3)**, **(e3)** are the cloud detection result image when **S** = 50; **(a4)**, **(b4)**, **(c4)**, **(d4)**, **(e4)** are the cloud detection result image when **S** = 100; **(a5)**, **(b5)**, **(c5)**, **(d5)**, **(e5)** are the cloud detection result image when **S** = 500. Table 3 shows the area proportion data of cloud area in cloud detection result image under different super-pixel sizes. Figure 3 shows the accuracy of cloud detection results under different super-pixel sizes through histogram and table, and Fig. 4 shows the algorithm operation speed under different super-pixel sizes through line chart.

In Table 3, we find that when **S** = 500, the detected cloud area is much smaller than the cloud detection results in the other two cases. Combined with the fusion images in Fig. 2 a2\b2, it is found that all cloud regions in the image, especially some smaller cloud regions, are not well detected, and there are many missed detection phenomena. Figure 3 also clearly shows that the accuracy is lower than the other two cases. The analysis shows that when the set value of **S** is large (the value of **K** is small), the total number of pixels in the super pixel will be more. Then some small cloud areas may become a super pixel with other ground objects, and then the average gray value of the super pixel may be less than the spectral threshold of the cloud boundary, so there may be missed detection.

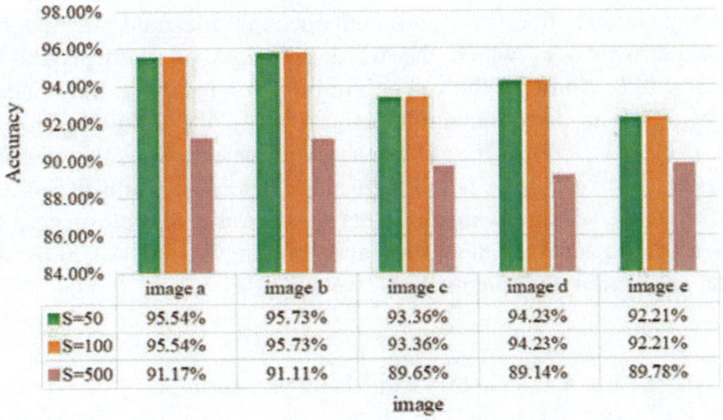

Fig. 3 Cloud detection accuracy under different super pixel sizes

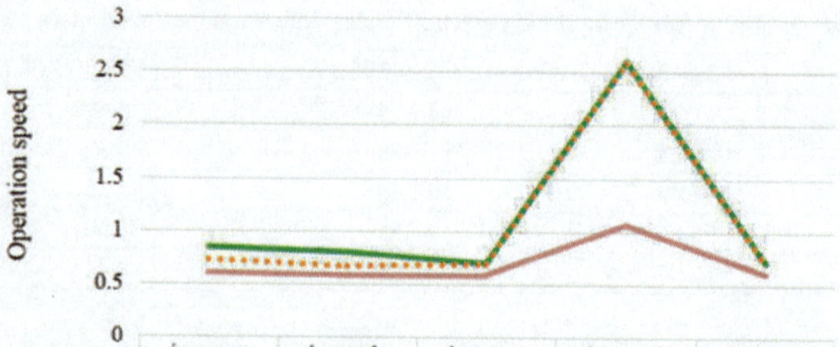

	image a	image b	image c	image d	image e
S=50	0.842	0.799	0.694	2.602	0.708
S=100	0.72	0.665	0.688	2.59	0.696
S=500	0.599	0.58	0.588	1.062	0.603

image

Fig. 4 Relationship between different super pixel size and operation speed

According to Fig. 2 and Table 3, when **S** = 50 and **S** = 100, the cloud detection results are almost the same. According to Fig. 4, the speed of the algorithm increases with the increase of the initial super pixel size. This shows that for the data of this experiment, **S** = 100 can meet the needs of cloud detection.

3.3 *Accuracy Evaluation*

So far, there is no uniform standard for cloud detection. The most commonly used method for cloud detection accuracy evaluation is to generate standard cloud image by visual interpretation, and to compare the cloud image identified by the algorithm with the standard cloud image to complete the accuracy evaluation. In this paper, we use the proposed algorithm and multispectral threshold method for cloud detection experiments, in which the fixed threshold of multispectral threshold method is **T** = 200. And take the visual interpretation results as the standard cloud detection results. The detection results are shown in the following figure, where **(a1)**, **(b1)**, **(c1)**, **(d1)**, **(e1)** in Fig. 5 are the cloud detection results of visual interpretation, **(a2)**, **(b2)**, **(c2)**, **(d2)**, **(e2)** are the cloud detection results of this algorithm, and **(a3)**, **(b3)**, **(c3)**, **(d3)**, **(e3)** are cloud detection results of multispectral threshold method. We will do accuracy evaluation and comparative analysis of the algorithm from visual evaluation and quantitative evaluation.

3.3.1 Comparative Analysis of Visual Interpretation

We selected some mountain area screenshots from GF-6 satellite remote sensing image. Figures 6 and 7 show the overlapping results of the three cloud detection

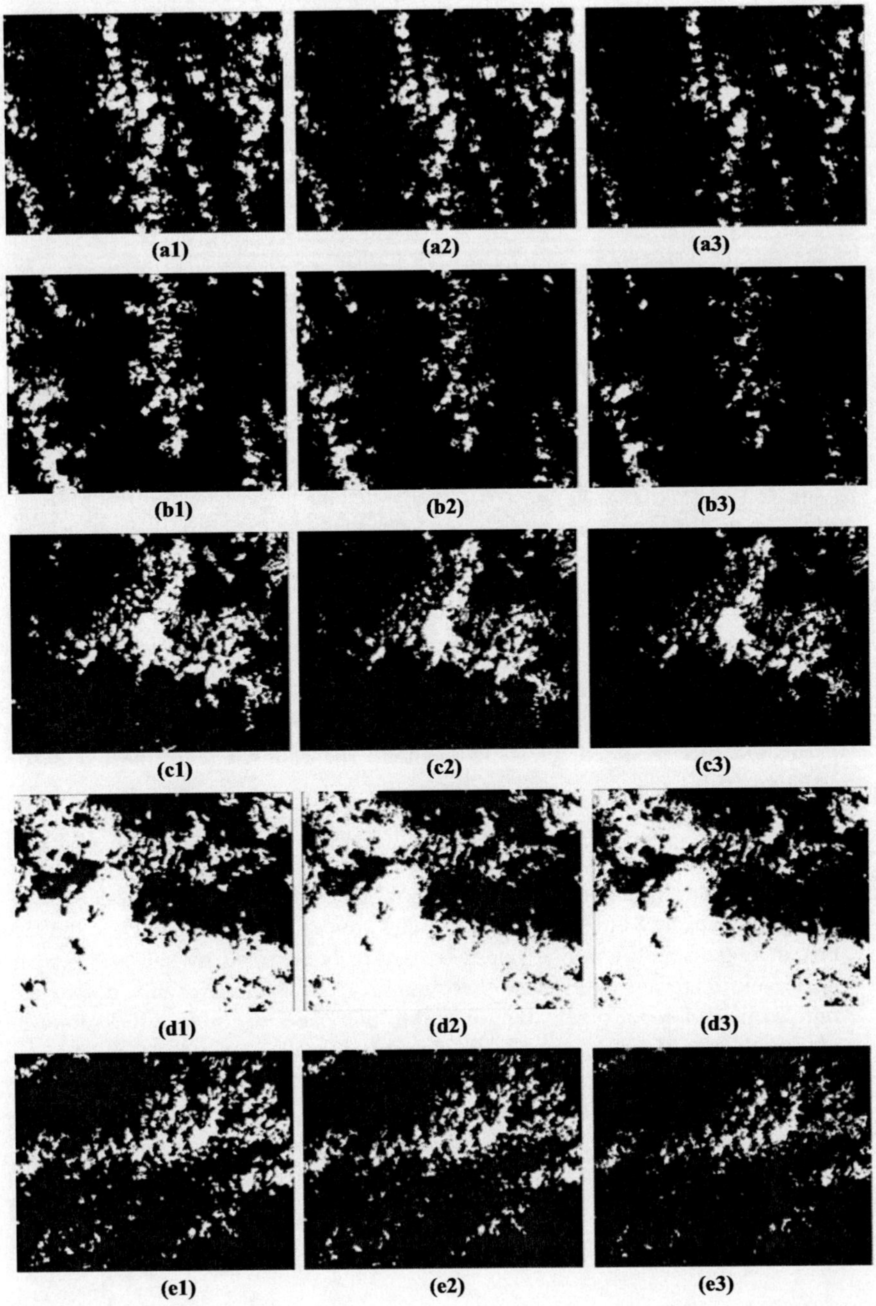

Fig. 5 The could detection result image of standard cloud image, algorithm in this paper, multispectral threshold method

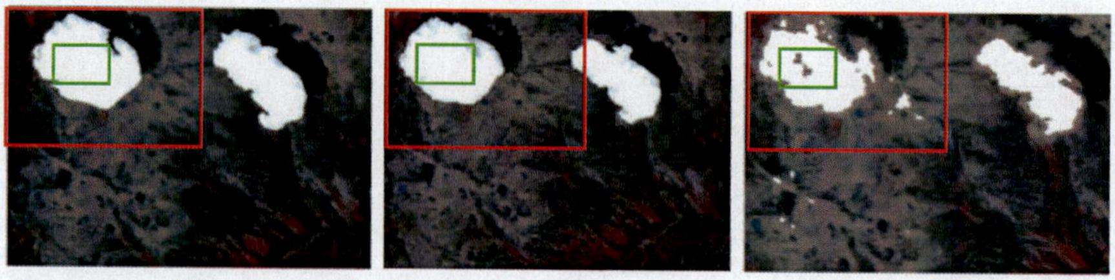

Fig. 6 Partial comparison 1

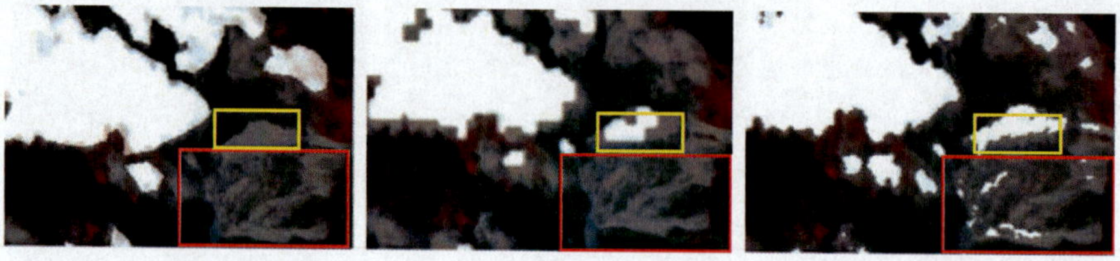

Fig. 7 Partial comparison 2

results and the original image, where **(1)** is the result of visual interpretation (i.e. the standard cloud image), **(2)** is the result of the algorithm in this paper, and **(3)** is the result of multispectral fixed threshold method. The following are the results of comparative analysis:

(1) According to Chapt. 2, in the visible and near-infrared bands, the reflectivity of mountain area is similar to that of cloud, and the reflectance is relatively high. The traditional fixed multispectral threshold method can easily misjudge a few pixels with spectral threshold close to cloud area as clouds, as shown in the red box in Figs. 6 and 7. Moreover, it is easy to be affected by salt and pepper noise, and there are many false detection areas in the image. But, due to the introduction of super pixels, the method in this paper can reduce the influence of small-scale highlight regions to a certain extent after proper super-pixel segmentation. However, as shown in the yellow box, for a large range of highlighted areas, false detection occurs. After that, it can be further improved in this respect.

(2) As shown in Fig. 6 (3), there is a "hole" in the cloud area detected by the multispectral threshold method, which has been improved in the method studied in this paper, as shown in Fig. 6 (2).

3.3.2 Comparative Analysis of Quantitative Evaluation

Use the results of visual interpretation as the standard cloud image, and the accuracy of the algorithm and multispectral threshold method are evaluated and compared by quantitative evaluation. The accuracy rate, false alarm rate and recall rate are used for quantitative evaluation. Their calculation formula is shown as formula (9).

$$\begin{aligned} Accuracy &= \frac{Na + Nd}{Na + Nb + Nc + Nd} \\ False\ alarm\ rate &= \frac{Nc}{Nc + Nd} \\ Recall &= \frac{Na}{Na + Nb} \end{aligned} \tag{9}$$

where, the meanings of **Na, Nb, Nc** and **Nd** are as follows (Table 4):

The accuracy rate represents the successful pixel ratio of true cloud and true non cloud in the algorithm results. The accuracy value represents the accuracy of the algorithm. The false alarm rate represents the ratio of true and non cloud pixels detected by the algorithm. The lower the false alarm rate, the more accurate the algorithm is. The recall rate represents the ratio of all pixels identified by the algorithm to recognize the true cloud pixels successfully. There is a negative correlation between the missing rate and the recall rate. The quantitative evaluation results are shown in Table 5:

Table 5 and Fig. 8 show the accuracy, false alarm rate and recall rate of cloud detection results of five high score six images under the two cloud detection methods, as well as their average values. It can be seen from the chart that the

Table 4 The letter meaning

	Cloud prime number detected by algorithm	Non cloud prime number detected by algorithm
True cloud image prime number	Na	Nb
True non cloud image prime number	Nc	Nd

Table 5 Quantitative evaluation table

Image	Multispectral threshold method			The algorithm in this paper		
	Accuracy (%)	False alarm rare (%)	Recall (%)	Accuracy (%)	False alarm rare (%)	Recall (%)
Image a	91.82	2.11	49.09	95.54	0.14	73.82
Image b	92.62	1.76	41.06	95.73	0.26	67.02
Image c	91.01	0.49	60.04	93.36	0.56	79.27
Image d	89.57	1.42	91.75	94.23	0.97	97.63
Image e	90.74	1.12	48.97	92.21	0.19	87.64
Average value	91.15	1.38	58.18	94.21	0.42	81.08

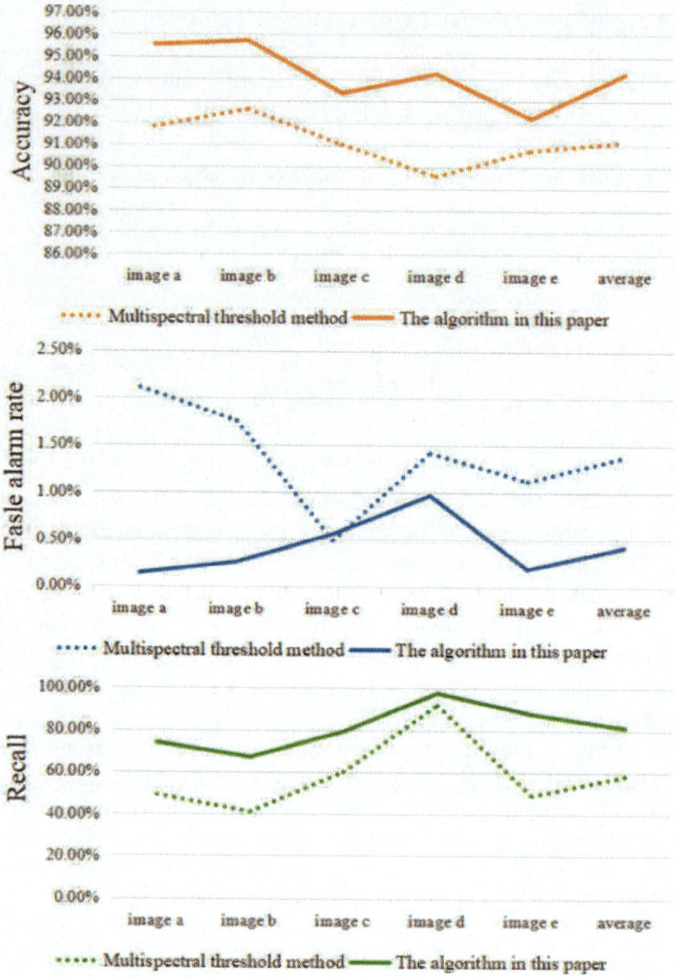

Fig. 8 Contrast line chart

accuracy of the proposed algorithm is better than that of the multispectral threshold method, with an average accuracy of 94.21%, while that of the multispectral threshold method is 91.15%. In terms of false alarm rate, the difference between this algorithm and multispectral threshold method is small. The average false alarm rate of this algorithm is 0.42%, which is slightly lower than that of multispectral threshold method by 1.38%. However, the recall rate of this algorithm is far less than that of multi spectral threshold method. The average recall rate of this algorithm is 81.08% when that of multi spectral threshold method is only 58.18%. The line chart clearly shows that the algorithm in this paper is better than the multi-spectral threshold method under the three quantitative evaluation. The results show that the accuracy of cloud detection is higher than that of multispectral threshold method and the detection accuracy is higher than 90% for the research of GF-6 wide range camera data.

4 Conclusions

In this work, we have introduced a cloud detection method which combine super pixel and dynamic threshold based on GF-6 wide camera data. Firstly, according to the band characteristics of GF-6 and the spectral characteristics of clouds and other ground objects, we choose the appropriate band combination for cloud detection. Secondly, we do histogram equalization to enhance the contrast of the image and increase the accuracy of subsequent threshold acquisition. Thirdly, using the OTSU method to dynamically obtain the appropriate cloud detection threshold of the image, and the threshold is limited between 160 and 230. If the threshold obtained by OTSU method is not in this range, the threshold is set to 200 to solve the problem that the appropriate threshold can not be obtained due to other abnormal conditions such as no cloud or all clouds in the image. Then, using SLIC algorithm to generate super-pixel objects, calculating spectral mean of super-pixel, replacing single pixel as unit pixel of subsequent processing. It increases the space and texture information, and yet reduces the image dimension. Finally, according to the threshold obtained by OTSU method, whether the super pixel object is cloud or not is judged to get the final cloud detection result. Through the experiment, we know that the cloud detection method is applicable to GF-6 wide camera data, and can effectively and accurately detect the cloud area in the image. In the comparative evaluation, the proposed method is approximately equal to or better than the multispectral threshold method in three indicators. In the accuracy evaluation, comprehensive visual evaluation and quantitative evaluation, the performance of this method is better, the average accuracy rate is 94.21%, and the detection accuracy rate is all higher than 90%.

Acknowledgements This work was supported by the National Natural Science Foundation of China (grant number: 91838303, 91638301). GaoFen-6 data were provided by the China Center for Resources Satellite Data and Application. This support was valuable.

References

1. Liu ZH, Wu YL (2017) A review of cloud detection methods in remote sensing images. Remote Sens Land Resour 29(4):6–12
2. Christodoulou CI, Michaelides SC, Pattichis CS (2003) Multi-feature texture analysis for the classification of clouds in-satellite i magery. IEEE Trans Geosci Remote Sens 41(11):2662–2668

3. Latry C, Panem C, Dejean P (2008) Cloud detection with SVM technique. Int Geosci Remote Sens Symp (IGARSS) 3:448–451
4. Ackerman S, Strabala K, Menzel WP et al (1998) Discriminating clear sky from clouds with MODIS. J Geophys Res Atmos 103(D24):32141–32157
5. Liu X, Jianmin X, Bingyu D (2005) Automatic cloud detection of GMS-5 image using dual channel dynamic threshold. Acta Sinica Sinica 04:434–444
6. Marais IVZ, Du Preez JA, Steyn WH (2011) An optimal image transform for threshold-based cloud detection using heteroscedastic discriminant analysis. Int J Remote Sens 32(6)
7. Li Z, Shen H, Li H et al (2017) Multi-feature combined cloud and cloud shadow detection in GaoFen-1 wide field of view imagery. Remote Sens Environ 191:342–358
8. Ren X-F, Malik J (2003) Learning a classification model for segmentation. In: Proceedings of the 9th IEEE international conference on computer vision, pp 10–17. IEEE Computer Society, Washington DC
9. Achanta R, Shaji A, Smith K, Lucchi A, Fua P, Süsstrunk S (2012) SLIC superpixels compared to state-of-the-art superpixel methods. IEEE Trans Pattern Anal Mach Intell 34 (11):2274–2282
10. Wang M, Cheng Y, Guo B, Jin S (2019) Parameters determination and sensor correction method based on virtual CMOS with distortion for the GaoFen6 WFV camera. Int J Photogrammetry Remote Sens 156:51–62
11. Zhang S, Dai H, Zhao Q, Yang Y (2017) Cloud detection algorithm for FY-3C/MERSI. Ecol Environ Monit Three Gorges 2(04):66–70
12. Shule GE, Shengyue DONG, Genyun SUN, Yongming DU, Ying LIN (2019) Cloud detection algorithm for images of visual and infrared multispectral imager. Aerosp Shanghai 36(S2):204–209
13. Quan-jun HE, Cheng LIU (2008) Improved algorithm of self-adaptive fire detection for MODIS Data. J Remote Sens 03:448–453
14. Cunjian YANG, Zijian ZHAO, Xiaolan REN, Jing NI, Qin WANG (2012) The analysis of the green vegetation cover change in western Sichuan based on GIS and remote sensing. Acta Ecol Sin 32(02):632–640

Research on Space-Ground Integrated Control Technology of Earth Observation Spacecraft

Zhang Fajia, Yang Tongzhi, Zhu Jia, Xing Nan, and Dang Jiancheng

Abstract With the improvement of spacecraft processing capacity, spacecraft has gradually possessed certain autonomous management capability. At present, users have put forward higher and higher expectations on the ease of use of earth observation spacecraft. One task often needs to schedule multi earth observation spacecraft of different types, scales and spectra. But the multi spacecraft task scheduling mainly relies on the cooperative mode of "spacecraft ground spacecraft", which has the disadvantages of poor timeliness, complex user interface and low cooperation efficiency, which cannot meet the needs of users. To solve this problem, this paper proposes an integrated autonomous satellite group management and control technology, which can coordinate the relationship between ground control and spacecraft autonomous management and control capability through hierarchical task planning, space resource synchronization and joint health assessment, so as to improve the joint control effect and realize the rapid response of spacecraft payload business, which can be used as a reference for future spacecraft mission management and control design.

Keywords Space-ground integrated control · Hierarchical task planning · Space resource synchronization · Joint health assessment

Z. Fajia · Y. Tongzhi (✉) · D. Jiancheng
Shanghai Institute of Satellite Engineering, Shanghai, China

Z. Jia
The 10th Research Institute of China Electronics Technology Group Corporation, Chengdu, China

X. Nan
China Xi'an Satellite Control Centre, Xi'an, China

L. Wang et al. (eds.), *Proceedings of the 7th China High Resolution Earth Observation Conference (CHREOC 2020)*, Lecture Notes in Electrical Engineering 757, https://doi.org/10.1007/978-981-16-5735-1_25

1 Introduction

With the development of space microwave remote sensing technology, especially the resolution of sensors has been improved by an order of magnitude in recent ten years. Since 1999, the United States has successively launched IKONOS, QuickBird and WorldView series of high-resolution remote sensing satellites, leading the development trend of high-resolution optical satellite remote sensing, forming a high-low orbit, multi-type, multi-scale, multi-spectrum earth observation spacecraft Group [1]. China has also constructed remote sensing satellite clusters represented by GF series and FY series, and commercial remote sensing satellite clusters represented by jilin-1 and gaojing-1. However, most spacecraft can only fly around the earth periodically in a fixed orbit. Due to the limitation of the observation range and maneuverability of spacecraft, a single spacecraft cannot continuously observe the designated area, so it is necessary to cooperate with multiple spacecraft to complete the observation task in a larger area [2]. In order to obtain the three-dimensional and multi-level information of a ground target or the comprehensive information of some ground targets, the joint observation of multiple spacecraft is often adopted, which puts forward higher requirements for spacecraft load, platform resource operation timeliness, spacecraft task state switching speed, and multi spacecraft resource coordination efficiency [3]. Due to the different orbit parameters and observation ranges of spacecraft, the existing regional decomposition method cannot effectively decompose the designated observation area according to different spacecraft parameters [4]. It is necessary to study a new space ground integrated joint management and control mode suitable for high-performance spacecraft group, build a space earth integrated joint management and control framework, and design a hierarchical management and control mode based on normal and abnormal conditions, To solve the core problems of intelligent interaction between satellite and ground, closed-loop collaborative management and control, integrated simulation of space and ground resources, and improve the efficiency of Spacecraft Earth observation.

2 Space-Ground Integrated Control Principle

With the improvement of spacecraft intelligent programmable ability, based on the current mode “ground programming and spacecraft execution”, the core business of ground control would be transplanted to spacecraft execution. So a space-based integrated joint management and control architecture consisting of space-based control system, ground-based management and control system, ground station network center and high-performance earth observation spacecraft group is constructed, which coordinates the relationship between ground control and spacecraft autonomous control capability, realizes autonomous and efficient cooperation between spacecraft, and improves the joint control effect. The hierarchical task management and control mechanism is adopted for the integrated management and control of the space-based control system, which can increase the real-time performance of the control task, optimize the task description mode, simplify the ground-based control process, improve the portability of the ground-based control system, and increase the flexibility of the ground-based control system deployment.

2.1 Hierarchical Control of Observation Tasks

Tasks are divided into user tasks, level I tasks and level II tasks. User tasks are directly associated with user requirements, which are high-level descriptions of tasks (including observation target type, location, time, frequency and other information). The level I tasks are tasks formed by analyzing the user's task requirements and can be executed by the space-based control system. According to the nature, they can be divided into macro observation tasks, subdivision observation tasks and emergency tasks. According to the complexity, they can be divided into single spacecraft mission and compound mission. The level I task is decomposed into the level II task, which can be performed by earth observation spacecraft. The task hierarchy is shown in Fig. 1.

As shown in Fig. 2, according to whether space-based control is involved, the hierarchical control mode is divided into normal control mode and abnormal control mode.

(1) Normal control mode

The ground-based management and control system receives the user's requirements, generates the level I task and send it to the space-based control system. The space-based control system carries out the level I task planning and sends the task response to the ground-based management and control system (including the

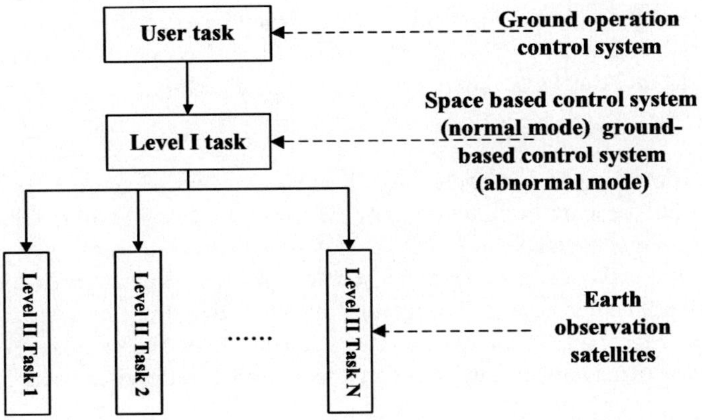

Fig. 1 Task hierarchy diagram

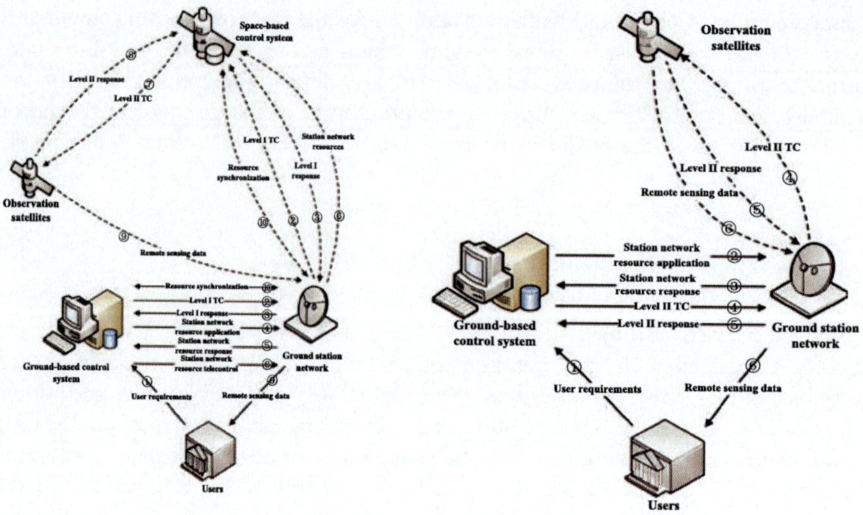

Fig. 2 Normal control mode (left) and abnormal control mode (right)

ground station network resource application requirements, etc.). The ground-based management and control system sends the ground station resource application to the ground station network center system, and the ground station network center system allocates resources and responds to the ground-based control system. The ground-based management and control system injects the ground station application resources into the space-based control system, and the space-based control system forms the level II tasks and broadcasts them to the earth observation spacecraft group. After receiving the level II tasks, the earth observation spacecraft sends response information to the space-based control system and carries out independent planning, which forms the work instruction package and control parameter package and executes instructions. The space-based control system and the ground-based management and control system realize resource synchronization through space-ground interaction. The earth observation spacecraft completes remote sensing tasks and transmits them to users through data link.

(2) Abnormal control mode

Abnormal control mode is a control mode when space-based control system is not available, and the control function only depends on ground-based control. The ground-based management and control system receives the user's requirements, plans user task and sends the ground station resource application to the ground station network center system. The ground station network center system allocates resources and responds to the ground-based management and control system. The ground-based management and control system forms the level II tasks and sends them to the earth observation spacecraft group. The spacecraft group receives the level II tasks, carries out independent planning and execution. The spacecraft completes missions and transmits sensing data to users through data link.

As shown in Fig. 3, the process of resource allocation and conflict resolution is shown. The ground management and control system classifies the task requirements into level I tasks and uploads them to the space-based control system. The space-based control system makes the decision on the type of single task and compound task, and decomposes it into a group of level II tasks. Then the types and lists of spacecraft resources are decided by spacecraft load parameter decision-making matrix and health status. According to the priority of tasks and the optimal principle, the observation resources and time window are allocated, the corresponding observation tasks are formulated, and the conflict resolution is completed. After the task planning and resource allocation, the satellite level II task is generated. According to the task plan, the ground-based management and control system applies for tracking and receiving resources from the ground station network, and optimizes the task according to the ground station feedback, so as to ensure the effective coordination of spacecraft observation and data reception. The observation task and the data reception plan are packaged to generate a level II task list, which is sent to the earth observation spacecraft group. According to the task list, the earth observation spacecraft group carries out level II task planning for the action sequence of payload, power supply, storage, data transmission, etc., and generates task instructions, which are executed in a closed loop.

2.2 *Task Planning Based on the Coordination of Space-Ground Resources*

In the traditional control mode, due to the lack of mission time distribution characteristics, spacecraft follows the most conservative resource constraints, which greatly limits the effectiveness. In order to achieve maximum efficiency, the ground-based management and control system or space-based control system needs to dynamically obtain the spacecraft's resource conditions (including energy, payload, storage, data transmission, attitude and orbit, etc.), calculate resource constraints in real time, and realize dynamic mission planning [5].

Through the asynchronous message communication between space and ground, the resource information synchronization is realized. Message interaction channel can rely on Beidou short message service. Message information includes sensor type, orbit, payload capacity and resource constraints of earth observation spacecraft group, mission planning algorithm, strategy, parameters and task evaluation results, ground station network resource operation status and meteorological information, etc. The ground resources are divided into various antennas, receiving resources, telemetry and control resources, etc.

In order to meet the requirements of resource allocation for different types of spacecraft, the task planning process based on space-ground resource collaboration

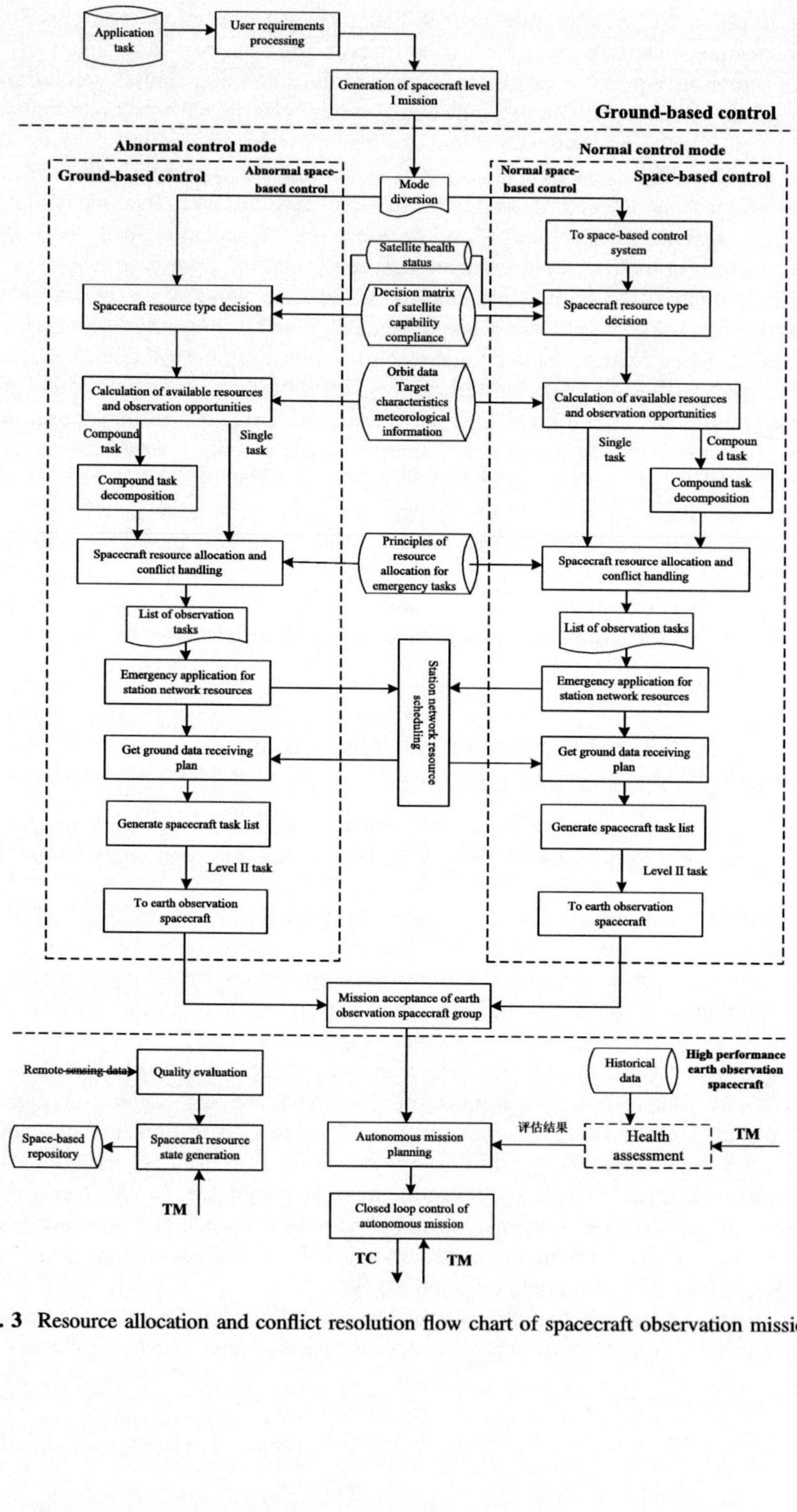

Fig. 3 Resource allocation and conflict resolution flow chart of spacecraft observation mission

is divided into four parts: resource information management, task sequencing management, observation transmission management and constraint detection management.

(1) Resource information management

The received level II tasks are instructive and resourceful, so that the tasks are compiled from the level II task sequence into the instruction sequence and resource information, so as to sort the tasks and manage the observation transmission.

Task resource allocation adopts time window tailoring mechanism. Each task has its own observation time window, and the spacecraft also has its own transmission time window. However, the spacecraft can only observe one task at the same time, and there are conflicts between the time windows. The observation or transmission time windows of different tasks may overlap. In order to avoid such conflicts, the optional time window of each activity must be trimmed [6], so that it can be removed from the available time period when other activities are arranged. Set the new task period as [Ts, Te], and the time window to be trimmed is [Ws, We]. The demonstration of the time window clipping algorithm is shown in Table 1.

(2) Task sequencing management

There are five steps in the task sequencing component group. Each step designs different sorting strategies based on heuristic method. The tasks are randomly selected according to the probability accumulation method, and a sorting sequence is obtained, and the task sorting results are submitted.

Table 1 Time window clipping

Comparison of time	Ts ≤ Ws	Ws < Ts < We	Ts ≥ We
Te ≤ Ws	The shooting time has no effect on the time window	The situation does not exist	The situation does not exist
Ws < Te < We	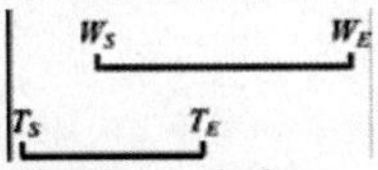 The time window [Ws, We] is cropped to [Te, We]	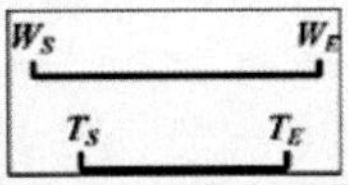 The time window [Ws, We] is cropped to [Ws, We] and [Ts, Te]	The situation does not exist
Te ≥ We	 The time window [Ws, We] was cut out	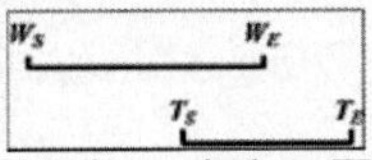 The time window [Ws, We] is cropped to [Ts, Te]	The shooting time has no effect on the time window

Step 1: sort tasks randomly;
Step 2: priority should be given to important or urgent tasks;
Step 3: priority should be given to tasks with short observation time;
Step 4: priority should be given to tasks with long observation time;
Step 5: priority should be given to tasks with early observation opportunity of the best imaging point.

(3) Observation transmission management

Observation activities and transmission activities have certain similarity. They both need time window and capacity constraint check, and then arrange activities in optional time window according to certain strategies. The observation arrangement component first judges the possibility of arranging observation activities for a given task (carrying out capacity constraint feasibility test), then determines the observation start time according to the time window selection strategy from the optional observation time period, and submits the constraint check to determine the actual observation start time of the task. The transmission scheduling component first judges the possibility of arranging a transmission activity for a specified task, then arranges the transmission activity according to the time window selection strategy within the optional transmission time period of a given task [7], and ensures that the echo is carried out after the observation, that is, the actual transmission start time is obtained.

Step 1: select the optional time window randomly;
Step 2: give priority to the earliest optional time window;
Step 3: the latest time window is preferred;
Step 4: give priority to the shortest optional time window;
Step 5: the longest optional time window is preferred.

(4) Constraint detection management

The storage space of spacecraft is limited, and the observation task information cannot exceed maximum solid storage, so it is impossible to include all tasks into the observation plan at the same time. Therefore, a constraint processing mechanism should be set up to maintain the storage space (such as updating the information, erasing expired tasks, etc.), so as to complete the observation and transmission more efficiently. Loop the above operation until the condition is met, and generate the final task scheme.

2.3 *Space-Ground Joint Health Assessment*

In order to realize the integrated management and control of space and ground, it is necessary to improve the ability of on orbit autonomous fault diagnosis and health assessment, and reduce the dependence on ground fault diagnosis system. Considering the complexity of fault diagnosis and the significant risk of misdiagnosis, it is necessary to carry out the joint health state assessment. Different disposal strategies are adopted according to the level of faults. Simple faults and low-risk faults are handled by spacecraft independently. Complex faults and high-risk faults are firstly pre diagnosed by spacecraft [8]. When anomalies are found independently, spacecraft interacts with the ground through space-based control system in time, and the ground is rechecked.

Spacecraft health assessment is divided into three levels: component level assessment, subsystem level assessment and spacecraft level assessment. Each level of health status assessment system includes assessment index, index weight and health status assessment level. Among them, the evaluation index is divided into core index and conventional index, component level evaluation index takes telemetry parameters and component level functional items as objects, system level assessment index takes health status of components/equipment and subsystem level functional items as objects, and spacecraft level assessment index takes health status of each subsystem and function items of spacecraft as objects [9]. The health status assessment of the high-level is obtained by low-level index weighted calculation based on the reliability theory (Fig. 4).

3 Design and Verification of Space-Ground Integrated Control System

3.1 System Design

As shown in Fig. 5, the integrated management and control system is mainly composed of space-based control system, ground-based management and control system, ground station network center and high-performance earth observation spacecraft group.

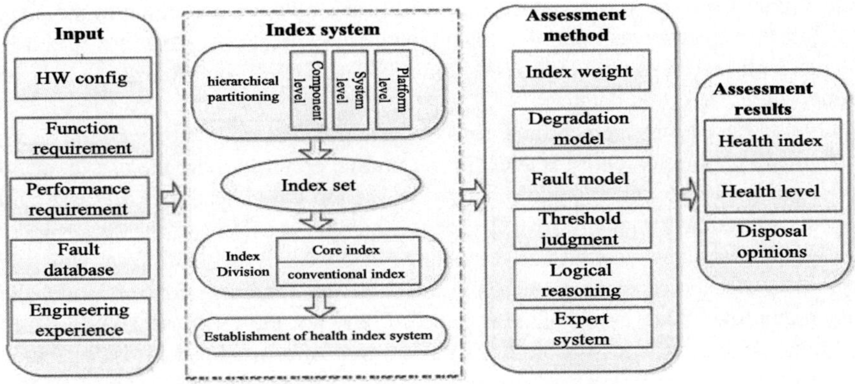

Fig. 4 Health assessment framework

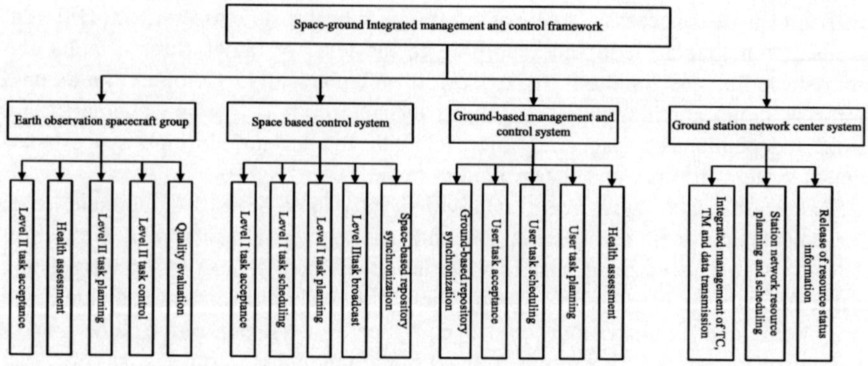

Fig. 5 Space-ground integrated management and control framework

Earth observation spacecraft group: it has certain autonomous health status assessment, level II task planning and closed-loop control, independent quality evaluation capability of payload data, and independent release capability of resource. Receive the level II tasks of space-based control system (normal mode) or ground management and control system (abnormal mode) and carry out task planning based on the results of independent health assessment, form instruction package and parameter package executed by each subsystem, and control the instruction implementation by closed-loop control. When remote sensing data is obtained, the quality evaluation can be carried out. The results of independent health status assessment and payload data quality evaluation are regularly released to space and ground management and control system.

Space based control system: it can receive the level I task of the ground-based management and control system, generate the level II task sequence executed by the spacecraft group, and coordinate the ground station network center system to form the ground station tracking and receiving plan. It broadcasts the level II task to the observation spacecraft group for execution, and the ground station network center system completes the spacecraft tracking and receiving work according to the plan. It automatically receives operation status of visible spacecraft which are stored in the space-based resource database, and automatically synchronized with the ground-based resource database.

Ground-based management and control system: it has different functions under normal and abnormal control modes. In the normal control mode, the user demand is automatically accepted. According to the ground-based resource database and task classification strategy, a level I task is generated and sent to the space-based control system. The ground-based resource database is synchronized with the space-based resource database automatically. In the abnormal control mode, the user requirements are automatically accepted, and according to the ground-based resource database, the task planning is carried out to form a level II task executed by the observation spacecraft group, which is sent to the operational spacecraft [10]. At the same time, it coordinates the ground station network center to track and receive data, carry out spacecraft health assessment and verify and confirm the evaluation results.

Ground station network center system: it can accept the station resources application and allocation of ground-based management and control system, complete spacecraft tracking and data receiving, and regularly release the ground station resource status to the ground-based management and control system.

The resource simulation component framework of the prototype system includes resource template, configuration parser and run container.

(1) Resource template: it defines the way of resource integration description, and describes resource attributes based on XML format. Resource templates are described hierarchically, and the low-level resources can be materialized on the basis of inheriting top-level resource templates. The description of a certain type of resource can be referenced by another resource, making it a part of the reference. The relationship between templates and definitions of different resources is shown in Fig. 6. A complete resource description template is formed by inheriting (or deriving low-level templates) from high-level templates and referencing other resource definitions, and the template is materialized as a resource that can be executed under parameter configuration [11]. In the top-level structure, resource templates are abstracted into the following five aspects:

   (1) The control interface is used to provide the access mode and control interface between the resource and the management and control system. The management and control system controls the behavior of resources and obtains the resource status through this interface.
   (2) Payload capacity is used to describe the payload and sensing capability. It mainly describes the core business capability and configurable scope of the resource.
   (3) Data transmission interface is the interface for sending or receiving its non-control data, which is generally used for sending or receiving payload data.

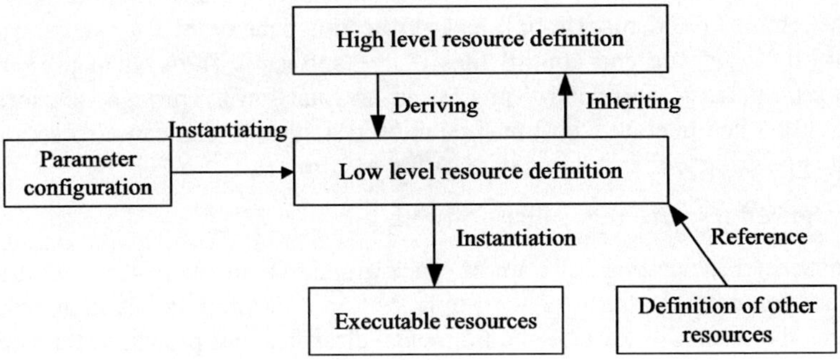

Fig. 6 Resource template definition and relationship

    (4) State description is used to describe the state of the resource itself, such as its space–time location, its health and integrity, its ability to perform tasks, its own task list and task planning.
    (5) Processing module, for the additional processing functions added by the resource, belongs to the description of the unique processing ability, such as processing payload data, decomposing tasks and compiling instructions.

(2) Configuration parser: the processing module plays the role of simulation run compiler. It is used to parse the resource and its template defined by XML text into code that can be run on the computing simulation system. The resource template not only describes the attributes of the system, but also the behavior of the system [12]. The extended Backus normal form (EBNF) is used to describe the syntax of resource behavior.

(3) Run container: the run container is a module to load the running and scheduling of resources. It ensures that different resources run in the same space–time framework, monitors the running status of each resource, and outputs it to users and necessary logs. The runtime container provides a standard spatiotemporal framework, and outputs it to each resource, so that each resource can refer to the spatiotemporal synchronization information.

3.2 Simulation Verification

The demonstration and verification architecture of the prototype system is shown in Fig. 7. The system is mainly composed of four core software: earth observation spacecraft management and control simulation system, space-based control prototype software, space-based resource integration scheduling software, guidance and control demonstration software.

(1) Guidance and control demonstration software

Guidance and control demonstration software is the core software of man–machine interface, data visualization, system management and operation maintenance. The user determines the demonstration and verification scheme of the system by configuring the guidance and control files of the software. The system presents the resources, operation and control process in a visual form. Through this software, system users can interact with the simulation test system, and can also set the test scheme and control the running state of the system.

(2) Integrated resource scheduling software

The integrated scheduling software of space-ground resources receives all kinds of user task requirements sent by the guidance and control demonstration software, analyzes the feasibility of user requirements, classifies and processes the accepted requirements. Then the software makes overall planning of multi tasks in combination with the status of space-ground resources, constraints of spacecraft use, task

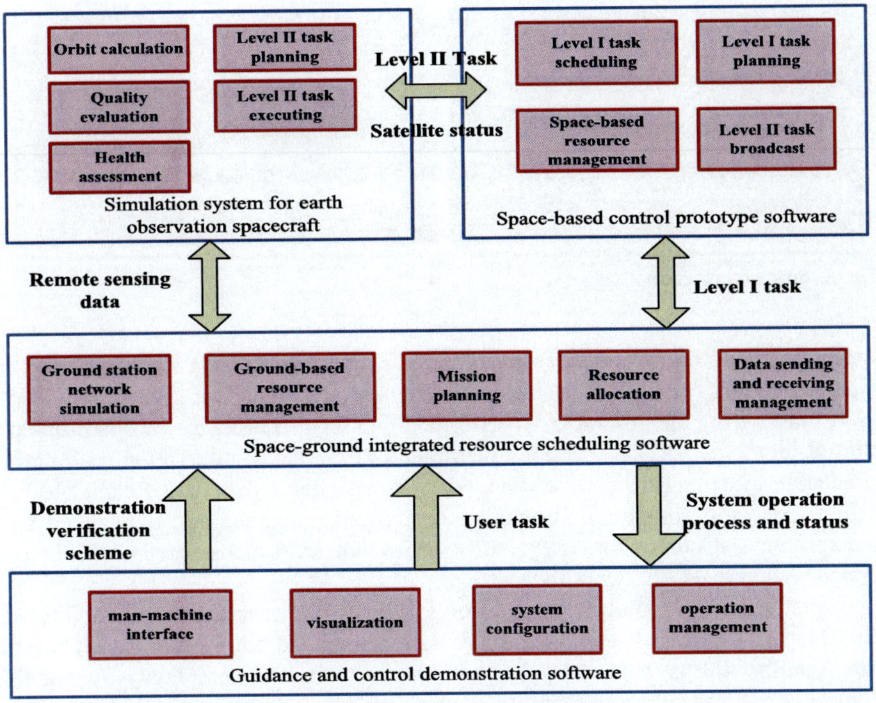

Fig. 7 Demonstration and verification system of prototype system

type and other factors, so as to form a level I task which is sent to the space-based control prototype. In the emergency mode, according to the real-time status of the earth observation spacecraft and other information, the user task requirements are refined to form the level II tasks which are sent to the simulation system of earth observation spacecraft.

(3) Simulation system for earth observation spacecraft

It is used to simulate the response and behavior of the earth observation spacecraft after receiving the level II mission. For different earth observation spacecraft, the software can access the parameters and indexes in the simulation database, simulate a scene and evaluate the payload capacity, imaging effect and health status according to the test scheme. The software system receives the level II tasks from the space-based control prototype software and the integrated resource scheduling software, and feeds back the detection results.

(4) Space-based control prototype software

Space-based control software is responsible for the transfer of mission data, collects the real-time status data of the simulation system of each earth observation spacecraft, and synchronizes it to the integrated resource scheduling software. The

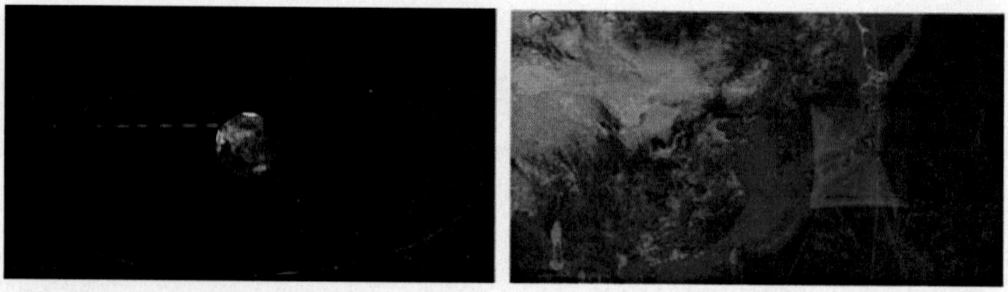

Fig. 8 Space based control diagram

space-based control prototype software is the nerve center of the whole prototype system, which is responsible for the centralized exchange of space-ground data. The level I tasks from the ground are transformed into level II tasks by the space-based control prototype software, and are distributed to the earth observation spacecraft simulation system. At the same time, the real-time status data of the earth observation spacecraft simulation system are centralized processed and transmitted by the space-based control prototype software, which are synchronized to the integrated resource scheduling software.

The effect of the prototype verification system is shown in Fig. 8. The ability to control the earth observation spacecraft group through the space-based inter-satellite link is verified by the prototype system. The complexity of ground control is reduced by more than 50%, which greatly improves the rapid response ability of the spacecraft group to Earth Observation Tasks.

4 Summary

This technology overcomes the shortcomings of the traditional centralized ground management and control system, such as lack of timeliness and agility of the management and control system. It adopts hierarchical task management mode, takes full account of the advantages of high orbit space-based control system, increases the mission real-time performance, reasonably allocates and organizes earth observation resources based on the synchronization of space-ground resources and joint health assessment, optimizes the task description mode, and simplifies the task description. It can improve the portability of the ground management and control system and increase the flexibility of the ground management and control system deployment. In the future, further research will be carried out in optimizing resource allocation, task dynamic adjustment, task sequence combination optimization, and evaluation model optimization.

References

1. Zhang S, Xu D (2016) The development of commercial remote sensing satellites in the United States. Geometrics 39(12):135–138
2. Han S-m, Beak S-w, Cho K-r (2008) Satellite mission scheduling using genetic algorithm. In: SICE annual conference, Tokyo, Japan
3. Lee J-H, Wang SM, Chung D (2012) Multi-satellite control system architecture and mission scheduling optimization. In: IEEE aerospace conference proceedings, New York, USA
4. Wang C, Chen H, Zhai B (2016) Satellite observing mission scheduling method based on case-based learning and a genetic algorithm. In: 2016 IEEE 28th international conference on tools with artificial intelligence, San Jose, CA, USA
5. Xing J, Li J, Zhu H (2010) Optimizing in advance: Shenzhou 8 space mission scheduling tryout architecture. In: The 5th international conference on computer science & education, Hefei, China
6. Liu R, Sheng M, Xu C (2017) Antenna slewing time aware mission scheduling in space networks. IEEE Commun Lett 21(3):516–519
7. Zhou D, Sheng M, Liu R (2018) Channel-aware mission scheduling in broadband data relay satellite networks. IEEE J Select Areas Commun 36(5):516–519
8. Yan Z, Zhao Y, Chen T (2010) Fault recovery based mission scheduling of AUV for oceanographic survey. In: 10th World congress on intelligent control and automation, Beijing, China
9. Xu J, Zhen H, Xu L (2012) Meta-synthesis information fusion for hybrid diagnostics of space avionics. J Aerospace Eng 228(2):219–235
10. Figueroa F, Underwood L, Walker MG (2019) NASA platform for autonomous systems (NPAS). In: AIAA SCITech 2019, San Diego, CA, United States
11. Xing J, Zhu H, Li L (2016) The unified scheduling language designed for the space mission scheduling platform. In: 2016 IEEE aerospace conference, Big Sky, MT, USA
12. Marshall A, Alqahtani S, Ridgway A (2015) Combining coordination with usage policies to improve mission scheduling resilience. In: IEEE 2015 resilience week, Philadelphia, PA USA

An Improved Ionospheric Estimation Algorithm for Low-Frequency SAR Interferometry

Beixin Qin, Yongsheng Zhang, Zaoyu Sun, and Xiaoxiang Zhu

Abstract Interferometric Synthetic Aperture Radar (InSAR) has been proven to successfully map surface deformations and monitor glacier movement. For space-borne InSAR systems working at low frequencies, the deformation measurement accuracy is strongly affected by the ionospheric effect, which can cause phase delay and lead to an obvious interference phase error. For compensation of the ionospheric effect, an improved ionospheric estimation algorithm based on the split spectrum is proposed. The theoretical basis and implementation process of the improved algorithm are described in detail in the article. In order to investigate the estimation accuracy of the proposed algorithm, simulation data with complex ionospheric changes are used and six different filtering algorithms are selected for optimization of the proposed algorithm. According to the results of the ionospheric estimation error, the improved algorithm has a smaller mean value and standard deviation than the traditional algorithm. Among the six filtering methods, the mean value and standard deviation of the ionospheric estimation result error of the Lee filtering method are much smaller than those of the original method. These results illustrate that the proposed algorithm improves the actual accuracy of ionospheric estimation and strongly certifies the effectiveness of the improved algorithm.

Keywords Interferometric synthetic aperture radar (InSAR) · Ionosphere estimation · Split-spectrum

1 Introduction

Space-borne interferometric synthetic aperture radar (InSAR) based on repeat-pass measures the surface deformation of the earth through different viewing angles [1–3]. However, the ionospheric effect introduces additional group delay and phase

B. Qin · Y. Zhang (✉) · Z. Sun · X. Zhu
College of Electronic Science and Technology, National University of Defense Technology, Changsha, China
e-mail: zyscn@163.com

L. Wang et al. (eds.), *Proceedings of the 7th China High Resolution Earth Observation Conference (CHREOC 2020)*, Lecture Notes in Electrical Engineering 757, https://doi.org/10.1007/978-981-16-5735-1_26

advance, resulting in estimation error of the interference phase, thereby affecting the accuracy of the interferometry. In order to ensure the imaging quality of the L-band space-borne InSAR system, the ionospheric effects must be compensated.

Currently, algorithms for compensating the ionospheric effect can be categorized into the range split-spectrum, group phase delay, and azimuth multi-aperture algorithms. The group phase delay algorithm adopts the characteristic that the ionospheric phase and group delay have a pair of equal but opposite signs to calculate the ionosphere [4]. The azimuth multi-aperture algorithm exploits [5] the proportional relationship between the differential azimuth offset and azimuth derivative. Brcic et al. [6, 7] conducted a theoretical analysis of the spectral split-spectrum algorithm, suggested dividing the two sub-bands, and provided the theoretical standard deviation in the case of BIOMASS, ALOSPALSAR, TerraSAR-L, and TerraSAR-X systems. However, this method is only sensitive to the azimuth and cannot recover the change in the range. Rosen et al. [8] proposed a multi-frequency split-spectrum technique, which was processed using real data, but its accuracy was not determined. De Zan et al. [9] confirmed that the two existing multi-looking schemes did not achieve the ideal performance limit, and proposed a two-stage averaging method. In [10, 11], the split-spectrum algorithm was described in detail for the first time with the theoretical value as the center frequency of the sub-bands and ALOS2 data were processed using mean and two-dimensional weighted large window Gaussian filtering.

This paper proposes a new compensation method for the ionospheric effect. The ionospheric dispersion characteristics and zero baseline interference characteristics are utilized to accurately estimate the ionosphere. In Sect. 2, the ionospheric effect on interferograms is introduced. Section 3 describes the implementation process of the new method in detail. Section 4 presents the simulation experiment. In this section, the six filtering algorithms, including median filtering, Gaussian filtering, local frequency filtering, rotary filtering, Lee filtering, and mean filtering are adopted and compared with the original method; the proposed algorithm improves the accuracy of the ionospheric estimation. Section 5 summarizes the findings and plans for future work.

2 Ionospheric Effect on Interferograms

Owing to the dispersive characteristics of the ionosphere, the ionospheric phase in the interferograms can be defined as

$$\Delta\varphi_{\text{io}} = \frac{4\pi K}{cf_0} \int n_e(z)dz = \frac{4\pi K}{cf_0} \cdot TEC, \quad (1)$$

where c is the speed of light, $K = 40.28\,\text{m}^3/\text{s}^2$, $TEC = \int n_e(z)dz$ is the difference between the total electron content (TEC) experienced by the signals, and f_0 is the carrier frequency.

The ionospheric effect leads to phase advancement; hence, the equation for dividing the phase in the interferogram into a dispersive phase $\Delta\varphi_{\text{io}}$ and a non-dispersive phase $\Delta\varphi_{no}$ is as follows:

$$\Delta\varphi = -\Delta\varphi_{\text{io}} + \Delta\varphi_{no}. \tag{2}$$

The split-spectrum method is based on this theory to eliminate the influence of the ionospheric effect during the interference process.

3 Proposed Approach for Split-Spectrum Method

The original split-spectrum method employs the theoretical value as the sub-band center frequency, but it is not sufficiently accurate. This paper proposes a new split-spectrum approach based on zero baseline interferometry characteristics, which utilizes more accurate sub-band center frequencies to improve the accuracy of ionospheric estimation.

3.1 *Theoretical Basis of the Proposed Approach*

Based on Sect. 2, the equation for ionospheric estimation using the split-spectrum method is

$$\Delta TEC = \frac{c}{4\pi K}\frac{f_H f_L}{f_H^2 - f_L^2}[\Delta\varphi_L f_H - \Delta\varphi_H f_L], \tag{3}$$

where f_L is the center frequency of the lower sub-bands,f_H is the center frequency of the higher sub-bands, $\Delta\varphi_L$ is the interferogram of the lower sub-bands, and $\Delta\varphi_H$ is the interferogram of the higher sub-bands.

It is noted that the center frequencies of the sub-bands have a significant influence on the result of ionospheric estimation. In the original split-spectrum method, the higher sub-band center frequency f_H and lower sub-band center frequency f_L are fixed theoretical values:

$$f_H = f_0 + B/3 \tag{4}$$

$$f_L = f_0 - B/3, \tag{5}$$

where B is the bandwidth. Nevertheless, the actual radar signal will be affected by various interferences, resulting in a deviation between the center frequency of the actual sub-band and the theoretical value. This introduces a non-negligible error in the estimation of the ionosphere by the spectrum division method.

As is well known, in the case of zero baseline interferometry, if the ionospheric effect is added, the interference phase is equivalent to the ionospheric phase. According to the dispersion characteristics of the ionosphere and Eq. (1), the center frequency of the sub-band can be derived as

$$f_i = \frac{c \cdot \Delta\varphi_{1i}}{4\pi K \cdot \Delta TEC_1}, \quad i = H, L, \tag{6}$$

where $\Delta\varphi_{1H}$ and $\Delta\varphi_{1L}$ are the interference phases of the higher and lower sub-bands under the zero baseline case.

By substituting the result of Eq. (6) into Eq. (3), a more accurate ionospheric estimation formula of the new method can be obtained:

$$\Delta TEC = \frac{c}{4\pi K} \frac{\frac{c \cdot \Delta\varphi_{1H}}{4\pi K \cdot \Delta TEC_1} \frac{c \cdot \Delta\varphi_{1L}}{4\pi K \cdot \Delta TEC_1}}{\left[\frac{c \cdot \Delta\varphi_{1H}}{4\pi K \cdot \Delta TEC_1}\right]^2 - \left[\frac{c \cdot \Delta\varphi_{1L}}{4\pi K \cdot \Delta TEC_1}\right]^2} \left[\Delta\varphi_L \frac{c \cdot \Delta\varphi_{1H}}{4\pi K \cdot \Delta TEC_1} - \Delta\varphi_H \frac{c \cdot \Delta\varphi_{1L}}{4\pi K \cdot \Delta TEC_1}\right], \tag{7}$$

3.2 *Implementation of the New Approach*

- First, a known ionosphere is added to the primary image as the secondary image, the registration and split-spectrum process are performed, the theoretical value is utilized as the sub-band center frequency, and gain interferograms. The center frequency error of the sub-band is compensated according to Eq. (6).
- Second, the registration and split-spectrum of the original primary image and secondary image are performed. We consider the results of the zero baseline interferometry estimator as the center frequency of the sub-bands and perform interference.
- Third, six different filtering methods are adopted for filtering, and the least cost flow method is applied to unwrap the filtering results. After the unwrapping error is corrected, the ionosphere is estimated.

4 Simulation Results

The simulation data are applied to verify the improved split-spectrum method. The space-borne InSAR works at the L-band and in the strip mode. The signal is in the form of chirp with a bandwidth of 103.96 MHz. The data parameters are illustrated in Table 1.

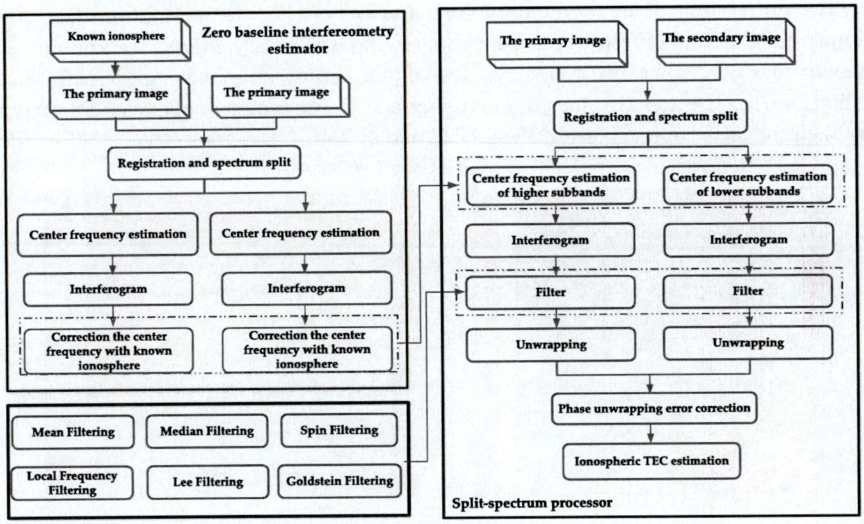

Fig. 1 Implementation of the proposed approach

Table 1 Radar system parameters

Parameters	Value
Carrier frequency	1.27 GHz
Bandwidth	103.96 MHz
PRF	3500 Hz
Signal pulse width	48 us
Azimuth sampling points	800
Range sampling points	1300

Fig. 2 Ionospheric difference between simulation data

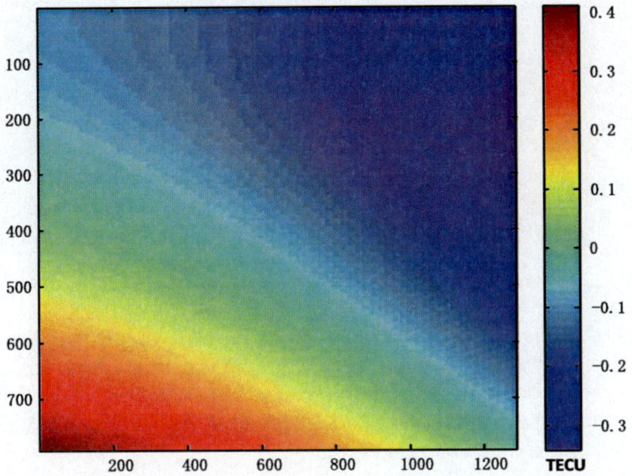

The correlation of the simulation data was 0.997. For the simulation data, the ionospheric difference between the primary and secondary images is known, as shown in Fig. 2. The theoretical values of the center frequency are 1.2353 and 1.3047 GHz. The actual values of the higher and lower center frequencies estimated by the proposed approach are 1.2354 GHz and 1.3046 GHz, respectively. When the

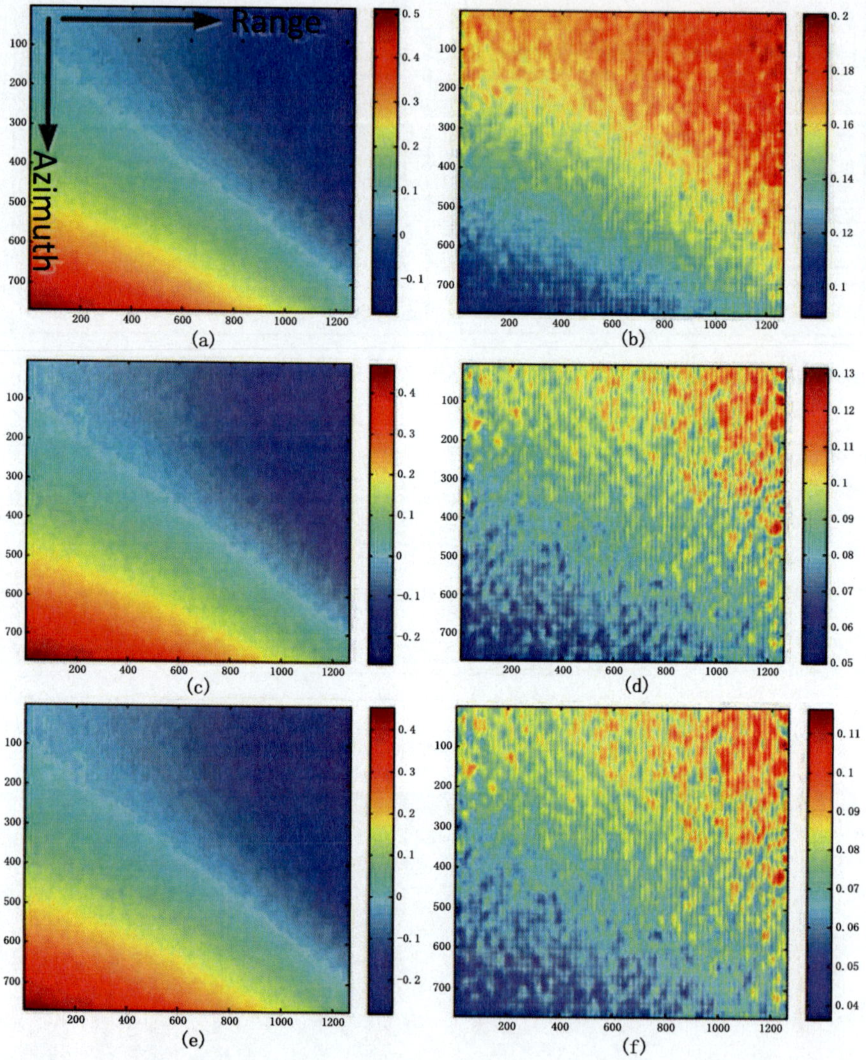

Fig. 3 Ionospheric estimation results of simulation data utilizing the zero baseline interferometry estimator with different filtering algorithms. **a**, **c**, and **e** are ionospheric estimation results; **b**, **d**, and **f** are ionospheric estimation errors. **a** and **b** are the original split-spectrum algorithm estimation results; **c** and **d** are spin filtering algorithm estimation results; **e** and **f** are local frequency filtering algorithm estimation results

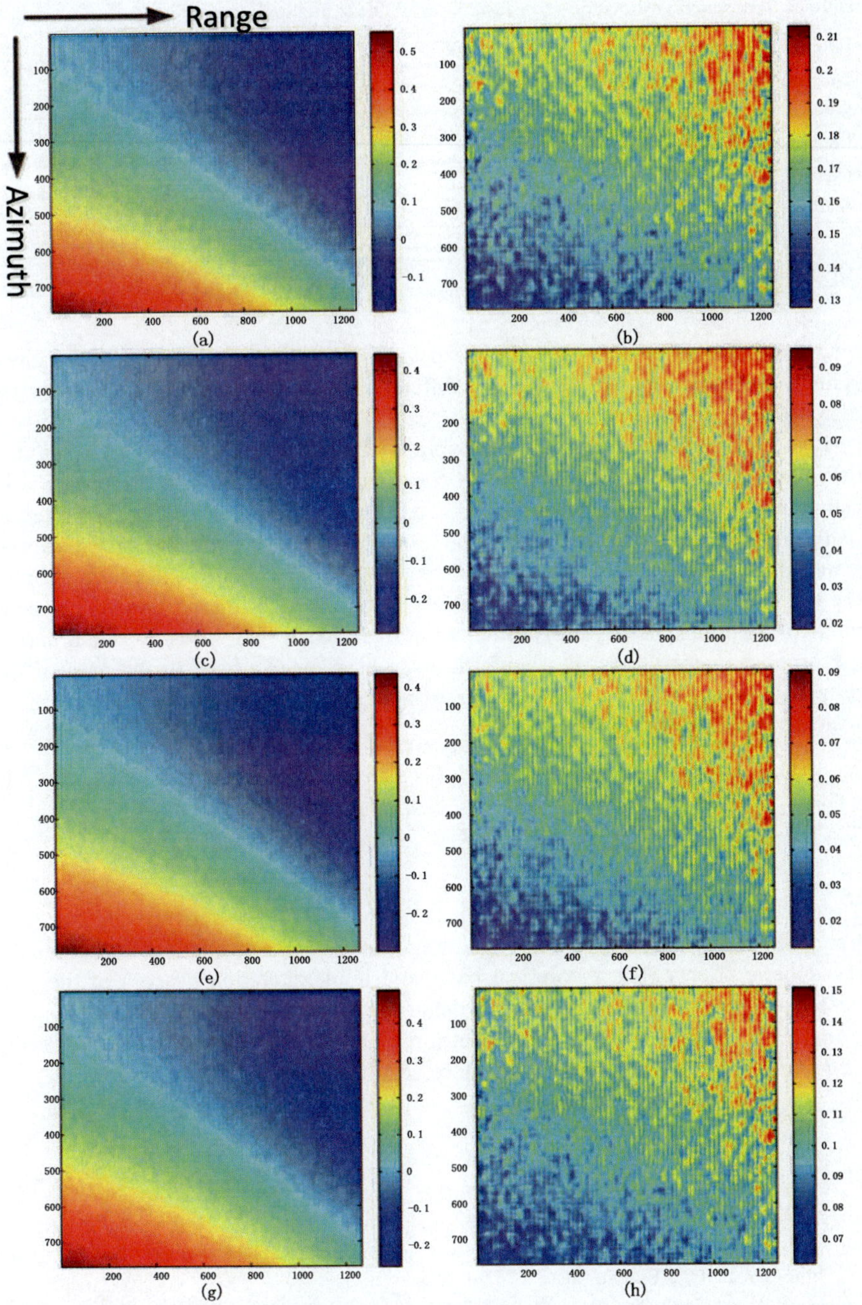

Fig. 4 Ionospheric estimation results of simulation data utilizing the zero baseline interferometry estimator with different filtering algorithms. **a**, **c**, **e**, and **g** are ionospheric estimation results; **b**, **d**, **f**, and **h** are ionospheric estimation errors. **a** and **b** are the mean filtering algorithm estimation results; **c** and **d** are median filtering algorithm estimation results; **e** and **f** are Lee filtering algorithm estimation results; **g** and **h** are Goldstein filtering algorithm estimation results

Table 2 Ionospheric estimation error results for different filtering algorithms

Filtering algorithm	Mean/TECU	Standard deviation/TECU
Original split-spectrum	0.1521	0.0197
Spin	0.0883	0.0113
Local frequency	0.0904	0.0110
Mean	0.1678	0.0115
Median	0.0567	0.0106
Lee	0.0503	0.0106
Goldstein	0.1027	0.0117

range resolution unit is 26, the proposed approach is followed to obtain the ionospheric estimation results under six different filtering methods. The results are displayed in Figs. 3 and 4.

Figs. 3 and 4 show that the original split-spectrum method has a larger error than the proposed approach for ionospheric estimation. The proposed approach can estimate the ionosphere more accurately, and the Lee filtering and median filtering methods have the smallest estimation errors. Comparing the mean value and standard deviation of the ionospheric estimation error in Table 2, the estimation of the proposed method is the most accurate when the Lee filter is used. The mean value of the ionospheric estimation error at this time is two-thirds less than that of the original method, and the standard deviation is nearly half. In the case of the proposed method for Lee filtering, the ionospheric estimation error is the smallest, which proves its effectiveness.

5 Summary

The ionospheric effects plague the space-borne InSAR system operating at low frequencies. This paper proposes a new approach that can effectively correct the ionospheric effects. The theoretical basis and implementation process of the proposed approach are introduced in detail and verified using simulation data. Compared with the original split-spectrum method, the proposed approach using Lee filtering has smaller errors and more accurate ionospheric estimation.

References

1. Gabriel AK, Zebker HA (1988) Crossed orbit interferometry: theory and experimental results from SIR-B. Int J Remote Sens 9:857–872
2. Zhang Q, Antoniou M, Chang W, Cherniakov M (2015) Spatial decorrelation in GNSS-Based SAR coherent change detection. IEEE Trans Geosci Remote Sens 53(1):219–228
3. Brcic R, Parizzi A, Eineder M (2011) Ionospheric effects in SAR interferometry: an analysis and comparison of methods for their estimation, Vancouver, Canada, pp 1497–1500

4. Meyer F, Bamler R, Jakowski N, Fritz T (2006) The potential of low-frequency SAR systems for mapping ionospheric TEC distributions. IEEE Trans Geosci Remote Sens Lett 3(4):560–564
5. Jung H, Lu Z, Zhang L (2013) Feasibility of along-track displacement measurement from Sentinel-1 interferometric wide-swath mode. IEEE Trans Geosci Remote Sens 51(1):573–578
6. Brcic R, Eineder M, Bamler R (2009) Interferometric absolute phase determination with TerraSAR-X wideband SAR data. In: 2009 IEEE radar conference, Pasadena, CA, pp 1–6. https://doi.org/10.1109/RADAR.2009.4977112
7. Brcic R, Parizzi A, Eineder M, Bamler R, Meyer F (2010) Estimation and compensation of ionospheric delay for SAR interferometry. In: Proceedings of IEEE IGARSS, pp 2908–2911
8. Rosen P, Hensley S, Chen C (2010) Measurement and mitigation of the ionosphere in L-band interferometric SAR data. In: Proceedings of IEEE radar conference, May 2010, pp 1459–1463
9. De Zan F, Prats-Iraola P, Rodriguez-Cassola M (2015) On the dependence of delta-k efficiency on multilooking. IEEE Geosci Remote Sens Lett 12(8):1745–1749. https://doi.org/10.1109/LGRS.2015.2424272
10. Gomba G, Parizzi A, De Zan F, Eineder M, Bamler R (2015) Toward operational compensation of ionospheric effects in SAR interferograms: the split-spectrum method. IEEE Trans Geosci Remote Sens 54(3):1446–1461. https://doi.org/10.1109/TGRS.2015.2481079
11. Gomba G, Rodríguez González F, De Zan F (2016) Ionospheric phase screen compensation for the Sentinel-1 TOPS and ALOS-2 ScanSAR modes. IEEE Trans Geosci Remote Sens 55(1):223–235. https://doi.org/10.1109/TGRS.2016.2604461

The Research of Multi-scale Effect on Remote Sensing Image Object Detection

Guangkuo Ma, Yifan Dong, Yun Su, Wei Xu, and Pingping Huang

Abstract Object detectors based on convolutional neural network (CNN) have already achieved state-of-the-art performance due to the powerful feature representation capabilities in computer vision tasks including image classification, object detection, and image segmentation. With the rapid development of remote sensing technology, the object detection of high-resolution optical remote sensing images has become an important part of remote sensing. On account of the large sight and complex background of the remote sensing images, along with the large size variation of the objects, the performance of the methods based on CNN is restricted. This paper will focus on the objects that multi-scale features are of the most obvious to analyze and discuss the impact of multi-scale effect on object detection. A new criterion for setting small objects is presented. Through the analysis of evaluation indicators, we conclude that the multi-scale effect decreases the detection accuracy of small objects under the same condition. Furthermore, a method is proposed to improve the accuracy of small objects detection.

Keywords CNN · Object detection · Remote sensing · Multi-scale

1 Introduction

Object detection has been an important research hotspot in both academic community and industry. For instance, a fundamental component of autonomous driving technology [1] that has been popular in recent years is object detection. In computer vision tasks which includes image classification [2], object detection [3],

G. Ma · Y. Dong (✉) · Y. Su · W. Xu · P. Huang
College of Information Engineering, Inner Mongolia University of Technology, Hohhot 010051, China
e-mail: yfdong@imut.edu.cn

G. Ma · Y. Dong · Y. Su · W. Xu · P. Huang
Inner Mongolia Key Laboratory of Radar Technology and Application, Hohhot 010051, China

L. Wang et al. (eds.), *Proceedings of the 7th China High Resolution Earth Observation Conference (CHREOC 2020)*, Lecture Notes in Electrical Engineering 757, https://doi.org/10.1007/978-981-16-5735-1_27

and instance segmentation, a number of methods based on convolution neutral network (CNN) [4–6] have achieved state-of-the-art performance in the natural sense images because CNN can extract high-level semantic feature from original images.

However, the progress of remote sensing image object detection is relatively slow, as a result of the complex background of satellite or aerial images, which are greatly affected by weather, clouds, light and other factors. In the meanwhile, object detection plays a pivotal role in the field of analysis and understanding of remote sensing images. Driven by this demand, lots of object detection approaches based on deep learning for high-resolution optical remote sensing images have been developed. In order to satisfy the requirements of deep learning algorithms, several object detection datasets, such as DIOR [7], DOTA [8], NWPU VHR-10, COWC, are proposed in the research area of remote sensing.

Nevertheless. The performance of object detection algorithms based on deep learning is limited seriously by the irreparable flaws in these datasets. In particular, the multi-scale variation problem and small objects. In natural scene images like MS COCO, the detection of small objects can be improved by copy-pasting small objects multiple times in each image containing small objects [9]. Such methods just add image preprocessing steps and the real matter that small objects disappear in the deep network is not tackled. In remote sensing images, Scale Adaptive Proposal Network (SAPNet) [10] model, is proposed to address the size issue of different kinds of objects. But the SAPNet is trained on pretrained ResNet-101 model for DOTA dataset, high-level semantic information can be required, while the deeper network layer is not friendly to small objects. At the same time, the objects are divided into three types: small, medium and large, according to different categories. For example, large instance categories have soccer ball field (SBF), ground track field (GTF), basketball court (BC), tennis court (TC) and baseball diamond (BD), medium ones have large vehicle (LV), bridge, harbor, storage tank (ST) and roundabout (RA), and small ones have helicopter (HC) ,swimming pool (SP), ship, plane and small-vehicle (SV). This standard of division is unreasonable because that the same class may contain small and large instances in the meanwhile, as a result of the huge size variation of the same object in different scenes.

In this letter, we will focus on the object instances that multi-scale features are of the most obvious to discuss the impact of multi-scale on object detection. A Large-scale Dataset for Object Detection in Aerial Images (DOTA) is used by the experiment. A new division pattern is proposed to divide the objects into three levels: large, medium and small, ResNet50 is used as feature extractor, instead of ResNet101, to avoid small objects disappear in the deep network. A state-of-the-art object detection method based on deep learning is used to train and test. Through the analysis of experiment results, it's a conclusion that the effect of large multi-scale makes the AP of small objects poor while the AP of large objects is relatively fair under the same condition. The detection accuracy of large instances is 67% higher than that of small instances in the new division pattern for DOTA. In order to overcome the multi-scale effect, a method is proposed and the experiment results shows the method achieve improvement under the same division way.

2 Related Works

Faster region-based convolutional network (Faster R-CNN) with a deep residual network (ResNet-50) backbone and feature pyramid network (FPN) framework is used to do this experiment.

2.1 *Faster R-CNN*

In the Faster R-CNN framework, feature extraction of images is accomplished through a series of convolution layers. Then, a large number of proposals are generated by region proposal network (RPN) which use a great many anchors [11] with five scales (32, 64, 128, 256, 512) and three aspect ratios (0.5, 1, 2) in the feature map. Finally, the proposals are processed as fixed dimensions by regions of interest (ROI) pooling layers to classify and regression in the next moment.

2.2 *ResNet-50*

With the deepening of neural network, the problem of gradient disappearance will appear. Deep Residual Network (ResNet) is proposed in order to overcome the above-mentioned problem. An identity mapping is added to convolution neural network in ResNet. It means the current output is passed directly to the next layer, at the same time, the gradient of the next layer is also transmitted directly to the network of the previous layer during backward process. ResNet-50 with 50 convolution layers and fully connection layers, as well as ResNet-101 with 101 convolution layers and fully connection layers are most frequently used, ResNet-50 is selected here, because that the features of small objects may disappear in deep convolution neural network, as shown in Table 1.

2.3 *FPN*

The core of Feature Pyramid network (FPN) [12, 13] is top-down, combined with feature extraction network which is named bottom-up to put the location information at the low-level and the semantic information at the high-level together. The experiments have shown that the performance of algorithms with FPN is greatly improved. Especially, in term of small objects, more information about the characteristics of small objects is obtained, as a result, the resolution of feature mapping is increased through FPN.

Table 1 ResNet-50 and ResNet-101

Layer name	Output size	50-layer	101-layer
Conv1	112 × 112	7 × 7, 64, stride 2	
Conv2_x	56 × 56	3 × 3 max pool, stride 2	
		$\begin{bmatrix} 1 & \times & 1,64 \\ 3 & \times & 3,64 \\ 1 & \times & 1,256 \end{bmatrix} \times 3$	$\begin{bmatrix} 1 & \times & 1,64 \\ 3 & \times & 3,64 \\ 1 & \times & 1,256 \end{bmatrix} \times 3$
Conv3_x	28 × 28	$\begin{bmatrix} 1 & \times & 1,128 \\ 3 & \times & 3,128 \\ 1 & \times & 1,512 \end{bmatrix} \times 4$	$\begin{bmatrix} 1 & \times & 1,128 \\ 3 & \times & 3,128 \\ 1 & \times & 1,512 \end{bmatrix} \times 4$
Conv4_x	14 × 14	$\begin{bmatrix} 1 & \times & 1,256 \\ 3 & \times & 3,256 \\ 1 & \times & 1,1024 \end{bmatrix} \times 6$	$\begin{bmatrix} 1 & \times & 1,256 \\ 3 & \times & 3,256 \\ 1 & \times & 1,1024 \end{bmatrix} \times 23$
Conv5_x	7 × 7	$\begin{bmatrix} 1 & \times & 1,512 \\ 3 & \times & 3,512 \\ 1 & \times & 1,2048 \end{bmatrix} \times 3$	$\begin{bmatrix} 1 & \times & 1,512 \\ 3 & \times & 3,512 \\ 1 & \times & 1,2048 \end{bmatrix} \times 3$
	1 × 1	Average pool, 1000-d fc, softmax	

2.4 Data Preparation

A Large-scale Dataset for Object Detection in Aerial Images (DOTA) is made up of 15 different categories, including ground track field (GTF), baseball diamond (BD), basketball court (BC), helicopter (HC), swimming pool (SP), soccer ball field (SBF), harbor, storage tank (ST), tennis court (TC), small vehicle (SV), large vehicle (LV), roundabout (RA), ship, bridge, and plane. As shown in Fig. 1. The images have a range of size from about 800 × 800 to 4000 × 4000 pixels, and every image has multiple objects of different scales, orientations and shapes in this dataset. We utilize the oriented annotation of DOTA in the experiment.

A series of 1024 × 1024 patches are cropped from the original images, erenow, they will be scaled with scaling factor of 0.5, 1, 1.5 in order to augment dataset, enhance training effect. Figure 2 reports the number of object instance per class after data enhancement. The object classes of ship and small vehicle have higher instance count, yet, the object classes of roundabout, ground track field, soccer ball field, and baseball diamond have lower instance count.

3 Experiment

The experiments were based on the deep learning framework Pytorch and implemented on a single CPU Inter Xeon W-2123 CPU at 3.60GHz server with two GPUs NVIDIA RTX 2080Ti. The computer operating system was Ubuntu 18.04.

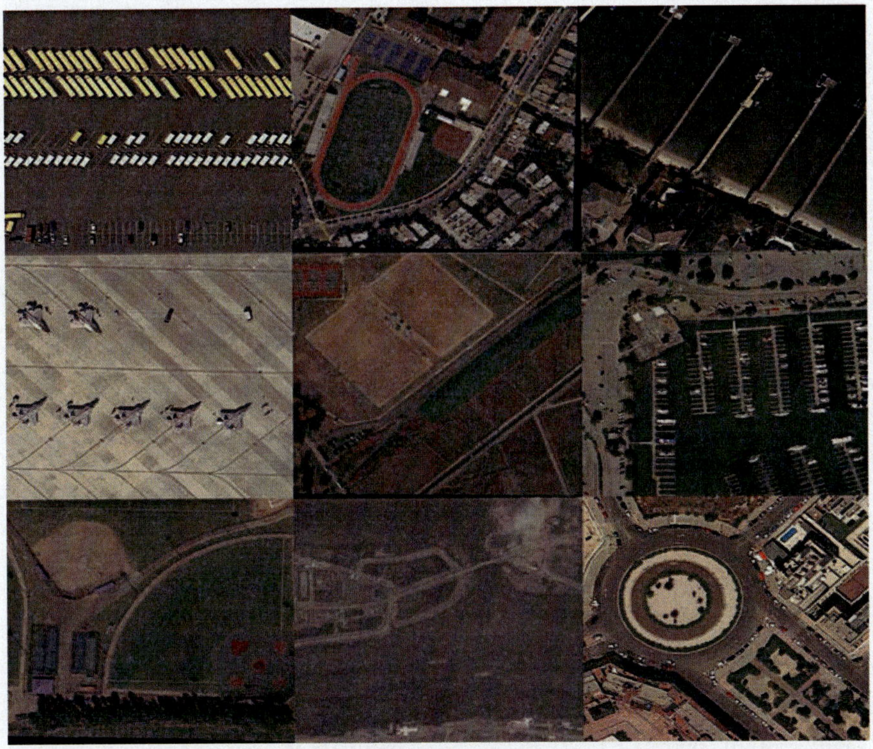

Fig. 1 Examples with small vehicle, ground track field, harbor, plane, soccer ball field, ship, baseball diamond, bridge, and roundabout in DOTA

The AP and mAP, which have been used in other CNN-based methods [14], is adopted to evaluate the performance of the object detection method in our experiment. The precision and recall are important evaluation indexes of object detection. And the precision can be formulated as

$$P = \frac{TP}{TP + FP} = \frac{TP}{all\,positive\,numbers} \times 100\%. \tag{1}$$

The recall can be calculated by

$$R = \frac{TP}{TP + FN} = \frac{TP}{all\,ground\,truths} \times 100\%. \tag{2}$$

where TP represents true positive numbers, FP and FN indicate false positive numbers and false negative numbers. If the intersection over union (IoU) [15] of a bounding box exceeds 0.5, it can be regarded as right, the IoU is defined as follows

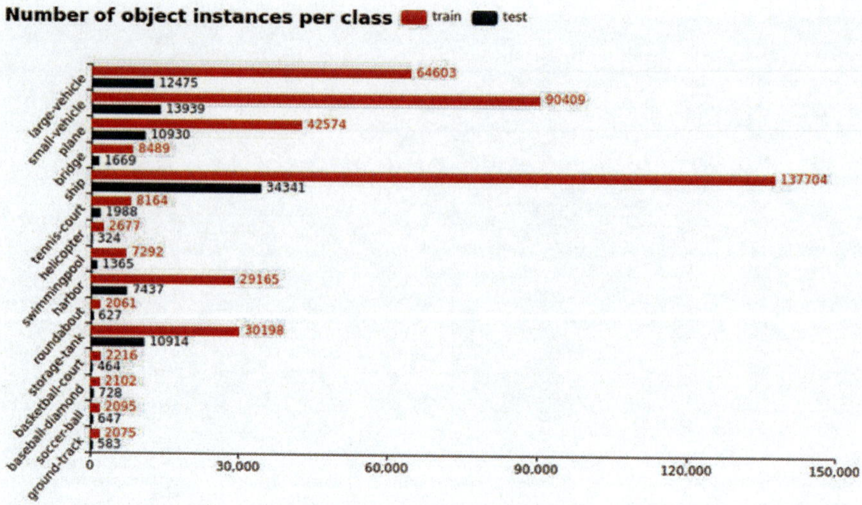

Fig. 2 Number of object instances per class

$$IoU = \frac{GT \cap DR}{GT \cup DR}. \tag{3}$$

where GT represents the ground truth bounding box, and DR is the detection results. The learning rate is set to be 0.005 and SGD is used to be as optimizer.

According to the strategy of division, which was mentioned earlier in the SAPNet framework. Large instance categories have soccer ball field (SBF), ground track field (GTF), basketball court (BC), tennis court (TC) and baseball diamond (BD), medium ones have large vehicle (LV), bridge, harbor, storage tank (ST) and roundabout (RA), and small ones have helicopter (HC) ,swimming pool (SP), ship, plane and small-vehicle (SV). In the testing images of DOTA-v1.0, the proposed method can achieve the best results in all small, medium, and large instance categories in terms of AP, the mAP has a giant improvement from 62.9 to 79.9% compared to SAPNet. For medium instance categories, the result of the proposed model is 22% higher than that of SAPNet. It can be seen from Table 2.

As mentioned earlier, this standard of division is unreasonable because that the same class may contain small and large instances in the meanwhile, as a result of

Table 2 Performance comparisons on DOTA

Methods	FRA (%)	MS-VANS [16] (%)	SAPNet (%)	The proposed (%)
AP_small	56.5	56.6	61.2	81.2
AP_medium	51.3	51.5	54.1	76.9
AP_large	74.2	73.4	73.4	81.8
mAP	60.6	60.5	62.9	79.9

the huge size variation of the same object in different scenes. The definitions of the small, medium, and large objects in MS COCO [17] are shown in Table 3. If the definitions are used in our experiment, the proportion of small object will be as high as 72.19%, this approach is not representative.

We proposed a new division pattern on the basis of characteristics of the remote sensing images and existing methods for remote sensing images. A longest side of a bunding box can be required through for vertices of the bounding box. The object is considered to be small object, when the longest side of the object is less than 15 pixels. Other details are shown in Table 4

Validation set of DOTA-v1.0 is used to calculate the precision rate and recall rate. Figure 3 shows the Recall-Precision-Curves of 15 categories objects in validation set.

Table 3 The definitions of the small, medium, large objects in MS COCO

	Min rectangle area	Max rectangle area
Small object	0 × 0	32 × 32
Medium object	32 × 32	96 × 96
Large object	96 × 96	∞ × ∞

Table 4 The definitions of the small, medium, large objects in DOTA

	Min rectangle area	Max rectangle area	Objects count (%)	AP (%)
Small object	0 × 0	15 × 15	26.88	22.62
Medium object	15 × 15	96 × 96	67.42	88.66
Large object	96 × 96	∞ × ∞	5.70	89.35

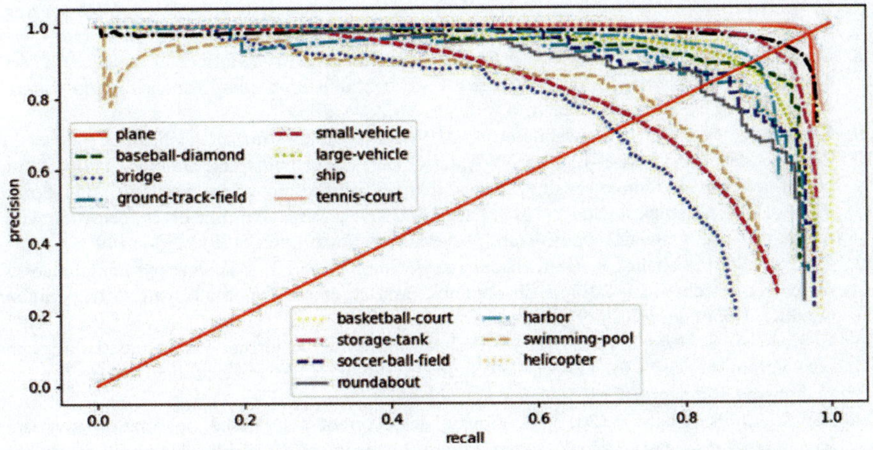

Fig. 3 The Recall-Precision-curves of 15 categories objects in DOTA validation set

4 Summary

In this paper, Faster R-CNN is combined with ResNet-50 and FPN to discuss the impact of multi-scale effect on object detection. Experimental results based on DOTA show that the proposed method significantly increases the mAP from 62.9 to 79.9% compared with the well-known SAPNet framework. In the meanwhile, the detection results of small, medium, and large object are higher than that of SAPNet framework. In addition, we proposed a new division pattern for optical remote sensing images, according to characteristics of remote sensing images and existing models.

Acknowledgements This work was supported in part by the Equipment Pre-Research Field Foundation (JZX7Y20190253041401) and the Project of Science and Technology Plan of Inner Mongolia Autonomous Region (2019GG138).

References

1. Hoffmann L, Elster C (2020) Deep neural networks for computational optical form measurements. J Sensors Sensor Syst 9(2)
2. Liu J-e, An F-P (2020) Image classification algorithm based on deep learning-Kernel function. Sci Program
3. He K (2020) Traffic sign identification based on target detection network. J. Softw Eng 23 (10):19–22
4. Zou Z, Shi Z (2016) Ship detection in spaceborne optical image with SVD networks. IEEE Trans Geosci Remote Sens 54(10):5832–5845
5. Chen X, Xiang S, Liu C-L, Pan C-H (2014) Vehicle detection in satellite images by hybrid deep convolutional neural networks. IEEE Geosci Remote Sens Lett 11(10):1797–1801
6. Tang T, Zhou S, Deng Z, Zou H, Lei L (2017) Vehicle detection in aerial images based on region convolutional neural networks and hard negative example mining. Sensors 17(2):336
7. Li K, Wan G, Cheng G, Meng L, Han J (2020) Object detection in optical remote sensing images: A survey and a new benchmark. ISPRS J Photogrammetry Remote Sens 159
8. Ding J, Xue N, Long Y, Xia G, Lu Q (2019) Learning RoI transformer for oriented object detection in aerial images. In: 2019 IEEE/CVF conference on computer vision and pattern recognition (CVPR), Long Beach, CA, USA, pp 2844–2853
9. Kisantal M, Wojna Z, Murawski J et al (2019) Augmentation for small object detection
10. Zhang S, He Gh, Chen H-B, Jing N, Wang Q (2019) Scale adaptive proposal network for object detection in remote sensing images. IEEE Geosci Remote Sens Lett 16(6):864–868
11. Ren S, He K, Girshick R, Sun J (2017) Faster R-CNN: towards real-time object detection with region proposal networks. IEEE Trans Pattern Anal Mach Intell 39(6):1137–1149
12. Lin T, Dollár P, Girshick R, He K, Hariharan B, Belongie S (2017) Feature pyramid networks for object detection. In: 2017 IEEE conference on computer vision and pattern recognition (CVPR), Honolulu, HI, pp 936–944
13. Li H, Miao S, Feng R (2020) DG-FPN: learning dynamic feature fusion based on graph convolution network for object detection. In: 2020 IEEE International conference on multimedia and expo (ICME), London, United Kingdom, pp 1–6
14. Zan L, Lu K, Chen Z (2019) Automatic detection of ship based on rotation invariant RetinaNet. In: Wang L, Wu Y, Gong J (eds) Proceedings of the 6th China high resolution

earth observation conference (CHREOC 2019). Lecture notes in electrical engineering, vol 657. Springer, Singapore. https://doi.org/10.1007/978-981-15-3947-3_39
15. Yan Z, Song X, Zhong H, Zhu X (2018) Object detection in optical remote sensing images based on transfer learning convolutional neural networks. In: 2018 5th IEEE International conference on cloud computing and intelligence systems (CCIS), Nanjing, China, pp 935–942
16. Wang C, Bai X, Wang S, Zhou J, Ren P (2019) Multiscale visual attention networks for object detection in VHR remote sensing images. IEEE Geosci Remote Sens Lett 16(2):310–314
17. Vinyals O, Toshev A, Bengio S, Erhan D (2017) Show and tell: lessons learned from the 2015 MSCOCO image captioning challenge. IEEE Trans Pattern Anal Mach Intell 39(4):652–663

Remote Sensing Road Segmentation Based on Feature Extraction Optimization and Skeleton Detection Optimization

Sijun Dong, Yu Zhao, and Zhengchao Chen

Abstract High-resolution remote sensing image segmentation is a mature application in many industrial-level image applications, such as those in the military, civil, and other fields. Scene analysis needs to be automated in high-resolution remote sensing images as much as possible. Nowadays, with the rise of deep learning algorithms, remote sensing image processing algorithms have made tremendous progress. Deep learning algorithms process unlabeled data by learning a certain amount of labeled data. We conducted a specific study on the road target with GF1 data collected in China, and the remote sensing image's resolution was 2 m (Jin et al. in J Anhui Agri Sci 43:358–362, 2015 [Jin et al. in J Anhui Agri Sci 43:358–362, 2015]). According to the observation of road features in remote sensing images, it still has a large number of small roads that are difficult to distinguish in the 2 m resolution GF1 remote sensing image. Due to the limitations of the downsampling calculation of the fully convolutional neural network, it is easy to lose a lot of information on small roads (Long et al. in Proceedings of the IEEE conference on computer vision and pattern recognition, pp. 3431–3440, 2015 [Long J, Shelhamer E, Darrell T (2015) Fully convolutional networks for semantic segmentation. In: Proceedings of the IEEE conference on computer vision and pattern recognition, 2015, pp 3431–3440]). Therefore, we have adjusted the feature extraction and backbone networ. We adopted EfficientNet (Tan and Le in Efficientnet: Rethinking model scaling for convolutional neural networks, 2019 [Tan M, Le QV (2019) Efficientnet: rethinking model scaling for convolutional neural networks. arXiv preprint arXiv:1905.11946]) as the skeleton network of the algorithm, and combined D-linkNet (Zhou et al. in CVPR workshops, pp. 182–186, 2018 [Zhou L, Zhang C, Wu M (2018) D-LinkNet: LinkNet with pretrained encoder and dilated convolution for high resolution satellite imagery road extraction. In: CVPR workshops, pp 182–186]) as our After preliminary training to

S. Dong · Y. Zhao · Z. Chen (✉)
Aerospace Information Research Institute, Chinese Academy of Sciences, Beijing 100094, China
e-mail: chenzc@aircas.ac.cn

Z. Chen
University of Chinese Acdemy of Sciences, Beijing 100049, China

L. Wang et al. (eds.), *Proceedings of the 7th China High Resolution Earth Observation Conference (CHREOC 2020)*, Lecture Notes in Electrical Engineering 757, https://doi.org/10.1007/978-981-16-5735-1_28

proposed D2-LinkNet, based on the road samples, we can get very good prediction results. According to the existing prediction results, there is still a certain difference in the fitting of our segmentation results to the groundtruth. To solve this problem, based on the extraction and analysis of the skeleton of the road prediction results and the integration of different prediction results, we proposed a skeleton optimization algorithm to optimize our prediction results for road samples. And on the one hand, it complements the segmentation of small roads. On the other hand, the road segmentation result fits the road boundary more closely. The final experiment results show that the optimization algorithm effectively improves the prediction accuracy of the network, and we have achieved better results compared with D-LinkNet in our large range remote sensing dataset.

Keywords Remote sensing image · Image semantic segmentation · Feature extraction · Road skeleton detection

1 Introduction

The land classification of high-resolution remote sensing images is of great significance for studying regional utilization, capturing the dynamic evolution of land use, and monitoring key resources. In this article, we conducted a road segmentation algorithm research for a wide range of scenes in eastern China, as shown in Fig. 1. High-resolution remote sensing image road segmentation is very helpful for urban planning, traffic planning, and automatic driving. This article selects large-scale remote sensing image data samples from 29 provinces in eastern China for experiments. Our samples cover most of China's provinces, and a lot of labeling work has been done in each province to ensure that our samples are sufficiently diverse. It can represent the road characteristics of remote sensing images across the country. Due to the limitation of the number of parameters, in the face of large-scale high-resolution image data, traditional remote sensing physics algorithms and machine learning algorithms usually cannot fit the data well. Since the introduction of deep learning in 2006, the powerful ability of convolutional neural networks (CNN) to solve complex problems has made great progress in the past 10 years, providing new methods for remote sensing land use mapping tasks [5].

Semantic segmentation of remote sensing images is also called land cover classification. By assigning a category to each pixel in the image, position information and semantic information can be obtained. With the development of deep learning, remote sensing image semantic segmentation based on deep learning has achieved great success. Since the first fully convolutional neural network proposed to solve the problem of image semantic segmentation, many classic image semantic segmentation algorithms have appeared in recent years. This paper uses the deep convolutional neural network method to extract and analyze the features of remote sensing images. Remote sensing image segmentation is different from natural scene image segmentation. There are many natural scene categories, and different object

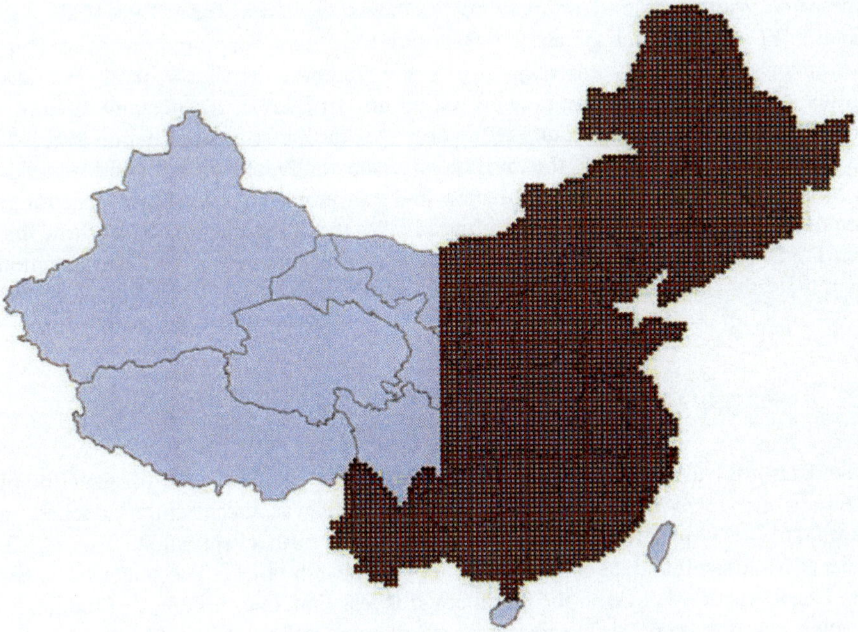

Fig. 1 Distribution range of land remote sensing image samples nationwide

categories usually have strong interdependence. For example, many objects will appear in the feature scene, which also leads us to use the scene information of the image to predict the object category. However, remote sensing images are different from natural scenes [6]. We usually do not need to recognize many target categories, or only a specific category in the remote sensing image scene, such as buildings or roads. This paper focuses on experiments on road samples from remote sensing images. The image features of road samples are quite different from other remote sensing impact samples. Because the road is long and may traverse the entire sample, a large field of view is required for feature extraction. And the road is thin, The proportion of foreground and background is very different, so the spatial information of the feature needs to be preserved. Therefore, expanding the field of view is not suitable for increasing the pooling layer, thereby destroying the spatial feature. Another benefit of expanding the field of view is that you can learn the connectivity and span characteristics of the road. LinkNet [7] is an effective semantic segmentation neural network, which takes advantage of skip connections [8], residual blocks, and encode-decode structure. D-LinkNet uses the pretrained encoder LinkNet as its backbone and has an additional expanded convolutional layer in the central part. The main highlight of D-LinkNet is to add expanded convolutional layers to expand the receptive field. According to the observation of the experimental sample data, our dataset is collected from a large-scale remote sensing image of eastern China. Therefore, our data samples are more complex compared with common land classification datasets such as the ISPRS Vaihingen dataset [9] and Inria Aerial Labeling Dataset [10]. Therefore, more powerful feature extraction capabilities are required in the feature extraction part. We use EfficientNet as the backbone network based on D-LinkNet. At the same time, the third block of D-LinkNet feature extraction uses the same center dilation part [11] as the fourth block. Usually, the third block in the backbone still has good semantic information, and this part of information plays an important role in our large range remote sensing image road classification. Finally, the experimental results show that our D2-LinkNet has achieved better classification results in the road segmentation of remote sensing images than D-LinkNet in this dataset.

2 Research Dataset Introduction

The dataset we used is GF-1 satellite data. The GF-1 satellite is the first satellite of China's high-resolution earth observation system. It was successfully launched on April 26, 2013, ushering in a new era of China's earth observation. This article selects GF-1 satellite data taken in 2015 as the research object. The selected images are based on sunny days, so the cloud cover is less than one percent, and the image is clear. At the same time, geometric correction, radiometric calibration, atmospheric correction, image enhancement, and other methods are used to preprocess the image to improve image quality. The label data studied in this paper is a road single-class binary data set with manual labeling, a total of 100,000 label images, the labels are evenly distributed in the provinces of the study area, the size of each label is 1024×1024, and the total data area is 419,000 km^2. In our dataset, we selected 1000 pictures of 1024×1024 size as our validation dataset, and we use this to evaluate the accuracy of the algorithm. In addition, this article uses data enhancement methods to synchronize remote sensing images and labels, including resize, panning and flipping to increase the total amount of training data and improve the generalization ability of the model [12].

3 Methodology

D-LinkNet uses ResNet34 [13] pretrained on the ImageNet dataset as its encoder. ResNet34 was originally designed for 256×256 medium-resolution image classification, but in this challenge, the task is to segment roads from 1024×1024 high-resolution satellite images. Considering the narrowness, connectivity, complexity, and long road span, it is important to increase the perception range of the features in the central part of the network and retain detailed information. Using a pooling layer can double the perception range of features, but it may reduce the resolution of the central feature map and reduce spatial information. The atrous convolution layer may be an ideal alternative to the pooling layer. D-LinkNet uses several atrous convolution layers stacked together with a skip-connection in the middle. In this way, the receptive field can be expanded without pooling calculation to lose image information, and all the information of the image is retained to the greatest extent, providing comprehensive image feature information for the decoding part. Because our national land remote sensing image road classification samples are different from some open source remote sensing image datasets, our data sample distribution range is very large, and the environment is also very complex, which increases the difficulty of feature extraction in our network structure. Based on the current sample environment, we proposed D2-LinkNet to process our large number of samples, as shown in Fig. 2.

The feature extraction module in D-LinkNet is ResNet34, which is not enough to fit our complex samples well, so we have made a complicated design based on D-LinkNet. First, the ResNet34 skeleton network is replaced with the EfficientNet. Compared with ResNet, EfficientNet proposed the compound model scaling algorithm, which achieves the goal of index improvement by comprehensively optimizing the network width, network depth, and resolution. It can achieve the same accuracy index and the existing classification network, greatly reducing the number of model parameters and calculations. So, a backbone network with more powerful feature extraction capabilities is proposed. Therefore, we adopted EfficientNet as our feature extraction backbone network. D-LinkNet proposed the center dilation part to optimize the extraction of the skeleton network to obtain high-level semantic

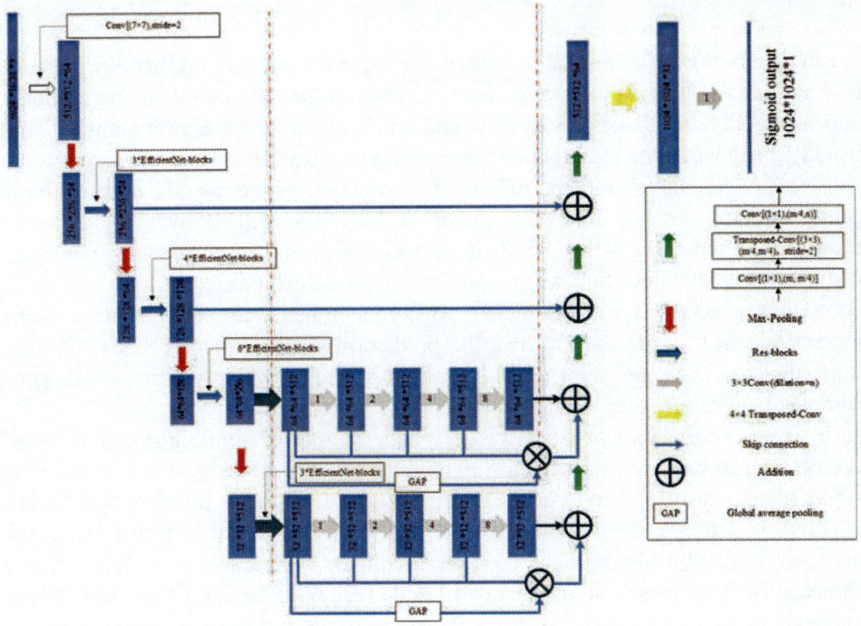

Fig. 2 D2-LinkNet network architecture

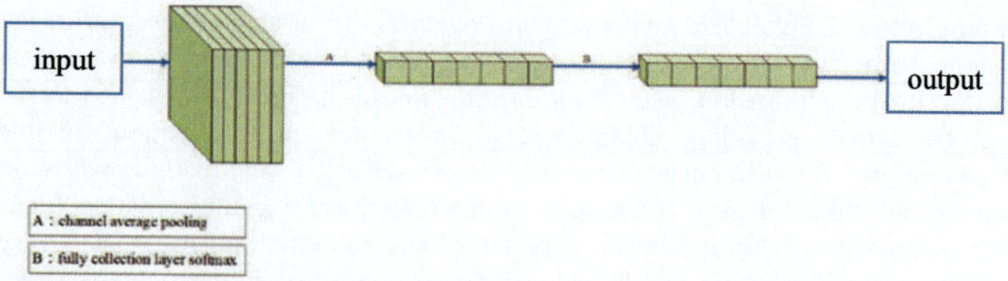

Fig. 3 Global average pooling

features. Through the observation of the slave modules, we found that the author used simple addition to combine the atrous convolution extraction with different atrous rates. In order to optimize the combination of information in this part of the feature map, we use the corresponding high-level semantic features extracted by the backbone network to perform global average pooling as a weight vector to monitor the combined feature map. We use this module to process high-level semantic features as the channel information for the addition of weighted supervision feature maps as shown in Fig. 3. In addition, we also use the center dilation part for processing in the third block of the feature extraction skeleton network.

4 Skeleton Detection Optimization

According to the experimental results of our algorithm, we found that our network still has certain flaws in the extraction of roads. Because the sample is too complex and large, there will be many narrow and small roads in the dataset samples, and usually small roads are easy to lose information in semantic segmentation of remote sensing images and cannot be effectively extracted. When we are doing sample prediction, we can get different degrees of prediction effects by using different thresholds. Prediction with a small threshold can be a thicker road segmentation result because the algorithm will recognize the area surrounding the road as a road, so the extracted result will be thicker, so it will be better for small roads, and the extraction result is not easy to lose the prediction results of small roads. Using a large threshold can get more refined prediction results, which can it the real road samples better, but some small roads may not be extracted.

In response to this situation, we designed an optimization algorithm for road prediction results based on skeleton extraction as shown in Fig. 4. First, the road segmentation results from A and B were obtained by using different thresholds. Road A is a coarse segmentation result, and Road B is a thin result. Then the skeleton extraction algorithm is used to obtain the skeleton part, skeleton A and skeleton B of the segmentation results of A and B. We take the difference between skeleton A and skeleton B, which is the road that is not detected by skeleton B. Therefore, we attach this difference set area to road B to roughly obtain a road

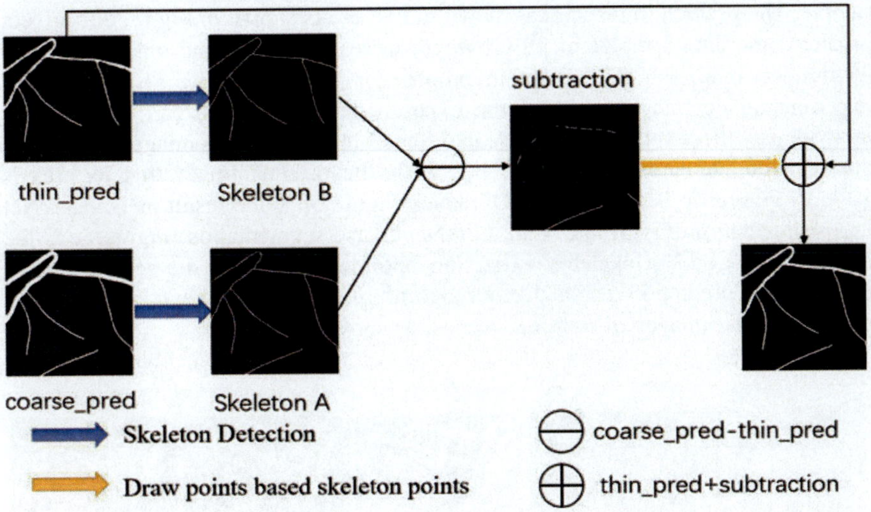

Fig. 4 Skeleton detection optimization

segmentation result after completion. Regarding this additional method, first of all, we can get the coordinate points of the skeleton of the difference set area. Based on each coordinate point, we draw a curve with a suitable width, so that the curve can be roughly regarded as a road which cannot be detected. This curve will be attached into road B, so we got the optimized result which has a more comprehensive identification of small roads. Through this method, we can not only obtain road prediction results with a suitable width, but also have a good prediction effect on small roads. In the actual situation, the road is blocked by trees in the remote sensing image, which leads to a disconnection in the extracted road results. This situation is difficult to achieve by optimizing the accuracy of the extraction algorithm. Therefore, we adopt the method of expansion and corrosion in mathematical morphology and realize the connection of adjacent open circuits by performing the closed operation of the road binary result first expansion and then corrosion. The results obtained through the above operations are more suitable for real road samples, and at the same time, due to our skeleton extraction optimization algorithm, there is no phenomenon that the thin roads cannot be extracted in the prediction results.

5 Experiment and Result

In this article, we conducted training and prediction on samples of 29 provinces in the land classification of our national remote sensing images. The experimental results show that our work can be excellent for a large-scale national land road sample. The prediction results are shown in Fig. 5. As shown in Fig. 5, our project predicted the data samples of 29 provinces in eastern China and obtained a road distribution map. For the algorithm proposed in this paper, we also conducted experimental comparisons. From the experimental results, we can see that our network can effectively extract roads in different remote sensing image scenes after optimization. As shown in Fig. 5, Fig. 5a is the original image that its area is 10 km^2. Figure 5b is the result of DlinkNet. Figure 5c is the result of D2-LinkNet fine segmentation. Figure 5d is D2-LinkNet coarse segmentation. Figure 5e is the result of D2-LinkNet skeleton extraction optimization. It can be seen from the experimental prediction results that our algorithm can effectively extract roads from remote sensing images of different scenes, as shown in Fig. 6.

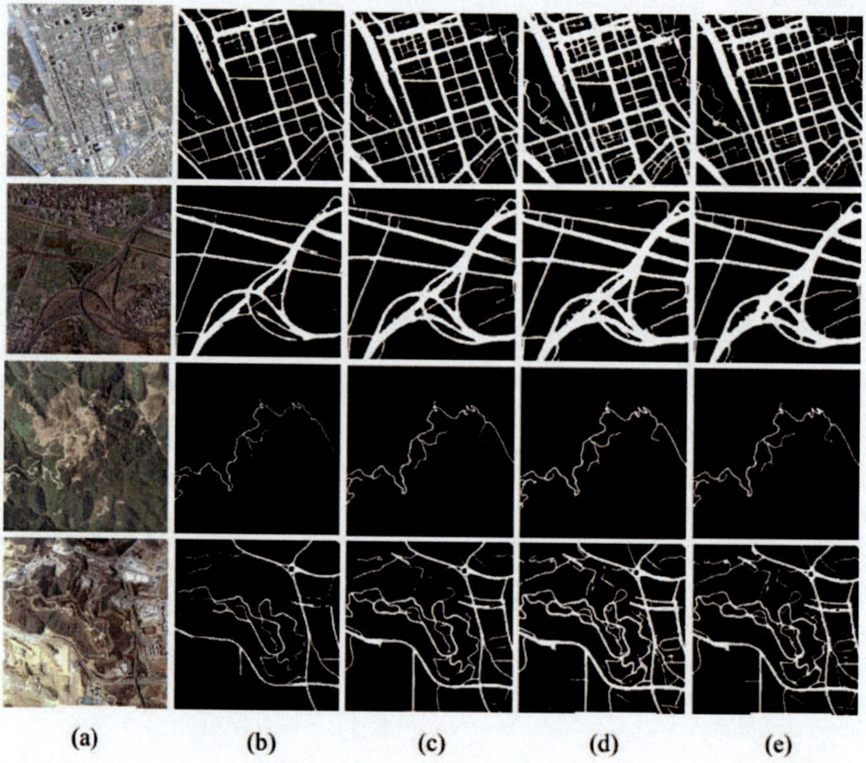

Fig. 5 Details in mapping results. **a** GF-1 images; **b** D-LinkNet results; **c** thin results; **d** coarse results; **e** D2-LinkNet results within skeleton detection optimization

Fig. 6 Distribution map of land classification roads in China

6 Summary

In this article, we proposed a D2-LinkNet network based on D-LinkNet to optimize the road classification results of our country's land remote sensing images. Finally, compared to the original D2-LinkNet, we have achieved obvious optimization results. We selected 1000 sample slices with more accurate marking results for calculation accuracy, and finally showed our accuracy and f1-score as shown in Table 1. At the same time, we also judge our road prediction effect based on the visualization of the large-scale prediction results. Through the observation of the visualization effect, our nationwide remote sensing image road prediction project worked well. Therefore, our national land road classification project can play a very important role in practical applications.

Table1 Accuracy and f1-score in validation dataset

Method	acc (%)	F1-score (%)
D-LinkNet	82.79	80.45
D2-LinkNet	84.64	82.07
D2-LinkNet*	86.26	85.74

*define skeleton detection optimization

References

1. Jin W, Haibo Z, Guobin L (2015) The application of GF-1 remote sensing image in the land use change monitoring. J Anhui Agri Sci 43(16):358–362
2. Long J, Shelhamer E, Darrell T (2015) Fully convolutional networks for semantic segmentation. In: Proceedings of the IEEE conference on computer vision and pattern recognition, 2015, pp 3431–3440
3. Tan M, Le QV (2019) Efficientnet: rethinking model scaling for convolutional neural networks. arXiv preprint arXiv:1905.11946
4. Zhou L, Zhang C, Wu M (2018) D-LinkNet: LinkNet with pretrained encoder and dilated convolution for high resolution satellite imagery road extraction. In: CVPR workshops, pp 182–186
5. Krizhevsky A, Sutskever I, Hinton GE (2012) Imagenet classification with deep convolutional neural networks. In: Advances in neural information processing systems, pp 1097–1105
6. Ding L, Tang H, Bruzzone L (2019) Improving semantic segmentation of aerial images using patch-based attention. arXiv preprint arXiv:1911.08877
7. Chaurasia A, Culurciello E (2017) Linknet: exploiting encoder representations for efficient semantic segmentation. In: 2017 IEEE visual communications and image processing (VCIP), pp 1–4
8. He K, Zhang X, Ren S, Sun J (2016) Identity mappings in deep residual networks. In: European conference on computer vision, pp 630–645
9. Rottensteiner F, Sohn G, Gerke M, Wegner JD, Breitkopf U, Jung J (2014) Results of the ISPRS benchmark on urban object detection and 3D building reconstruction. ISPRS J Photogramm Remote Sens 93:256–271
10. Maggiori E, Tarabalka Y, Charpiat G, Alliez P (2017) Can semantic labeling methods generalize to any city? The inria aerial image labeling benchmark. In: 2017 IEEE international geoscience and remote sensing symposium (IGARSS), pp 3226–3229
11. Chen L-C, Papandreou G, Kokkinos I, Murphy K, Yuille AL (2017) Deeplab: semantic image segmentation with deep convolutional nets, atrous convolution, and fully connected crfs. IEEE Trans Pattern Anal Mach Intell 40(4):834–848
12. Zhao H, Shi J, Qi X, Wang X, Jia J (2017) Pyramid scene parsing network. In: Proceedings of the IEEE conference on computer vision and pattern recognition, pp 2881–2890
13. He K, Zhang X, Ren S, Sun J (2016) Deep residual learning for image recognition. In: Proceedings of the IEEE conference on computer vision and pattern recognition, pp 770–778

Capability Requirement Analysis of Space-Based Ocean Reconnaissance and Surveillance System Based on Improved QFD Method

Min Wan and Yan Hou

Abstract The capability requirement analysis of space-based ocean reconnaissance and surveillance system is the basic work of system design, which has an important influence on the construction and optimization of system capability. By summarizing the concept and application status of QFD (Quality Function Deployment) method, and combining with the characteristics of space-based ocean reconnaissance and surveillance system capability requirement analysis, an improved QFD method based on fuzzy theory and GRA (Grey Relational Analysis) is proposed. Based on the analysis of space-based ocean reconnaissance and surveillance task and capability composition, the HoQ (House of Quality) model is established by using the improved QFD method. Through case analysis, the importance ranking of each capability is obtained. The research results verify the feasibility of the improved QFD method, and can also provide reference for the design and optimization of space-based ocean reconnaissance and surveillance system.

Keywords Space based ocean reconnaissance and surveillance · Capability requirement analysis · Quality function deployment (QFD) · House of quality (HoQ) · Fuzzy theory · Grey relational analysis (GRA)

1 Introduction

Space-based reconnaissance and surveillance is a military activity of carrying out reconnaissance and surveillance in outer space on the platform of various spacecraft using equipment such as photoelectric remote sensors, radar or radio receivers [1]. As one of the national strategic reconnaissance means, space-based reconnaissance and surveillance has the characteristics of wide reconnaissance range, high recon-

M. Wan (✉) · Y. Hou
Space Engineering University, Beijing 101416, China
e-mail: wanmin08006328@163.com

M. Wan
PLA Unit 95685, Kunming 650500, China

L. Wang et al. (eds.), *Proceedings of the 7th China High Resolution Earth Observation Conference (CHREOC 2020)*, Lecture Notes in Electrical Engineering 757, https://doi.org/10.1007/978-981-16-5735-1_29

naissance efficiency, not limited by national boundaries and geographical conditions, and good reconnaissance effect. In many modern wars and local conflicts, it has demonstrated a strong strategic deterrent effect and a supporting capability multiplier role. Now it has become the focus of research and development of major military powers.

Ocean reconnaissance and surveillance is an important application direction of space-based reconnaissance and surveillance, which mainly aims at detecting ships, submarines, aircraft carrier formations and other targets. With the upgrading of modern warfare countermeasures, the demand for battlefield target information and situation information is increasing day by day. At the same time, the improvement of the performance of shipborne electronic equipment and the maneuverability of warship makes the target anti-reconnaissance capability greatly improved, which puts forward higher requirements for space-based ocean reconnaissance and surveillance capability.

Military demand analysis is an activity to study and demonstrate the demand basis, content and specification of military system construction [1]. Military requirement analysis is the basic work to guide the construction and development of equipment and form the combat capability of the system. In order to build a strong space-based ocean reconnaissance and surveillance capability, the first thing to do is to conduct accurate military demand analysis, and use scientific analysis methods to get accurate analysis results, so as to better guide the construction and development of space-based ocean reconnaissance and surveillance system.

Based on the traditional QFD method, this paper proposes an improved QFD method which combines the fuzzy theory and the grey relational analysis. By analyzing the task and capability structure of the space-based ocean reconnaissance and surveillance system, combined with the data of literature search and expert consultation, the HoQ model of the analysis problem is constructed, and the results of relevant examples are given. The research results can provide reference for the design and optimization of space-based ocean reconnaissance and surveillance system.

2 Overview of QFD Method Concept and Application

QFD method is a multi-level deductive analysis method which decomposes, unfolds and transforms the customer or market demand gradually through the intuitive graph form quantitative scoring table and matrix combination [2]. QFD method originated from Japanese manufacturing industry in 1960s, and was first proposed and used by Shigeru Mizuno and Yoji Akao. The QFD method directly meets the needs of customers, reflecting the comprehensive quality management concepts of "customer first", "source grabbing", "system planning", "quantitative management", and "multidisciplinary collaboration" [3]. It has the characteristics of visual, wide application and strong operability in the analysis and operation process. Because of these characteristics, QFD method is applied to the food industry

[4], financial services [5], and investment Decision [6], administrative countermeasures [7], automation management [8] and other different application areas and scope continue to expand.

In the aspect of military requirement analysis, reference [9] introduced a QFD-based method for mapping weapon demand in the demonstration of weapon demand. Reference [10] used the QFD method to model, analyze and evaluate the military requirements of the shipboard missile electromagnetic vertical catapult system, clarified the key functions of the system, and provided guidance and support for the system scheme design. Reference [11] proposed a requirement analysis method combining QFD and combat simulation for the problem of ship equipment requirements, gave the design process and operation steps, and obtained the ranking of the importance of ship equipment combat performance. Reference [12] took the electronic blue army antimissile defense combat drill as an example, and discussed the application of QFD method in the combat task of the combat force. Reference [13] studied the application of QFD method in defense missions, and took the long-range air strike cost-benefit analysis as an example to verify the practicability of QFD technology.

In terms of aerospace applications, reference [14] used the QFD method to analyze the requirements of space-based space surveillance capabilities and methods, and obtained the importance order of various surveillance capabilities, providing guidance for the priority development order of space-based space surveillance methods. Reference [15] proposed a new multi-dimensional structure QFD analysis model based on the two-dimensional house of quality, the system house, and applied it to the design and analysis of large and complex space systems. Reference [16] aimed at the problem that the relevant work of aerospace test for new space platform is still in the exploratory stage, and the QFD method is used to analyze the capacity requirements for aerospace tests.

In the application of QFD method, the determination of the functional relationships between customer requirements and engineering characteristics are always the difficulty and focus. The above QFD-based demand analysis research mainly uses the digital sequence method of expert scoring to determine the functional relationships. The method is simple and intuitive but has large subjectivity and errors. The objective method is to use the competitive evaluation data obtained from the investigation or experiment to determine the relationships through the corresponding solution. Using this as a guide, in order to reduce the bias caused by subjective evaluation, many scholars have carried out corresponding research. Reference [17] proposed a method for determining the functional relationships based on rough set theory, but it is only applicable to the situation where a large amount of sample data can be obtained. Reference [18, 19] used the multi-attribute decision-making method based on competitive evaluation to determine the functional relationships in the HoQ model, and achieved more true and reliable results in the case. This objective method can reduce subjective effects and achieve more accurate results, but the method relies on a large amount of accurate data. In reality, due to competition confidentiality and test costs, The scale of data obtained is often small, the distribution is not typical, the numerical value is uncertain and fuzzy.

Aiming at the problem of small data samples and atypical distributions, reference [20] proposed a method using grey relational analysis to determine the functional relationships. Reference [21] applied the orthogonal experiment method to the determination the functional relationships of QFD, taking engineering characteristics as influencing factors, and customer needs as evaluation indicators, so that the problem was considered as a multi-factor and multi-index experimental design problem. Aiming at the problem of data uncertainty and fuzziness, reference [22] used the method of fuzzy weighted average based on α-cut set to calculate the importance of technical characteristics, and used the average level cut set to defuzzify to get an accurate value. According to the value, the technical characteristics were sorted and resources were allocated. In reference [23], the multi-attribute value theory combined with fuzzy regression and fuzzy optimization theory was used to establish a model to comprehensively solve the QFD problem. Reference [24] proposed a method for determining the functional relationships based on asymmetric fuzzy linear regression by expanding symmetrical triangular fuzzy numbers into asymmetric triangular fuzzy numbers and asymmetric trapezoidal fuzzy numbers, combining least square regression and fuzzy linear regression, and said that it has better application effect in problems with uncertainty, qualitative and fuzzy relations.

At present, there are few countries and international organizations with complete space-based ocean reconnaissance and surveillance capability, and the index parameters that represent the capability of the system are mostly military secrets. Therefore, the information obtained is not only fuzzy but also small in sample data. In order to overcome the above shortcomings, this paper proposes an improved QFD method based on fuzzy theory and grey relational analysis for the demand analysis of space-based ocean reconnaissance and surveillance capability. Fuzzy theory is used to solve the uncertainty and fuzziness of data, and grey relational analysis is introduced to determine the relationships matrix in QFD, so as to solve the situation of atypical data distribution and less sample data.

3 Improved QFD Method Based on Fuzzy-Gray Relational Analysis

The core of the QFD method is an analysis model. The analysis model uses a series of matrices to quantitatively analyze the relationship between customer needs and engineering characteristics by expanding the form of a chart. At present, the most common analysis model of the QFD method is the House of Quality (HoQ) model [25].

A typical house of quality is mainly composed of Customer requirements and importance (left wall), engineering characteristics (ceiling), engineering characteristics autocorrelation matrix (roof), relationships matrix (room), importance of engineering characteristics (floor), technology competitiveness evaluation matrix

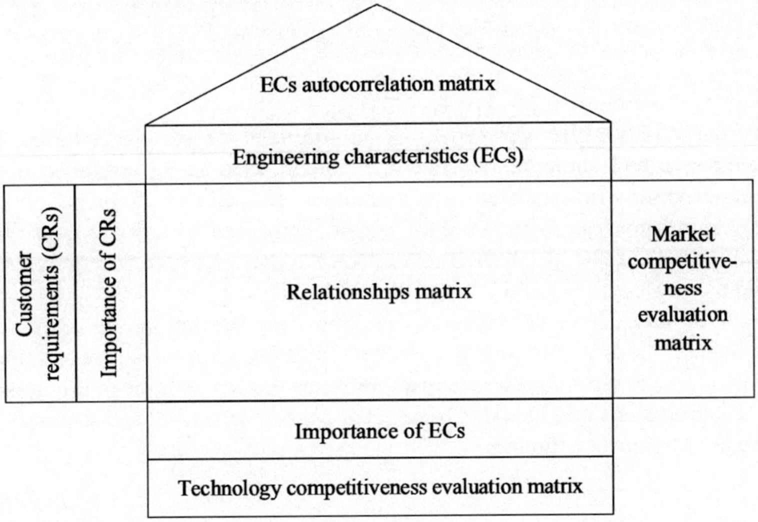

Fig. 1 Structural composition of HoQ

(basement), market competitiveness evaluation matrix (right wall), as shown in Fig. 1.

According to the general analysis steps of the QFD method and the HoQ modeling process, combined with the method steps of grey relational analysis, the implementation steps of the improved QFD method designed for the space-based ocean reconnaissance and surveillance capability requirement are as follows.

3.1 Data Collection

Collect and analyze the task requirements of space-based ocean reconnaissance and surveillance system, determine the main tasks, record them as set $\boldsymbol{M}$, get expert evaluation through expert consultation, use analytic hierarchy process (AHP) to get the task requirement importance matrix $\boldsymbol{P}$, and determine the "left wall" of HoQ, where $\boldsymbol{P} = \left[p_{ij}\right]_{l\times 1}$, and l is the number of tasks.

Based on the collection of mission information and technical data of the countries or organizations that have established the space-based ocean reconnaissance and surveillance system, combined with expert consultation, the triangular fuzzy number is used to evaluate the ability of the reference system to complete the corresponding tasks, and the fuzzy evaluation matrix W is obtained to determine the "right wall" of HoQ, where $\boldsymbol{W} = \left[\hat{w}_{ij}\right]_{l\times n}$ and n is the number of reference systems.

For the fuzzy evaluation of multiple experts, the triangular fuzzy number operation rules are used for processing, as shown in Eq. (1):

$$\hat{w}_{ij} = \frac{1}{x} \otimes \left(\hat{w}_{ij}^1 \oplus \hat{w}_{ij}^2 \oplus \cdots \oplus \hat{w}_{ij}^x \right) \tag{1}$$

where x is the number of experts, $i \in [1, l], j \in [1, n]$.

Sort out and analyze the capability requirement of the space-based ocean reconnaissance and surveillance system, record it as a set $\boldsymbol{Q}$, establish a corresponding capability index system, and determine the "ceiling" of HoQ. Collect and consult the information of the reference system, combined with expert consultation, determine the "basement" of HoQ. Here, the number of capacity requirements is denoted as m.

For the ability index with accurate numerical representation, the relevant data can be directly used; for the ability index without accurate numerical representation, the triangular fuzzy number is used to represent, the lower limit is the most pessimistic estimate, the middle value is the most possible estimate, and the upper limit is the most optimistic estimate.

3.2 Standardization of Index Data

There are usually two types of attributes for engineering characteristics: benefit and cost. The benefit type is the attribute that the larger the index value is, the better. The cost type is the attribute that the smaller the index value is, the better. For different indicator attributes, the corresponding data needs to be standardized. Combined with the engineering characteristics of space-based ocean reconnaissance and surveillance capability index, the max min normalization method in reference [24] is used to process the technology competitiveness evaluation matrix $\boldsymbol{C}$, where $\mathbf{C} = [\hat{c}_{ij}]_{m \times n}, \hat{c}_{ij} = \left(c_{ij}^l, c_{ij}^m, c_{ij}^u\right)$.

For the benefit indicator, mark its characteristic as "+", which can be calculated by the following formula:

$$\begin{cases} z_{ij}^l = \frac{c_{ij}^l - \min}{\max - \min} \\ z_{ij}^m = \frac{c_{ij}^m - \min}{\max - \min} \\ z_{ij}^u = \frac{c_{ij}^u - \min}{\max - \min} n \end{cases} \tag{2}$$

For the cost indicator, mark its characteristic as "–", which can be calculated by the following formula:

$$\begin{cases} z_{ij}^l = \frac{\max - c_{ij}^l}{\max - \min} \\ z_{ij}^m = \frac{\max - c_{ij}^m}{\max - \min} \\ z_{ij}^u = \frac{\max - c_{ij}^u}{\max - \min} \end{cases} \tag{3}$$

In (2) and (3), min and max are the minimum and maximum values of the project's engineering characteristics, and $\boldsymbol{Z}$ is the standardized technology competitiveness evaluation matrix, where $\mathbf{Z} = [\hat{z}_{ij}]_{m\times n}$, $\hat{z}_{ij} = \left(z_{ij}^{l}, z_{ij}^{m}, z_{ij}^{u}\right)$.

3.3 Calculation of Grey Relational Degree

Based on the data obtained in the above steps, the grey relational analysis method is used to get the relationships matrix, which is to determine the "room" of HoQ.

First define the distance d between two triangular fuzzy numbers $\hat{a} = (a^l, a^m, a^u)$ and $\hat{b} = (b^l, b^m, b^u)$:

$$d(\hat{a},\hat{b}) = \sqrt{\frac{(a^l - b^l)^2 + (a^m - b^m)^2 + (a^u - b^u)^2}{3}} \quad (4)$$

Corresponding to the task CRi of item i, $i \in [1, l]$, a gray association set $\boldsymbol{X}^i$ is established, where $\boldsymbol{X}^i = \{X_0^i, X_1^i, \ldots, X_m^i\}$, X_0^i is the reference sequence, where $X_0^i = [\hat{w}_{i1}, \hat{w}_{i2}, \ldots, \hat{w}_{in}]^{\mathrm{T}}$, $X_1^i, \ldots, X_m^i$ is the comparison sequence, where $X_m^i = [\hat{z}_{1m}, \hat{z}_{2m}, \ldots, \hat{z}_{nm}]^{\mathrm{T}}$.

Use Eq. (4) to solve the distance dxy between the comparison sequence and the reference sequence, where $d_{xy} = d\left(X_y^i(x), X_0^i(x)\right)$, $x \in [1, n]$, $y \in [1, m]$.

According to the grey relational analysis theory [26], the correlation coefficient is calculated as follows:

$$\xi_{xy}^{i} = \frac{\min_x \min_y (d_{xy}) + \rho \max_x \max_y (d_{xy})}{d_{xy} + \rho \max_x \max_y (d_{xy})} \quad (5)$$

In the formula, ρ is the resolution coefficient, $\rho \in (0, 1)$. If ρ is smaller, the difference of correlation coefficient will be larger and the differentiation will be stronger. Generally, ρ is 0.5.

According to the correlation coefficient, the grey relational degree of each sequence relative to the reference sequence is calculated:

$$r_y^i = \frac{1}{n}\sum_{x=1}^{n} \xi_{xy}^{i} \quad (6)$$

Further, the relationships matrix $\boldsymbol{R}$ of HoQ is as follows:

$$\boldsymbol{R} = \left[r_y^i\right]_{l\times m} \tag{7}$$

3.4 Rank the Importance of Capability Requirement

Use the task requirement importance matrix $\boldsymbol{P}$ and the relationships matrix $\boldsymbol{R}$ to solve the capability requirement importance proportion, that is, determine the "floor" of HoQ.

$$\boldsymbol{F} = [f_i]_{1\times m} = \boldsymbol{P}^{\mathrm{T}} \times \boldsymbol{R} \tag{8}$$

Normalize equation (8):

$$\boldsymbol{F}' = [f_i']_{1\times m} = f_i / \sum_{i=1}^{m} f_i \tag{9}$$

where $\boldsymbol{F}'$ is the importance matrix of capability requirement.

4 Example Analysis

4.1 Task of Space-Based Ocean Reconnaissance and Surveillance System

According to the current application status of the space-based ocean reconnaissance and surveillance system, combined with the requirements of space-based ocean reconnaissance and surveillance information for future maritime operations, corresponding to the work flow of the space-based ocean reconnaissance and surveillance system [27], through research, data review, expert consultation, The following four tasks are summarized, as shown in Fig. 2.

(1) Search and discovery targets. According to the user's reconnaissance requirements, a scheme is developed to dispatch the reconnaissance satellite to detect and search the specific sea area or target, so as to obtain the electromagnetic radiation signal, image information and other characteristic data of the possible target.

(2) Identify and confirm targets. The electronic and image information transmitted by the reconnaissance satellite is processed to obtain the target characteristic data, and the target is identified and confirmed by automatic data matching and manual research.

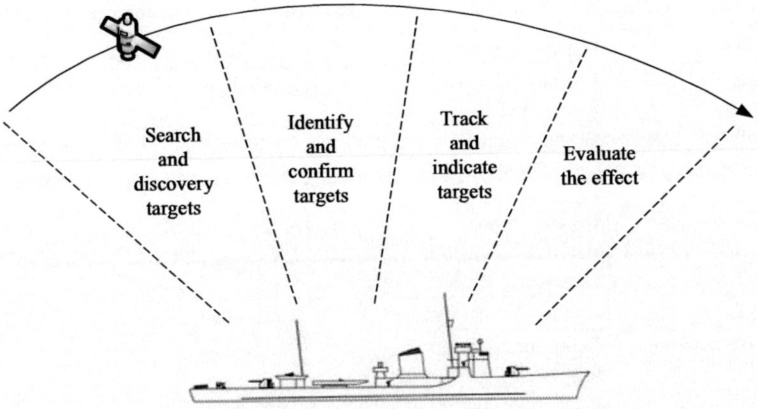

Fig. 2 Task of space-based ocean reconnaissance and surveillance system

(3) Track and indicate targets. After identifying and confirming the target, according to the user's requirements, the key sensitive target shall be tracked and monitored to generate the real-time situation information of the sea battlefield. When it is necessary to strike the target, the target position, course, speed, target characteristic parameters and other information shall be transmitted to the combat unit to guide and direct the strike target.
(4) Evaluate the effect. After the guided strike is completed, the reconnaissance satellite is dispatched to take photos of the target, and the strike effect is evaluated through research and calculation to determine whether the combat effect is achieved.

4.2 Ability Composition

The requirement of space-based ocean reconnaissance and surveillance capability is mainly reflected in the ability to use the space-based ocean reconnaissance and surveillance system to carry out fast, accurate and continuous reconnaissance and surveillance on the target. Starting from the six capabilities of information acquisition capability, relay transmission capability, task control capability, data processing capability, mobile protection capability and support capability, this paper makes a detailed analysis of space-based ocean reconnaissance and surveillance capability, gives 23 capability indicators, and gives data sources and data characteristics for reference of corresponding indicators, as shown in Table 1.

Table 1 The requirement of space-based ocean reconnaissance and surveillance capability

Capability	Capability index	No	Index data	Unit	Characteristic
Information acquisition capability	Imaging information resolution	1	Resolution	m	−
	Electromagnetic spectrum coverage	2	Frequency range	GHz	+
	Load continuous working time	3	Operating hours	h	+
	Reconnaissance coverage	4	Coverage rate	%	+
	Average revisit interval	5	Time interval	h	−
Relay transmission capability	Relay transmission capacity	6	Bandwidth	GHz	+
	Relay transmission time efficiency	7	Time delay	s	−
	Relay transmission quality	8	Bit error rate	10–8	−
	Relay coverage	9	Coverage rate	%	+
Task control capability	Degree of requirement integration	10	Integration degree	%	+
	Mission planning efficiency	11	Efficiency	%	+
	Emergency response time	12	Response time	h	−
Data processing capability	Target extraction rate	13	Extraction time	min	−
	Target recognition rate	14	Accuracy	%	+
	Target positioning accuracy	15	Error distance	km	−
	Degree of information fusion	16	Fusion degree	/	+
	Production time of information products	17	Production time	h	−
Mobile protection capability	Mobile networking speed	18	Networking time	h	−
	Anti-interference capability	19	Failure rate	%	−
	Anti-collision capability	20	Force	N	+
Support capability	Emergency launch capability	21	Launch time	h	−
	Accuracy of measurement and control	22	Measurement error	m	−
	Personnel technical support level	23	Degree	/	+

4.3 Model of HoQ

According to the tasks and capacity requirements obtained from the above analysis, the improved QFD analysis method is used to find the functional relationships between tasks and capabilities, and the importance of capabilities is sorted to guide the improvement direction of future system construction and application.

First, the QFD team members invited three experts in the field of space reconnaissance to consult on the importance of the task. The AHP method was used to comprehensively process the results of the consultation to obtain the task importance ranking matrix $\boldsymbol{P}$, $\boldsymbol{P} = [0.4245, 0.2975, 0.0824, 0.1956]^{\mathrm{T}}$.

Then, referring to the construction of the two generations of ocean surveillance satellite system of United States and Russia, as well as their application in wars and regional conflicts, The QFD team members consulted the data and consulted three experts, taking 0–1 as the scale, and using Eq. (1) for comprehensive processing, the fuzzy market competition evaluation matrix $\boldsymbol{W}$ and the technology competition evaluation matrix $\boldsymbol{C}$ were given in Tables 2 and 3.

Then, the capability requirement factors of different attributes are normalized by using Eqs. (2) and (3), and then the grey relational analysis method is used to solve the association matrix $\boldsymbol{R}$ of HoQ by using Eqs. (4)–(7).

$$\boldsymbol{R} = [r_{ij}^{i}]_{l \times m} = \left[\begin{array}{cccccccc}
0.3020 & 0.6917 & 0.4529 & 0.7044 & 0.6280 & 0.2898 & 0.3611 & 0.2887 \\
0.3975 & 0.7210 & 0.4055 & 0.6485 & 0.6674 & 0.2755 & 0.4464 & 0.3768 \\
0.2805 & 0.5179 & 0.8240 & 0.6326 & 0.5338 & 0.5903 & 0.3490 & 0.2707 \\
0.3437 & 0.7869 & 0.4405 & 0.8272 & 0.7518 & 0.2908 & 0.3882 & 0.3278 \\
0.3127 & 0.3509 & 0.3456 & 0.3943 & 0.4440 & 0.3139 & 0.4989 & 0.6530 \\
0.4132 & 0.4727 & 0.4650 & 0.5436 & 0.5504 & 0.4164 & 0.7284 & 0.7066 \\
0.2885 & 0.3051 & 0.3051 & 0.3366 & 0.4029 & 0.2844 & 0.3867 & 0.6427 \\
0.3511 & 0.4032 & 0.3985 & 0.4586 & 0.4824 & 0.3590 & 0.6312 & 0.6308 \\
0.3530 & 0.4294 & 0.4061 & 0.2711 & 0.3469 & 0.3014 & 0.4095 & \\
0.4723 & 0.4400 & 0.4891 & 0.2607 & 0.4680 & 0.3970 & 0.5687 & \\
0.3178 & 0.4970 & 0.5158 & 0.5311 & 0.3073 & 0.2784 & 0.3469 & \\
0.4109 & 0.4540 & 0.4458 & 0.2728 & 0.3977 & 0.3434 & 0.4974 &
\end{array}\right]_{4\times 23}$$

So far, the HoQ model of QFD analysis case of space-based ocean reconnaissance and surveillance capability requirements has been established.

Table 2 The fuzzy market competition evaluation matrix ***W***

	US1	US2	RUS
Task1	(0.37,0.43,0.50)	(0.64,0.67,0.77)	(0.37,0.40,0.50)
Task2	(0.47,0.50,0.57)	(0.73,0.77,0.80)	(0.47,0.53,0.60)
Task3	(0.17,0.27,0.57)	(0.37,0.37,0.47)	(0.17,0.23,0.30)
Task4	(0.50,0.53,0.63)	(0.63,0.67,0.73)	(0.37,0.40,0.50)

Table 3 The technology competition evaluation matrix ***C***

No	1	2	3	4	5	6	7	8
US1	(1,4,6)	(4,5,7)	(1,1.5,3)	(80,82,85)	(4,4.5,5)	(0.2,0.5,1)	(10,50,100)	(1,5,10)
US2	(0.2,0.5,2)	(5,7,9)	(1.5,2,3)	(80,85,90)	(3,3.5,4)	(0.5,1,2)	(5,20,30)	(0.1,0.3,0.5)
RUS	(3,5,9)	(3,5,6)	(0.5,1,2)	(60,65,70)	(5,6,7)	(0.2,0.5,1)	(10,20,80)	(1,5,10)
Min	0.01	0.1	0.2	50	0.1	0.1	0.01	0.01
Max	50	10	5	100	10	10	100	100
No	9	10	11	12	13	14	15	16
US1	100	(80,85,90)	(80,85,90)	(2,3,4)	(10,15,20)	90	4	(0.2,0.3,0.5)
US2	100	(85,90,95)	(90,92,95)	(1.5,1.5,2)	(2,5,10)	94	2	(0.5,0.6,0.8)
RUS	(70,80,100)	(80,90,95)	(80,85,90)	(2,3,4)	(5,10,20)	95	2.5	(0.5,0.6,0.6)
Min	33	0	0	0.1	0.1	0	0.1	0.01
Max	100	100	100	12	30	100	10	1
No	17	18	19	20	21	22	23	
US1	(1,2,2.5)	(24,36,48)	(0.4,0.5,0.6)	0.3	(24,36,48)	(2,4,5)	70	
US2	(0.2,0.5,1)	(12,18,24)	(0.7,0.7,0.8)	0.5	(18,24,30)	(0.5,2,3)	90	
RUS	(1,2,3)	(24,24,48)	(0.2,0.3,0.3)	0.3	(24,48,62)	(1.5,3,5)	75	
Min	0.1	2	0.01	0.1	2	0.01	0	
Max	8	48	1	10	240	50	100	

4.4 Results and Analysis

According to the established HoQ model, the importance proportion of each capability requirement is obtained by using Eq. (8), and the importance matrix $\boldsymbol{F}'$ of each capability requirement is obtained by normalizing by using Eq. (9).

$$\boldsymbol{F}' = [0.0326, 0.0683, 0.0453, 0.0684, 0.0636, 0.0301, 0.0379, 0.0311, 0.0337, 0.0381, 0.0376, 0.0433, 0.0465, 0.0340, 0.0566, 0.0643, 0.0385, 0.0429, 0.0434, 0.0281, 0.0378, 0.0326, 0.0454]$$

The generated histogram is shown in Fig. 3.

From Fig. 3, it can be seen that the capability requirements of space-based ocean reconnaissance and surveillance have different weights on the completion of system tasks. The proportion of information acquisition capability is 27.82%, the proportion of relay transmission capability is 13.28%, the proportion of task control capability is 11.90%, the proportion of data processing capability is 23.98%, the proportion of mobile protection capability is 11.44%, and the proportion of support capability is 11.58%.The five most influential indicators are electromagnetic spectrum coverage, reconnaissance coverage, average revisit interval, target positioning accuracy, and degree of information fusion.

By analyzing the results of Fig. 3, the following inferences can be drawn.

(1) Information acquisition capability is the basis and prerequisite for the formation of space-based ocean reconnaissance and surveillance systems and the completion of tasks. Data processing capability is the key focus of completing space-based ocean reconnaissance and surveillance tasks. Relay transmission capability, task control capability, mobile protection capability, support capability are important supports for the system to complete the task, and have an important impact on the success or failure of the task.

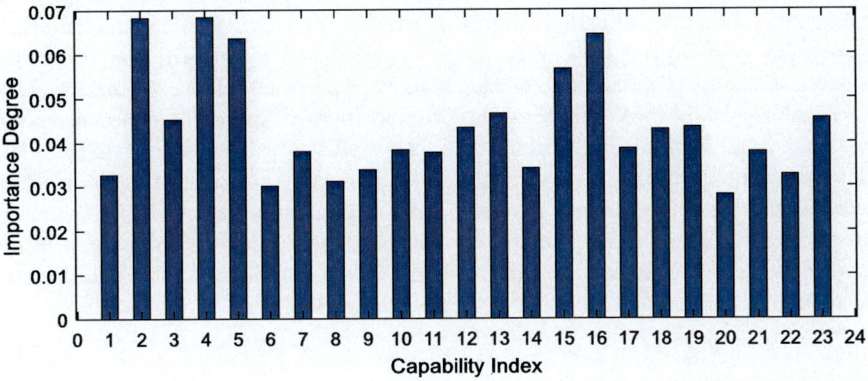

Fig. 3 Importance distribution of capacity requirement

(2) The electromagnetic spectrum coverage is an indicator that reflects the satellite's electromagnetic reconnaissance capability. Its largest proportion indicates the basic position of electromagnetic reconnaissance methods in acquiring marine target information and is the key to forming the system's capabilities. In terms of satellite load selection and performance design, the index of electromagnetic spectrum coverage can be specifically improved to better improve the system capability.
(3) Reconnaissance coverage and average revisit interval reflect the performance of satellite constellation, which can be improved by optimizing constellation design to improve the system capability.
(4) Target positioning accuracy is a key index reflecting the system's ability to complete reconnaissance and surveillance tasks, which has an important impact on combat activities. It can be improved by optimizing data processing algorithm and other methods.
(5) The degree of information fusion reflects the comprehensive utilization of different types of data information. Information fusion can not only improve the accuracy of information, but also expand the information dimension, so it has a high degree of importance. In order to improve the system's ability, the research on the degree of information fusion should be further strengthened to improve the level of information fusion.

The situation and inference reflected in Fig. 3 are consistent with the existing reality and cognition, which can provide reference for the construction and optimization of space-based ocean reconnaissance and surveillance system.

5 Conclusions

The space-based ocean reconnaissance and surveillance system is an important source of information for combat operations such as offshore operations, offshore defense, and precision marine strikes. It is also an important force to support the victory of modern naval battles. In this paper, the improved QFD method based on fuzzy theory and grey relational analysis is proposed to analyze and study the system capability requirements. The analysis steps and methods are given. Based on the analysis of the task and capability composition of space-based ocean reconnaissance and surveillance system, the HoQ model is established and simulated. The importance ranking of each capability is obtained. At the same time, this paper verifies the feasibility and effectiveness of the proposed improved QFD method through an example. This method can be used not only for system capability requirements analysis, but also for decision analysis problems such as equipment system construction and technical system optimization.

References

1. Wang ZY (2009) Encyclopedia of Chinese military-military space technology, Beijing. Encyclopedia of China Publishing House, China
2. Chan LK, Wu ML (2002) Quality function deployment: a literature review. Eur J Oper Res 143(3):463–497
3. Liu HE, Zhang LP (2000) Quality function deployment theories and methods-review on research progress. Syst Eng 18(2):1–6
4. Costa AIA, Dekker M, Jongen WMF (2000) Quality function deployment in the food industry: a review. Trends Food Sci Technol 11(9):306–314
5. Wang BJ, Huang HQ, Xu SN (2011) Study on service quality improvement in commercial banks based on fuzzy QFD. Sci-Technol Manage 13(2):120–123
6. Dinçer H, Yüksel S, Martinez L (2019) Balanced scorecard-based analysis about European energy investment policies: a hybrid hesitant fuzzy decision-making approach with quality function deployment. Expert Syst Appl 115:152–171
7. Yu LY, Dai YQ (2013) Research on the influence factors of regional scientific and technological innovation coopetition relationship based on fuzzy QFD and fuzzy TOPSIS method. Sci Technol Progress Policy 30(2):47–52
8. Cordeiro EC, Barbosa GF, Trabasso LG (2016) A customized QFD (quality function deployment) applied to management of automation projects. Int J Adv Manuf Technol 87 (5):2427–2436
9. Zhang BZ, Guo QS (2012) Theory and method of army weapon equipment requirement demonstration, Beijing. National Defense Industry Press, China
10. Li WB, Yang DW, Qi XW (2019) Military requirement research in weapon system for shipborne based on QFD. Ship Electron Eng 39(7):1–3
11. Xu YP, Shi FL, Yang F, Zhu YF (2010) Requirement analysis method in naval battleships design based on QFD and operational simulation techniques. Syst Eng Theory Pract 30 (1):167–172
12. Sheng DB, Sheng D, He P (2015) Requirement analysis for army forces drilling based on QFD. Command Control Simul 37(2):67–73
13. Nicholds B, Bil C, Marzocca P et al (2018) Case study of quality function deployment method in defence missions. In: Proceedings of 2018 AIAA aerospace sciences meeting, Kissimmee, Florida, USA, 2018, no. 1753
14. Wang JJ, Yu XH (2009) Research on military requirements for space-based space surveillance capabilities and equipments based on QFD. Military Oper Res Syst Eng 23 (2):56–60
15. Li YS, Ling SM, Li WZ (2011) QFD and FMECA technology and application for large and complex space system, Beijing. China Aerospace Publishing House, China
16. Liu SM, Feng SX (2015) QFD requirement analysis of space test on new space platform. Military Opera Res Syst Eng 2:52–56
17. Li YL, Tang JF, Chin KS et al (2012) A rough set approach for estimating correlation measures in quality function deployment. Inf Sci 189:126–142
18. Park T, Kim KJ (1998) Determination of an optimal set of design requirements using house of quality. J Oper Manag 16(5):569–581
19. Li YL, Zhi XG, Han Y (2011) Determination of functional relationship measures in product planning based on competitive evaluations. China Mech Eng 22(13):56–62
20. Li CL, Gao QS (2008) Research on functional relationships for QFD based on grey correlation degree. Mach Des Res 24(4):10–12
21. Li CL, Gao QS (2009) Orthogonal experiment method of determining functional relationships in quality function deployment. Ind Eng Manage 14(1):24–27
22. He Z, Zhao Y, Ma YH (2008) Method for rating technical characteristics in fuzzy QFD. J Tianjin Univ 41(5):631–634

23. Kim KJ, Moskowitiz H, Dhing RAA et al (2000) Fuzzy multi-criteria models for quality function deployment. Eur J Oper Res 121(2):504–518
24. Fung RYK, Chen Y, Tang J (2006) Estimating the functional relationships for quality function deployment under uncertainties. Fuzzy Sets Syst 157(1):98–120
25. Chan LK, Wu ML (2002) Quality function deployment: a comprehensive review of its concepts and methods. Qual Eng 15(1):23–35
26. Yazdani M, Kahraman C, Zarate P et al (2019) A fuzzy multi attribute decision framework with integration of QFD and grey relational analysis. Expert Syst Appl 115:474–485
27. Wan M, Hou Y, Wang Q (2019) Research on the process and technical system of space-based reconnaissance and surveillance for maritime moving targets. In: Proceedings of 2019 IEEE 3rd advanced information management, communicates, electronic and automation control conference (IMCEC), Chongqing, China, 2019, pp 285–289

Remote Sensing Scene Classification with Multi-task Learning

Tengfei Gong, Xiangtao Zheng, and Xiaoqiang Lu

Abstract Remote sensing scene classification is a fundamental task for understanding the high-resolution remote sensing images and has achieved great progress thanks to the deep neural networks. However, the deep neural networks heavily depend on the large-scale training samples, while sufficient manually annotated data is often unavailable. In this paper, a Multi-Task Scene Classification Network (MTSCN) is proposed to address the label consumption dilemma by using annotated information from other related tasks. The MTSCN simultaneously conducts multiple tasks so that scene classification task can leverage information shared by other tasks. The MTSCN consists of a shared branch and multiple task-specific branches with each for one task. First, the shared branch extracts shared features for all tasks. Then, the task-specific branch adapts shared features into task-specific features to finish the classification task. Such architecture allows complementary information shared among tasks as well as avoids task interference by selecting specific information for a different task. The method is evaluated on two multi-task learning scenarios with respect to accuracy and the number of network parameters. The results show that the proposed method is efficient not only in terms of data but also in terms of network parameters.

Keywords Remote sensing scene classification · Multi-task learning · Attention · Parameter efficient

T. Gong · X. Zheng (✉) · X. Lu
Key Laboratory of Spectral Imaging Technology CAS, Xi'an Institute of Optics and Precision Mechanics, Chinese Academy of Sciences, Xi'an, China
e-mail: zhengxiangtao@opt.cn

T. Gong
University of Chinese Academy of Sciences, Beijing, China

L. Wang et al. (eds.), *Proceedings of the 7th China High Resolution Earth Observation Conference (CHREOC 2020)*, Lecture Notes in Electrical Engineering 757, https://doi.org/10.1007/978-981-16-5735-1_30

1 Introduction

Remote sensing scene classification tries to identify a high-resolution remote sensing image with a semantic category [1, 2]. It serves as the basis of automatic image understanding and has a variety of applications, such as urban planning, and resource management [3–5]. Recently, deep neural networks have achieved high performance on remote sensing scene classification task [6]. However, the deep neural network strongly depends on a large amount of training data, while sufficient manually annotated data is often unavailable [7].

A typical way of circumventing the label consumption dilemma is to leverage information from other related tasks. It is possible in many cases to access labeled images from other tasks, which provides opportunities to better finish the current task. Multi-Task Learning (MTL) has displayed remarkable success in the field of leveraging related information [8]. Multiple tasks are simultaneously learned by sharing relevant information across tasks. For example, Volpi and Tuia [9] consider that tasks can provide complementary information and propose to joint learning of semantic segmentation and semantic boundary detection. Zhang et al. [10] explore the intrinsically hierarchical relationships between land cover and land use classification tasks, and propose a joint learning framework for the two tasks.

This paper focuses on remote sensing scene classification task (target task) when annotated data is limited. Some related tasks (auxiliary tasks) act as a regularizer for the target task. Both the target task and auxiliary tasks are simultaneously considered in a deep neural network. The complementary information from auxiliary tasks is shared among tasks to improve target classification. In this paper, two scenarios of using information from related tasks are explored. The first scenario uses semantic segmentation as an auxiliary task. The second scenario uses other scene classification tasks as auxiliary tasks.

A Multi-Task Scene Classification Network (MTSCN) is proposed to allow information shared among tasks while avoiding task inference. The MTSCN consists of a shared branch and multiple task-specific branches with each for one task. First, the shared branch exploits an existing network to extract deep representations as shared features for multiple tasks. Then, the task-specific branch adapts shared features into task-specific features according to the executed task. The task-specific branch includes a task head and multiple attention modules. The task head takes an image of the current task as input and outputs a task description. The attention modules take the task description and shared features as input, and generate attention masks to select task-specific features from shared features. Multiple attention modules are used to achieve different attention-aware features. Experimental results show that the proposed methods can achieve superior accuracy without a large amount of training data.

The rest of this paper is organized as follows: In Sect. 2, the related works on remote sensing scene classification and multi-task learning are briefly reviewed. In Sect. 3, the proposed method is discussed in detail. In Sect. 4, experiments are provided. Section 5 summarizes the paper.

2 Related Work

2.1 Remote Sensing Scene Classification

Remote sensing scene classification takes an image as input and outputs a scene category for the image. A classifier is trained on a labeled dataset with a supervised learning paradigm and tested on an unlabeled dataset. Recently, deep neural networks based methods have become the leading way for remote sensing scene classification. Deep neural networks can capture the complex structure characteristics of remote sensing images and obtain good feature representation. For example, Penatti et al. [6] use a convolutional neural network (CNN) as the feature extractor and express the remote sensing image with the neuronal output of the last convolutional layer. Zheng et al. [1] extract CNN activations and perform multi-scale pooling to learn multi-scale invariant scene representation. Castelluccio et al. [7] fine-tune the pre-trained CNN with labeled remote sensing images, and then directly use the CNN for classification.

Generally, it is not feasible to train a new CNN for a scene classification task as it requires a large amount of labeled data. Therefore, most of the works use existing CNN pre-trained on ImageNet [11]. There are also some works which transfer classifier trained on other remote sensing datasets to classify current dataset [12, 13]. Domain adaptation is used to reduce the distribution discrepancy between two datasets [14]. This paper uses multi-task learning which extracts complementary information from related tasks to promote scene classification.

2.2 Multi-Task Learning

Multi-task learning is a paradigm that learns multiple related tasks jointly, in which shared information among tasks is leveraged to improve generalization performance. Currently, multi-task learning in remote sensing mainly focuses on the relationship within one task, i.e. multiple sub-tasks in one task are learned simultaneously instead of separately. For example, in the vehicle instance segmentation task, vehicle region segmentation and semantic boundary detection are learned jointly [15]. In the road extraction task, road detection and centerline extraction are optimized simultaneously [8]. Though some techniques learn multiple sub-tasks jointly instead of learning them separately, there is very limited research focusing on multi-task learning among different tasks. Liu et al. [16] propose that learning aerial image retrieval and scene classification at the same time can benefit from each other and lead to a good generalization capability. To achieve this, similarity distribution learning, center metric learning, and classification are incorporated into one CNN.

These methods generally assume that the multiple tasks are well correlated and use a shared representation from a single shared backbone for all tasks. However, if tasks are not well correlated, there may be interference when using exactly the same features. In this paper, attention modules are proposed to filter out the task-specific information so that task-shared information can be further explored.

3 Method

Given a target task defined over dataset $\{(X_t, Y_t)\}_{t=1}$ and T auxiliary tasks defined over datasets $\{(X_t, Y_t)\}_{t=2}^{T+1}$, the goal is to learn a map from input image $x_1 \in X_1$ to label $y_1 \in Y_1$, i.e., to classify the target task, with the information provided by the auxiliary tasks. To this end, MTSCN is proposed with the framework shown in Fig. 1. We first introduce the architecture of the proposed network, then extensively illustrate the attention module, and finally explain the multi-task classification procedure.

3.1 Architecture Design

The proposed MTSCN completes the target task and auxiliary tasks in one framework simultaneously. The network consists of two parts: a shared branch, and multiple task-specific branches. Firstly, the shared branch extracts shared features for all tasks. This setting allows information to be shared among tasks. The target task can leverage information from auxiliary tasks. Secondly, multiple task-specific branches are incorporated to execute the multiple tasks, with each branch for one task. The task-specific branch adapts shared features to task-specific features

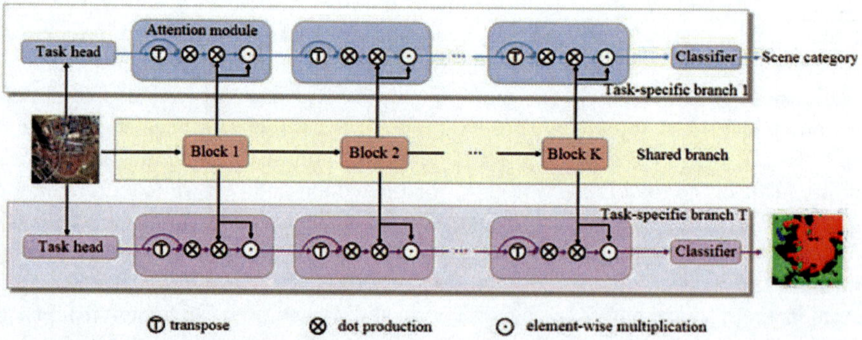

Fig. 1 Framework of the proposed method. The MTSCN is composed of a shared branch and multiple task-specific branches with each for one task. The shared branch extracts shared features for all tasks. Each task-specific branch selects task-specific features from shared features with attention modules, and then use the features for the executed task. Multiple attention modules are used to learn task-specific features progressively

according to the executed task. Attention modules are used to learn task-specific features, in which the features important to the task are emphasized, and the irrelevant ones are ignored.

Concretely, the shared branch is a deep neural network which is stacked by K convolutional blocks. Given input image x_t from the t-th task, the shared branch outputs intermediate layer feature at each block. These features are referred to as shared features $\{F_t^k\}_{k=1}^K$, where F_t^k denotes feature output by the k-th block.

The task-specific branch is constructed by a task head, and K attention modules, where each attention module latches onto one block of the shared branch. Firstly, for the t-th task, given image x_t, the task head outputs a task description h_t. Secondly, the attention modules take as input task description h_t as well as shared features $\{F_t^k\}_{k=1}^K$, and generate attention masks $\{m_t^k\}_{k=1}^K$. The masks are then applied to features of the corresponding block in the shared branch, with one mask per block. The task-specific feature of the k-th block f_t^k is obtained by element-wise multiplication of the attention mask m_t^k with the shared feature F_t^k. K attention modules are used to achieve different attention-aware features. After a sequence of attention modules, the output of the last attention module is used for final classification.

3.2 Attention Module

We now present the procedure of learning task-specific features with attention modules in the task-specific branch. Given the shared feature F_t^k at the k-th block, the task-specific feature f_t^k in this block is obtained by applying a same size attention mask m_t^k to the shared feature:

$$f_t^k = F_t^k \odot m_t^k \quad (1)$$

where $\odot$ identifies element-wise multiplication.

The attention mask m_t^k is generated by the k-th attention module in the t-th task-specific branch, which transfers feature maps into attention mask. The attention mask is formulated as:

$$m_t^k = h_t^{k-1} \otimes \left(h_t^{k-1}\right)^T \otimes \tilde{F}_t^k \quad (2)$$

where $\otimes$ identifies dot production. h_t^{k-1} is a task-specific feature. For the first attention module, h_t^{k-1} is the task description h_t output by the task head. For the subsequent attention module, $h_t^{k-1} = E(f_t^{k-1})$, where $E(\cdot)$ is an extractor and indicates a convolutional operation over the task-specific feature learned from the previous block. $\tilde{F}_t^k$ is a feature from the k-th block of the shared branch, but from a different layer from the shared feature F_t^k.

Each channel of the high-level feature h_t^{k-1} can be regarded as a unique response of a class, and different semantic responses are related to each other. By mining the interdependence between channel mappings, the interdependent feature mapping can be emphasized and the feature representation of specific semantics can be improved. Therefore, $h_t^{k-1} \otimes (h_t^{k-1})^T$ gets the relationship between channels. Through the dot production on $h_t^{k-1} \otimes (h_t^{k-1})^T$ and $\tilde{F}_t^k$, the channel relationship and the characteristics of the shared branch are further integrated into the attention mask.

3.3 Multi-task Classification and Optimization

The task-specific feature f_t^k output by the last attention module is passed to a task-specific classifier, which is designed based on the specific task. For the scene classification task, an average pooling layer following a single fully connected layer is used. For the auxiliary task such as semantic segmentation, a classifier symmetric to the shared branch is constructed to form an encoder-decoder structure as SegNet [17].

For the scene classification task, each image is assigned to a scene category. The standard cross-entropy loss is applied. For the semantic segmentation task, each pixel in an image is assigned to a label. The pixel-wise cross-entropy loss is applied for each predicted category label. The proposed method is optimized to perform multi-task learning in a unified framework. The loss function is set to the combination of multiple task-specific losses:

$$L = \sum_{t=1}^{T+1} L_t \tag{3}$$

where $T+1$ is the number of tasks. L_t is the loss of the t-th task.

The method can use semantic segmentation as an auxiliary task, where semantic segmentation takes as input images the same as the scene classification task. The loss of simultaneously conducting scene classification and semantic segmentation is:

$$\begin{aligned} L(X, Y_{1:2}) = & - \sum_{(x,y_1)\in(X,Y_1)} y_1 \log \hat{y}_1 \\ & - \sum_{(x,y_2)\in(X,Y_2)} \sum \frac{1}{WH} \sum_{w=1}^{W} \sum_{h=1}^{H} y_2(w,h) \log \hat{y}_2(w,h) \end{aligned} \tag{4}$$

where X is the input data for two tasks, Y_1 and Y_2 are the corresponding ground-truth for scene classification and semantic segmentation tasks, respectively. The first term of the formula is the standard cross-entropy loss for the scene classification task. $\hat{y}_1$ is the predicted scene label for image x. The second term is the pixel-wise cross-entropy loss for semantic segmentation task. $\hat{\mathrm{y}}_2$ is the semantic segmentation prediction. W and H are the width and height of image x.

The proposed method can also incorporate other scene classification tasks as auxiliary tasks, where different tasks are defined over different datasets. The loss of simultaneously performing multiple scene classification tasks using multiple datasets is:

$$L(\mathrm{X}_{1:T+1}, \mathrm{Y}_{1:T+1}) = \sum_{t=1}^{T+1} - \sum_{(\mathrm{x}_t, \mathrm{y}_t) \in (\mathrm{X}_t, \mathrm{Y}_t)} \mathrm{y}_t \log \hat{\mathrm{y}}_t \quad (5)$$

where $(\mathrm{X}_t, \mathrm{Y}_t)$ is dataset corresponding to the t-th task. Each image x_t in X_t has one scene label y_t and one prediction $\hat{\mathrm{y}}_t$. Multiple scene classification tasks are optimized simultaneously.

We jointly optimize the weights of the shared branch and multiple task-specific branches. The problem of incorporating useful information from auxiliary tasks is transformed into the problem of learning weights in MTSCN, which is solved through end-to-end training. At test time, when a new image is input to the multi-task network, the shared branch first extracts the shared features, and then the task-specific branch is followed, selectively choosing needed features to complete the executed task.

4 Experiments

In this section, the proposed method is evaluated on two MTL scenarios. The first scenario uses semantic segmentation as an auxiliary task. We refer to this scenario as Pixel-Scene Classification and explain it in Sect. 4.1. The second scenario uses other scene classification tasks as auxiliary tasks. This scenario is referred to as Multiple-Scene Classification and is explained in Sect. 4.2.

4.1 *Pixel-Scene Classification*

This section firstly introduces the datasets used for semantic segmentation and scene classification tasks. Then, a comparison with baselines and state-of-the-art multi-task learning methods is conducted. Finally, an analysis on the attention module of the proposed method is performed.

(1) **Datasets**

Two datasets: GID-MTL5 and GID-MTL15 [18] are used in the experiments. Every image in each dataset contains pixel-level land-cover labels and image-level scene label. The image is cropped from GID datasets [19] and set to a size of 224 × 224 pixels. The categories of the scene are the same as the categories of pixel-level annotations, except that pixel-level annotations further contain an undefined category. GID-MTL5 contains five land-cover categories. GID-MTL15 contains fifteen land-cover categories. The contained categories, the number of images per category, and some sample images of GID-MTL5 are shown in Fig. 2. The same information for GID-MTL15 is shown in Fig. 3.

(2) **Model configurations & Metrics**

The proposed MTSCN is implemented in the Pytorch environment. VGG-16 [20] network is used as the shared branch. A classifier symmetric to VGG-16 is constructed for semantic segmentation. An average pooling of [7 × 7] kernels followed by a single fully connected layer is used as the scene classifier. Task head is a convolutional layer with batch normalization and non-linear activation. The extractor $E(\cdot)$ contains a convolutional layer with [3 × 3] kernels, following by a pooling layer. Four extractors are used in the proposed network. The number of parameters of the shared branch, semantic segmentation classifier, scene classifier, task head for each task, and four extractors are shown in Table 1. Adam optimizer is used with the learning rate set to 10^{-4}. The batch size is set to 2. 80% of the dataset is used for training and 20% of the dataset is used for testing. The entire network is trained from scratch.

Fig. 2 Sample images of GID-MTL5. Each group displays an image corresponding to a scene category, and the corresponding segmentation map. The annotations below list the category, the number of scene images contained in the category, and the color of the pixel-level label

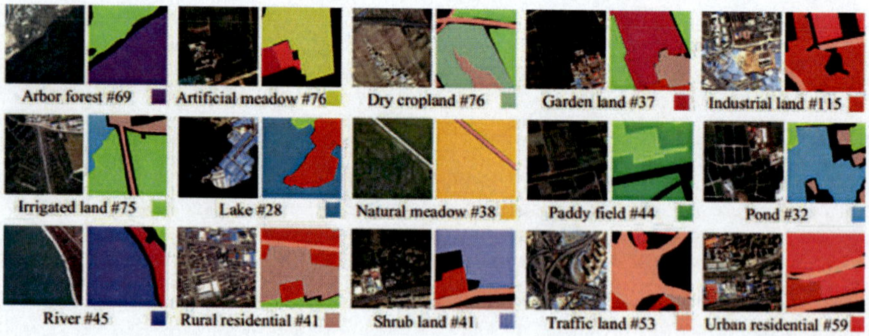

Fig. 3 Sample images of GID-MTL15. Same annotations as above

Table 1 Number of parameters of each component of MTSCN with backbone constructed on VGG-16. No consideration is given to the subtle difference in the number of parameters between GID-MTL5 and GID-MTL15. The parameters of the scene classifier and task head are few and are therefore ignored

Components	Shared branch	Segmentation classifier	Scene classifier	Task head	Extractors
Parameters	15.49M	9.44M	~0M	~0M	6.27M

The method is evaluated with respect to accuracy and the number of network parameters. For the semantic segmentation, the classification accuracy is assessed on all pixels with mean Intersection over Union (mIoU) and Pixel Accuracy (PA). For scene classification, an image is assigned to a scene class, and the Overall Accuracy (OA) is used to assess the performance. For all the evaluation metrics: mIoU, PA and OA, the higher values indicate better classification performance.

(3) **Comparison Experiments**

This section evaluates the effectiveness of the proposed method. Firstly, we validate that using information from related tasks with multi-task learning can improve scene classification accuracy when training data is limited. Four variants of MTSCN are designed as the baselines. Secondly, MTSCN is compared with state-of-the-arts to validate the superiority of the proposed method. Comparisons are conducted on GID-MTL5 and GID-MTL15 datasets. All comparison methods use the same shared branch as MTSCN. Performances are evaluated with respect to mIoU, PA, OA, and the number of parameters. Experimental results are shown in Table 2. The explanations are presented as follows:

Scratch identifies using SegNet and VGG-16 trained from scratch to finish the prediction tasks. Each network is trained independently. The SegNet used for semantic segmentation task is composed of the shared branch and the segmentation classifier. VGG-16 used for scene classification task is composed of the shared branch and the scene classifier. Therefore, the parameter of this baseline is 41.32M. Since different tasks are trained independently, no mutually complementary information is provided to each other. This variant provides a benchmark on GID-MTL5 and GID-MTL15 datasets.

Finetune means that SegNet and VGG-16 are firstly pre-trained on the ImageNet and then fine-tuned using the remote sensing images. The other setting is the same as the variant of *Scratch*. Semantic segmentation and scene classification tasks are trained independently. This variant serves two purposes. First, it assesses the suitability of pre-training on ImageNet. Second, by comparing *Finetune* with multi-task methods, we can assess whether incorporating related tasks is effective and whether the performance is better than that of *Finetune*.

Data-Augmentation means that semantic segmentation annotations are used to train the scene classification network. It is achieved by sharing SegNet and VGG-16 with a same backbone. Specifically, a shared branch is used to extract features for two tasks. A segmentation classifier and a scene classifier complete the

Table 2 Experimental results of pixel-scene classification on GID-MTL5 and GID-MTL15 datasets

Method	Parameters	GID-MTL5			GID-MTL15		
		Segmentation		Classification	Segmentation		Classification
		mIoU	PA	OA	mIoU	PA	OA
Scratch	41.32M	18.97%	59.89%	88.75%	8.53%	39.35%	78.12%
Finetune	41.32M	23.49%	64.61%	90.52%	9.38%	44.19%	78.30%
Data-augmentation	24.93M	33.03%	74.66%	91.67%	11.09%	50.78%	74.37%
Scratch-mask	53.86 M	27.41%	69.92%	86.04%	7.97%	38.67%	75.00%
Cross-stitch [21]	41.32M	29.51%	75.77%	**95.42%**	10.37%	54.88%	80.63%
Uncertainty-weights [22]	24.93M	32.87%	76.38%	94.58%	10.80%	49.12%	77.50%
MTSCN (Ours)	31.26M	**33.41%**	**77.70%**	95.12%	**13.45%**	**60.02%**	**81.17%**

corresponding tasks, respectively. Therefore, the parameter of this architecture is 24.93M. This variant is actually a basic MTL method that provides a baseline of incorporating information from the related task.

Scratch-Mask identifies each task trained from scratch individually and attention masks are applied to the features of shared branch. It is a variant of *MTSCN* without the benefits of multi-task learning. Concretely, for semantic segmentation task, shared branch, segmentation classifier, task head and extractors are used. For scene classification, the shared branch, scene classifier, task head and extractors are used. The parameter of *Scratch-Mask* is 53.86M. Compared with *Scratch*, this method has higher accuracy on GID-MTL5 dataset, but similar accuracy on GID-MTL15 dataset. The higher accuracy of *Scratch-Mask* on GID-MTL5 dataset is because it can make full use of network parameters in a simple way to complete the simple task. However, for more complex tasks, MTL methods are required for better results.

Cross-Stitch Network [21] is a state-of-the-art MTL method. It contains multiple standard feed-forward networks with each for one task. Multiple cross-stitch units are utilized to allow features in corresponding layers to be shared across tasks. The parameters of cross-stitch units are negligible. Therefore, the parameters of *Cross-Stitch Network* are the same as *Scratch*, which is 41.32M. The accuracy of this method is superior to *Scratch* due to information sharing among tasks.

Uncertainty-Weights [22] uses the same architecture as *Data-Augmentation*, but balances feature sharing by weighting the multi-task learning losses. Task uncertainty is used as a dynamic weight instead of manually choosing the optimal weight. The parameter of this method is 24.93M. Compared with *Data-Augmentation*, this method has improved performance on the scene classification task, but has similar performance on semantic segmentation task.

MTSCN is the proposed method. Two tasks use a shared branch. Each task further uses attention modules to select the needed features. Therefore, the network is composed of the shared branch, segmentation classifier, scene classifier, task heads for two tasks, and extractors. The number of parameters is 31.26M. The performance of *MTSCN* is outstanding. The mIoU and PA of the semantic segmentation task conducted on GID-MTL5 dataset are 33.41 and 77.70%, which are highest among all comparison methods. The OA of scene classification on GID-MTL5 dataset is 95.12%, which is also competitive. On the GID-MTL15 dataset, mIoU, PA and OA are 13.45%, 60.02% and 81.17%, respectively, which outperform all the other methods.

By comparing *Scratch, Finetune*, and *Data-Augmentation*, it can be found that both *Finetune* and *Data-Augmentation* can increase the scene classification accuracy. But the *Data-Augmentation* may sometimes be distracted by additional information. For example, on the GID-MTL15 dataset, the scene classification accuracy is lower than that of the benchmark. But the precision of the segmentation task has been improved obviously. The proposed method *MTSCN*, which uses the attention mask, enables the executed task to make full use of useful information and avoid task interference. Moreover, by comparing the proposed method with state-of-the-art methods, it can be found that *MTSCN* obtains the optimal performance.

In terms of network parameters, compared with methods that directly use a shared branch, MTSCN uses a few more parameters, but improves the performance significantly. In the meantime, compared with the methods using completely separate networks, MTSCN has a substantial reduction in the number of parameters, but keeps the high performance maintained. The superiority will be more obvious when there are more tasks performed simultaneously. The superiorities come from two aspects: First, task-specific features are learned for different tasks, thereby avoiding task interference. Second, instead of assigning an independent network for each task, constructing a task-specific branch with attention modules only needs a small number of parameters.

(4) **Qualitative Results**

Except for the above quantitative results, the qualitative results of the proposed *MTSCN* on GID-MTL5 are also presented in Fig. 4. Specifically, several semantic segmentation results and learned attention masks for different tasks are visualized.

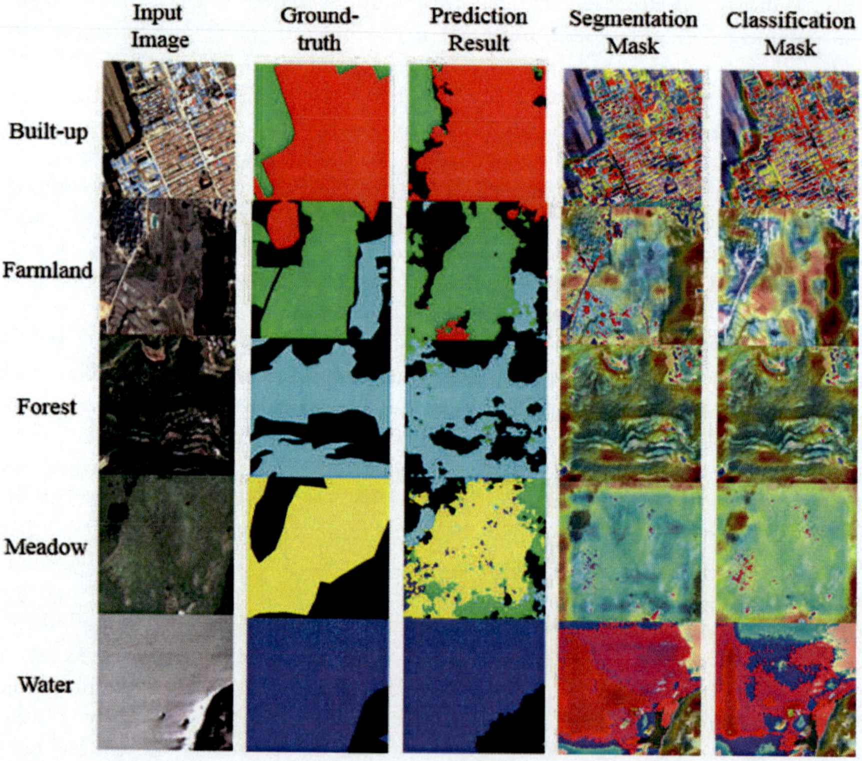

Fig. 4 Visualization of semantic segmentation and attention masks on GID-MTL5 dataset. Each row shows results for one of five scene categories. The input image, semantic segmentation ground-truth map, predicted segmentation maps from MTSCN, the attention masks for semantic segmentation task and scene classification task, are showed respectively

The third column in Fig. 4 is the prediction results of the semantic segmentation task. The fourth and fifth columns are the learned attention masks for semantic segmentation and scene classification tasks, respectively. All the attention masks come from the same channel of the last attention module. It can be seen that the attention masks of different tasks are different. Each mask picks out features that are important for the current task, and filters out features that are uninformative. The results validate the effectiveness of the proposed method.

4.2 Multiple-Scene Classification

In this section, the proposed method is evaluated in the MTL scenario where other scene classification tasks are used as auxiliary tasks. We first introduce the used datasets, and then compare the proposed method with state-of-the-arts.

(1) **Datasets**

Three heterogeneous remote sensing scene classification datasets, Merced [1], AID [2], and NWPU [23] are used to build MTL scene classification datasets. Merced dataset includes 21 aerial scene categories, and each category contains 100 RGB images. Every image has a size of 256 × 256 pixels and a resolution of 0.3 m. All the images are collected from the United States Geological Survey National Map. AID consists of 10,000 RGB images in 30 scene categories collected from Google Earth imagery. These images have a large scale of 600 × 600 pixels with resolutions ranging from 0.5 to 8 m. NWPU contains 45 scene classes and includes 700 RGB images per class. Every image has a size of 256 × 256 pixels extracted from Google Earth imagery. 12 common categories shared by the three datasets are selected: *airfield, anchorage, beach, dense residential, farm, flyover, forest, game space, parking space, river, sparse residential, and storage cisterns*. We further select 100 images per category per dataset to evaluate the proposed method.

(2) **Model configurations & Comparisons**

Wide Residual Network (WRN) [24] with a depth of 28, widening factor of 4 is used as the shared branch. Each task has an individual classifier same as the classifier in WRN. Four attention modules are used in each task. Three extractors are used throughout the network. The batch size is set to 32. Other settings are the same as the experiment in Sect. 4.1.

Consistent with the prior experiments, the proposed method is compared with multiple baselines and state-of-the-art multi-task learning methods. Performance is evaluated with respect to the number of parameters and OA. The experimental results are shown in Table 3. The explanations are presented as follows:

Scratch trains three separate WRNs to complete scene classification tasks from scratch, with each network trained independently. The number of parameters is 17.58M. The average OA is 82.36%.

Table 3 Experimental results (number of parameters and OA) on the MTL scenario of multiple-scene classification

Method	Parameters	Merced	AID	NWPU	Average
Scratch	17.58M	90.84%	82.05%	74.20%	82.36%
Finetune	17.58M	91.19%	82.43%	77.62%	83.74%
Data-augmentation	5.87M	90.26%	82.99%	79.88%	84.37%
Cross-stitch [21]	17.58M	91.88%	83.20%	75.78%	83.62%
DEN [25]	7.35M	90.62%	84.38%	77.60%	84.20%
MTSCN (Ours)	6.84M	**94.14%**	**86.61%**	**83.05%**	**87.93%**

Finetune means fine-tuning the pre-trained WRNs on remote sensing images. Three tasks are finished separately. The parameter is the same as that of *Scratch* which is 17.58M. The average OA is 83.74%.

Data-Augmentation is a MTL method that uses a shared WRN for three tasks except for the classifiers. The number of parameters is 5.87M, which is parameter efficient. The average OA is 84.37% which is higher than *Scratch* due to the information sharing among multiple tasks.

Cross-Stitch Network is a state-of-the-art MTL method that has been mentioned above. The number of parameters is 17.58M. The average OA is 83.62%.

Deep Elastic Network (DEN) [25] is a state-of-the-art MTL method that dynamically selects a model from multiple candidate models to execute a task. A selector is used to choose which model to use. The number of parameters is 7.53M. The average OA is 84.20%.

MTSCN is the proposed method. The number of parameters is 6.84M, which is comparative among the comparison methods. The average OA is 87.93% which is the highest. It can be seen that the proposed method has a significant parameter reduction compared to *Scratch*. This effect will be more pronounced when more tasks are executed simultaneously.

5 Summary

In this paper, a high performance as well as parameter efficient multi-task learning network is proposed to finish the remote sensing scene classification task. The proposed method uses information from other related tasks to alleviate the label consumption problem. Concretely, the proposed method completes multiple tasks simultaneously. A shared branch is used to extract features for all tasks, but multiple task-specific branches are used to allow every task to learn task-specific features adaptively. Such a strategy allows useful information shared among tasks and avoids interference among tasks. Two MTL scenarios: Pixel-Scene Classification, and Multiple-Scene Classification are evaluated. Experimental results validate the effectiveness of the proposed method with respect to classification accuracy and number of parameters.

References

1. Zheng X, Yuan Y, Lu X (2019) A deep scene representation for aerial scene classification. IEEE Trans Geosci Remote Sens 57:4799–4809
2. Xia G, Hu J, Hu F, Shi B, Bai X, Zhong Y, Zhang L, Lu X (2017) Aid: a benchmark data set for performance evaluation of aerial scene classification. IEEE Trans Geosci Remote Sens 55:3965–3981
3. Gao H, Yuan Y, Zheng X (2019) Remote sensing road extraction by refining road topology. In: Proceedings of the 6th China high resolution earth observation conference, pp 187–197
4. Chen P, Chen Z, Yang X, Li B, Zhang B (2019) High-resolution land-use mapping in Beijing-Tianjin-Hebei region based on convolutional neural network. In: Proceedings of the 6th china high resolution earth observation conference, pp 213–222
5. Zhang Y, Ran H, Peng Y, Zheng Y (2019) Automatic prediction of landslides over InSAR techniques and differential detection using high-resolution remote sensing images: application to Jinsha River. In: Proceedings of the 6th China high resolution earth observation conference, pp 471–486
6. Penatti A, Nogueira K, dos Santos J (2015) Do deep features generalize from everyday objects to remote sensing and aerial scenes domains? In: Proceedings of IEEE computer vision and pattern recognition, Workshops, pp 44–51
7. Castelluccio M, Poggi G, Sansone C, Verdoliva L (2015) Land use classification in remote sensing images by convolutional neural networks. Acta Ecol Sin 28:627–635
8. Lu X, Zhong Y, Zheng Z, Liu Y, Zhao J, Ma A, Yang J (2019) Multi-scale and multi-task deep learning framework for automatic road extraction. IEEE Trans Geosci Remote Sens 57:9362–9377
9. Volpi M, Tuia D (2018) Deep multi-task learning for a geographically-regularized semantic segmentation of aerial images. ISPRS J Photogramm Remote Sens 144:48–60
10. Zhang C, Sargent I, Pan X, Li H, Gardiner A, Hare J, Atkinson P (2019) Joint deep learning for land cover and land use classification. Remote Sens Environ 221:173–187
11. Deng J, Dong W, Socher R, Li L, Li K, Li F (2009) Imagenet: a large-scale hierarchical image database. In: Proceedings of IEEE conference on computer vision and pattern recognition, pp 248–255
12. Lu X, Gong T, Zheng X (2020) Multisource compensation network for remote sensing cross-domain scene classification. IEEE Trans Geosci Remote Sens 58:2504–2515
13. Othman E, Bazi Y, Melgani F, Alhichri H, Alajlan N, Zuair M (2017) Domain adaptation network for cross-scene classification. IEEE Trans Geosci Remote Sens 55:4441–4456
14. Bashmal L, Bazi Y, Alhichri H, AlRahhal M, Ammour N, Alajlan N (2018) Siamese-GAN: learning invariant representations for aerial vehicle image categorization. Remote Sens 10:351–369
15. Mou L, Zhu X (2018) Vehicle instance segmentation from aerial image and video using a multitask learning residual fully convolutional network. IEEE Trans Geosci Remote Sens 56:6699–6711
16. Liu Y, Han Z, Chen C, Ding L, Liu Y (2020) Eagle-eyed multitask cnns for aerial image retrieval and scene classification. IEEE Trans Geosci Remote Sens 58:6699–6721
17. Badrinarayanan V, Kendall A, Cipolla R (2017) Segnet: a deep convolutional encoder-decoder architecture for image segmentation. IEEE Trans Pattern Anal Mach Intell 39:2481–2495
18. [Online]. Available: https://github.com/cheer00/GID-MTL.git
19. Tong X, Xia G, Lu Q, Shen H, Li S, You S, Zhang L (2020) Land-cover classification with high-resolution remote sensing images using transferable deep models. Remote Sens Environ 237:111322
20. Simonyan K, Zisserman A (2015) Very deep convolutional networks for large-scale image recognition. In: Proceedings of international conference on learning representations, pp 1–10

21. Misra I, Shrivastava A, Gupta A, Hebert M (2016) Cross-stitch networks for multi-task learning. In: Proceedings of IEEE conference on computer vision and pattern recognition, pp 3994–4003
22. Cipolla R, Gal Y, Kendall A (2018) Multi-task learning using uncertainty to weigh losses for scene geometry and semantics. In: Proceedings of IEEE conference on computer vision and pattern recognition, pp 7482–7491
23. Cheng G, Han J, Lu X (2017) Remote sensing image scene classification: benchmark and state of the art. Proc IEEE 105:1865–1883
24. Zagoruyko S, Komodakis N (2016) Wide residual networks. In: Proceedings of the British machine vision conference, pp 87.1–87.12
25. Ahn C, Kim E, Oh S (2019) Deep elastic networks with model selection for multi-task learning. In: Proceedings of IEEE international conference on computer vision, pp 6528–6537

Top-of-Atmosphere Radiance Image Simulation from the Visible to Thermal Infrared

Dong Liu, Yanbing Dong, and Hongxia Mao

Abstract Top-of-Atmosphere (TOA) radiance image simulation is very useful for the definition of new observation systems, the optimization of optical instruments and the design of data processing or scientific algorithms. A physical based simulation model from visible to thermal infrared incorporating the coupling between heterogeneous non-Lambertian surface and atmosphere is proposed in this paper. It is necessary for multi-angle optical remote sensing of natural land. The simulation examples in optical domain and thermal infrared domain are analyzed. At last, we suggest the further investigation about the acquisition of high spatial and high spectral resolution surface Bidirectional Reflectance Distribution Function (BRDF), which helps the proposed model applied to various scene generation demands.

Keywords TOA radiance image · Image simulation · Visible to thermal infrared · Non-Lambertian

1 Introduction

The end-to-end simulation tool of remote sensing images is playing more and more important role for earth observation plans. It helps the definition of future observation systems with new concepts, the optimization of optical instruments, as well as the design of data processing algorithms in quantitative way [1]. The model of Top-of-Atmosphere (TOA) radiance image simulation is the major part of an end-to-end tool. It generates radiance image at imaging sensor which is input to instrumental model for digital number scene generation. Therefore, a variety of physical based TOA radiance image simulation models have been developed for remote sensing systems in recent years [2–5]. However, the couple between atmosphere and heterogeneous non-Lambertian land surface has not been established very well due to its complexity. Generally, the assumptions are introduced at

D. Liu (✉) · Y. Dong · H. Mao
Science and Technology On Optical Radiation Laboratory, Beijing 100854, China
e-mail: ldwl2009@126.com

L. Wang et al. (eds.), *Proceedings of the 7th China High Resolution Earth Observation Conference (CHREOC 2020)*, Lecture Notes in Electrical Engineering 757, https://doi.org/10.1007/978-981-16-5735-1_31

the beginning of modeling. For example, many models treat land surface as Lambertian object, which simplifies the model significantly, because there is no application based on the exploitation of directional information for their observation plans. With the development of multi-directional remote sensing of land, it is necessary now to build the model that precisely accounts for the radiation effects in highly heterogeneous non-Lambertian land surface to support the construction of these sensors.

In this paper, we propose a new model for TOA radiance image simulation that incorporates the coupling between atmosphere and heterogeneous non-Lambertian surface in physical consistent way. To calculate TOA radiance from visible to thermal infrared, the radiative transfer processes of solar radiation and terrestrial thermal radiation are involved in the model simultaneously which are both important in the mid-infrared spectral range. The algebraic method employing MODTRAN is established to efficiently calculate the atmospheric parameters that are independent with the surface optical properties. The surface bidirectional reflectance, emissivity and temperature images from the existing satellite measurements are utilized to generate the realistic land texture pattern. In the following parts, the methodology of the model is described first; several TOA radiance image simulation examples are presented then; and a brief summary is given at last.

2 Methodology

2.1 Modeling Theory of TOA Radiance Image Simulation

The radiative transfer of successive orders of radiation interactions is adopted to build the physical model for TOA radiance image simulation [6]. It has been widely used as the basic principle in the satellite signal interpretation and image simulation, which is because the surface optical parameters can be separated from the atmospheric parameters directly in the successive orders of scattering to help extract the surface information from coupled measurements or accelerate the image simulation. Figures 1, 2, 3 and 4 present the decomposed processes that contribute to the one-pixel TOA radiance, that is, solar radiation has no interaction and once interaction with the surface through the atmosphere (Fig. 1); and solar radiation has multiple interaction with the surface (Fig. 2); and thermal radiation has no interaction and once interaction with the surface (Fig. 3); and thermal radiation has multiple interaction with the surface (Fig. 4). Note that thermal emission from the surface is regarded as once interaction here.

The mathematical expressions of thirteen radiative transfer processes are given respectively. It is necessary to clarify the notation at first. In Figs. 1, 2, 3 and 4, (x, y) denotes the coordinate of target pixel on the surface which is signed by red line; (ξ, η) denotes the coordinate of background pixel that is around the target one. Blue circle indicates atmospheric scattering of radiation; orange circle indicates

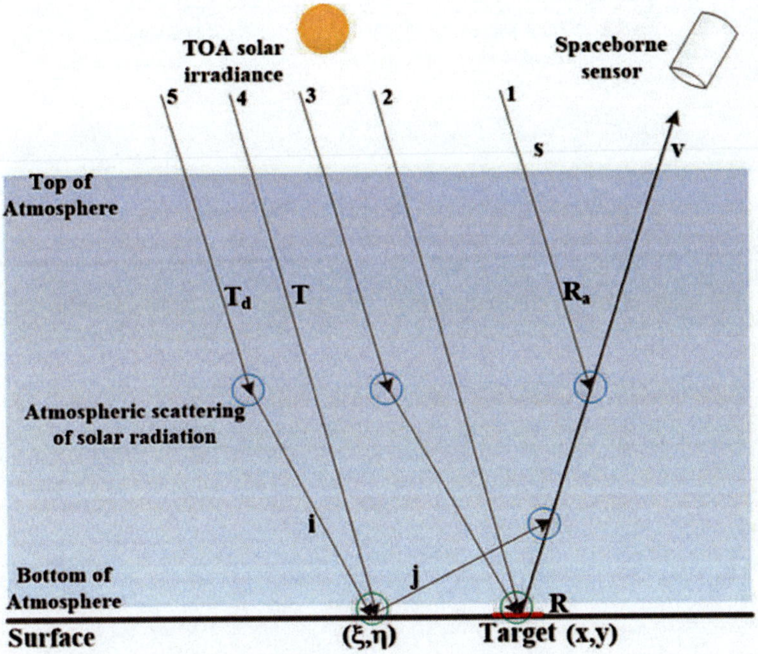

Fig. 1 The radiative transfer processes that solar radiation is scattered by atmosphere only and interacted with surface once

atmospheric thermal emission; green circle indicates surface reflection of radiation; and red circle indicates surface thermal emission. Then, s used as subscript denotes the solar incident direction; v denotes the viewing direction for observing the surface target from the satellite; i and j denote the differential direction of atmospheric downward diffuse radiation and surface upward diffuse radiation, respectively.

In Fig. 1, the radiance contribution from process 1 can be expressed as

$$L_1 = L_s\mu_s \cdot R_a(\theta_s, \theta_v, \varphi_{sv}) \tag{1}$$

where R_a is atmospheric intrinsic reflectance that denotes the magnitude of pure atmospheric scattering of solar radiation; L_s denotes TOA solar irradiance; θ_s is the solar zenith angle (SZA) and θ_v is the viewing zenith angle (VZA); φ_{sv} is the relative azimuth angle (RAA) between solar direction and viewing direction; μ_s is cosine of θ_s. The wavelength subscript is omitted for simplicity.

The contribution from process 2 can be expressed as

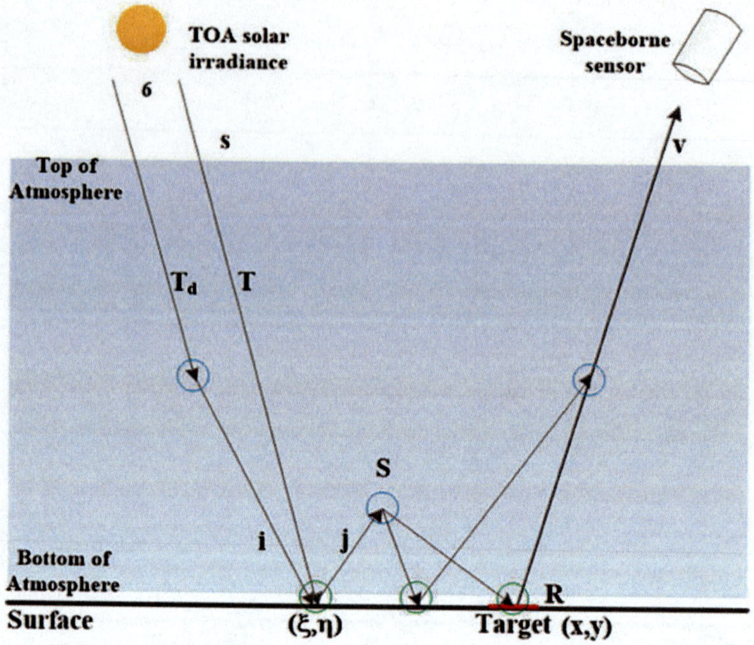

Fig. 2 The radiative transfer processes that solar radiation is interacted with surface in two and multiple times

$$L_2 = L_s\mu_s T(\theta_s) \cdot \frac{R(x, y; \theta_s, \theta_v, \varphi_{sv})}{\pi} \cdot T(\theta_v) \tag{2}$$

where R denotes the bidirectional reflectance that would have obvious spatial distribution and angular distribution for heterogeneous non-Lambertian surface; T (θ_s) and $T(\theta_v)$ are the direct transmittances in the solar direction and viewing direction respectively.

The contribution from process 3 can be obtained as

$$L_3 = L_s\mu_s \int_0^{2\pi} \int_0^1 P(\theta_s, \theta_i, \varphi_{si}) \cdot \frac{R(x, y; \theta_i, \theta_v, \varphi_{iv})}{\pi} \mu_i d\mu_i d\varphi_i \cdot T(\theta_v) \tag{3}$$

Here, P is introduced to characterize atmospheric diffuse transmission property. Generally, angular integral in Eq. (3) has to been computed numerically for each pixel of imaging region, which will be remarkable computational burden if the image contains millions of pixels. The angular separation approximation is implemented in our model, so that L_3 becomes

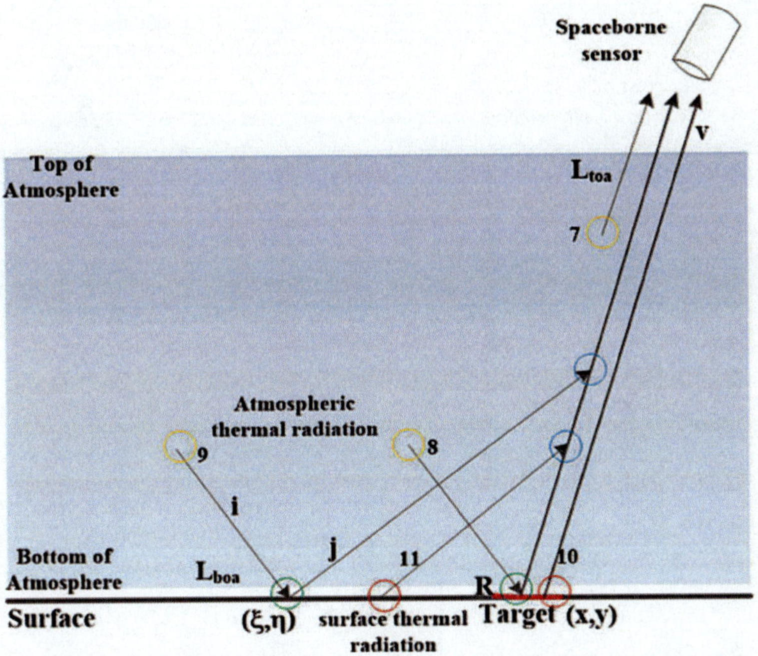

Fig. 3 The radiative transfer processes that atmospheric thermal emission and surface thermal emission have no interaction or once interaction with surface

$$
\begin{aligned}
L_3 &= L_s \mu_s T_d(\theta_s) \cdot \frac{A(x, y; \theta_v)}{\pi} \cdot T(\theta_v) \\
T_d(\theta_s) &= \int_0^{2\pi} \int_0^1 P(\theta_s, \theta_i, \varphi_{si}) \mu_i d\mu_i d\varphi_i \\
A(x, y; \theta_v) &= \frac{1}{\pi} \int_0^{2\pi} \int_0^1 R(x, y; \theta_i, \theta_v, \varphi_{iv}) \mu_i d\mu_i d\varphi_i
\end{aligned}
\tag{4}
$$

where $T_d(\theta_s)$ is the total atmospheric diffuse transmission in the θ_s direction that is determined by atmospheric condition only. $A(x, y, \theta_v)$ is the target pixel albedo in the θ_v direction. It should be noted that Zibordi et al. [7] has proven atmospheric angular diffuse transmission does not deviate from uniform distribution too much in most situations, so that the approximation above can be treated as a basic assumption for our model.

Process 4 and 5 describe that solar radiation transmitted through atmosphere directly and diffusely to surface is reflected by the surroundings and then diffusely transmitted into viewing direction. To express these two processes, the atmospheric point spread function M is introduced, so that the contribution 4 can obtained as

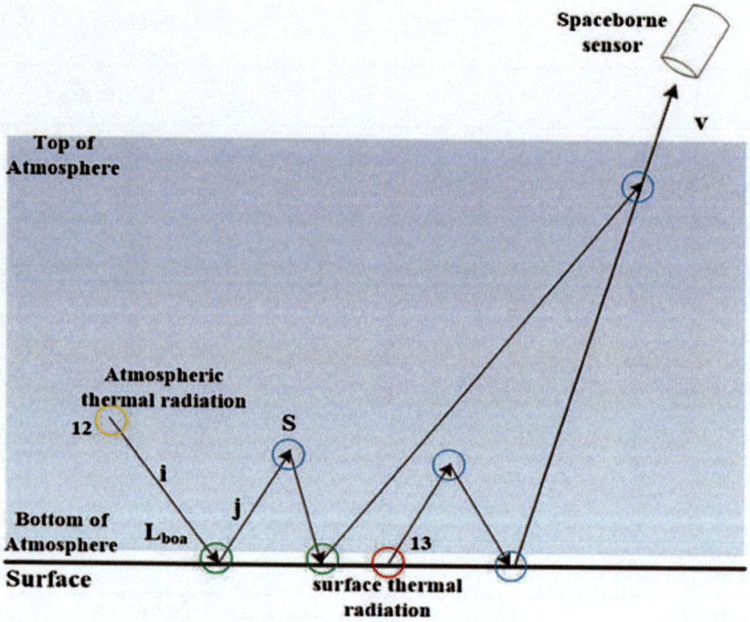

Fig. 4 The radiative transfer processes that atmospheric thermal emission and surface thermal emission have multiple interactions with surface

$$L_4 = L_s\mu_s T(\theta_s) \cdot \int_{-\infty}^{+\infty}\int_{-\infty}^{+\infty} \frac{A(\xi,\eta;\theta_s)}{\pi} \cdot M(\xi,\eta;x,y)d\xi d\eta \tag{5}$$

The angular separation approximation has been applied in Eq. (5). Using the normalized atmospheric point spread function, L_4 can be expressed as

$$\begin{aligned} L_4 &= L_s\mu_s T(\theta_s) \cdot \frac{\langle A(\theta_s)\rangle}{\pi} \cdot T_d(\theta_v) \\ M(\xi,\eta;x,y) &= T_d(\theta_v) \cdot f(\xi,\eta;x,y) \\ \langle A(\theta_s)\rangle &= \int_{-\infty}^{+\infty}\int_{-\infty}^{+\infty} A(\xi,\eta;\theta_s) \cdot f(\xi,\eta;x,y)d\xi d\eta \end{aligned} \tag{6}$$

where $T_d(\theta_v)$ is the total atmospheric diffuse transmission in the θ_v direction. $\langle A(\theta_s)\rangle$ illustrates the adjacency effect from process 4 that mixes radiation signals from surroundings into remote sensing signal of target region. Actually, it is a convolution that atmosphere filter is applied to surface and centered on the target pixel, which must be integrated pixel by pixel for heterogeneous surface. To speed up this computation, Fast Fourier Transform (FFT) is implemented in the integral and atmospheric point spread function is parameterized as

$$
\begin{aligned}
f(\xi,\eta;x,y) &= \frac{dF(r)}{dr}\cdot\frac{1}{2\pi r},\quad r=\sqrt{(\xi-x)^2+(\eta-y)^2} \\
F(r) &= \frac{\alpha_R\tau_R F_R(r)+\alpha_A\tau_A F_A(r)}{\alpha_R\tau_R+\alpha_A\tau_A},\quad \alpha_R=\frac{1}{2},\ \alpha_A=\frac{5}{6}
\end{aligned} \tag{7}
$$

where τ_R and τ_A are atmospheric molecular optical thickness and aerosol optical thickness respectively. It means that atmospheric point spread function can be calculated pixel by pixel efficiently when two atmospheric optical thicknesses are derived.

Similarly, the contribution 5 can be obtained as

$$
\begin{aligned}
L_5 &= L_s\mu_s T_d(\theta_s)\cdot\frac{\langle A_w\rangle}{\pi}\cdot T_d(\theta_v) \\
\langle A_w\rangle &= \int_{-\infty}^{+\infty}\int_{-\infty}^{+\infty} A_w(\xi,\eta)\cdot f(\xi,\eta;x,y)\,d\xi d\eta
\end{aligned} \tag{8}
$$

where A_w is the surface white-sky albedo and $\langle A_w\rangle$ denotes the adjacency effect from process 5.

It has been proven that the contribution from multiple interactions with surface is always very few percent of total target remote sensing signal. Consequently, the expression of process 6 in Fig. 2 can be remarkably simplified and given as

$$
L_6 = \frac{1}{\pi}L_s\mu_s T^{\downarrow}(\theta_s)\cdot\frac{\langle A_w\rangle^2 S}{1-\langle A_w\rangle S}\cdot T^{\uparrow}(\theta_v) \tag{9}
$$

Here, $T^{\downarrow}(\theta_s)$ and $T^{\uparrow}(\theta_v)$ are sums of atmospheric direct and diffuse transmission in solar and viewing direction respectively.

In Fig. 3, L_{toa} and L_{boa} are introduced to characterize atmospheric thermal radiation property. L_{toa} is the radiance of atmospheric thermal emission received at the top of atmosphere, of which the photon has no interaction with surface. L_{boa} is the radiance received at the bottom of atmosphere from the same source as L_{toa}. Thus, the contribution from process 7 is

$$
L_7 = L_{toa}(\theta_v) \tag{10}
$$

Based on derivation in the processes that downward diffuse radiation from solar source is reflected by surface into viewing direction, contribution from downward thermal radiation can be derived in similar way. Therefore, the contribution from process 8 can be expressed as

$$
\begin{aligned}
L_8 &= E_{boa} \cdot \frac{A(x, y; \theta_v)}{\pi} \cdot T(\theta_v) \\
E_{boa} &= \int_0^{2\pi} \int_0^1 L_{boa}(\theta_i) \cdot \mu_i d\mu_i d\varphi_i
\end{aligned}
\tag{11}
$$

where E_{boa} is atmospheric downward thermal flux at the bottom of atmosphere. The contribution from process 9 can be expressed as

$$
L_9(x, y; \theta_v) = E_{boa} \cdot \frac{<A_w>}{\pi} \cdot T_d(\theta_v) \tag{12}
$$

where $\langle A_w \rangle$ has same meaning as those in Eqs. (8) and (9); and it demonstrates the adjacency effect from surrounding surface reflection of atmospheric thermal radiation here.

Process 10 describes that thermal emission from the target region is transmitted directly into sensor. The contribution can be expressed as

$$
\begin{aligned}
L_{10} &= \varepsilon(x, y; \theta_v) L_B(x, y) \cdot T(\theta_v) \\
L_B(x, y) &= \frac{2hc^2}{\pi\lambda^5} \frac{1}{(\exp(\frac{hc}{\lambda k T(x,y)}) - 1)}
\end{aligned}
\tag{13}
$$

where L_B is the blackbody radiance with apparent temperature of the target region which can be calculated by Planck function as shown in Eq. (13); and ε is its emissivity ($\varepsilon = 1 - A$). Process 11 describes that thermal emission from the surrounding is transmitted diffusely into sensor, so that the contribution can be given as

$$
\begin{aligned}
L_{11} &= \langle \varepsilon_w L_B \rangle \cdot T_d(\theta_v); \quad \varepsilon_w = 1 - A_w \\
\langle \varepsilon_w L_B \rangle &= \int_{-\infty}^{+\infty} \int_{-\infty}^{+\infty} \varepsilon_w(\xi, \eta) L_B(\xi, \eta) \cdot f(\xi, \eta; x, y) d\xi d\eta
\end{aligned}
\tag{14}
$$

Here, $\langle \varepsilon_w L_B \rangle$ demonstrates the adjacency effect from surrounding thermal emission.

At last, the contributions from multiple interactions of atmospheric and surface thermal emission with surface in Fig. 4 can be obtained following the derivation of Eq. (9), that is

$$
L_{12} = \frac{E_{boa}}{\pi} \cdot \frac{\langle A_w \rangle^2 S}{1 - \langle A_w \rangle S} \cdot T^{\uparrow}(\theta_v) \tag{15}
$$

$$L_{13} = \langle \varepsilon_w L_B \rangle \cdot \frac{\langle A_w \rangle S}{1 - \langle A_w \rangle S} \cdot T^{\uparrow}(\theta_v) \tag{16}$$

To sum all the contributions above, the total TOA radiance of target pixel is derived as

$$L(x, y) = L_{SS} + L_{GR} + L_{PT} + L_{SE} \tag{17}$$

with

$$\begin{aligned}
L_{SS} =& L_s \mu_s R_a + \frac{L_s \mu_s}{\pi} \cdot T_d(\theta_v) \left[T(\theta_s) \langle A(\theta_s) \rangle + T_d(\theta_s) \langle A_w \rangle + T^{\downarrow}(\theta_s) \frac{< A_w >^2 S}{1 - < A_w > S} \right] \\
L_{GR} =& \frac{L_s \mu_s}{\pi} T(\theta_v) \cdot \left[T(\theta_s) R + T_d(\theta_s) A(\theta_v) + T^{\downarrow}(\theta_s) \frac{\langle A_w \rangle^2 S}{1 - \langle A_w \rangle S} \right] \\
& + \frac{E_{boa}}{\pi} T(\theta_v) \cdot \left(A(\theta_v) + \frac{\langle A_w \rangle^2 S}{1 - \langle A_w \rangle S} \right) + \langle \varepsilon_w L_B \rangle T(\theta_v) \cdot \frac{\langle A_w \rangle S}{1 - \langle A_w \rangle S} \\
L_{PT} =& L_{toa}(\theta_v) + \frac{E_{boa}}{\pi} T_d(\theta_v) \langle A_w \rangle \cdot \frac{1}{1 - \langle A_w \rangle S} + \langle \varepsilon_w L_B \rangle T_d(\theta_v) \cdot \frac{1}{1 - \langle A_w \rangle S} \\
L_{SE} =& \varepsilon(\theta_v) L_B \cdot T(\theta_v)
\end{aligned} \tag{18}$$

According to Eqs. (17) and (18), the total TOA radiance consists of four terms, that is, first term denotes atmospheric scattering of solar radiation into sensor no matter whether it is reflected by surface or not; second term denotes target surface region reflection of solar radiation and terrestrial thermal radiation directly transmitted into sensor; third term denotes atmospheric path thermal radiation into sensor which is from atmospheric thermal emission or from atmospheric scattering of terrestrial thermal emission; and fourth term denotes target surface region thermal emission directly transmitted into sensor. These terms can be calculated separately using a surface-atmosphere coupling stratified radiative transfer model, like MODTRAN. The difference is that each term is expressed in the analytical form here with surface optical parameters and atmospheric optical parameters which are independent with surface. It means that the atmospheric parameters in our model just need to be calculated once for the imaging region where atmosphere status and observation geometry keep constant.

2.2 Numerical Computation Method

The algebraic method is established here to calculate the atmospheric parameters employing MODTRAN [8]. The uniform Lambertian surface is assumed for MODTRAN run with albedo R_O 0, 0.5, and 1 respectively. $L_{SS}(R_O)$, $L_{GR}(R_O)$, $L_{PT}(R_O)$, $L_{SE}(R_O)$, $L_{DR}(R_O)$, $L_{GT}(R_O)$, $L_{AT}(R_O)$ and DEPTH are generated from tape7 of MODTRAN outputs under the condition of specified atmospheric status and observation geometry. Here, L_{DR} is the direct reflected radiance of target surface region; and L_{GT} is the total target surface radiance which is sum of L_{GR} and L_{SE}; and L_{AT} is the total atmosphere radiance which is sum of L_{SS} and L_{PT}. DEPTH is the optical depth of atmosphere in the viewing direction. Accordingly, the atmospheric parameters under the specified conditions can be obtained by algebraic operation of these MODTRAN outputs, i.e.

$$\begin{aligned} T(\theta_v) &= \exp(DEPTH) \\ T(\theta_s) &= \frac{\pi L_{DR}(1)}{L_s\mu_s \exp(DEPTH)} \end{aligned} \tag{19}$$

$$\begin{aligned} R_a &= \frac{L_{SS}(0)}{L_s\mu_s} \\ S &= \frac{\Delta L_{GT}(1) - 2\Delta L_{GT}(0.5)}{\Delta L_{GT}(1) - \Delta L_{GT}(0.5)} \\ T_d(\theta_v) &= \frac{\Delta L_{AT}(1)}{\Delta L_{GT}(1)} T(\theta_v) \\ T_d(\theta_s) &= \frac{\pi}{L_s\mu_s}(L_{SS}(1) - L_s\mu_s R_a)\frac{1-S}{T_d(\theta_v)} - T(\theta_s) \end{aligned} \tag{20}$$

$$\begin{aligned} L_{toa}(\theta_v) &= L_{PT}(0) - T_d(\theta_v)L_B \\ E_{boa} &= (L_{PT}(1) - L_{toa}(\theta_v))\frac{\pi(1-S)}{T_d(\theta_v)} \end{aligned} \tag{21}$$

where $\Delta L_{GT}(1)$ is $L_{GT}(1)$ minus $L_{GT}(0)$; $\Delta L_{GT}(0.5)$ is $L_{GT}(0.5)$ minus $L_{GT}(0)$; and $\Delta L_{AT}(1)$ is $L_{AT}(1)$ minus $L_{AT}(0)$. Note that TOA solar irradiance L_s is also gained from tape7 output TOA_SUN. Then, another MODTRAN run with no aerosol is needed to derive atmospheric molecular optical thickness τ_R, i.e.

$$\tau_R = \mu_v DEPTH(clear) \tag{22}$$

where DEPTH(clear) denotes the optical depth of atmosphere with no aerosol in the viewing direction. The total atmospheric optical thickness τ can be derived in the same way using MODTRAN output with aerosol. The atmospheric aerosol optical thickness τ_A is τ minus τ_R. Thus, ten atmospheric parameters in our model are calculated with four times MODTRAN runs. This method can also be utilized to generate an off-line Look-Up Table (LUT) of each atmospheric parameter.

On the other hand, the surface images of bidirectional reflectance, albedo and white-albedo, emissivity and apparent temperature required by our model are obtained from satellite measurements. Generally, directional reflectance can be calculated according to Bidirectional Reflectance Distribution Function (BRDF) retrieved from multi-angle reflectance when solar and viewing geometry are defined; and emissivity can be derived from albedo directly. It should be noted that if the surface must be treated as Lambertian approximately where BRDF is not available, the Lambertian albedo can be used in the model as inputs of bidirectional reflectance, albedo and white-albedo, and so does emissivity. The model would become its Lambertian version automatically in this situation.

3 Examples and Discussion

Several examples of TOA radiance image simulation in the optical and thermal infrared domain are conducted for a test region in northwestern China to show the effectiveness of our model. Middle latitude summer atmospheric profiles and rural aerosol type in MODTRAN are chosen. The surface elevation is assumed to be 0 km except the example for the effect of elevation change. The size of subzone of adjacency effect integral is fixed to 21*21 pixels here.

3.1 Radiance Image Simulation in the Optical Domain

The MODIS MCD43 data is used for input of surface directional reflectance and derived directional emissivity of test region, and MODIS MOD21 data is used for input of surface apparent temperature. Though surface emissivity and apparent temperature have negligible influence on TOA radiance in the optical domain, they should be involved to ensure the completeness of input information.

Figure 5 shows the surface bidirectional reflectance and its corresponding TOA radiance images for three bands of MODIS when the SZA and VZA are both set as 10° and RAA is set as 0°. Band VIS is MODIS Band4 (545–565 nm); Band NIR is MODIS Band2 (841–876 nm); and Band SWIR is MODIS Band6 (1628–1652 nm). It is obvious that the pattern of reflectance images is significantly different between three bands, which is the main reason for the difference of TOA radiance images. The difference of absolute value of TOA radiance images is also attributed to the difference of TOA solar irradiance and atmospheric scattering property.

To demonstrate the influence of non-Lambertian surface, the TOA radiance images with non-Lambertian surface and TOA radiance images with Lambertian surface are compared under the condition of different viewing zenith angle.

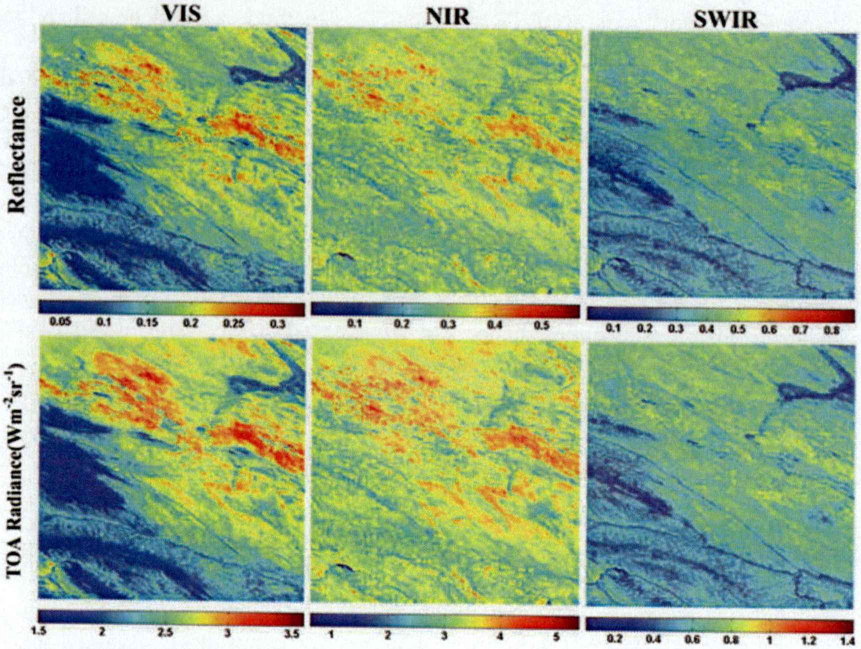

Fig. 5 The surface bidirectional reflectance images (first row) and TOA radiance images (second row) for Band VIS, Band NIR and Band SWIR

Figure 6 shows that realistic non-Lambertian surface reflectance images have clear variability with different VZA, so that TOA radiance becomes smaller with reflectance smaller. The white albedo of test region from MODIS MCD43 is assumed to be Lambertian surface reflectance (not shown in Fig. 6). It can be seen that TOA radiance images with Lambertian surface have obviously different variability comparing to those with non-Lambertian surface. The TOA radiance becomes larger with VZA increase, which could be attributed to the increased atmospheric scattering.

3.2 *Radiance Image Simulation in the Thermal Infrared Domain*

Since it is difficult to achieve the surface BRDF in the thermal infrared domain at present, we show the examples of thermal infrared radiance image simulation using MODIS MOD21 data. The surface emissivity and apparent temperature are both obtained from it, and albedo is derived from the surface emissivity. Figure 7 shows the surface emitted thermal radiance that is not attenuated by atmosphere yet and its corresponding TOA radiance at 8 μm and 12 μm using MODIS data of Band 29

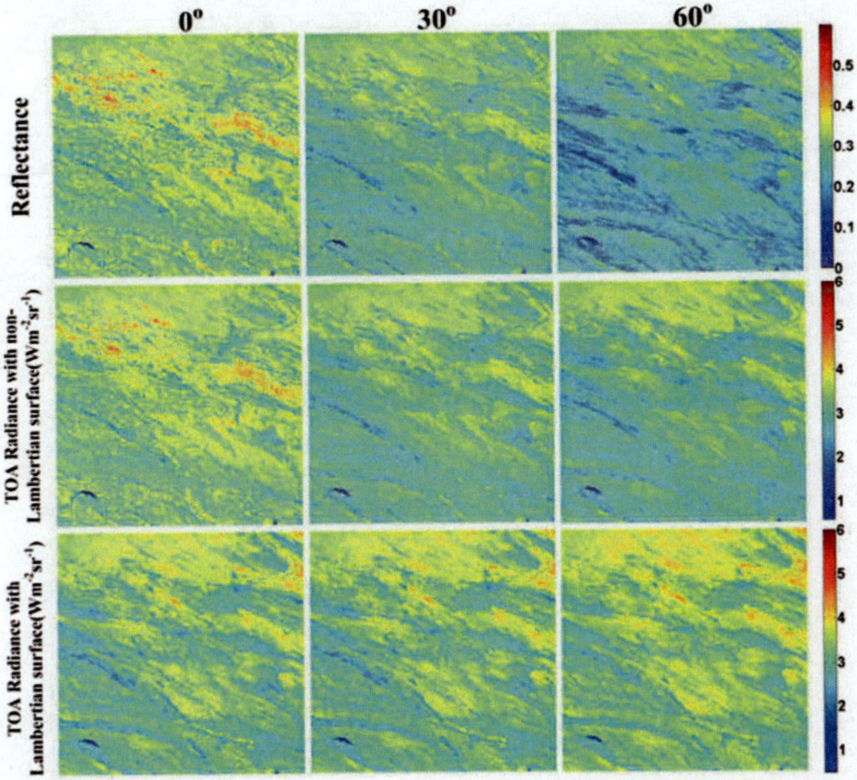

Fig. 6 The surface bidirectional reflectance images (first row), TOA radiance images (second row) and TOA radiance images under the assumption of Lambertian surface (third row) under the condition of three VZA (i.e. 0°, 30°, 60°)

and Band 32. It is visual that TOA radiance is slightly smaller than surface radiance, which means that atmospheric thermal emission does not fill in the loss from the surface radiance decrease. Besides, the radiance at 12 μm is obviously larger than the radiance at 8 μm due to stronger emission in the larger wavelength domain.

Figure 8 shows the surface emitted thermal radiance image and its corresponding TOA radiance images. The TOA radiance image with viewing zenith angle 0° and elevation 1 km visually decreases due to atmospheric absorption. The TOA radiance image with viewing zenith angle 60° and elevation 1 km decreases more substantially due to longer atmospheric path. However, the TOA radiance image with viewing zenith angle 0° and elevation 5 km is almost same as the surface emitted thermal radiance image, which demonstrate that atmospheric influence on surface thermal remote sensing with high elevation is negligible at some extent.

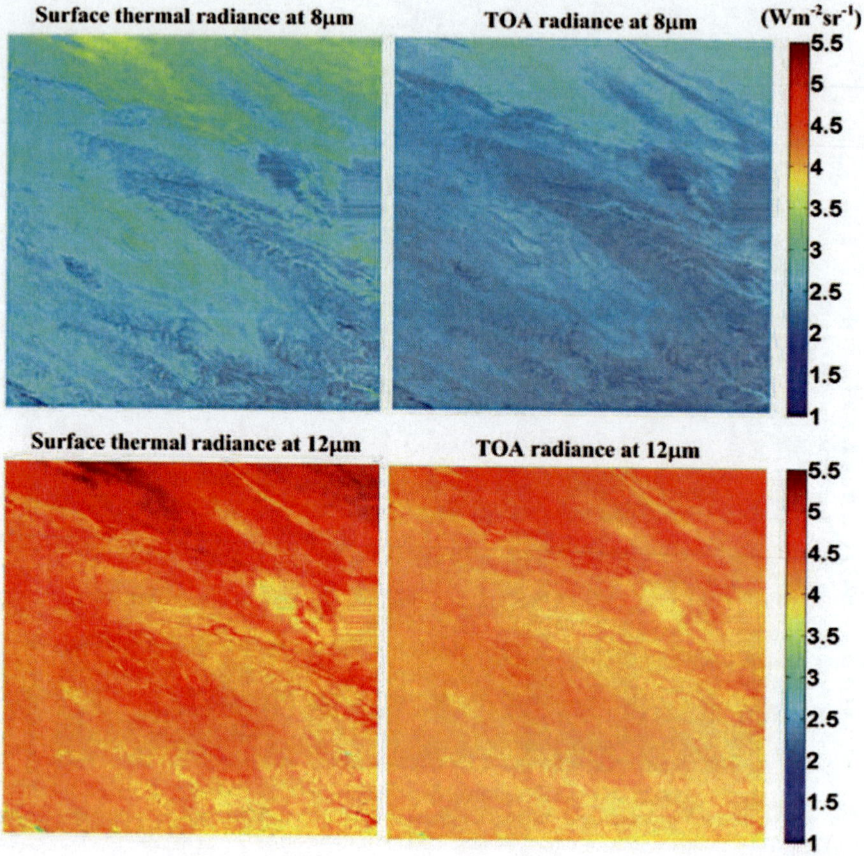

Fig. 7 The surface thermal radiance and its corresponding TOA radiance at 8 μm and 12 μm

3.3 Discussion

The acquisition problem of thermal infrared BRDF image reminds us that the method of surface bidirectional reflectance and emissivity retrieval with high spatial resolution and high spectral resolution ranging from visible to thermal infrared must be investigated further to enable the TOA image simulation in a variety of remote sensing applications. BRDF/Albedo retrieval of surface, like MODIS BRDF/Albedo product adopted here, has the fixed spatial resolution and spectral range. MODIS BRDF/Albedo product can generate daily BRDF, albedo and white-sky albedo maps with 0.5 km resolution in seven spectral bands ranging from visible to short wave infrared for the specified region. But when higher spatial resolution, higher spectral resolution and different spectral range are required, MODIS data will be incapable. Utilizing very high spatial resolution multi-angle reflectance

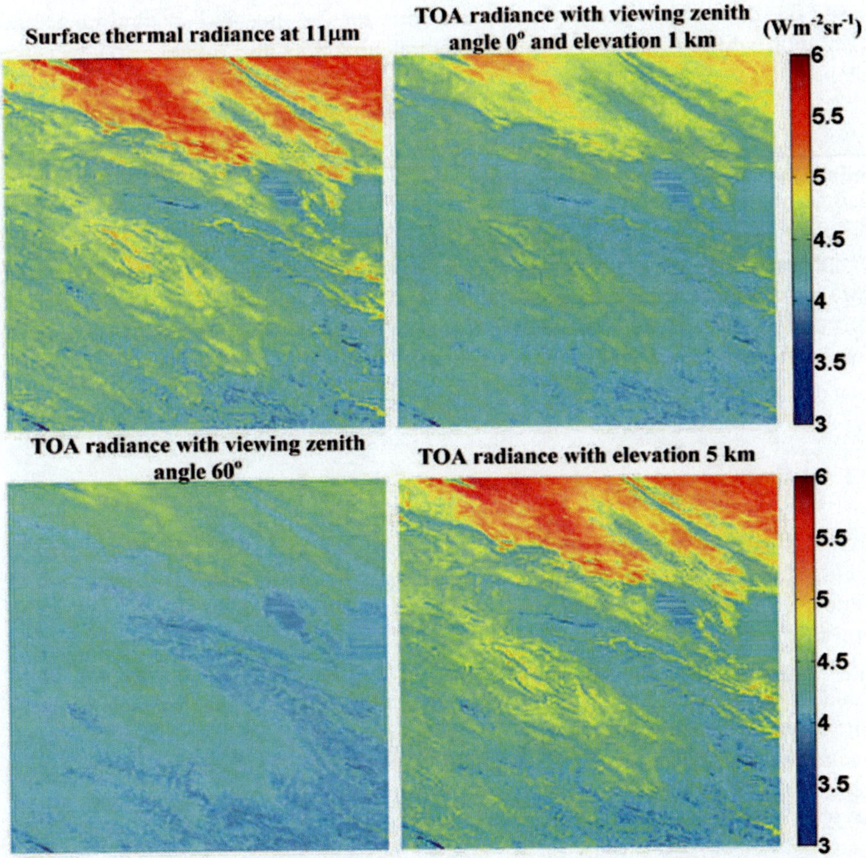

Fig. 8 The surface thermal radiance and its corresponding TOA radiance with different viewing zenith angles and surface elevations

measurements to retrieve surface BRDF seems to be a straightforward way to solve spatial resolution problem, such as Sentinel-2 with spatial resolution up to 10 m [9]. The lower spatial resolution map can be obtained by resampling the original data. Though, the spectral problem still exists because Sentinel-2 just supports multi-spectral imaging with moderate spectral resolution; and a definite fact based on the existing remote sensing technique is that spectral resolution and spatial resolution are contradictory for the imager.

Verhoef et al. [10] proposed a promising way to solve the spectral resolution problem, that is, the geophysical parameters of surface (such as soil moisture, leaf area index, fraction of brown leaves, chlorophyll content and dry matter) in each pixel are retrieved first by the biophysical land BRDF model (such as coupled soil-leaf-canopy model) based on multispectral measurements or multi-angle measurements; then spectral BRDF of the pixel is calculated by the BRDF model based on the retrieved geophysical parameters for the specified spectral range and resolution which is limited only by spectral library in the model. However, the accuracy of geophysical retrieval still needs to be improved due to the complexity of the biophysical BRDF model. Note that the kernel-driven BRDF model used for MODIS may be regarded as the intermediate form for the retrieval of geophysical parameters, because the kernels are derived from the approximation of BRDF geometric optics modeling and radiative transfer modeling to linearize the inversion problem; and kernel coefficients are the results of combination of the several geophysical parameters. Therefore, if the BRDF kernel coefficients can be expressed by the surface geophysical parameters in more explicit way, retrieval of spectral BRDF will be achieved based on kernel-driven model, which suggests a further surface BRDF model investigation.

4 Summary

In this paper, we propose a physical based TOA radiance image simulation model incorporating the coupling of heterogeneous and non-Lambertian surface with atmospheric radiative transfer consistently. The model is built based on successive orders of radiation interactions, which involves the processes of solar radiation and terrestrial thermal radiation. Adjacency effect of heterogeneous surface and directional effect of non-Lambertian surface are demonstrated in our model completely. The algebraic method employing MODTRAN is established for ten atmospheric parameters calculation with the specified atmospheric condition and solar-viewing geometry. The surface BRDF/Albedo, emissivity and apparent temperature of MODIS have been used for simulation of this paper. Several examples have been analyzed in the optical domain and thermal infrared domain. At last, the possible acquisition methods of atmospheric and surface inputs have been discussed for variety of remote sensing scene simulations. A feasible way to generate the high spatial and high spectral resolution surface BRDF is suggested for further investigation.

References

1. Segl K et al (2012) EeteS—The EnMAP end-to-end simulation tool. IEEE J Sel Topics Appl Earth Observ Remote Sens 5
2. Schott JR et al (1999) An advanced synthetic image generation model and its application to multi/hyperspectral algorithm development. Can J Remote Sens 25
3. Börner A et al (2001) Sensor: A tool for the simulation of hyperspectral remote sensing systems. ISPRS J Photogramm Remote Sens 55
4. Guanter L, Segl K, Kaufmann H (2009) Simulation of optical remote sensing scenes with application to the EnMAP hyperspectral mission. IEEE Trans Geosci Remote Sens 47

5. Segl K et al (2015) S2eteS: an end-to-end modeling tool for the simulation of sentinel-2 image products. IEEE Trans Geosci Remote Sens 53
6. Tanre D, Herman M, Deschamps PY (1983) Influence of the atmosphere on space measurements of directional properties. Appl Opt 21
7. Zibordi G, Voss KJ Geometrical and spectral distribution of sky radiance: comparison between simulation and field experiments. Remote Sens Environ 27
8. Berk A et al (1998) MODTRAN cloud and multiple scattering upgrades with application to AVIRIS. Remote Sens Environ 65
9. Drusch M et al (2012) Sentinel-2: ESA's optical high resolution mission for GMES operational services. Remote Sens Environ 120
10. Verhoef W, Bach H (2003) Simulation of hyperspectral and directional radiance images using coupled biophysical and atmospheric radiative transfer models. Remote Sens Environ 87

Application of Improved RBF Neural Network in Remote Sensing Image Restoration

Yunsen Wang, Yong Wang, Daqiang Feng, Jing Yang, and Ke An

Abstract Neural network is widely used in remote sensing image restoration. To avoid huge time and space consumption of conventional neural network models, a novel remote sensing image restoration algorithm base on improved RBF (Radial Basis Function) neural network is proposed. The training data set could be organized by degrading the high-quality remote sensing images or promoting degraded images' quality by other necessary methods. And the number of hidden layer is decided by the size of training data at the condition of small training sample scale. To accelerate convergence speed of RBF neural network during training process, conjugate gradient descent method is adopted to realize the weight parameters' iterative correction. To further reduce calculating time, this paper proposes a matrix factorization algorithm to realize matrix's parallel arithmetic. Simulations and experiments indicate that the improved RBF neural network model could acquire relatively approving remote sensing image restoration results and time overhead.

Keywords Remote sensing image restoration · RBF neural network · Conjugate gradient descent · Matrix factorization

1 Introduction

Remote sensing data is always degraded by different kinds factors in space environment [1–3]. To obtain high-quality images, traditional method attempts to design space camera with long focal length and large aperture, which would increases research costs. And remote sensing image restoration plays an important role when it is difficult to suppress the vibration of optical platform on spacecraft [4].

Traditional restoration methods generally adopt filter algorithms according to the degradation principle or mathematical statistics theory, such as inverse filtering [5],

Y. Wang (✉) · Y. Wang · D. Feng · J. Yang · K. An
Shandong Institute of Space Electronic Technology, Yantai 264670, Shandong, China
e-mail: yswang513@163.com

L. Wang et al. (eds.), *Proceedings of the 7th China High Resolution Earth Observation Conference (CHREOC 2020)*, Lecture Notes in Electrical Engineering 757, https://doi.org/10.1007/978-981-16-5735-1_32

Richardson-Lucy filtering [6, 7], Wiener filter [8], Laplacian constraint [9, 10]. Most of these methods are implemented based on PSF (Point Spread Function), or MTF (Modulation Transfer Function) [11]. But the PSF and MTF are unable to obtained occasionally.

In recent years, the deep learning methods have brought rapid development to various fields of computer image and vision, including restoration of degraded images [12], especially CNN (Convolution Neural Network). Sun et al. firstly adopted CNN model to deblur degraded images [13]. Then this algorithm is optimized in literature [14]. Another popular model is Hopfield network. Zhou et al. firstly adopted Hopfield neural network to realize image restoration by using optimization method [15]. And this method is improved by many scholars from the network model and the convergence rate [16, 17]. Additionally, there are many other popular methods, such as GAN [18], U-net [19], residual net [20] and so on. But all of these methods' connection weights are very huge, which would cost huge storage and computing resources.

On the basis of previous studies, a novel remote sensing image restoration algorithm base on improved radial basis function neural network (RBFNN) is proposed. The training data set could be organized by degrading the high-quality remote sensing images or promoting degraded images' quality by other necessary methods. To reduce operational time, the number of RBFNN's hidden layer equals to the size of training data at the condition of small training sample scale. To accelerate convergence speed of RBFNN during training process, conjugate gradient descent method is adopted to realize the weight parameters' iterative correction. To further reduce calculating time, this method adopts matrix factorization algorithm to realize matrix's parallel arithmetic. Experiments results indicate that the improved RBF neural network model could acquire relatively approving remote sensing image restoration results and time overhead.

2 Image Degradation Model and RBF Neural Network

2.1 Remote Sensing Image Degradation and Restoration Models

To build restoration model, remote sensing image degradation and restoration elements are introduced firstly, as shown in Fig. 1. $f(x, y)$ represents the original remote sensing image, $d(x, y)$ is the degraded image, $p(x, y)$ is the motion blur function (Point Spread Function, PSF), and $n(x, y)$ represents the noise model.

To simplify operating process, the PSF $p(x, y)$ is set up as a linear and space-unchanging model. So the description of degradation process could be expressed by

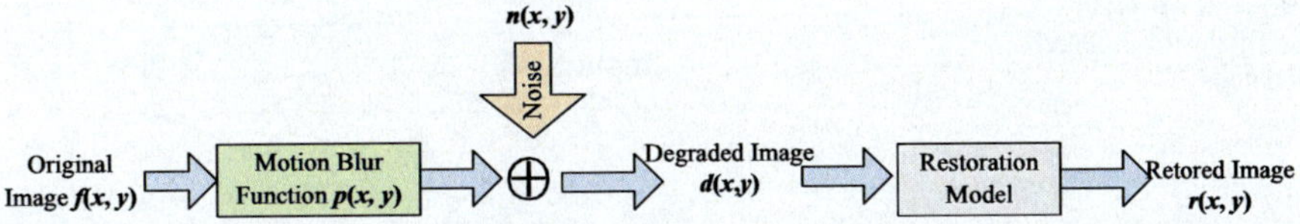

Fig. 1 Remote sensing image's degradation and restoration models

$$d(x, y) = f(x, y) * p(x, y) + n(x, y) \tag{1}$$

And the degradation model's Fourier transform can be described as

$$D(u, v) = F(u, v) * P(u, v) + N(u, v) \tag{2}$$

Therefore, the Fourier transform of original image could be calculated by

$$F(u, v) = \frac{D(u, v) - N(u, v)}{P(u, v)} \tag{3}$$

In Fig. 1, *r*(*x*, *y*) represents the recovered remote sensing image. The restoration model's effect is to fulfill the following equation.

$$F(u, v) = R(u, v) \tag{4}$$

where *R*(*u*, *v*) is the Fourier transform of *r*(*x*, *y*).

2.2 *RBF Neural Network*

The RBF neural network is a type of forward network model, as show in Fig. 2. RBFNN has three layers: an input layer, a hidden layer with a non-linear RBF activation function and a linear output layer. RBF neural network has good performance and global approximation, so that RBFNN could be free from local minima problem.

The structure of RBFNN shows in Fig. 2 is *m*-*h*-*n*. The network has *m* input neurons (X_1, X_2 … X_m), *h* hidden neurons (N_1, N_2 … N_h) and *n* output neurons (Y_1, Y_2 … Y_n). The size of *m* and *n* is determined by realistic applications. And the value of *h* is decided by the number of training samples.

In the remote sensing image restoration algorithm, the size of input nodes equals to the output nodes, $m = n$. So the network's output *y* is a vector with *m* components, determined in terms of the input vector *x* with *m* components.

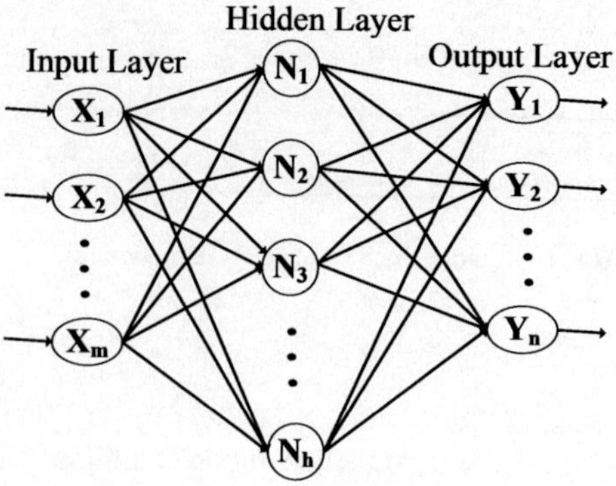

Fig. 2 Two-layer feed-forward RNFNN structure

$$y_i = \sum_{j=1}^{h} [w_{ij}\varphi_j(x) + \theta_{wi}], \quad i = 1, 2 \ldots m \tag{5}$$

where $\varphi(\ )$ is the radial basis function, w_{ij} is the interconnection weight between hidden-layer and output layer. And the threshold offset is denoted by θ_{wi}.

There are many types of radial basis function used by RBFNN. In the proposed algorithm, Gaussian activation function is chosen as the radial basis function, which is described by

$$\varphi_j(x) = \exp\left\{-\frac{|x - c_i|}{2\sigma^2}\right\} \tag{6}$$

where c is the center and the σ is the variance. For simplicity, the centers and variances are pre-defined and fixed.

3 Design of Remote Sensing Image Restoration with RBF Neural Network

RBFNN has strong capability of non-linear mapping. Through learning the internal mapping relationship between input remote sensing image and output sample, the non-linear relationship between original image and degraded image could be built. And this relationship could realize degraded images' restoration process. The design of RBF restoration model is shown in Fig. 3.

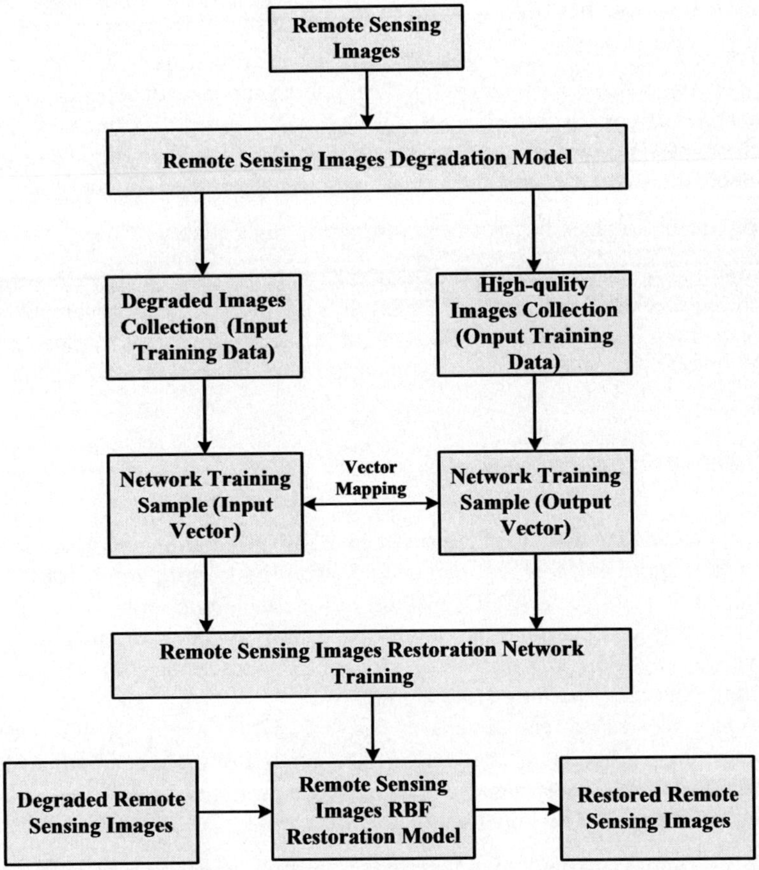

Fig. 3 Design of RBF restoration model

3.1 Collection of Training Data

The training sample data could be expressed by $\{(x(i), y(i))\}_{i=1}^{h}$, which has h samples. And the training sequence of one round is organized by $\{(x(1), y(1)), (x(2), y(2)), \ldots (x(h), y(h))\}$. Though continuous training, the RBFNN's weights would be adjusted incessantly until the network tends to be stable.

To reduce computing time, the hidden layer's number of RBFNN is set as the number of training samples, because the scale of training data is relatively small.

For the problem of remote sensing image restoration, the original high-quality image is unknown, but the motion model of the aerial camera could be acquired. Therefore, there are two ways to obtain the training sample data.

(1) Original images has been restored to high quality

If the original remote sensing sample has been restored to high level, the train sample data could be obtained directly. One training sample can be expressed as (*X*, *Y*), where *X* is original degraded image, *Y* is high-quality image. At the case of lacking enough samples, generative adversarial network (GAN) would be adopted to grow in the number of samples, so that the approaching results can become more accurate.

(2) Low quality images haven't been restored to high quality

If low quality images haven't been restored to ideal status, classical Wiener filter method could be applied to upgrade the image's quality, for the motion model of the aerial camera is known. Equation (3) is employed to restore those degraded images. And other handles could also been adopted to make the images more distinct.

3.2 Network Learning Rules

Gradient descent method is widely used in RBFNN's training process. But low convergence speed and low accuracy would affect the learning results of RBFNN. Conjugate gradient decent (CGD) method has the quadratic termination property, which could apace determine the width of the basis function through supervised iterative learning. So CGD method is adopted to improve the RBFNN's learning algorithm. And the procedure steps are as flows.

Assume the size of remote sensing image is $W*H$, where W and H are the image's width and height, respectively. So the number of input and output neurons m equals to W*H. The terminal circulating error is set as τ. And the maximum alternative time is N. The initial alternative time $t = 1$.

Step 1 Apply Gaussian radial function to compute output results of RBFNN at time t.

$$\varphi_j(x) = \exp\left\{-\frac{|x - c_i|}{2\sigma^2}\right\}, j = 1, 2, \ldots h \tag{7}$$

$$y_t(x) = \sum_{j=1}^{h} w_{kj}\varphi_j, k = 1, 2, \ldots m \tag{8}$$

Step 2 Compute the efficiency function of RBFNN.

$$E = \sqrt{\frac{1}{Nmh}\sum_{t=1}^{N}\sum_{k=1}^{m}[y_k(x) - y(k)]^2} \tag{9}$$

If $E \leq \tau$, terminate training, else go to Step 3.

Step 3 Calculate the radial function's width σ_{ji} (t) by non-supervised learning. And revise link weights $w_{kj}(t)$ by supervised learning. If $3 \leq t \leq N$,

$$\sigma_{ji}(t) = \sigma_{ji}(t-1) + \eta g_{\sigma_{ji}(t)}(t-1) + \alpha[\sigma_{ji}(t-1) - \sigma_{ji}(t-2)] \quad (10)$$

$$w_{kj}(t) = w_{ji}(t-1) + \eta g_{w_{kj}(t)}(t-1) + \alpha[w_{kj}(t-1) - w_{kj}(t-2)] \quad (11)$$

where η is the learning speed, and α is the momentum factor.

Step 4 Calculate derivative of function E(t) with respect to σ_{ji} (t) and $w_{kj}(t)$ by CGD method to obtain search direction.

$$g_{\sigma_{ji(t)}}(1) = -\frac{\partial E}{\partial \sigma_{ji}(1)} \quad (12)$$

$$g_{\sigma_{ji(t)}}(t-1) = -\frac{\partial E}{\partial \sigma_{ji}(t-1)} + g_{\sigma_{ji(t)}}(t-2)\left\|\frac{\partial E}{\partial \sigma_{ji}(t-1)}\right\|^2 \Big/ \left\|\frac{\partial E}{\partial \sigma_{ji}(t-2)}\right\|^2 \quad (13)$$

$$g_{w_{kj(t)}}(1) = -\frac{\partial E}{\partial w_{kj}(1)} \quad (14)$$

$$g_{w_{kj(t)}}(t-1) = -\frac{\partial E}{\partial w_{kj}(t-1)} + g_{\sigma_{kj(t)}}(t-2)\left\|\frac{\partial E}{\partial w_{kj}(t-1)}\right\|^2 \Big/ \left\|\frac{\partial E}{\partial w_{kj}(t-2)}\right\|^2 \quad (15)$$

Step 5 If $t > N$, terminate training, else $t = t + 1$, return to Step 1.

3.3 Calculating Optimization of RBFNN

The RBFNN's input and output could be transformed to the matrix form below:

$$Y = W^T \Psi(X) \quad (17)$$

where $\Psi(X) = [\varphi_1, \varphi_2, \ldots \varphi_h]^T$ and W is the weights matrix.

The data scale of a remote sensing image is very large. To reduce calculating time, the proposed restoration method adopts a parallel calculating architecture to partition the large matrix into several small matrixes. This procedure is described as follows.

The output number of the training samples is known. In this application, the size of output array equals to the size of the input array. Assume $F(X)$ represents the

RBFNN's mapping non-linear function, and $\boldsymbol{E}$ is the output matrix, the RBFNN can be described as

$$\boldsymbol{F}(\boldsymbol{X}) = \boldsymbol{E} \tag{18}$$

Then the problem becomes how to calculate a matrix $\boldsymbol{W}$, which can satisfy the following equation.

$$\mathbf{\Phi}^T\mathbf{W} = \mathbf{E}^T \tag{19}$$

where $\boldsymbol{\Phi}^T = \begin{bmatrix} \varphi_{11}\varphi_{12}\cdots\varphi_{1M} \\ \varphi_{21}\varphi_{22}\cdots\varphi_{2M} \\ \cdots \\ \varphi_{N1}\varphi_{N2}\cdots\varphi_{NM} \end{bmatrix}$, and $\boldsymbol{\Phi}^T$ represents the activation array. Then $\boldsymbol{W}$ can be calculated by

$$\hat{\mathbf{W}} = (\mathbf{\Phi}^T)^{\#}\mathbf{E}^T \tag{20}$$

Here ()$^{\#}$ is the pseudo-converse operation of an array. The variance could be chosen as

$$\sigma_i^2 = \sigma^2 = \frac{d_{\max}^2}{2N} \tag{21}$$

If the training samples are chosen as the centers, Eq. (15) can be expressed as

$$\hat{\mathbf{W}} = \mathbf{\Phi}^{-1}\mathbf{E}^T \tag{22}$$

To reduce the calculation time, a matrix factorization algorithm is applied. Each decomposed matrix is distributed to different computing core (GPU, CPU or DSP).

Stipulate $\mathbf{\Phi} = \begin{bmatrix} \mathbf{A} & \mathbf{B} \\ \mathbf{C} & \mathbf{D} \end{bmatrix}$. The inverse matrix of $\mathbf{\Phi}$ could be calculated by

$$\mathbf{\Phi}^{-1} = \begin{bmatrix} (\mathbf{A} - \mathbf{B}\mathbf{D}^{-1}\mathbf{C})^{-1} & -\mathbf{A}^{-1}\mathbf{B}(\mathbf{D} - \mathbf{C}\mathbf{A}^{-1}\mathbf{B})^{-1} \\ -\mathbf{D}^{-1}\mathbf{C}(\mathbf{A} - \mathbf{B}\mathbf{D}^{-1}\mathbf{C})^{-1} & (\mathbf{D} - \mathbf{C}\mathbf{A}^{-1}\mathbf{B})^{-1} \end{bmatrix} \tag{23}$$

In Eq. (9), $(\mathbf{A} - \mathbf{B}\mathbf{D}^{-1}\mathbf{C})^{-1}$ and $(\mathbf{D} - \mathbf{C}\mathbf{A}^{-1}\mathbf{B})^{-1}$ could be calculated by Woodbury Formula.

$$(\mathbf{A} - \mathbf{B}\mathbf{D}^{-1}\mathbf{C})^{-1} = \mathbf{A}^{-1} + \mathbf{A}^{-1}\mathbf{B}(\mathbf{D} - \mathbf{C}\mathbf{A}^{-1}\mathbf{B})^{-1}\mathbf{C}\mathbf{A}^{-1} \tag{24}$$

$\mathbf{A}^{-1}$ and $\mathbf{D}^{-1}$ could be calculated in parallel, as can $(\mathbf{D} - \mathbf{C}\mathbf{A}^{-1}\mathbf{B})^{-1}$ and $(\mathbf{A} - \mathbf{B}\mathbf{D}^{-1}\mathbf{C})^{-1}$.

4 Experimental Results and Analysis

To validate the performance of the proposed restoration model, a training software implementation is developed in a graphics workstation, and the final restoration algorithm is realized on an embedded GPU.

300 doubles of remote sensing image with the size 5000×5000 are chosen as training data. All of the images were taken by the same remote-sensing camera. So the original remote sensing data are degraded by the same point spread function and noise function, and these data are restored to ideal status by many steps, such as image calibration and Wiener filter, which cost quite a long time. The average recovery time of each remote sensing image is 36.3 seconds in a graphic workstation. Another 100 original images are used to validate the restoration effects of the proposed RBF method.

4.1 Restoration Results

Figure 4 shows the restoration results of four remote sensing images using the proposed RBF model. The left four images are original degraded data, and the right four images are restored data. From the comparison of the four groups' images, it is obvious that the contours and details of the restored images are more distinct, and the surface features appear more prominent.

To further illustrate the restoration results of the proposed RBF model, the rectangular areas in the first group's images above are enlarged as Fig. 5. As we can see from the contrastive data, the left remote sensing image is degraded relatively serious, and object outlines are not distinct by observation of human eye. While the right restored image seems more clear, and the contour profiles of roads, bridges and fields in the area appear more distinguishable.

In addition, statistics suggest that the average processing time of 100 image is 13.6 seconds when operating in Nvidia Tegra X2, and the restored time reduce to 3.4 seconds when a graphic workstation is chosen as processing hardware. For a remote sensing image of size 5000×5000, the processing time is relatively acceptable.

4.2 Comparative Experiments

To further show the superiority of the proposed RBFNN restoration algorithm, another two frequently-used methods were implemented for comparison with the RBFNN model. These two methods are Wiener filter and inverse filtering, respectively. The original degraded remote sensing image and restoration results are shown in Fig. 6.

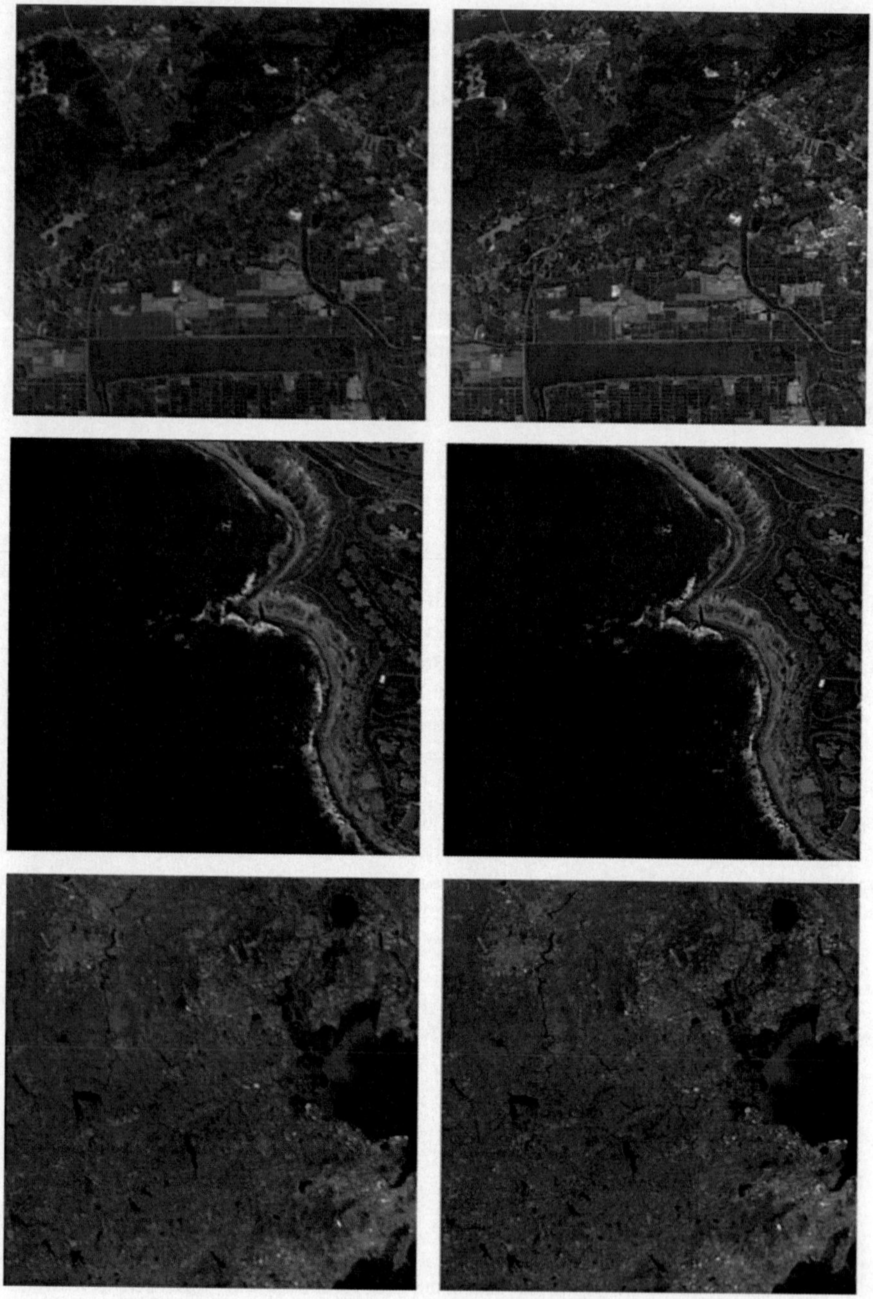

Fig. 4 Restoration results of four groups' images using RBF model

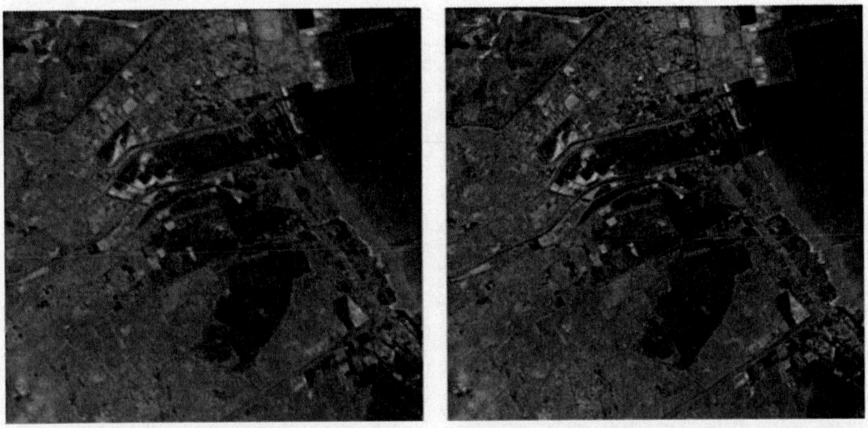

Fig. 4 (continued)

Fig. 5 Local enlarged results of the first group's images

Table 1 shows the parameter comparisons of the results using different restoration methods. From the PSNR (peak signal to noise ratio) and contrast statistics, it is obvious that the proposed RBFNN model has the best restoration result, as shown in Fig. 6b. Compared with original image, the results has more clear clear lines or contours. While using inverse filter method, new noise is easy to introduce into the images, as shown in Fig. 6d, so the PSNR and contrast would be dropped. Wiener filter method is built on the statistics minimization principle, but the power spectrum ratio is hard to estimate. So the restoration quality is not satisfactory, as shown in Fig. 6c.

Table 1 Comparison of the results of image restoration using different methods

	PSNR	Contrast
Original image	45.54	124.72
Wiener filter with RBFNN	64.21	8.27
Conventional Wiener filter	45.98	47.54
Inverse filter	21.12	54.28
RBRNN	75.86	142.62

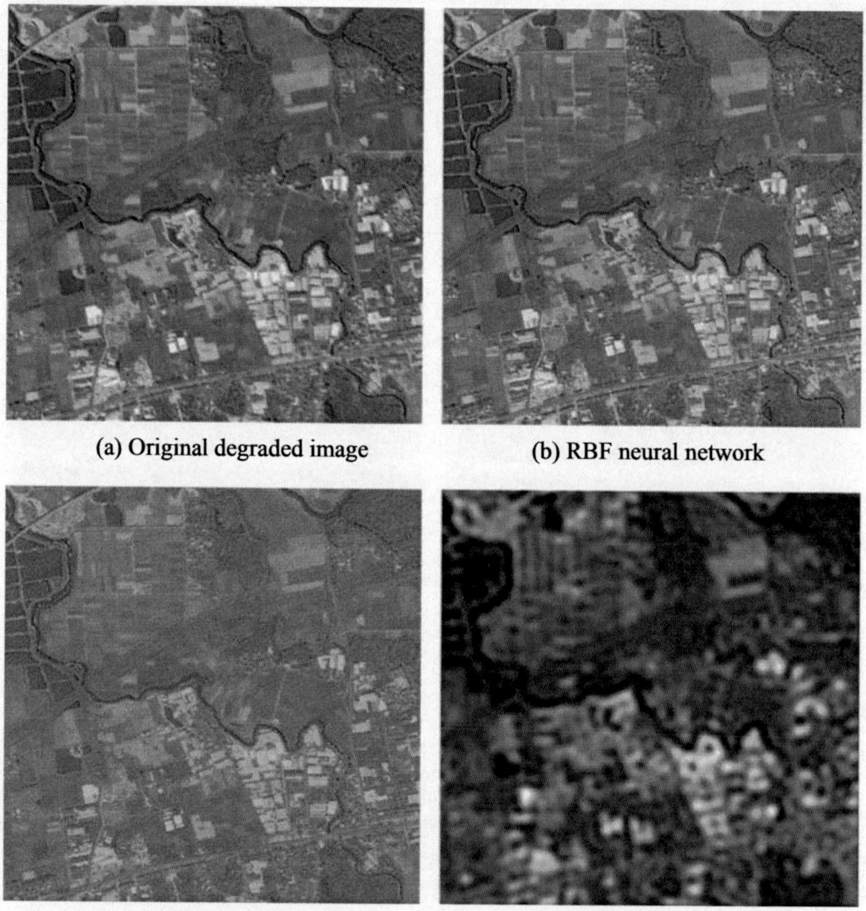

(a) Original degraded image (b) RBF neural network

(c) Conventional Wiener filter (d) Inverse filter

Fig. 6 Results of restoration using different recovery methods. **a** Original degraded image. **b** RBF neural network. **c** Conventional Wiener filter. **d** Inverse filter

5 Conclusions

This paper presents a novel remote sensing image restoration algorithm base on improved radial basis function neural network (RBFNN). Compared to traditional restoration methods, the proposed method has two main advantages: (I) The RBF neural network method could obtain more satisfactory restoration results when the motion-blurred direction and the blur extent are unknown; (II) To reduce the the operational time in practice, this paper proposes a conjugate gradient descent method to realize the weight parameters' iterative correction. And to further reduce calculating time, this method adopts matrix factorization algorithm to realize matrix's parallel arithmetic. Experiments results indicate that the improved RBF neural network model could acquire relatively approving remote sensing image restoration results and time overhead.

References

1. Jia T, Shi Y, Zhu Y, Wang L (2016) An image restoration model combining mixed L1/L2 fidelity terms. J Vis Commun Image R 38:461–473
2. Cai X, Chan R, Zeng T (2013) A two-stage images segmentation method using a convex variant of the Mumford-Shah model and thresholding. SIAM J Imag Sci 6:368–390
3. He L, Cui G, Feng H, Xu Z, Li Q, Chen Y (2015) The optimal code searching method with an improved criterion of coded exposure for remote sensing image restoration. Optics Commun 338:540–550
4. Ye Z (2017) Optical remote sensing motion degration simulation and image restoration technology, Zhejiang University
5. Wu X, Wu J, Zhang H (2011) Research on image restoration techniques based on inverse filtering algorithm, Information. Technology 35:183–185
6. Richardson WH (1972) Bayesian-based iterative method of image restoration. J Opt Soc Am 62:55–59
7. Lucy L (1974) An iterative technique for the rectification of observed distributions. Astron J 79:745–753
8. Khan M, Nizami IF, Majid M (2019) No-reference image quality assessment using gradient magnitude and wiener filtered wavelet features. Multimedia Tools Appl 78:85–14509
9. Levin A, Fergus R, Durand F et al (2007) Image and depth from a conventional camera with a coded aperture. ACM Trans Graphic 26:70
10. Krishnan D, Fergus R (2009) Fast image deconvolution using hyper-Laplacian priors. In: Proceedings of neural information processing systems (NIPS), pp 1033–1041
11. Wang Y, Mu W, Du X, Ma C, Shen X (2020) Remote sensing image on-board restoration based on adaptive wiener filter. Lecture Notes in Electrical Engineering 657:271–282
12. Liu L, Li S, Lai S (2019) Advance of neural network in degraded image restoration. J Graph 40:213–224
13. Sun J, Cao W, Xu Z et al (2015) Learning a convolutional neural network for non-uniform motion blur removal. In: Proceedings of the 2015 IEEE Conference on Computer Vision and Pattern Recognition. IEEE Press, New York, pp 769–777
14. Nah S, Kim TH, Lee KM (2017) Deep multi-scale convolutional neural network for dynamic scene deblurring. In: 2017 IEEE conference on computer vision and pattern recognition (CVPR). IEEE Press, New York, p 3

15. Zhou YT, Chellappa R (1988) Image restoration using a network. IEEE Trans Acoust Speech Signal Process 36(1988):1141–1151
16. Han Y, Wu L (2002) Image restoration using a modified Hopfield neural network of continuous state change. Signal Process 20:431–435
17. Perry SW, Guan L (2000) Weight assignment for adaptive image restoration by neural networks. IEEE Trans Neural Netw 11:156–170
18. Kupyn O, Budzan V, Mykhailych M et al (2018) DeblurGAN: blind motion deblurring using conditional adversarial networks. In: 2018 IEEE/CVF conference on computer vision and pattern recognition, pp 8183–8192
19. Tang J, Wang K, Zhang W et al (2020) A deep learning-based image restoration method in optical synthetic aperture imaging system. Acta Optica Sinica 40:2111001
20. Li Y, Xu Q, Li K (2020) New method of residual dense generative adversarial networks for image restoration. J Chin Comput Syst 41:830–836

Distributed HRWS MIMO-SAR Linear Vertical Baseline Estimation and Compensation

Pengcheng Li, Zaoyu Sun, Yongsheng Zhang, Feng He, Shenjing Wang, and Zhen Dong

Abstract The satellite formation configuration directly determines the functions of the distributed SAR system and has an important impact on system performance. The most practical formation configuration for realizing high resolution wide swath imaging is satellite formation along track. However, during the formation flight, the distributed array changes with time, so there is a linearly changing vertical baseline. The linearly changing vertical baseline will cause mismatch of the steering vector and cause the azimuth ambiguity resolution to fail. Based on the satellite formation configuration along track, this paper studies the vertical baseline variation and its influence on the ambiguity resolution along the track and converts the vertical baseline into a phase term for estimation and compensation. Through experimental analysis, by estimating and compensating the vertical baseline, the ambiguity caused by the mismatch of the steering vector caused by the linearly varying vertical baseline can be effectively solved.

Keywords Distributed synthetic aperture radar · Satellite formation · Vertical linear baseline · High-resolution wide-swath imaging · Multiple-Input Multiple-Output (MIMO)

1 Introduction

The multiple transmitting and receiving antennas of the distributed spaceborne MIMO-SAR system are respectively installed on small satellites arranged along track, and the transmitting and receiving antennas are separated from each other in the space position. The combination of multiple transmission and multiple reception and digital beam forming technology achieves high resolution and wide-swath Imaging. The satellite formation along track is not ideally arranged along track

P. Li · Z. Sun · Y. Zhang (✉) · F. He · S. Wang · Z. Dong
College of Electronic Science and Technology, National University of Defense Technology, Changsha, China
e-mail: zhangyongsheng@nudt.edu.cn

L. Wang et al. (eds.), *Proceedings of the 7th China High Resolution Earth Observation Conference (CHREOC 2020)*, Lecture Notes in Electrical Engineering 757, https://doi.org/10.1007/978-981-16-5735-1_33

during the flight. It has a baseline that changes linearly with time in the vertical track direction. The linear baseline in the direction of the vertical track will cause mismatch of the steering vector, resulting in the failure of azimuth ambiguity resolution.

For the estimation and compensation of the vertical baseline measurement error of the distributed SAR, An array auto-calibration technique [1] based on the space-time-frequency characteristics of echoes is proposed, which can correct the array errors with high precision and high efficiency. A new 3-D baseline error estimation method for distributed small satellites is proposed in this paper [2]. This method separates the two kinds of baselines through image registration in the field of the image and estimates the two kinds of baseline errors by image registration and subspace decomposition, respectively. Aiming to eliminate approximation error arising from the first-order Taylor series expansion of the position-error exponential function, some researchers have developed an improved array error estimation method [3] by compensating the position error estimates obtained at each iteration. A joint SAR imaging algorithm with Doppler ambiguity removal is proposed, which can be used for high resolution wide swath SAR imaging in low pulse repetition frequency (PRF) for distributed micro satellite constellation. The algorithm [4] is capable of correcting the vertical baseline effectively, not only for the range invariant component and range variant component, but also for the phase related to residual baseline in vertical combined with Doppler ambiguity removal. This letter [5] has described the parametric position error model and a novel calibration method for large aperture arrays in a GEO SAR system. Compared with the conventional method, the proposed algorithm can decrease the computational complexity and obtain the more accurate estimation of the position errors simultaneously.

The above research is based on the vertical baseline estimation and compensation based on the constant term of the array vertical baseline. The measurement error of the vertical baseline is considered, and the linear variation of the vertical baseline during satellite formation flight is not considered. In order to improve the accuracy of phase error estimation and improve the quality of synthetic aperture radar (SAR) image, this paper studies the law of array changes with time during satellite formation flight and the influence of linear vertical baseline on azimuth ambiguity based on satellite formation along track. It is of great significance in practical applications.

2 Signal Model

The small satellites in the satellite formation along track have the same ground track [6]. In order to obtain highly correlated radar echo data, a compact formation configuration is adopted. To prevent collisions between small satellites, the distance between satellites is generally set at a hundred meters level.

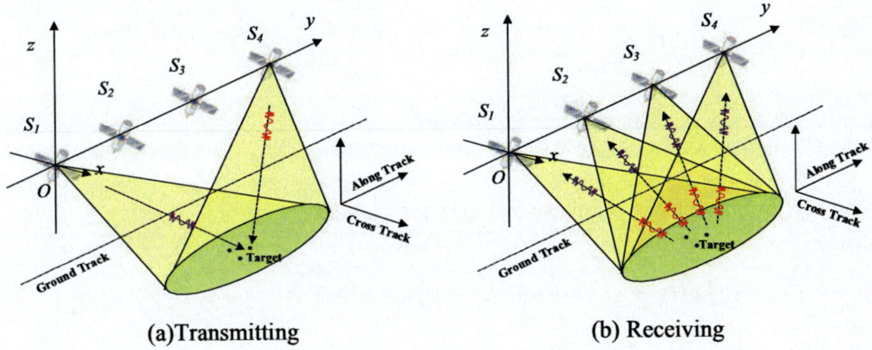

Fig. 1 Distributed SAR formation along track. **a** Transmitting, **b** Receiving

The multiple small satellites of the distributed MIMO-SAR system transmit and receive radar signals simultaneously, which can obtain more equivalent phase centers than the actual receiving and transmitting antennas, which improves the freedom of the system and helps to achieve high-resolution and wide-swath imaging. Take the "two-transmit and four-receive" as an example, that is, two satellites at the beginning and the end transmit radar signals, and four small satellites receive echo signals at the same time.

Figure 1 is the distributed SAR signal model of formation along track. Define the coordinate system as follows: y-axis is in the direction of the small satellite velocity vector; the x-axis is along the orbital angular momentum vector; the z-axis direction is from the center of the earth to the satellite. Suppose there are M small satellites along the course, S_1 is the reference satellite located at the origin of the coordinate, satellite S_1 and satellite S_4 transmit the short-time translational orthogonal waveform signals [7], and all small satellites receive the signals at the same time, and the echo signals are separated in the range direction [8], 8 equivalent phase centers can be obtained, and one of the equivalent phase centers coincides. The number of equivalent phase centers is 7. It is assumed that the separated echo signal has been obtained.

2.1 Ideal Signal Model

The position of the mth small satellite is $(0, d_m, 0)$ in the ideal condition. The equivalent phase center position is located at the center of the transmitting and receiving antenna, denoted by $(0, d_m/2, 0)$. The ground target position is (x_n, y_n, z_n). For the convenience of the following description, let $\widetilde{d_m} = d_m/2$.

The SAR echo signal received by the mth equivalent phase center can be expressed as

$$s_m(t) = \iiint_{(x,y,z)} \Delta(x,y,z) \cdot G_m\left(t - \frac{y - \widetilde{d_m}}{v_a}\right) \cdot p\left(\tau - \frac{2R_m(t)}{c}\right) \cdot \exp\left(-j\frac{4\pi R_m(t)}{\lambda}\right) dxdydz \quad (1)$$

where G represents the double-ways antenna pattern.

$$R_m(t) \approx \sqrt{\left(v_a \cdot t + \widetilde{d_m} - y\right)^2 + x^2 + z^2 + \frac{\widetilde{d_m}^2}{2\sqrt{x^2+z^2}}} \quad (2)$$

Further,

$$\begin{aligned} s_m(t) &= \iiint_{(x,y,z)} \exp\left(-j\frac{2\pi \widetilde{d_m}^2}{\lambda\sqrt{x^2+z^2}}\right) \cdot \Delta(x,y,z) \cdot G_m\left(t - \frac{y - \widetilde{d_m}}{v_a}\right) \\ &\quad \cdot p\left(\tau - \frac{2R_m(t)}{c}\right) \cdot \exp\left(-j\frac{4\pi R(t + \widetilde{d_m}/v_a)}{\lambda}\right) dxdydz \\ &= \exp\left(-j\frac{2\pi \widetilde{d_m}^2}{\lambda\sqrt{x^2+z^2}}\right) \cdot s\left(t + \widetilde{d_m}/v_a\right) \end{aligned} \quad (3)$$

2.2 Modified Signal Model

The simulation analysis of the formation along track reveals that the formation of small satellites that formed along track is not strictly arranged during the flight but has a vertical baseline. Moreover, the value of the distance between adjacent satellites is not fixed. Assuming that the equivalent phase centre of the mth small satellite under non-ideal conditions is $(\delta x_m(t), \widetilde{d_m} + \delta y_m(t), \delta z_m(t))$.

$$R_m(t) \approx \sqrt{\left(v_a t + \widetilde{d_m} + \delta y_m(t) - y\right)^2 + x^2 + z^2} - \frac{1}{2}\left(\frac{x \cdot \delta x_m(t)}{\sqrt{x^2+z^2}} + \frac{z \cdot \delta z_m(t)}{\sqrt{x^2+z^2}}\right) \quad (4)$$

where,

$$\begin{aligned} \delta r_m(t) &= -\frac{1}{2}\left(\frac{x \cdot \delta x_m(t)}{\sqrt{x^2+z^2}} + \frac{z \cdot \delta z_m(t)}{\sqrt{x^2+z^2}}\right) \\ &= -\frac{1}{2}\left(\delta x_m(t) \cdot \sin\theta + \delta z_m(t) \cdot \cos\theta\right) \end{aligned} \quad (5)$$

where θ indicates the incident angle.

The vertical baseline is equivalent to the phase term expressed as follows:

$$\begin{aligned}\varphi_m(t) &= -\frac{4\pi\delta r_m(t)}{\lambda} \\ &= 2\pi\left(\frac{\delta x_m(t)}{\lambda}\sin\theta + \frac{\delta z_m(t)}{\lambda}\cos\theta\right)\end{aligned} \tag{6}$$

The corrected echo signal of the mth equivalent phase center is expressed as follows:

$$Z_m(t) = \exp(j\varphi_m(t)) \cdot s\left(t + \widetilde{d_m}\Big/v_a\right) \tag{7}$$

Transform formula (7) into the range Doppler domain, we have

$$Z_m(f_a) = \exp(j\varphi_m) \cdot \sum_{k=0}^{K-1} S(f_a + kf_s) \cdot \exp(j2\pi(f_a + kf_s) \cdot t_m) \tag{8}$$

where,

$$t_m = \widetilde{d_m}\Big/v_a \tag{9}$$

The echo signal matrix form of the equivalent phase center is

$$\mathbf{Z}(f_a) = \mathbf{T}(f_a)\mathbf{S}(f_a) \tag{10}$$

where **T** is the steering vector matrix.

3 Analysis of Formation Along Track

Taking four satellite formations as an example for analysis, that is, M = 4, the satellite orbit parameters are shown in Table 1. The satellite orbit parameters are only slightly different in the ascension of the mean anomaly at epoch and longitude of the ascending node, and other orbital parameters are the same.

In an orbital period, the vertical baseline variation trend between adjacent satellites is the same. According to the satellite orbit parameters shown in Table 1, take satellite 1 (as a reference satellite) and satellite 2 as examples for simulation analysis.

It can be seen from Figs. 2 and 3 that the vertical baseline changes linearly during a cycle of formation. The vertical baseline in the x-axis direction changes linearly within the range of $[-4 \times 10^{-4}, 4 \times 10^{-4}]$, and the vertical baseline in the

Table 1 Satellite formation orbital parameters along track

Parameters	Satellite 1	Satellite 2	Satellite 3	Satellite 4
a [m]	7,119,030	7,119,030	7,119,030	7,119,030
e	2.010e − 05	2.010e − 05	2.010e − 05	2.010e − 05
i [°]	97	97	97	97
Ω [deg]	0	1.1061e − 4	2.2121e − 4	3.1812e − 4
ω [deg]	0	0	0	0
M [deg]	0	−0.0016	−0.0032	−0.0048

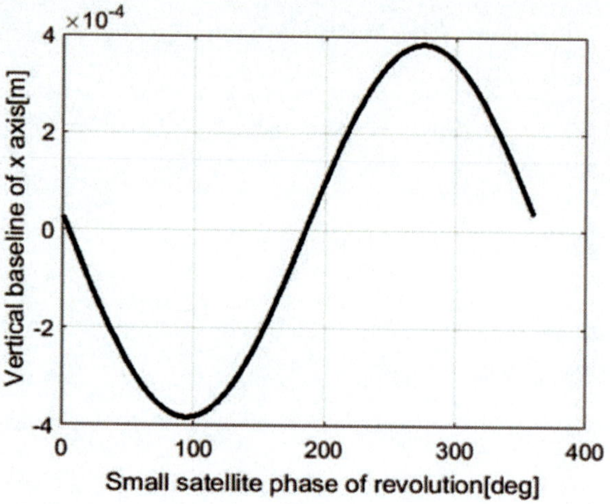

Fig. 2 Vertical baseline in the x-axis direction

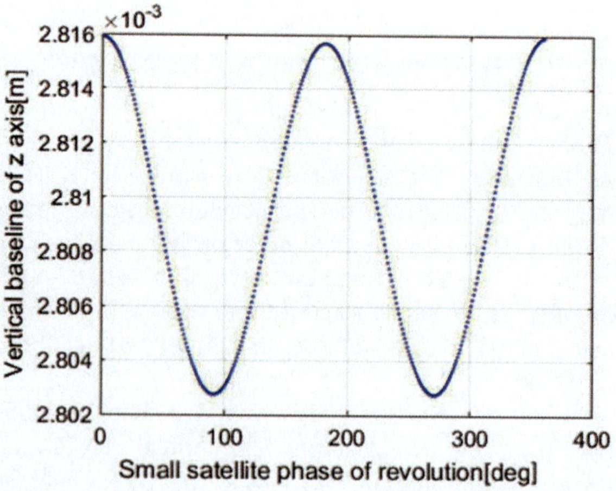

Fig. 3 Vertical baseline in the z-axis direction

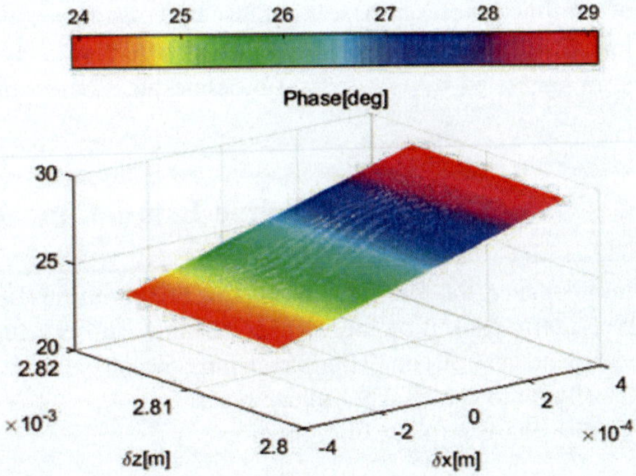

Fig. 4 Phase term generated by the vertical baseline

z-axis direction changes linearly within the range of $[-2.802 \times 10^{-3}, 2.816 \times 10^{-3}]$. Next, we will study the impact of the linear vertical baseline on the system reconstruction performance, and use azimuth ambiguity to signal ratio (AASR) [9] to evaluate the signal reconstruction effect. Based on Eq. (6), the simulation analysis of the influence of the vertical baseline in the x-axis direction and the vertical baseline in the z-axis direction on the phase term.

It can be seen from Fig. 4 that the phase term changes linearly in the x-axis direction, and basically does not change in the z-axis direction. Although the vertical baseline length in the z-axis direction is an order of magnitude larger than the vertical baseline length in the x-axis direction. Therefore, the linear vertical baseline in the x-axis direction makes a major contribution to the phase. The phase term produced by the linear vertical baseline has an effect on AASR as shown in Fig. 5. We can see that AASR has dropped by nearly 2 dB. Therefore, the influence

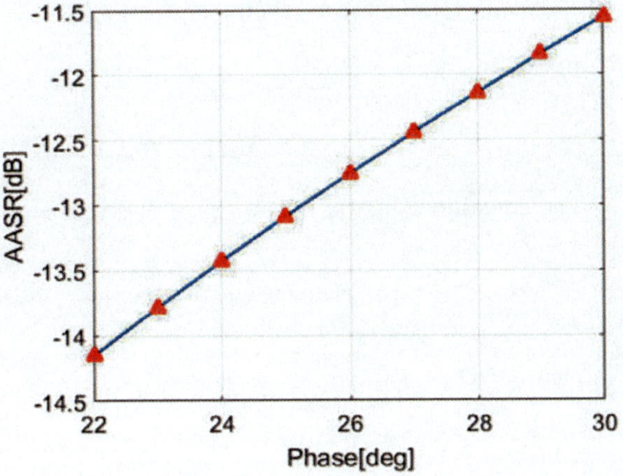

Fig. 5 The relationship between the phase term and AASR

of the linear vertical baseline must be considered, although the magnitude of the linear vertical baseline is 10^{-4}. Therefore, before signal reconstruction, the phase term generated by the vertical baseline must be estimated and compensated.

4 Linear Vertical Baseline Estimation Method

In this paper, the Orthogonal Subspace algorithm [10] is used to estimate the linear phase term caused by the vertical baseline, and the digital beam forming is used for signal reconstruction. This paper puts the phase term caused by the linear vertical baseline into the channel phase error.

The phase error can be given as

$$\xi_m = e^{j(\varphi_m(t) + \phi_m)} \tag{11}$$

ϕ_m is the phase constant term, $\varphi_m(t)$ is the linear phase converted from the linear vertical baseline. The estimation of the linear vertical baseline can be transformed into the phase term for estimation, that is, the steering vector $\xi\mathbf{T}(f_a)$ with the linear vertical baseline is estimated, and the modified signal model can be obtained as

$$\mathbf{Z}(f_a) = \xi\mathbf{T}(f_a)\mathbf{S}(f_a) \tag{12}$$

The reconstruction filter is given as follows:

$$\mathbf{W}(f_a) = \mathbf{T}'(f_a)\left(\mathbf{T}'(f_a)^H\mathbf{T}'(f_a)\right)^{-1} \tag{13}$$

where

$$\mathbf{T}'(f_a) = \xi\mathbf{T}(f_a) = [\,\xi\mathbf{t}f_a \quad \xi\mathbf{t}(f_a + f_s) \quad \cdots \quad \xi\mathbf{t}(f_a + (k-1)f_s)\,] \tag{14}$$

We have

$$\hat{\mathbf{S}}(f_a) = \mathbf{W}(f_a)^H\mathbf{Z}(f_a) \tag{15}$$

The covariance matrix of the signal is given as follows

$$\mathbf{R}(f_a) = \sum_{i=1}^{2M-1} \xi_i \mathbf{v}_i \mathbf{v}_i^H \tag{16}$$

where ξ_i and $\mathbf{v}_i$ are the eigenvectors corresponding to the eigenvalues of $R(f_a)$, and $\xi_1 \geq \xi_2 \geq \ldots \geq \xi_{2M-1}$.

The signal subspace feature vector is $\mathrm{V}_S = [\mathrm{v}_1 \quad \mathrm{v}_2 \quad \ldots \quad \mathrm{v}_K]$. The noise subspace feature vector is $\mathbf{V}_N = [\mathbf{v}_{K+1} \quad \mathbf{v}_{K+2} \quad \ldots \quad \mathbf{v}_{2M-1}]$, Both are orthogonal.

Define a cost function as follows

$$J = \mathbf{\Lambda}^H \mathbf{R}(f_a)\mathbf{\Lambda} \tag{17}$$

where

$$\mathbf{\Lambda} = [\xi_1 \quad \xi_2 \quad \cdots \quad \xi_{2M-1}]^T \tag{18}$$

$$\mathbf{R}(f_a) = \sum_{k=1}^{K} \sum_{n=K+1}^{2M-1} \mathbf{T}_k^{\mathbf{H}}(f_a)\left(\mathbf{v}_n \mathbf{v}_n^{\mathbf{H}}\right)\mathbf{T}_k(f_a) \tag{19}$$

$$\mathbf{T}_k(f_a) = \mathrm{diag}(\xi \mathbf{t}(f_a + (k-1)f_s)) \tag{20}$$

The objective function to estimate the phase error is given as follows

$$\begin{cases} \min\limits_{\widehat{\Lambda}} J \\ s.t\, \widehat{\Lambda}^{\mathbf{H}} g_1 = 1 \end{cases} \tag{21}$$

where g_1 is the normalization factor. Calculate the objective function to obtain the estimated value $\widehat{\Lambda}$ of the phase error and compensate.

5 Measured Data Analysis and Verification

The "two-transmit and four-receive" mode of distributed four satellites has 7 separate equivalent phase centers. In theory, the system can solve 7 times ambiguity.

This paper uses the RADARSAT-1 satellite data in the Vancouver area for analysis and verification. The original data PRF is 1256.98 Hz, the Doppler bandwidth is 706.2 Hz, and the azimuth is not ambiguous.

In order to simulate the signals received by multiple satellites are ambiguity, the original data is down-sampled 7 times, which is equivalent to the PRF of the signal transmitted by each small satellite is 179.57 Hz. The phase term caused by the linear vertical baseline is added to the data obtained by down-sampling, and the influence of the linear vertical baseline on the SAR image is introduced. Since the linear vertical baseline causes the phase term to have the space-variant characteristic, the phase term estimated value $\widehat{\Lambda}$ is compensated in blocks in the distance dimension Table 2 [2].

Table 2 Radar main parameters

Parameters	Value
Carrier frequency [GHz]	5.3
Wavelength [m]	0.05657
Incident angle [deg]	40.15
Doppler center frequency [Hz]	−6900
Sensor velocity [m/s]	7062
Number of satellites	4
PRF [Hz]	1256.98
Doppler bandwidth [Hz]	706.2
Range sampling rate [MHz]	32.317
Distance between adjacent satellites [m]	123.6

The following gives the unambiguous azimuth spectrum of the original data as shown in Fig. 6a. The azimuth spectrum of the signal after down-sampling 7 times for each satellite is shown in Fig. 6b. Here, digital beamforming technology is used to reconstruct the down-sampled data, and the reconstructed spectrogram is shown in Fig. 6c. Under ideal conditions, the signal spectrum is well reconstructed and the ambiguity is resolved. When the influence of the linear vertical baseline is added to the down-sampled data, the signal spectrogram is very different from the original image spectrogram as shown in Fig. 6d. The reconstruction algorithm fails, and the spectrum of the original data cannot be reconstructed. Then the phase term caused by the linear vertical baseline is estimated and compensated, and the reconstructed azimuth spectrum is shown in Fig. 6e. We can see that the reconstructed spectrum is consistent with the original data spectrum. Therefore, the compensation of the linear vertical baseline is very important for unambiguous imaging.

Then use the Chirp Scaling algorithm to image the processed data. The following figures are some of the characteristic areas of Vancouver. Figure 7a is the imaging of raw data. Figure 7b is the image after down-sampling by 7 times, the marked part has obvious azimuth ambiguity. Figure 7c is the image after adding the influence of the vertical baseline. It can be seen from the figure that there is ambiguity in the azimuth. This is because the linear vertical baseline causes the steering vector mismatch and the reconstruction filter fails. Figure 7d is the image after reconstruction of the signal after the phase term estimation and compensation. It can be seen from the figure that the phase term caused by the linear vertical baseline is well estimated and compensated, and the image without ambiguity is obtained.

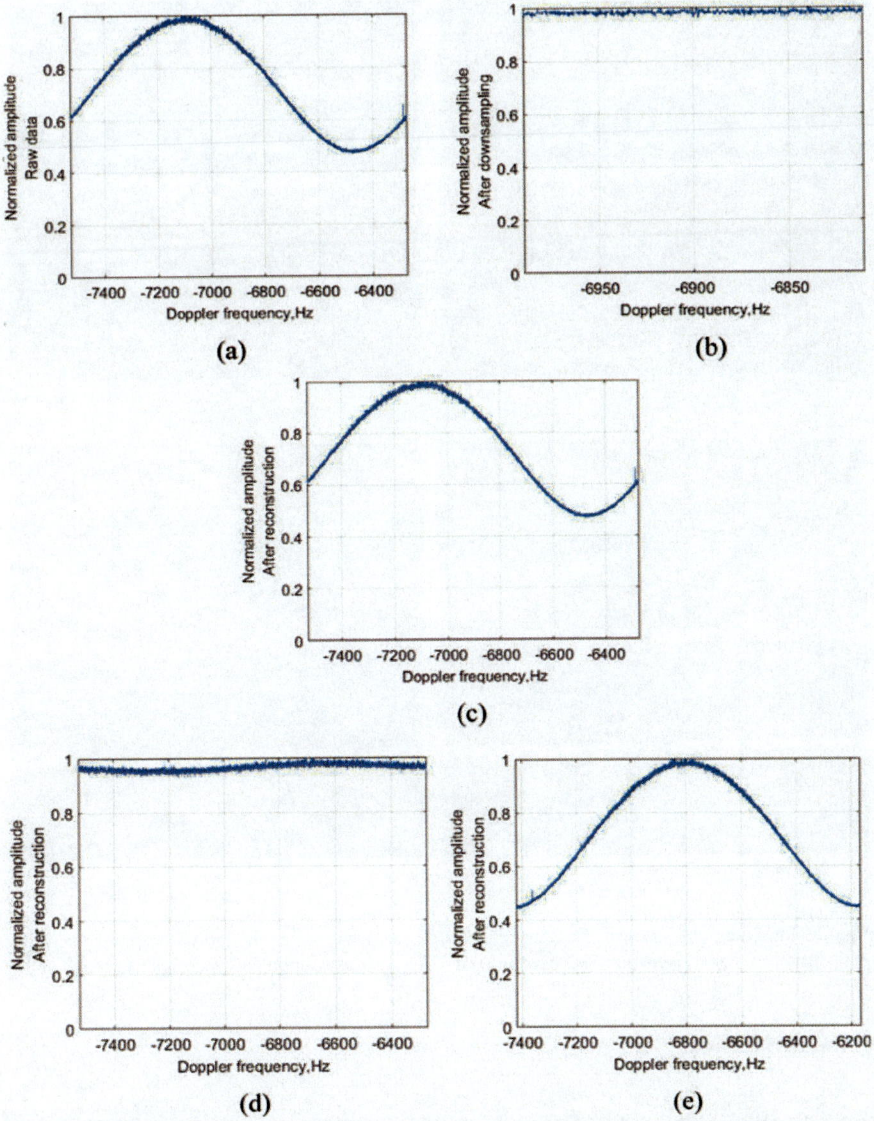

Fig. 6 Radar data spectrogram. **a** Raw data. **b** Down-sampling. **c** After reconstruction (no linear vertical baseline). **d** After reconstruction (with linear vertical baseline). **e** After reconstruction (after compensation)

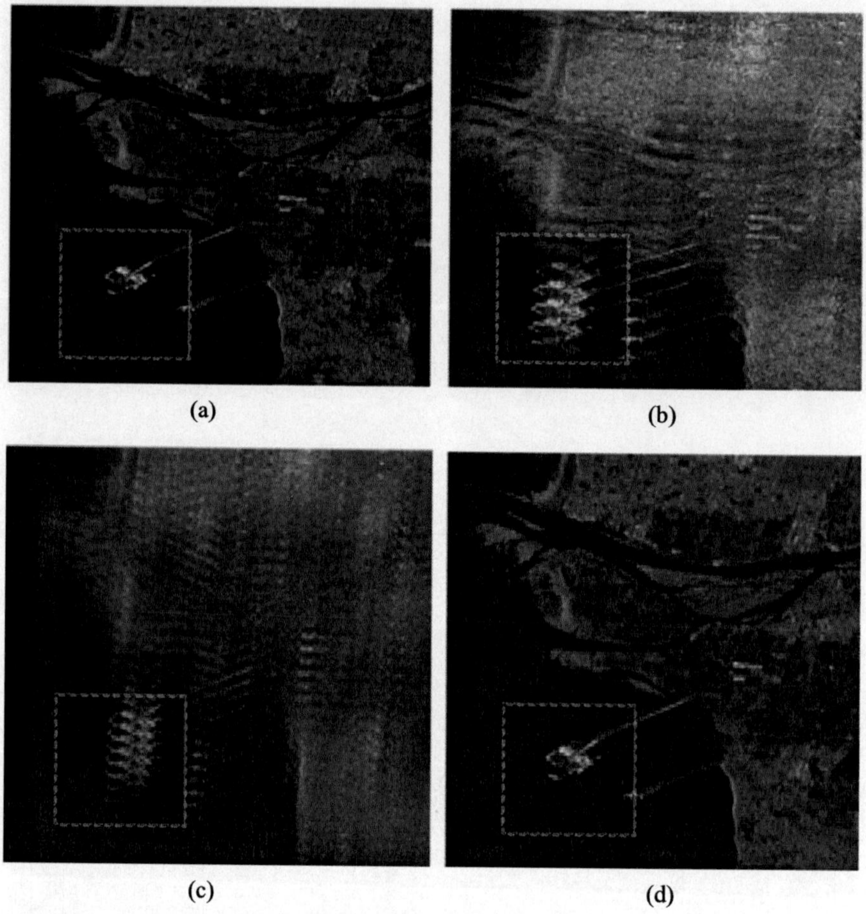

Fig. 7 Imaging with Chirp Scaling algorithm. **a** Raw data. **b** Downsampling. **c** After reconstruction (with linear vertical baseline). **d** After reconstruction (after compensation)

6 Conclusion

Based on the configuration of satellite formation along the course, this paper studies the law of vertical baseline change and its influence on ambiguity resolution. This paper puts the phase term caused by the linear vertical baseline into the channel phase error, and estimates and compensates the overall phase error term. The actual measurement data is used for analysis and verification, which effectively solves the azimuth ambiguity caused by the mismatch of the steering vector caused by the linear vertical baseline.

Acknowledgements This work was supported by the National Natural Science Foundation of China under Grant 61771478.

References

1. Li Z, Wang H, Bao Z et al (2004) Approach to array error estimation for distributed micro-satellite SAR systems. J Syst Eng Electron 26(9):1159–1161
2. Xu Q, Liao G, Liu Y et al (2008) 3-D baseline error estimation method for distributed small satellites. J J Xidian Univ 35(4):668–672
3. Liu A, Liao G, Xu Q et al (2011) An improved array-error estimation method for constellation SAR systems. J IEEE Geosci Remote Sens Lett 9(1):90–94
4. Zhang L, Su T, Liu Z et al (2015) Joint Doppler ambiguity removal SAR imaging algorithm for distributed micro-satellite. J Syst Eng Electron 37(3):544–552
5. Huo L, Liao G, Yang Z et al (2018) An efficient calibration algorithm for large aperture array position errors in a GEO SAR. J IEEE Geosci Remote Sens Lett 9:1–5
6. Xi X, Wang W, Gao Y (2003) Fundamentals of near-earth spacecraft orbit. National University of Defense Technology Press, Changsha, pp 254–256
7. Krieger G, Younis M, Huber S et al (2012) MIMO-SAR and the orthogonality confusion. In: International geoscience and remote sensing symposium, pp 1533–1536
8. Wang S, Sun Y, He F, Sun Z, Li P, Dong Z (2020) DBF processing in range-doppler domain for MWE SAR waveform separation based on digital array-fed reflector antenna. Remote Sens 12(19):3161
9. Ma X, Sun Z, Dong Z et al (2014) Signal reconstruction performance analysis of azimuth multi-channel SAR. J Electron Inf Tech 36(3):545–551
10. Li Z, Bao Z, Wang H et al (2006) Performance improvement for constellation SAR using signal processing technique. J IEEE Trans 42:436–452

Fast Three-Dimensional Imaging Method of Aerospace Targets Based on Linear Array Radar

Zhang Yi, Yu Anxi, Jin Guanghu, and Dong Zhen

Abstract Distributed Array Radar (DAR) uses space diversity to obtain better detection performance. As a new configuration of DAR, Linear Array Radar (LAR) can achieve a higher frame rate and higher dimensional imaging of aerospace targets. Based on the real-virtual aperture hybrid mode, LAR obtain the three-dimensional image of the target through space–time dimensions. The imaging process using the three-dimensional Back Projection (BP) algorithm accurately reconstruct the three-dimensional image of the target, but the amount of calculation is huge. In order to solve this problem, the Fast Factorized Back Projection (FFBP) algorithm is extended to three-dimensional space. And in the Cartesian coordinate system, a detailed theoretical derivation of the subaperture coarse image interpolation method is made. Through the imaging experiment and imaging quality evaluation of ideal point targets, it is verified that the fast imaging algorithm has good imaging quality. It has only a small performance loss compared with the traditional three-dimensional BP algorithm. The imaging processing time is also greatly reduced. In addition, this paper conducts imaging experiments based on satellite model simulation data generated by Radarbase. The results show that the algorithm can reconstruct the geometric configuration of the target well, which verifies the effectiveness of this algorithm.

Keywords Distributed array radar · Linear array radar · Aerospace targets · Three-dimensional imaging · Fast factorized back projection algorithm

Z. Yi · Y. Anxi · J. Guanghu (✉) · D. Zhen
College of Electronic Science and Technology, National University of Defense Technology, Changsha, Hunan 410073, P. R. China
e-mail: guanghujin@nudt.edu.cn

L. Wang et al. (eds.), *Proceedings of the 7th China High Resolution Earth Observation Conference (CHREOC 2020)*, Lecture Notes in Electrical Engineering 757, https://doi.org/10.1007/978-981-16-5735-1_34

1 Introduction

Distributed Array Radar (DAR) [1–3] can use space diversity to obtain better detection performance. As a new configuration of distributed array radar, Linear Array Radar (LAR) can achieve higher frame rate and higher dimensional imaging of aerospace targets. It can provide more favorable support for target classification and recognition. LAR is based on the real-virtual aperture hybrid mode, which can obtain the three-dimensional image of the target through space-time dimensions. Because the distribution of aerospace targets in space is objectively three-dimensional, obtaining a three-dimensional image of the target is crucial for accurately depicting the geometric configuration of the target and accurately identifying the target category and model.

In array radar imaging, the Back Projection (BP) [4] algorithm is a widely used time-domain imaging algorithm. It has no approximation in principle and is an accurate imaging algorithm. The algorithm is suitable for multiple imaging modes of any array configuration. The classic BP algorithm finds the radar echo amplitude corresponding to the imaging area on the profile obtained by each element. When the number of grid divisions in the imaging area is large, the calculation amount of the BP algorithm will be very large. To meet the requirements of imaging resolution, the number of grids is usually large. When the BP algorithm is extended to three dimensions, the impact of this computational burden on the efficiency of the algorithm will be more obvious.

In order to solve the problem of the large amount of calculation of the BP algorithm, a series of improved BP algorithms was proposed. Yegulalp [5] first proposed a Fast Back Projection (FBP) algorithm. The algorithm is based on the principle of local approximation and divides the full aperture into multiple subapertures. In the subaperture polar coordinate system, each subaperture uses the BP algorithm to image the target separately, and then coherently accumulates to obtain the final image. On the basis of FBP, Ulander [6] processed the subaperture coarse image merging process hierarchically, and proposed a Fast Factorized Back Projection (FFBP) algorithm [7–9]. This algorithm reduces the number of interpolation operations in the image merging process, and further accelerates the imaging processing speed of the BP algorithm. The FFBP algorithm has been widely recognized because of its excellent imaging quality and good imaging efficiency. In recent years, related research has been conducted. Yang Zemin proposed a fast back projection algorithm based on polar coordinate format. In view of the low efficiency of the BP algorithm when the FFBP algorithm is used to obtain the coarse image in the subaperture, the Polar Format Algorithm (PFA) was proposed to obtain the coarse subaperture image. The algorithm achieved higher efficiency in the curve track SAR imaging experiment.

To achieve efficient three-dimensional imaging of aerospace targets, this paper applies the local approximation principle and subaperture classification recursion to the three-dimensional BP algorithm based on LAR. Combined with the LAR observation geometry, coarse image interpolation fusion is realized in the rectangular coordinate system. The simulation experiments are carried out on the point target simulation echo data and Radarbase-based satellite model simulation data to verify the effectiveness and efficiency of the algorithm.

2 Three-Dimensional BP Imaging Algorithm Based on LAR

An point target is taken as an example to illustrate the principle of LAR's three-dimensional BP imaging of aerospace targets. The observation geometry of LAR three-dimensional imaging is shown in Fig. 1.

To obtain a high resolution in the range, the radar transmit signal is a Linear Frequency Modulation (LFM) signal.

$$p(t_f) = rect\left(\frac{t_f}{T_p}\right) \exp\left(j\pi K_r t_f^2\right) \tag{1}$$

where, t_f is the fast time, T_p is the pulse width, and K_r is the frequency of the LFM signal.

Assume that the position of LAR element antenna phase center is p_i, $i = 1, 2, \cdots, M$. The cumulative time of the observation target T is t_s. Then the signal model of the target echo is

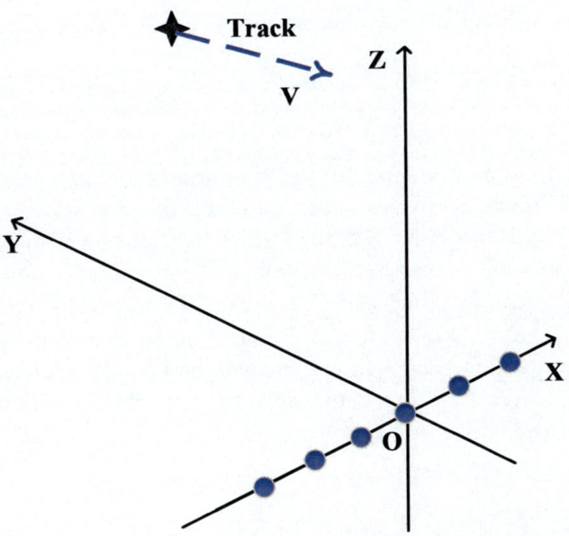

Fig. 1 Observation geometry of LAR three-dimensional imaging

$$s(t_s, p_i, t_f) = rect\left(\frac{t_f - \frac{2R(t_s, p_i)}{c}}{T_p}\right) \exp\left(j\pi K_r \left(t_f - \frac{2R(t_s, p_i)}{c}\right)^2\right) \exp\left(-j2\pi f_c \frac{2R(t_s, p_i)}{c}\right) \tag{2}$$

where, t_s is the slow time, $R(t_s, p_i)$ is the range between array element p_i and target T at time t_s, f_c is the signal center frequency, and c is the speed of light.

Operate range pulse compression on the echo signal, and the matched filter function of the system is

$$s_r(t_f) = p * (-t_f) = rect\left(\frac{t_f}{T_p}\right) \exp\left(-j\pi K_r t_f^2\right) \tag{3}$$

To facilitate calculation, matched filtering is generally performed in the frequency domain. The signal after matched filtering is

$$s(t_s, p_i, t_f) = A \sin c\left[B\left(t_f - \frac{2R(t_s, p_i)}{c}\right)\right] \exp\left(-j2\pi f_c \frac{2R(t_s, p_i)}{c}\right) \tag{4}$$

where, the $\sin c$ function is $\sin c(a) = \frac{\sin(\pi a)}{\pi a}$, A is the amplitude of the target position signal after pulse range compression, and B is the signal bandwidth.

Perform phase compensation on the signal after pulse range compression. The compensation phase is

$$Ph = \exp\left(j2\pi f_c \frac{2R(t_s, p_i)}{c}\right) \tag{5}$$

Then the array element profile can be expressed as

$$P(t_s, p_i, t_f) = A \sin c\left[B\left(t_f - \frac{2R(t_s, p_i)}{c}\right)\right] \tag{6}$$

At a slow time, for different array element antenna phase center positions, the distance between the target and the antenna phase center is known. The processed signal model is the profile, which is a function of fast time, and the function reaches its peak at the target position.

The imaging area G is divided into a number of grids $g_j(x_{g_j}, y_{g_j}, z_{g_j}) j = 1, 2, \cdots N$, $N = a \times b \times c$. a, b, and c are the number of grids divided into the imaging area in the real aperture, azimuth and distance directions, respectively.

At a slow time t_s, the range from the phase center of the array element antenna to the grid is

$$R(p_i, g_j) = \left\| x_{g_j} - x_{p_i}, y_{g_j} - y_{p_i}, z_{g_j} - z_{p_i} \right\|_2 \tag{7}$$

The corresponding radar echo delay is expressed as

$$\tau = \frac{2R(p_i, g_j)}{c} \tag{8}$$

Interpolate the profile of the array element at that time to obtain the echo value of the array element as

$$s(g_j) = \int \left(\sum_{i=1}^{M} S(t_s, p_i, \tau) \right) dt_s \tag{9}$$

Operate the above processing on all grids in the entire imaging area to obtain the three-dimensional image. It can be observed from (9) that LAR obtains a three-dimensional image of the target through space-time dimensions observation.

3 Three-Dimensional FFBP Algorithm in Rectangular Coordinate System

3.1 Local Approximation Principle of Three-Dimensional FFBP Algorithm

As shown in Fig. 2, suppose there are two adjacent grids $g_1(x_{g_1}, y_{g_1}, z_{g_1})$, $g_2(x_{g_2}, y_{g_2}, z_{g_2})$, and two adjacent array elements $p_1(x_{p_1}, y_{p_1}, z_{p_1})$,$p_2(x_{p_2}, y_{p_2}, z_{p_2})$.

$$\begin{cases} x_{g_2} = x_{g_1} + \Delta x_g \\ y_{g_2} = y_{g_1} + \Delta y_g \\ z_{g_2} = z_{g_1} + \Delta z_g \end{cases} \tag{10}$$

$$\begin{cases} x_{p_2} = x_{p_1} + \Delta x_p \\ y_{p_2} = y_{p_1} + \Delta y_p \\ z_{p_2} = z_{p_1} + \Delta z_p \end{cases} \tag{11}$$

According to (7), we can obtain

$$R(p_1, g_1) = \left\| x_{g_1} - x_{p_1}, y_{g_1} - y_{p_1}, z_{g_1} - z_{p_1} \right\|_2 \tag{12}$$

$$R(p_2, g_2) = \left\| x_{g_2} - x_{p_2}, y_{g_2} - y_{p_2}, z_{g_2} - z_{p_2} \right\|_2 \tag{13}$$

Fig. 2 Local approximation principle

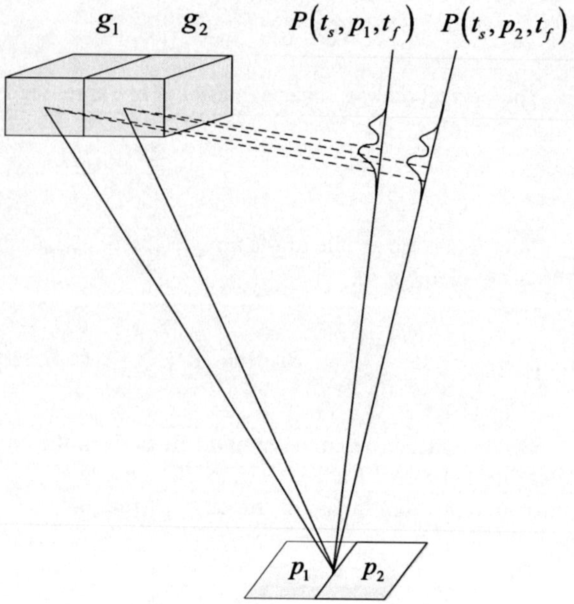

Then,

$$R(p_2, g_2) = \left\| (x_{g_1} - x_{p_1}) - (\Delta x_g - \Delta x_p), (y_{g_1} - y_{p_1}) - (\Delta y_g - \Delta y_p), (z_{g_1} - z_{p_1}) - (\Delta z_g - \Delta z_p) \right\|_2 \tag{14}$$

The selected grid and array elements are in close positions, and the spacing is much smaller than the distance resolution. Thus,

$$R(p_1, g_1) \approx R(p_2, g_2) \tag{15}$$

According to (8), we can obtain

$$\tau_{11} \approx \tau_{22} \tag{16}$$

When the array element is in the adjacent position, its profile is approximately equal.

$$P(t_s, p_1, t_f) \approx P(t_s, p_2, t_f) \tag{17}$$

In addition, we determine that $\tau_{11} \approx \tau_{22}$, so

$$P(t_s, p_1, \tau_{11}) \approx P(t_s, p_2, \tau_{22}) \tag{18}$$

The echo values of adjacent elements in the adjacent grid are approximately equal. According to this principle, the adjacent array elements are divided into subapertures, and the adjacent grids are merged to reduce the number of grids. First, the coarse image is generated by the subaperture, and the fine image is obtained by interpolation of the coarse image.

3.2 *Interpolation of Three-Dimensional FFBP Algorithm in Cartesian Coordinate System*

3.2.1 Array Element Profile Interpolation to Generate Coarse Subaperture Images

A rectangular coordinate system $X - Y - Z$ is established in the imaging area and the grid is divided. The profile of the array element generated by the pulse range compression is distributed along the phase center of the array element and the direction of the target line of sight. Therefore, it is necessary to project the grid coordinates onto the line of sight of the array element, and obtain the grid echo value through interpolation (Fig. 3).

Point the X axis of the grid coordinate system $X - Y - Z$ to the line of sight of the array element through coordinate system transformation. The projection of the grid points $T(x_t, y_t, z_t)$ on the line of sight of the array element can be regarded as coordinate x_t'' on axis X'' after the coordinate system transformation. The coordinate system transformation is divided into two steps (Fig. 4).

First, a certain angle γ is rotated with the Z axis as the rotation axis. At this time, the Z axis coordinate does not change, and the coordinate in the $X - Y$ plane changes to the $X' - Y'$ plane. The coordinates in the $X - Y$ plane are expressed in polar coordinates (ρ, θ) as

$$\begin{cases} x = \rho \cos\theta \\ y = \rho \sin\theta \end{cases} \tag{19}$$

After the coordinate system $X - Y$ is rotated by a certain angle γ to the coordinate system $X' - Y'$, it can be expressed in polar coordinates $(\rho, \theta - \gamma)$ as

$$\begin{cases} x' = \rho\cos(\theta - \gamma) = \rho\cos(\theta)\cos(\gamma) + \rho\sin(\theta)\sin(\gamma) \\ y' = \rho\sin(\theta - \gamma) = \rho\sin(\theta)\cos(\gamma) - \rho\cos(\theta)\sin(\gamma) \end{cases} \tag{20}$$

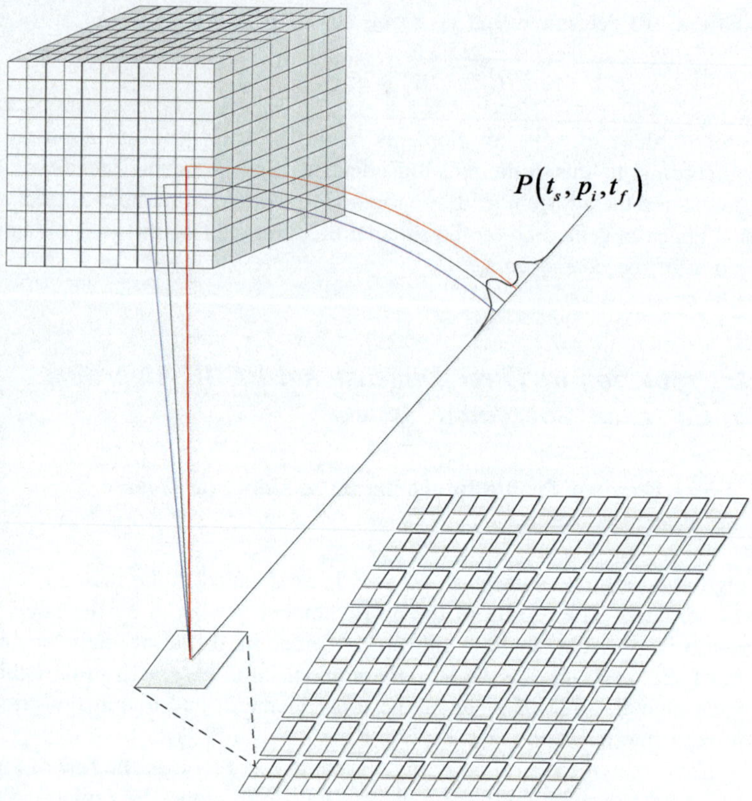

Fig. 3 Profile interpolation

Substituting (19) into (20), we obtain

$$\begin{cases} x' = x\cos(\gamma) + y\sin(\gamma) \\ y' = y\cos(\gamma) - x\sin(\gamma) \end{cases} \tag{21}$$

In addition

$$z' = z \tag{22}$$

Then the transformation relationship from coordinate system $X - Y - Z$ to coordinate system $X' - Y' - Z'$ can be expressed as

$$\begin{bmatrix} x'_t \\ y'_t \\ z'_t \end{bmatrix} = \begin{bmatrix} \cos\gamma & \sin\gamma & 0 \\ -\sin\gamma & \cos\gamma & 0 \\ 0 & 0 & 1 \end{bmatrix} \begin{bmatrix} x_t \\ y_t \\ z_t \end{bmatrix} \tag{23}$$

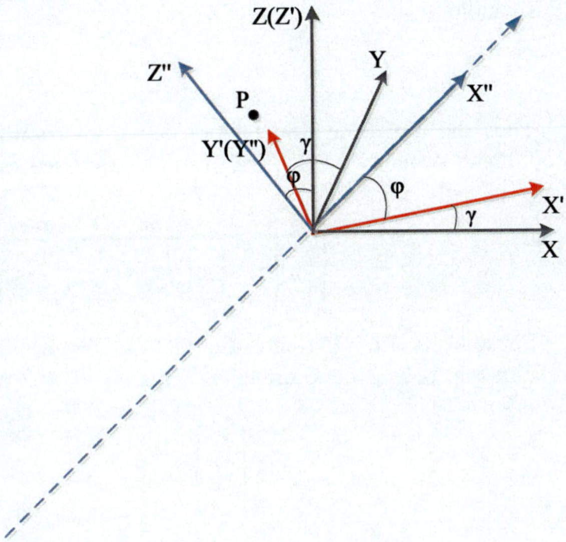

Fig. 4 Coordinate system transformation diagram

where, t is the time.

Second, transform from the $X' - Y' - Z'$ coordinate system to the $X'' - Y'' - Z''$ coordinate system, and rotate a certain angle φ with the Y' axis as the rotation axis. At this time, the Y' axis coordinate does not change, and the coordinate in the $X' - Z'$ plane changes to the $X'' - Z''$ plane. The coordinates in the $X' - Z'$ plane are expressed in polar coordinates (ρ, θ) as

$$\begin{cases} x' = \rho \cos \theta \\ z' = \rho \sin \theta \end{cases} \tag{24}$$

After the coordinate system $X' - Z'$ is rotated by a certain angle φ to the coordinate system $X'' - Z''$, it can be expressed in polar coordinates $(\rho, \theta - \varphi)$ as

$$\begin{cases} x' = \rho \cos(\theta - \varphi) = \rho \cos \theta \cos \varphi + \rho \sin \theta \sin \varphi \\ z' = \rho \sin(\theta - \varphi) = \rho \sin \theta \cos \varphi - \rho \cos \theta \sin \varphi \end{cases} \tag{25}$$

Substituting (24) into (25), we obtain

$$\begin{cases} x'' = x' \cos \varphi + z' \sin \varphi \\ z'' = z' \cos \varphi - x' \sin \varphi \end{cases} \tag{26}$$

In addition

$$y'' = y' \tag{27}$$

Then the transformation relationship from coordinate system $X' - Y' - Z'$ to coordinate system $X'' - Y'' - Z''$ can be expressed as

$$\begin{bmatrix} x_t'' \\ y_t'' \\ z_t'' \end{bmatrix} = \begin{bmatrix} \cos\varphi & 0 & \sin\varphi \\ 0 & 1 & 0 \\ -\sin\varphi & 0 & \cos\varphi \end{bmatrix} \begin{bmatrix} x_t' \\ y_t' \\ z_t' \end{bmatrix} \tag{28}$$

Substituting Eq. (23) into Eq. (28) to obtain the coordinate transformation relationship between coordinate systems $X - Y - Z$ and $X'' - Y'' - Z''$, we obtain

$$\begin{aligned} \begin{bmatrix} x_t'' \\ y_t'' \\ z_t'' \end{bmatrix} &= \begin{bmatrix} \cos\varphi & 0 & \sin\varphi \\ 0 & 1 & 0 \\ -\sin\varphi & 0 & \cos\varphi \end{bmatrix} \begin{bmatrix} x_t' \\ y_t' \\ z_t' \end{bmatrix} \\ &= \begin{bmatrix} \cos\varphi & 0 & \sin\varphi \\ 0 & 1 & 0 \\ -\sin\varphi & 0 & \cos\varphi \end{bmatrix} \begin{bmatrix} \cos\gamma & \sin\gamma & 0 \\ -\sin\gamma & \cos\gamma & 0 \\ 0 & 0 & 1 \end{bmatrix} \begin{bmatrix} x_t \\ y_t \\ z_t \end{bmatrix} \\ &= \begin{bmatrix} \cos\varphi\cos\gamma & \cos\varphi\sin\gamma & \sin\varphi \\ -\sin\gamma & \cos\gamma & 0 \\ -\sin\varphi\cos\gamma & -\sin\varphi\sin\gamma & \cos\varphi \end{bmatrix} \begin{bmatrix} x_t \\ y_t \\ z_t \end{bmatrix} \end{aligned} \tag{29}$$

The projection x_t'' of the sampling point $T(x_t, y_t, z_t)$ in the sampling direction can be expressed as

$$x_t'' = \cos\varphi\cos\gamma x_t + \cos\varphi\sin\gamma y_t + \sin\varphi z_t \tag{30}$$

3.2.2 Subaperture Coarse Image Interpolation to Generate Fine Image

Suppose the coordinate of the upper level grid is $T_f(x_f, y_f, z_f)$, and the projection on the line of sight of the array element is

$$x_f'' = \cos\varphi\cos\gamma x_f + \cos\varphi\sin\gamma y_f + \sin\varphi z_f \tag{31}$$

For the next grid point $T(x_t, y_t, z_t)$, (x_t, y_t) is known. When T has the same projection as T_f

$$\cos\varphi\cos\gamma x_t + \cos\varphi\sin\gamma y_t + \sin\varphi z_t = \cos\varphi\cos\gamma x_f + \cos\varphi\sin\gamma y_f + \sin\varphi z_f \tag{32}$$

Then

$$z_f = z_t + \cot\varphi\cos\gamma\left(x_t - x_f\right) + \cot\varphi\sin\gamma\left(y_t - y_f\right) \tag{33}$$

After the coordinates of the next level grid points are obtained, interpolation can be performed on the coarse image to obtain the echo values of the next level grid points.

3.3 Analysis of Calculation Amount of Three-Dimensional BP and FFBP Algorithm

The LAR-based three-dimensional BP imaging algorithm and the three-dimensional FFBP imaging algorithm are analyzed separately. The most basic operations of the algorithm can be divided into complex addition operations and complex multiplication operations. The complexity of complex multiplication operations is much higher than that of complex addition operations. For three-dimensional BP and three-dimensional FFBP algorithms, the amount of calculation is mainly concentrated in the interpolation processing part using complex multiplication operations. Therefore, when using interpolation methods with the same precision and the same principle, the number of interpolation processing required by the algorithm can objectively reflect the amount of calculation.

For ease of analysis, suppose the number of array elements is M, and the number of space–time dimensions array elements formed by the relative motion between LAR and the target is $M \times M$, $t_s \times PRF = M$. The subaperture size is $m \times m$. The observation duration is t_s, and the pulse repetition frequency is PRF. The grid of the imaging area is divided into $N \times N \times N$.

For the classic three-dimensional BP imaging algorithm, the number of interpolations required for imaging = the number of elements × image pixels is

$$M \times M \times N \times N \times N = M^2N^3 \tag{34}$$

The three-dimensional FFBP imaging algorithm needs to be discussed in two steps. First, for the subaperture process of generating coarse images, the interpolation times of subaperture coarse images = number of aperture elements × coarse image pixels, as

$$m \times m \times \frac{N}{M/m} \times \frac{N}{M/m} \times N = \frac{m^4N^3}{M^2} \tag{35}$$

Every two subaperture coarse images were merged as a group. Interpolation times of subimage fusion = number of subapertures × of new level image pixels, as

$$\left(\frac{M}{m}\right)^2 \times \frac{2N}{M/m} \times \frac{2N}{M/m} \times N = 4N^3 \tag{36}$$

We repeat the interpolation and fusion process to obtain the final full-resolution image. The number of fusions is $\log_2\left(\frac{M}{m}\right)$. Then the total FFBP imaging interpolation times are expressed as

$$\frac{m^4 N^3}{M^2} + 4N^3 \times \log_2\left(\frac{M}{m}\right) = \left[\frac{m^4}{M^2} + 4\log_2\left(\frac{M}{m}\right)\right] N^3 \tag{37}$$

To visually compare the calculation amount of the two algorithms, it is assumed that the imaging area grid is divided into 512 × 512 × 512. When the subaperture size is 2 × 2, the calculation amount of the algorithm changes with the number of LAR elements as shown in Fig. 5. Evidently, as the number of elements increases, the calculation amount of the three-dimensional BP algorithm increases rapidly, while the calculation amount of the three-dimensional FFBP algorithm increases steadily and slowly and is much lower than the calculation amount of the three-dimensional BP algorithm.

When the number of LAR elements remains unchanged and the size of the subaperture changes, the calculation curve of the algorithm is as shown in Fig. 6.

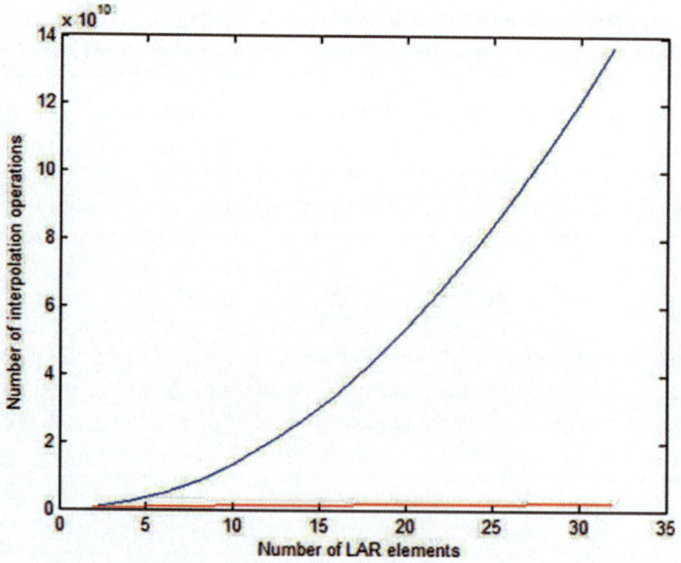

Fig. 5 BP and FFBP algorithm calculation amount changes with number of LAR array elements

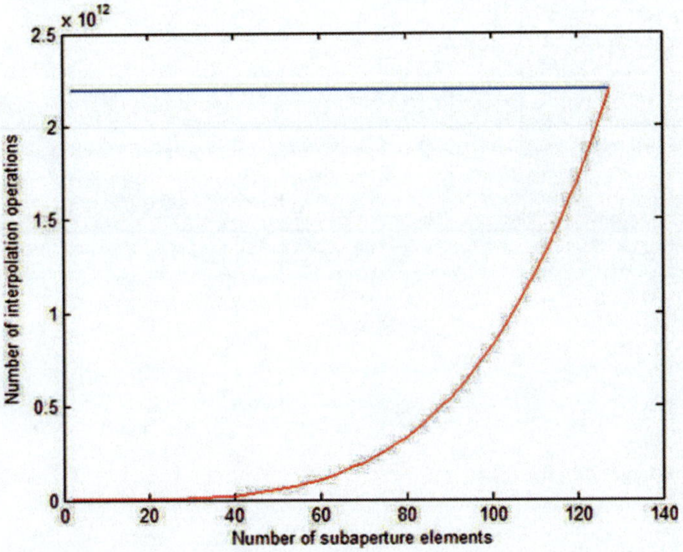

Fig. 6 BP and FFBP algorithm calculation of volume change curve with subaperture size

Table 1 Point target imaging simulation experiment parameter settings

System parameters	Value	System parameters	Value
Array length [m]	50	Signal frequency modulation rate [GHz/s]	2000
Number of elements	16	Pulse repetition frequency [Hz]	16
Fusion base	2	Sampling rate [MHz]	125
Target vertical array velocity component [m/s]	50	Scene size [m]	20 × 20 × 20
Synthetic aperture time [s]	1	Grid size [m]	0.25 × 0.25 × 0.25
Pulse width [μs]	5	Theoretical resolution of real aperture [m]	1.3941
Signal bandwidth [MHz]	100	Theoretical resolution of azimuth [m]	1.3941
Center frequency [GHz]	16	Theoretical range resolution [m]	1.3281

It can be observed that the smaller the number of elements constituting the subaperture, the better the acceleration performance of the FFBP algorithm. The number of layers of image fusion also increases. When the number of subaperture elements is the number of full aperture elements, the FFBP algorithm is equivalent to the BP algorithm.

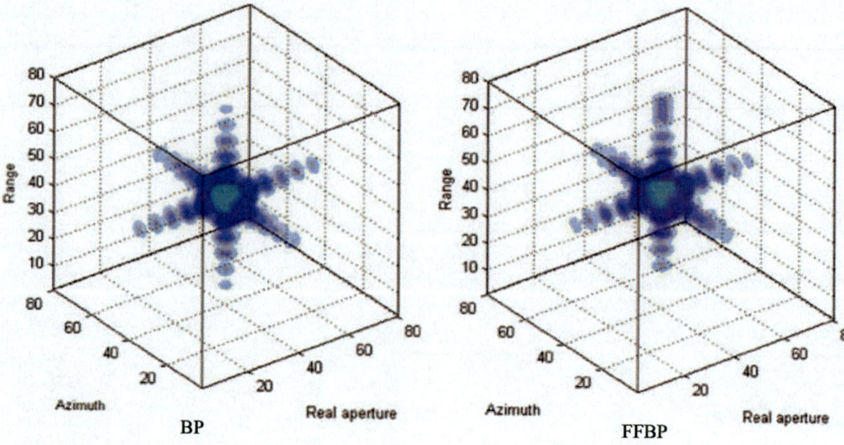

Fig. 7 Point target imaging result

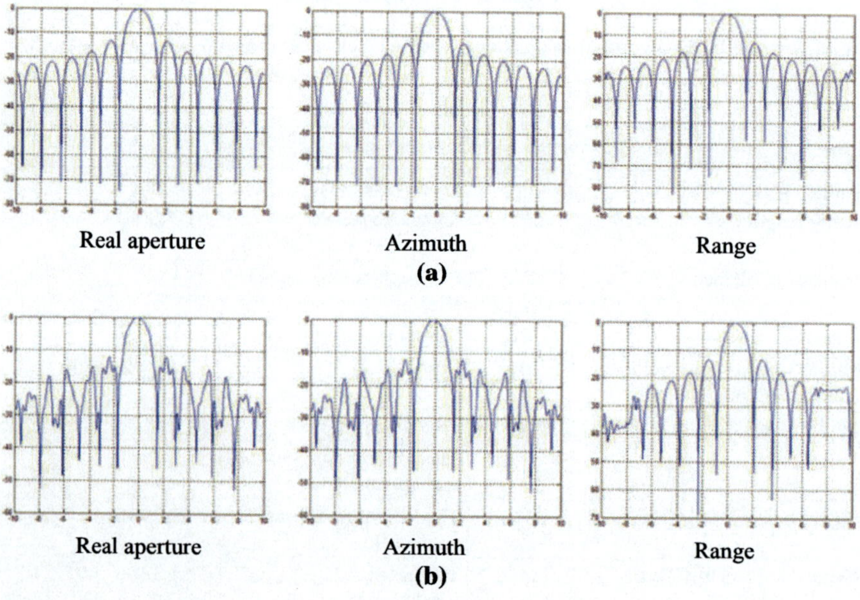

Fig. 8 Point target profile

Table 2 Point target simulation performance analysis

	Resolution [m]			Resolution [dB]			Time[s]	Speedup ratio
	Real aperture	Azimuth	Range	Real aperture	Azimuth	Range		
BP	1.4014	1.4014	1.3242	−13.15	−13.15	−13.26	120.1937	6.6
FFBP	1.4443	1.4443	1.3252	−12.02	−12.02	−13.24	18.2158	

4 Simulation Experiment

To verify the effectiveness and efficiency of the algorithm, simulation experiments were first performed on the ideal point target. The key experimental parameter settings are listed in Table 1.

The imaging result of the point target is shown in Fig. 7.

We perform the imaging performance analysis through the point target profile (Fig. 8).

The main performance indicators are listed in Table 2.

It is observed that the three-dimensional FFBP imaging algorithm has only a small loss in image quality, and the image quality loss is mainly caused by interpolation errors. However, its imaging speed is significantly accelerated, and the imaging time of the traditional three-dimensional BP algorithm is approximately 6.6 times that of the three-dimensional FFBP imaging algorithm.

Table 3 Satellite model imaging simulation experiment parameter setting

System parameters	Value	System parameters	Value
Number of elements in synthesized two-dimensional array	64 × 64	Meshing	232 × 256 × 256
Fusion base	2	Theoretical resolution of real aperture [m]	0.1411
Signal bandwidth [GHz]	0.58	Theoretical resolution of azimuth [m]	0.1411
Center frequency [GHz]	9.785	Theoretical range resolution [m]	0.2584
Grid size [m]	0.06 × 0.06 × 0.064		

Satellite model

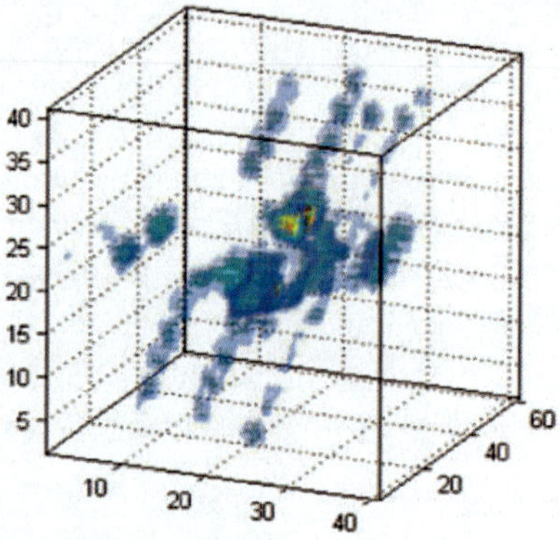

Three-dimensional BP algorithm imaging results

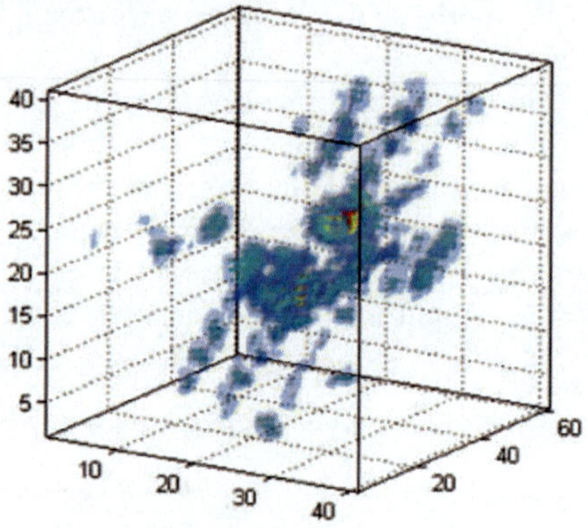

Three-dimensional FFBP algorithm imaging results

Fig. 9 Satellite model simulation data imaging results

To better verify the effectiveness and efficiency of the algorithm, this paper also uses satellite model simulation echo data based on Radarbase for imaging. The key experimental parameter settings are listed in Table 3.

The three-dimensional FFBP imaging result is shown in the Fig. 9. Further, the result is compared with the imaging results of the three-dimensional BP algorithm.

It can be observe from a comparison with the three-dimensional BP imaging algorithm that the imaging quality of the three-dimensional FFBP algorithm does not significantly decrease. Compared with the imaging time, the traditional three-dimensional BP imaging algorithm consumes 40,323.16 s, whereas the three-dimensional FFBP imaging algorithm consumes 278.86 s. The speedup ratio reaches 144.6. Evidently, the imaging processing speed of the three-dimensional FFBP algorithm is faster, and it is a more efficient three-dimensional imaging algorithm for LAR aerospace targets.

5 Summary

This paper explained the algorithm principle of the LAR system for three-dimensional BP imaging of aerospace targets through space-time two-dimensional observation and highlighted the problem of excessive calculation. To extend the FFBP algorithm to three dimensions, the local approximation principle of the three-dimensional FFBP algorithm was analyzed in detail, and a calculation method of three-dimensional interpolation and fusion in a rectangular coordinate system was proposed. Three-dimensional FFBP imaging based on LAR was realized. Through experiments on point targets and satellite model echo simulation data based on Radarbase, it was verified that the three-dimensional FFBP imaging algorithm based on LAR has high imaging accuracy for aerospace targets and a good reconstruction effect on the geometric configuration of the target. Compared with the classic three-dimensional BP imaging algorithm, the amount of calculation is significantly reduced. This algorithm has a high practicability.

References

1. Wu M-Q (2006) The development of digital array radar in China. In: International conference on radar, Shanghai, China, pp 5–9
2. Clem PG, Rodriguez M, Voigt JA, Ashley CS (2001) US Patent 6,231,666
3. Cantrell B (2002) Development of a digital array radar (DAR). IEEE Aerospace Electron Syst Magazine 17(3):22–27
4. Wang Y-D, Xi P, Xue W (2010) Fast filter back projection algorithm based on hexagonal grid. In: International conference on electrical and control engineering, Wuhan, China, pp 964–969
5. Yegulalp AF (1999) Fast backprojection algorithm for synthetic aperture radar. In: The Record of the 1999 IEEE Radar Conference, Waltham, Massachusetts, pp 60–65
6. Ulander LMH, Hellsten H, Stenstrom G (2003) Synthetic-aperture radar processing using fast factorized back-projection. IEEE Trans Aerospace Electron Syst 39(3):760–776
7. Vu VT, Sjogren TK, Pettersson MI (2012) SAR imaging inground plane using fast backprojection for mono- and bistatic cases. In: 2012 IEEE Radar Conference, Atlanta, USA, pp 184–189
8. Durand R, Ginolhac G, Thirion-Lefevre L et al (2011) Back projection version of subspace detector SAR processors. IEEE Trans Aerospace Electron Syst 47(2):1489–1497
9. Albuquerque M, Prats P, Scheiber R (2008) Applications of time-domain back-projection SAR processing in the airborne case. In: 7th European conference on synthetic aperture radar, Friedrichshafen, Germany, pp 1–4

Edge Detection of SAR Images Based on Shearlet

Zengguo Sun, Guodong Zhao, Weirong Chen, Robertas Damaševičius, and Marcin Woźniak

Abstract Synthetic aperture radar (SAR) has the characteristics of all-weather imaging and is widely used in various fields. However, owe to coherent imaging mechanism, speckle appears in SAR images, which seriously affects the interpretation of images. Due to merits of the strong directional sensitivity and the optimal sparsity, shearlet is used to construct an edge detector for SAR images. The detector firstly converts the constructed even symmetric shearlet into odd symmetry by Hilbert transform, then determines the main direction of the edge according to the odd symmetric shearlet coefficient of the SAR image, and calculates the possibility of edge existence according to both odd and even symmetric shearlet coefficient. Experiments show that, in comparison with the traditional difference-based edge detectors, the proposed edge detector has stronger ability of anti-speckle interference, and can better detect the edge information in SAR images.

Keywords Synthetic aperture radar images · Speckle · Edge detection · Shearlet

Z. Sun
Key Laboratory of Modern Teaching Technology, Ministry of Education, Xi'an, China

Z. Sun · G. Zhao
School of Computer Science, Shaanxi Normal University, Xi'an, China

W. Chen
China Centre for Resources Satellite Data and Application, Beijing, China

R. Damaševičius (✉)
Department of Applied Informatics, Vytautas Magnus University, Kaunas, Lithuania
e-mail: robertas.damasevicius@vdu.lt

M. Woźniak
Faculty of Applied Mathematics, Silesian University of Technology,
Kaszubska 23, 44-100 Gliwice, Poland

L. Wang et al. (eds.), *Proceedings of the 7th China High Resolution Earth Observation Conference (CHREOC 2020)*, Lecture Notes in Electrical Engineering 757, https://doi.org/10.1007/978-981-16-5735-1_35

1 Introduction

This work is supported by the National Natural Science Foundation of China (no. 61102163), the Fundamental Research Funds for the Central Universities (no. GK201903085), and the Key Laboratory of Land Satellite Remote Sensing Application Center, Ministry of Natural Resources of the People's Republic of China (no. KLSMNR-202004).

Synthetic aperture radar (SAR) is a high-resolution coherent imaging radar [1], which can work at any time and under all weather conditions. It is widely used in agriculture, disaster monitoring, geology, ocean monitoring, and other fields. Since the basic structure of object is generally outlined by its edge, the edge detection is of great significance to practical applications of SAR images [2].

Due to the coherent imaging mechanism, speckle appears in SAR images, which degrades the image quality seriously [3]. So the traditional difference-based edge detector cannot be applied to the edge detection of SAR images. Traditional edge detectors are based on difference, such as Prewitt, Sobel, and Log detectors [4, 5]. They treat the gray differentiation as a representation of the edges and perform edge detection by constructing a differential operator that is sensitive to pixel gray level changes. Although this method can detect the edge information of the image, it cannot overcome the speckle effect of the SAR image [6, 7], and it will detect the false edge mixed with the real edge.

In recent years, multi-scale geometric analysis (MGA) tools have been widely used in the field of image processing [8]. Among them, shearlet is the latest development of multi-scale geometric analysis with strong directional sensitivity and the optimal sparsity, so it can capture edge information more efficiently [9]. Therefore, the edge detection of SAR images using shearlet can better suppress the influence of speckle and improve the detection accuracy. In this paper, an edge detector based on shearlet is proposed, and its efficiency is demonstrated by edge detecting experiments on real SAR images.

2 Shearlet Introduction

Easley proposed a new multi-scale geometric analysis method based on the traditional affine system theory: shearlet. Shearlet are constructed by a special form of affine system with synthetic expansion. When dimension $n = 2$, the affine system with compound expansion is defined as [10–12]

$$\Psi_{AB}(\psi) = \left\{\psi_{j,l,k}(x) = |\det A|^{j/2}\psi\left(B^l A^j x - k\right) : j, l \in \mathbf{Z},\ k \in \mathbf{Z}^2\right\} \quad (1)$$

where $\psi \in L^2\left(\mathbf{R}^2\right)$, j, l, and k are scale, shear and translation parameters, A and B are invertible matrices, $|\det B| = 1$, A^j is a scale transformed matrix, B^l is associated with a geometric transformation that maintains an area constant. When $A = A_0 = \begin{pmatrix} 4 & 0 \\ 0 & 2 \end{pmatrix}$ is the anisotropic expansion matrix and $B = B_0 = \begin{pmatrix} 1 & 1 \\ 0 & 1 \end{pmatrix}$ is the shear matrix, the equation above becomes shearlet.

In current MGA tools, shearlet is demonstrated to have strong directional sensitivity and the optimal sparsity, so it is very appropriate for capturing edge information accurately and completely [10, 12]. Due to such merits, an edge detector is constructed for SAR images based on shearlet.

3 Edge Detection Based on Shearlet

Based on the characteristics of multi-scale, directionality and sparsity of shearlet [13], we firstly perform shearlet transformation on SAR images to obtain the transformed coefficients, and then use different shearlet coefficients to obtain edge information by the relationship between edge position and direction. Thse image processed by the shearlet transform is subjected to morphological processing to obtain a final edge detection image.

3.1 Shearlet Transformation

According to (1), the symmetrical shearlet is constructed by using the tensor product of Mexican straw hat wavelet and Gaussian wavelet [14], and the SAR image is processed by shearlet transform to obtain the shearlet coefficient of SAR image under even symmetric shearlet. Then, a complex shearing system is created by modifying the frequency band structure of the shearlet, and it produces local geometric information and some desirable features. In order to ensure that the real part of the constructed function is even symmetric (such as the cosine function) and the imaginary part is odd symmetric (such as sine function), the Hilbert transform [15, 16] is used to convert the even symmetric function into odd symmetric function. Let ψ^{even} be an even symmetric shearlet and be the real part of the function, then the complex shearlet is defined as follows

$$\psi = \psi^{even} + i\psi^{odd} := \psi^{even} + iH(\psi^{even}) \tag{2}$$

where ψ^{odd} represents an odd symmetric shearlet and the function $H(\cdot)$ represents the Hilbert transform [17]. The transformation formula is given by

$$H(f)(t) = \lim_{a->\infty} \int_{-a}^{a} \frac{f(\tau)}{t-\tau} d\tau \tag{3}$$

The expressions of the even symmetric shearlet and the odd symmetric shearlet are given as follows:

$$\psi_{j,k,x}^{even}(\cdot) = 2^{\frac{3j}{2}}\psi^{even}(S_k A_j(\cdot - x)) \tag{4}$$

$$\psi_{j,k,x}^{odd}(\cdot) = 2^{\frac{3j}{2}}\psi^{odd}(S_k A_j(\cdot - x)) \tag{5}$$

where $j \in \mathbb{N}_0$, $|k| < \left\lceil 2^{\frac{j}{2}} \right\rceil$ and $x \in \mathbb{Z}^2$ are scale, shear and translation parameters, $S_k = \begin{pmatrix} 1 & k \\ 0 & 1 \end{pmatrix}$ is the shear matrix and $A_j = \begin{pmatrix} 2^j & 0 \\ 0 & 2^{j/2} \end{pmatrix}$ is the anisotropic expansion matrix.

3.2 *Optimal Direction and Edge Existence Possibility Calculation*

A simple and intuitive way to calculate the likelihood of the presence of edges in a SAR image is to maximize all possible directions [18]. However, experiments have shown that a more satisfactory result can be achieved by pre-selecting the main direction of the edge. The primary direction can be determined by the maximum coefficient associated with the odd symmetric shearlet. That is to say, for each point in the two-dimensional plane, we will select the value with the largest absolute value among all the coefficients related to the odd symmetric shearlet, and select the corresponding shearlet direction as the main direction. Finally, based on the direction selected, we calculate the possibility of the edge of the pixel using the even symmetric shearlet coefficients and the odd symmetric shearlet coefficients obtained in the previous step [14].

First, we calculate the odd symmetric shearlet coefficients at each pixel according to (5). For the scale parameter $j \in \{J_{\min}, \ldots, J_{\max}\}$, the shear parameter $|k| \leq \left\lceil 2^{\frac{j}{2}} \right\rceil$ and the translation parameter $x \in \mathbb{R}^2$, the shearlet direction is taken as the optimal direction for the edge by selecting the maximum of the absolute values of all the coefficients. The optimal direction $k^*_{\psi^{even},\psi^{odd}}(f,x) \in \left\{ -\left\lceil 2^{\frac{J_{\min}}{2}} \right\rceil, \ldots, \left\lceil 2^{\frac{J_{\min}}{2}} \right\rceil \right\}$ of the edge at each pixel point $x \in \mathbb{R}^2$ is calculated by

$$k^*_{\psi^{even},\psi^{odd}}(f,x) = \underset{\tilde{k} \in \left\{ -\left\lceil 2^{\frac{J_{\min}}{2}} \right\rceil, \ldots, \left\lceil 2^{\frac{J_{\min}}{2}} \right\rceil \right\}}{\arg\max} \; \max_{j \in \{J_{\min}, \ldots, J_{\max}\},\, k \in \left\{ -\left\lceil 2^{\frac{j}{2}} \right\rceil, \ldots, \left\lceil 2^{\frac{j}{2}} \right\rceil \right\},\, \left| \tilde{k} - k \left\lceil 2^{\frac{J_{\min}}{2}} \right\rceil \left\lceil 2^{\frac{j}{2}} \right\rceil^{-1} \right| \leq \frac{1}{2}} \left| \left\langle f, \psi_{j,k,x}^{odd} \right\rangle \right| \tag{6}$$

The probability of the existence of an edge at this pixel is calculated by

$$\tilde{E}_{\psi_{even},\psi_{eodd}}(f,x) = \frac{\left|\sum_{j=J_{\min}}^{J_{\max}} \langle f, \psi_{j,k_j,x}^{odd}\rangle\right| - \sum_{j=J_{\min}}^{j=J_{\max}} \left|\langle f, \psi_{j,k_j,x}^{even}\rangle\right|}{(J_{\max} - J_{\min} + 1) \max_{j\in\{J_{\min},\dots,J_{\max}\}} \left|\langle f, \psi_{j,k_j,x}^{odd}\rangle\right| + \varepsilon} \tag{7}$$

where f is the image matrix, $S_k = \begin{pmatrix} 1 & k \\ 0 & 1 \end{pmatrix}$ is the shear matrix, and $A_j = \begin{pmatrix} 2^j & 0 \\ 0 & 2^{j/2} \end{pmatrix}$ is the anisotropic expansion matrix. ε ($\varepsilon > 0$) in the above equation is the minimum value approaching zero to ensure that the denominator will not be zero. In order to make the edge measurement probability value of each pixel point between 0 and 1, we finally set the following mapping relationship

$$E_{\psi_{even},\psi_{odd}}(f,x) = \max(\tilde{E}_{\psi_{even},\psi_{odd}}(f,x), 0) \tag{8}$$

It produces a value between 0 and 1 that is used to indicate the likelihood of the existence of an "edge" at the location. That is to say, a metric of 0 means that the position has no edge, and 1 means that there is definitely an edge.

3.3 *Morphological Operation*

Use of shearlet can ensure the detection accuracy of the edge, but usually the edge detected is thicker, and the false edges caused by speckle also appear. In this section, mathematical morphology [19] is used to further optimize the results of edge detection by shearlet, including removing false edges and edge refinement operations [20, 21].

3.3.1 Removing False Edges

A binary edge image is usually obtained from Sect. 3.2. It is found that there are false edges caused by speckle in the binary image. Experiments have found that the number of pixels in the connected region of the false edge is smaller than that of the normal edge, which can be eliminated by an appropriate method.

We firstly mark the connected domain in the binary edge image obtained [22]. This means that, in an image composed of white pixels (usually represented by "255" in the grayscale image and by "1" in the binary image) and black pixels (usually represented by "0"), the pixel sets having pixel values "1" or "255" adjacent to each other are marked with different numbers [23]. And the number of white pixels is counted in different connected domains.

In SAR images, the false edges caused by speckle usually corresponds to small connected areas. The threshold is set, and the portion where the number of pixels in each connected region of the image is less than the threshold is assumed to be a false edge. The pixel values in these connected regions are all set to 0 to achieve the purpose of eliminating false edges.

3.3.2 Edge Refinement

The purpose of the refinement is to keep the width of the edge one pixel wide [24] and to remove some pixels along the edges of image until the width of edge reaches a single pixel. The result of refinement is called the skeleton of the original image, and the skeleton has the same topological shape and connectivity as the original image. We use the edge refinement operations based on mathematical morphology to refine edges. Hit-or-miss transformation is the theoretical basis for the edge refinement based on mathematical morphology [25]. Its basic idea is to extract all the pixels in the image that match the given neighborhood structure.

Let $f(x, y)$ be the input image, $b(x, y)$ be the structural element, and $b(x, y)$ be composed of two disjoint parts $b_1(x, y)$ and $b_2(x, y)$, the intersection of $b_1(x, y)$ and $b_2(x, y)$ is an empty set, and the union is the structural element $b(x, y)$, so the definition of $f(x, y)$ hit by $b(x, y)$ is given as follows.

$$f * b(x, y) = (f \odot b_1) \cap (f^c \odot b_2) \tag{9}$$

where $\odot$ represents the morphological erosion. Further, we can know that the definition of edge refinement is given as follows

$$f \otimes b(x, y) = f - f * b(x, y) \tag{10}$$

Let's set up a series of structural elements as follows

$$\{b(x, y)\} = \{b_1(x, y), b_2(x, y) \cdots b_n(x, y)\} \tag{11}$$

Then image $f(x, y)$ is represented by the structural element refinement as

$$f \otimes \{b\}(x, y) = (\cdots((f \otimes b_1) \otimes b_2) \cdots) \otimes b_n \tag{12}$$

In the experiment, we find that skeleton offset occurs in some places of the result of hit-or-miss transformation due to the existence of edge burrs. In order to make the edge detection result more accurate, we need to deal with the edge burrs before the edge is refined. The specific method is to firstly perform morphological dilation on the image, so that the edges of the image tend to be smooth, and then perform morphological erosion.

In a two-dimensional Euclidean space Z^2, let $f(x,y)$ be the input image and $b(x,y)$ be the structural element. The morphological dilation of $f(x,y)$ with $b(x,y)$ is defined as follows:

$$f \oplus b(x,y) = \max\left\{ \begin{array}{c} f(x-s, y-t) + b(s,t) | (x-s), \\ (y-t) \in D_f, (s,t) \in D_b \end{array} \right\} \quad (13)$$

The above two-step morphological processing is simple and easy to implement, and the edge can be refined and a large number of pseudo edges can be removed, and finally a binary image with few false edges and sharp edges is obtained.

The shearlet-based edge detector is shown in Fig. 1. Based on the characteristics of multi-scale, directionality and sparsity of shearlet [26–30], we perform shearlet transformation firstly on SAR images to obtain transformation coefficients. Secondly, in order to obtain the even symmetric shearlet and the odd symmetric shearlet, the Hilbert transform can be used to realize the characteristics of the even and odd function transformation. By the Hilbert transform, the even symmetric shearlet constructed using tensor product of Mexican straw hat wavelet and Gaussian wavelet is converted into the odd symmetric shearlet. Thirdly, the main

Fig. 1 Flow chart of the proposed edge detector

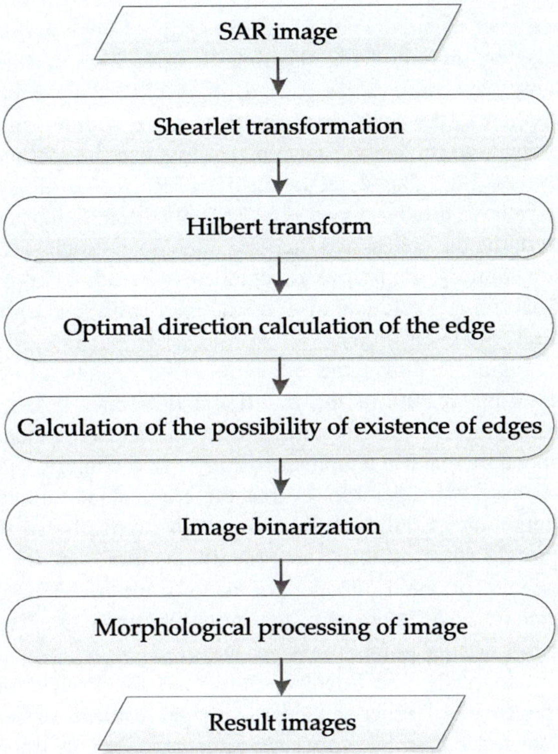

direction of the edge is determined according to the odd symmetric shearlet coefficient of the SAR image, and the possibility of the edge on the pixel is calculated according to the even symmetric shearlet coefficient and the odd symmetric shearlet coefficient. Because the shearlet has the best sparsity and strong directional positioning, the edge information in the result is guaranteed to have higher precision. Finally, aiming at the problem of false edges and rough edges, the morphological processing is used to optimize the detected results, including removing false edges, removing edge burrs, and edge refinement operations, leading to the final detected edges.

4 Experimental Results

This section uses two SAR images to verify the proposed edge detector. Figure 2a is a SAR image to be detected, and Fig. 2e is the corresponding result of edge detection. Obviously, the use of shearlet with optimal sparsity and strong directional positioning can effectively detect the edge information in the image and ensure the detection accuracy and continuity of the edge information [12].

In order to get the result of Fig. 2e, we firstly use shearlet directly to calculate the best direction of edge and the possibility of existence of edge. Figure 2b is the result of calculating the best direction of edge and the possibility of existence of edge by directly using shearlet. However, we find that the detection results inevitably have false edges caused by speckle. This is because the shearlet is directly applied to the SAR image with speckle without any preprocessing. Secondly, in order to eliminate the false edges, we use the method mentioned in Sect. 3.3.1 to process Fig. 2b and get the result of Fig. 2c. Thirdly, morphological dilation is used to remove the burrs, and Fig. 2d is obtained. Finally, morphological erosion is used to refine the edges, and the final edge detection result shown in Fig. 2e is obtained. In summary, the proposed edge detector leads to high accuracy of edge position for SAR images, due to use of shearlet with the optimal sparsity and the strong directional sensitivity.

Figure 3 shows the edge detection results of another SAR image, and the detection results of the traditional difference-based operator are given for comparison. The Sobel operator and the Prewitt operator are differential template operators that use templates to convolve the image and extract edges. The difference of these two operators is that the element of Sobeltemplate has a corresponding weighting in the 4-neighbor position of pixels, so the Sobel operator suppresses speckle stronger than the Prewitt operator, but the edge detected is wider. The essence of the Log operator is to smooth the original image with a Gaussian function, and then use the non-directional Laplacian operator to extract the zero-crossing points for edge detection.

Obviously, the Sobel operator and the Prewitt operator cannot overcome the speckle interference in the SAR image, resulting in much false edges. Therefore, the detected edges by Sobel operator and Prewitt operator are unclear. There is a

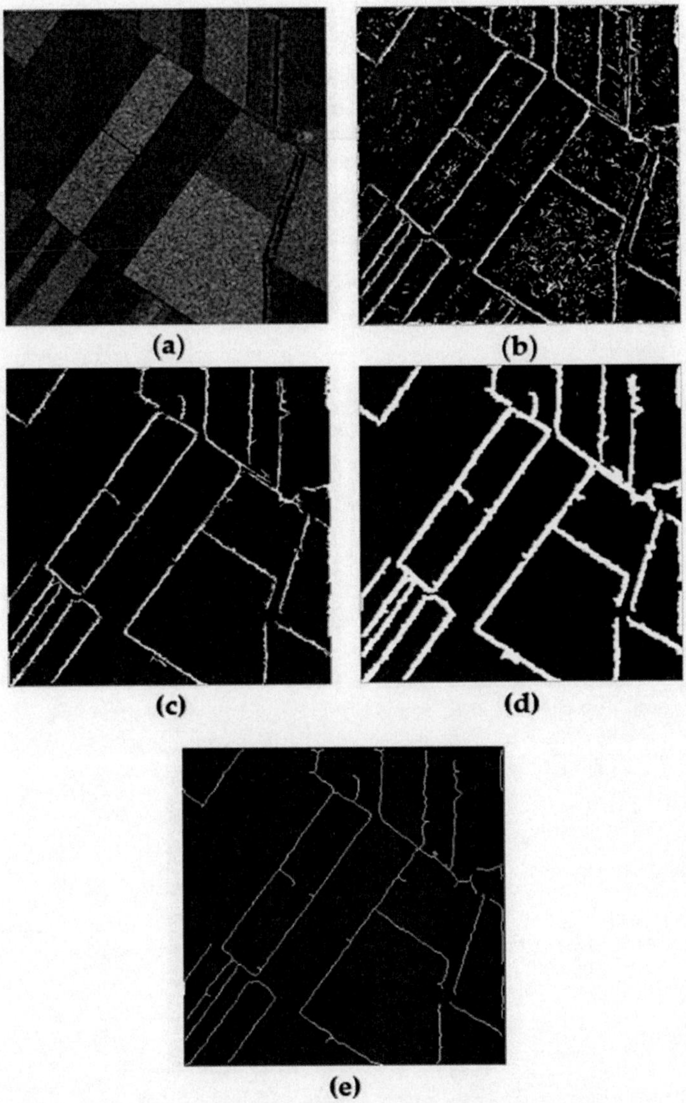

Fig. 2 The proposed edge detection results: **a** original image, **b** detection result by shearlet, **c** removal of false edges, **d** edge expansion result, **e** edge refinement result

serious rupture in the edges detected by the Log operator, and the edges trend can be barely seen as a whole. These three traditional edge detectors are based on difference, so the interference of speckle is extremely serious. The proposed edge detector makes full use of the optimal sparsity and strong directional localization of shearlet. At the same time, morphological processing, including false edge removal,

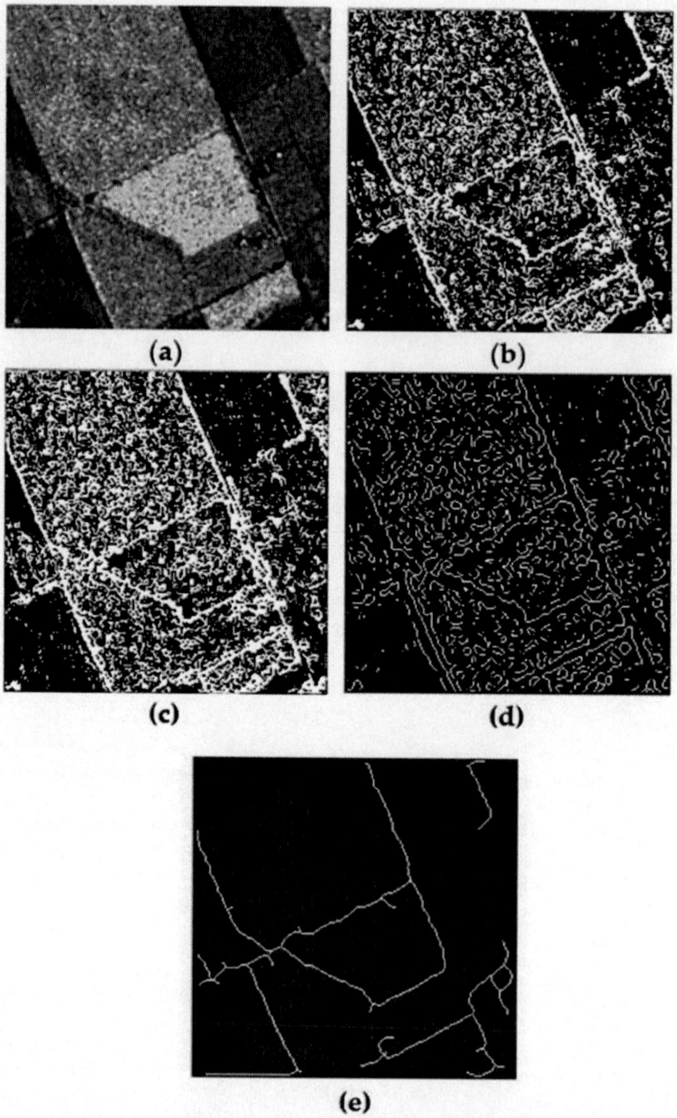

Fig. 3 Experimental results of different edge detectors: **a** original image, **b** Sobel operator, **c** Prewitt operator, **d** Log operator, **e** the proposed edge detector

edge burr removal and edge refinement operations, is performed to optimize the detection results to obtain ideal edge detection results. In short, the results of the proposed edge detector are more accurate and the detected edges are more complete and smooth, which is very suitable for the edge detection of SAR images under speckle interference.

5 Conclusions

In view of the shortcomings of low detection accuracy and sensitivity to speckle for the traditional difference-based edge detectors, we propose an edge detector for SAR images based on shearlet. The proposed edge detector makes full use of the characteristics of the optimal sparsity and strong directional position of shearlet, and the detection result is so well. The proposed detector first converts the constructed even symmetric shearlet into odd symmetry using Hilbert transform, then determines the main direction of the edge and the possibility of edge existence according to the shearlet coefficient of the SAR image. Finally, the morphological processing is used to optimize the detection result. Experiments show that the proposed detector is suitable for edge detection of SAR images, and it can ensure that the edges detected are continuous and smooth, and the position of edges is consistent with the original image.

References

1. Shao W, Yuan X, Sheng Y, Sun J, Zhou W, Zhang Q (2018) Development of wind speed retrieval from cross-polarization chinese Gaofen-3 synthetic aperture radar in typhoons. Sensors 18:412–427
2. Wang J, Bi J, Wang L, Wang X (2018) A non-reference evaluation method for edge detection of wear particles in ferrograph images. Mech Syst Signal Process 100:863–876
3. Li J, Huang S, Peng Y, Zhang W (2012) A novel method to configure the parameters of the bilateral filtering for synthetic aperture radar images speckle reduction. ActaPhysicaSinica. 61:9501–9505
4. Reisenhofer R, Kiefer J, King EJ (2016) Shearlet-based detection of flame fronts. Exp Fluids 57:41–54
5. Yang D, Lu C, Zhang J, Yang K, Chen X (2013) Edge detection of SAR images based on ROEWA and Hough transform. J Electron Measur Instr 27:543–548
6. An Q, Pan Z, You H (2018) Ship detection in gaofen-3 SAR images based on sea clutter distribution analysis and deep convolutional neural network. Sensors 18:34–354
7. Shao W, Sheng Y, Sun J (2017) Preliminary assessment of wind and wave retrieval from chinese Gaofen-3 SAR imagery. Sensors 17:1705–1717
8. Kutyniok G, Labate D (2012) Shearlets: Multiscale analysis for multivariate data, Birkhäuser Basel, New York, USA, pp 30–31
9. Yuan Y, Xie L, Zhu Y, Wang S, Zhuang Z (2017) SAR image de-noising using local properties analysis and discrete non-separable shearlet transform. In: 2017 10th International Congress on Image and Signal Processing, BioMedical Engineering and Informatics (CISP-BMEI) IEEE, Shanghai, China, October 14, 2017
10. Easley G, Labate D, Lim WQ (2008) Sparse directional image representation using the discrete shearlet transform. Appl Comput Harmon Anal 25:25–46
11. Liu S, Hu S, Xiao Y (2013) SAR image de-noising based on complex shearlet transform domain Gaussian mixture model. Acta Aeronautica et Astronautica Sinica. 34:173–180
12. Lim WQ (2010) The discrete shearlet transform: a new directional transform and compactly supported shearlet frames. IEEE Trans Image Process 19:1166–1180
13. Huang Z, Ding M, Zhang X (2017) Medical image fusion based on non-subsampled shearlet transform and spiking cortical model. J Med Imaging Health Inf 7(2017):229–234

14. King EJ, Reisenhofer R, Kiefer J, Lim W-Q, Li Z, Heygster G (2015) Shearlet-based edge detection: flame fronts and tidal flats. Appl Digital Image Process XXXVIII:9599
15. Yan B, Yang J (2004) Analysis and research on Hilbert transform. J Electric Electron Educ 26:27–29
16. Zheng J (2014) Local feature scale decomposition method and its application in mechanical fault diagnosis. Hunan University, Changsha, China
17. Selesnick IW (2001) Hilbert transform pairs of wavelet bases. IEEE Signal Process Lett 8:170–173
18. Ding W, Wang W, Zhang X, Su L (2010) Extracting straight lines from building image based on edge orientation image. Acta Optica Sinica. 30:2904–2911
19. Wisaeng K, Sa-Ngiamvibool W (2018) Improved fuzzy C-means clustering in the process of exudates detection using mathematical morphology. Soft Comput 22:2753–2764
20. Bouaynaya N, Charif-Chefchaouni M, Schonfeld D (2007) Theoretical foundations of spatially-variant mathematical morphology part I: binary images. IEEE Trans Pattern Anal Mach Intell 30:823–836
21. Wang H, Zhan G (2009) Research and application of edge detection operator based on mathematical morphology. Comput Eng Appl 45:223–226
22. Huang X (2010) Locally univalent harmonic mappings with linearly connected image domains. Chin Ann Math 5:625–630
23. Étienne B, Frédéric N, Millon G, Ruan S (2008) Binary-image comparison with local-dissimilarity quantification. Pattern Recogn 41:1461–1478
24. Bouchara F, Ramdani S (2009) Subpixel edge refinement using deformable models. J Opt Soc Am A Optics Image Sci Vis 4:820–832
25. Jin L (2012) Separation and application of strong electromagnetic interference based on mathematical morphology. Central South University, Changsha, China
26. Reisenhofer R (2014) The complex shearlet transform and applications to image quality assessment. Technische Universität Berlin Fachgruppe Angewandte Funktionalanalysis, Berlin, Germany
27. Zhou B, Polap D, Wozniak M (2019) A regional adaptive variational PDE model for computed tomography image reconstruction. Pattern Recogn 92:64–81
28. Xia X, Marcin W, Fan X, Damasevicius R, Li Y (2019) Multi-sink distributed power control algorithm for Cyber-physical-systems in coal mine tunnels. Comput Netw 161:210–219
29. Wei W, Houbing S, Wei L, Peiyi S, Athanasios V (2017) Gradient-driven parking navigation using a continuous information potential field based on wireless sensor network. Inf Sci 408:100–114
30. Wang X, Qi Y, Wang Z et al (2019) Design and implementation of SecPod: a framework for virtualization-based security systems. IEEE Trans Dependable Secure Comput 16:44–57

Research on Hysteresis Control Method of Fast Steering Mirror in Satellite Laser Communication

Yu Zhiliang, Li Xue, Yang Mingliang, and Fang Zhiyi

Abstract In the precise pointing system of satellite laser communication, the piezoelectric fast steering mirror (FSM) is used in the system. In the free state, the hysteresis between the input voltage and the output displacement of FSM is obvious, which has a serious impact on the precision positioning accuracy. Therefore, aiming at the hysteresis problem of the fast steering mirror in satellite laser communication, a phenomenal hysteresis mathematical model based on Bouc-Wen operator is established by analyzing the hysteresis mechanism of the piezoelectric fast steering mirror. An improved nonlinear particle swarm optimization (PSO) parameter identification method is proposed, which greatly improves the optimization efficiency of particle swarm optimization. On this basis, a feedforward inverse compensation linearization method is proposed. In order to verify the effectiveness of the proposed model, a fast steering mirror system experiment is established. The experimental results show that the improved PSO parameter identification method proposed in this paper can accurately identify the parameters in the model, and the proposed feedforward linearization method can improve the linearity of piezoelectric ceramic actuator by more than 90%, which can meet the actual index requirements of the fast steering mirror.

Keywords Fast steering mirror · Hysteresis · Bouc-Wen · Feedforward linearization

1 Introduction

At present, satellite laser communication technology has become a hot topic in the world. Compared with microwave communication technology, laser communication technology has the advantages of large communication throughput, strong reliability and good confidentiality [1–5]. The realization of satellite laser

---

Y. Zhiliang (✉) · L. Xue · Y. Mingliang · F. Zhiyi
Aerospace System Engineering Shanghai, Shanghai, China
e-mail: yzlhit@163.com

L. Wang et al. (eds.), *Proceedings of the 7th China High Resolution Earth Observation Conference (CHREOC 2020)*, Lecture Notes in Electrical Engineering 757, https://doi.org/10.1007/978-981-16-5735-1_36

communication link requires acquisition, pointing and tracking (APT) technology. The realization of these technologies depends on the fast steering mirror, which can adjust the incident and outgoing beams continuously [6–9]. Therefore, the fast steering mirror plays an important role in the laser communication. Piezoelectric ceramic actuator is used to drive the fast steering mirror, which has the advantages of small volume, large output force, high precision, wide bandwidth and fast response speed. However, due to the nonlinear hysteresis between the input voltage and output displacement of piezoelectric ceramics, the positioning accuracy of the fast steering mirror is seriously affected, and even the system is unstable, which leads to the interruption or loss of laser communication links.

The nonlinear modeling and control method of piezoelectric actuator has become a hot topic in the field of high precision control of fast steering mirror. At present, the mathematical description models of hysteresis can be roughly divided into the following two categories: physical model and phenomenal model. Physical model is based on the first principle, the relationship between energy and strain stress, such as domain wall model, homagenezed energy model, Jiles Atherton model [10], etc. However, the physical model is usually based on microscopic polarization mechanisms. Affected by the micro-materials and hysteresis characteristics, the physical model is not universal, and the realization of the physical model requires large-scale driving equipment, which is not conducive to practical application. Different from the physical model, the phenomenon model is independent of the physical mechanism of hysteresis. The phenomenon model is based on the relationship between the input and output of the actuator, such as Preisach model [11], Prandtl Ishlinskii model [12], Bouc-Wen model [13], Maxwell model [14], Duhem model [15] etc. Liu and Su [16] proposed an improved Prandtl Ishlinskii model based on play operator, which can not only describe rate independent hysteresis, but also rate dependent hysteresis. Wei et al. [17] proposed a composite control method combining sliding mode control with neural network control. In this method, the output of radial function of neural network is used as the input voltage of sliding mode control method, and the compensator of hysteresis network is used to compensate the output displacement, so that the system finally converges to the sliding mode surface. Cheng et al. [18] put forward a prediction method for nonlinear model. In this method, the hysteresis is equivalent to a nonlinear control object, and then the hysteresis nonlinear autoregressive moving average model is obtained by using multi-layer neural network, and then the hysteresis trajectory is predicted and optimized according to the tracking target, and the numerical analytical solution is obtained by solving the method. Wang et al. [19] proposed an improved model and parameter identification method based on Bouc-Wen operator, which has high efficiency of parameter identification, simple identification process and model prediction error less than 5.79%.

In order to solve the hysteresis problem, in addition to establishing the hysteresis mathematical model, another method is to reduce the nonlinear hysteresis and improve the linearization between input and output. For example, high-precision displacement sensor and control algorithm are used for closed-loop feedback control of piezoelectric actuator. Liu et al. [20] used the self-tuning neural network

PID feedback method to control the piezoelectric actuator, which greatly improved the tracking performance of the system. Wang et al. [21] realized the high-precision positioning of piezoelectric ceramic actuator on the basis of ellipse model and digital PID feedback method. Feng et al. [22] combined Prandtl Ishlinskii hysteresis model with the transmission function of piezoelectric ceramic actuator system, adaptive backstepping feedback controller is designed, and the high precision tracking effect is obtained by simulation experiment. Lai et al. [23] designed a Multi-mode variable structure feedback control method for the hysteresis characteristics of piezoelectric ceramic actuators, by introducing an exponential term, the static error free of the system was reduced, and a switching surface between multimode was designed, which not only overcame the chattering problem in the switching process, but also accelerated the convergence speed of the system.

Although the feedback method can improve the control accuracy of hysteresis, in practical engineering, the feedback control method increases the structure size, system complexity, complex control algorithm and many parameters. Another direct and effective method is the open-loop feedforward inverse compensation control method, that is, the hysteresis mathematical model of the piezoelectric actuator is obtained, and the next operator voltage of the piezoelectric actuator will be predicted and compensated according to the mathematical model, so as to reduce and eliminate the influence of nonlinear hysteresis characteristics. The open-loop inverse compensation method does not need complex feedback sensors and composite control algorithm. Therefore, the open-loop inverse compensation method is easy to implement in engineering.

Therefore, based on the above analysis, in this paper a Bouc-Wen based mathematical model is employeed for the hysteresis characteristics of the piezoelectric actuator with fast steering mirror, and an improved PSO algorithm is proposed for the nonlinear parameter identification problem in this model. The identification method not only improves the parameter identification efficiency, but also solves the problem of falling into the dead zone in the process of particle swarm optimization. On this basis, a feedforward inverse compensation linearization method is proposed. Finally, a fast steering mirror experimental system is established to verify the proposed method.

2 Operating Principle of Fast Steering Mirror

The operating principle of the laser communication terminal is shown in Fig. 1. Since the divergence angle of the satellite laser beam is small, when the beacon light of the other side is captured, the terminal will introduce the beacon light into the lens barrel, reflect it to the secondary mirror through the primary mirror, and then reflect it to the fast steering mirror. At this time, the fast steering mirror adjusts the angle of the incident light through continuous position deflection, and introduces the beam into the collimating mirror system CCD receives, thus obtains the other party's position information. In the same way, when the signal light is

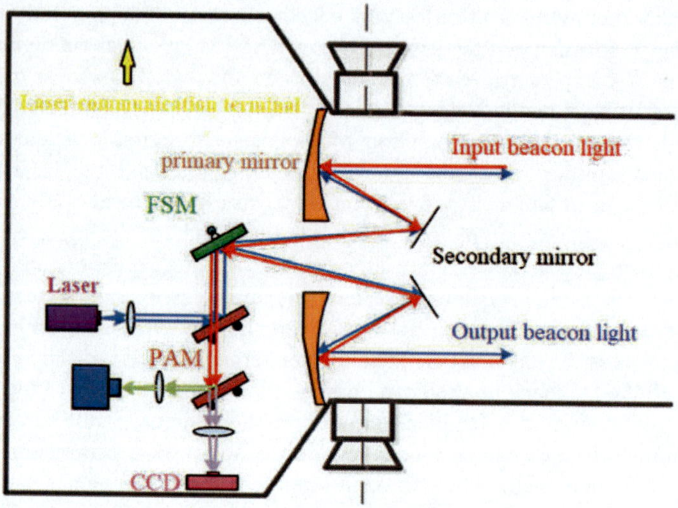

Fig. 1 Schematic diagram of laser terminal communication

transmitted from the terminal, the laser transmitter emits the signal light. The signal light is reflected to the secondary mirror through the adjustment of the fast steering mirror, and then reflected from the secondary mirror to the primary mirror, and then shot out of the lens barrel. It can be seen that the fast steering mirror plays an important role in a complete communication process.

The actuator of the fast steering mirror is a piezoelectric ceramic actuator. However, the piezoelectric ceramic can't directly act on the mirror body, so it needs to drive the mirror body to deflect through flexible hinge and rigid platform. The structure is shown in Fig. 2.

The deflection angle of the fast steering mirror can be calculated by the parameters in the figure. If the piezoelectric actuator is an ideal linear output characteristic, then the ratio of angle to displacement will be a constant. However, due to the hysteresis characteristics of piezoelectric ceramics, there is a nonlinear characteristic between the control input and the displacement output, which further leads to the angle and input of the fast steering mirror The voltage difference is also nonlinear, so that the desired input voltage of the piezoelectric actuator is given, but the actuator does not reach the desired deflection angle, which seriously affects the positioning accuracy of the fast steering mirror, which makes the optical path unable to enter and emit according to the expected angle. Therefore, Hysteresis Modeling and control of piezoelectric actuator are needed.

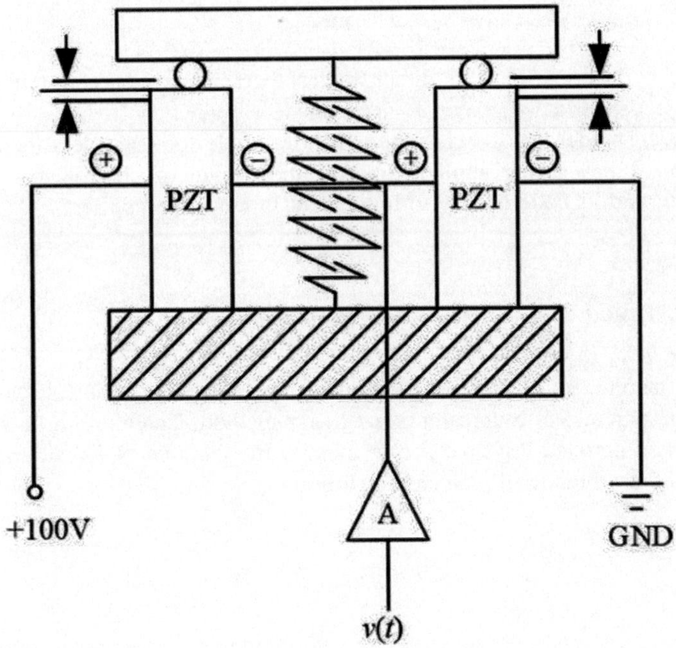

Fig. 2 Schematic diagram of FSM operating principle

3 Hysteresis Modeling and Parameter Identification of FSM

3.1 Hysteresis Modeling

The control method of hysteresis mainly is described that, the first is to establish an accurate mathematical model of hysteresis. The accuracy of the hysteresis model affects the description of the hysteresis behavior of the system; the second is the accuracy of parameter identification in the model. This is because the parameters have a decisive role in the accuracy of the model. If the parameter deviation is large, the hysteresis mathematical model is difficult to accurately control the hysteresis behavior description will also affect the inverse compensation, which makes the prediction voltage of inverse compensation have a large deviation. The following will be the hysteresis modeling analysis. The output of piezoelectric actuator can be expressed by the following formula:

$$\begin{cases} x(t) = \alpha ku(t) + (1-\alpha)kz(t) \\ \dot{z} = A\dot{x} - \beta|\dot{x}||z|^{n-1}z - \lambda\dot{x}|z|^{n} \end{cases} \tag{1}$$

$$\alpha = rand() \tag{2}$$

There $\alpha \in (0, 1)$; A, β, λ are dimensionless parameters; $x(t)$ is output displacement of piezoelectric actuator; z is intermediate variable.

Bouc-Wen model has the characteristics of simple form and less identification parameters. It has strong applicability for the design of controller in practical application, and its code is easy to implement in embedded system.

3.2 *Improved PSO Parameter Identification*

Since the parameters of Bouc-Wen model are nonlinear model, it is difficult to identify the parameters by conventional linear method. Therefore, in this paper is proposed an improved PSO method to identify the parameters for the model. The objective fitness function is selected as follows:

$$f(X) = \frac{1}{N}\sum_{i=1}^{N}(y_i - \tilde{y}_i) \tag{3}$$

where, N is the total number of data points, y_i is the model output value, $\tilde{y}_i$ is the actual measurement value.

The traditional PSO method has the problem that particle swarm optimization falls into local optimum and can't jump out, which leads to the parameter unable to find the optimal solution. Therefore, in this paper is proposed an improved rate update method to solve this problem:

$$\begin{cases} V_i^{t+1} = \omega V_i^t + c_1 r_1^t (P_{besti}^t - X_i^t) + \cdots \\ c_2 r_2^t (G_{best}^t - X_i^t) \\ X_i^{t+1} = X_i^t + V_i^{t+1} \end{cases} \tag{4}$$

There $r_i = \text{rand}_i(), i = 1, 2, \cdots$, ω is the inertia factor of particle swarm optimization, c_1 and c_2 are local acceleration coefficient and global acceleration coefficient respectively.

We know that inertia weight factor affects the whole particle search situation. Large inertia factor is conducive to global large-scale search, small inertia factor is conducive to small-scale search. Conventional PSO generally choose constant inertia factor, which will lead to limitations in search. Therefore, in this paper we change the constant inertia weight to dynamic inertia weight, as follows:

$$\omega_i = \omega_{\min} \exp\left(\frac{\omega_{\max} - \omega_{\min}}{T}(T - t_i)\right) \tag{5}$$

where $i = 1, 2, \ldots, T$, $\omega_{\max}$ is the maximum weight factor, $\omega_{\min}$ is the minimum weight factor, t_i is the ith particle, and T is the number of iterations. The improved dynamic weighting factor formula can be expressed as:

$$
\begin{aligned}
V_i^{t+1} = {} & \omega_{\min} \exp\left(\frac{\omega_{\max} - \omega_{\min}}{T}(T - t_i)\right) V_i^t + c_1 r_1^t (P_{besti}^t - X_i^t) \\
& + c_2 r_2^t (G_{best}^t - X_i^t) + c_3 r_3^t (Q_{besti}^t - X_i^t)
\end{aligned}
\tag{6}
$$

c_3 is the correlation acceleration coefficient, r_3^t is the random number between 0 and 1, Q_{besti}^t is the historical position information randomly selected, the parameter identification process is as follows:

$$
\begin{aligned}
V_i^{t+1} = {} & \omega_{\min} \exp\left(\frac{\omega_{\max} - \omega_{\min}}{T}(T - t_i)\right) V_i^t + c_1 r_1^t (P_{besti}^t - X_i^t) \\
& + c_2 r_2^t (G_{best}^t - X_i^t) + c_3 r_3^t (Q_{besti}^t - X_i^t)
\end{aligned}
\tag{7}
$$

The improved PSO flow chart is as follows (Fig. 3).

Step 1 set the total number of operators;
Step 2 initialize particle parameters, set population size, iteration times, acceleration coefficient and other initialization values;
Step 3 initialize the velocity and position of each particle in the population;

$$
\begin{cases}
X_i = X_{\min} + \text{rand}(x)(X_{\max} - X_{\min}) \\
V_i = V_{\min} + \text{rand}(x)(V_{\max} - V_{\min})
\end{cases}
\tag{8}
$$

Step 4 calculate the fitness of particles according to formula 3;
Step 5 find the optimal position of individual and global;

$$
pbest_i(t+1) = \begin{cases}
pbest_i(t), f(X_i(t+1)) \geq f(pbest_i) \\
X_i(t+1)), f(X_i(t+1)) < f(pbest_i) \\
gbest(k+1) = \min(pbest_i(k+1))
\end{cases}
\tag{9}
$$

Step 6 update the speed and position according to formula (8) and Eq. (9);
Step 7 if the termination conditions are met, stop the optimization process. If not, return to step 4 to continue the optimization.

4 Feedforward Linearization of FSM

The feedforward linearization of FSM process is shown in Fig. 4. The expected input signal $r(t)$ is transformed into the intermediate control input signal $u(t)$

Fig. 3 Structure diagram of improved PSO

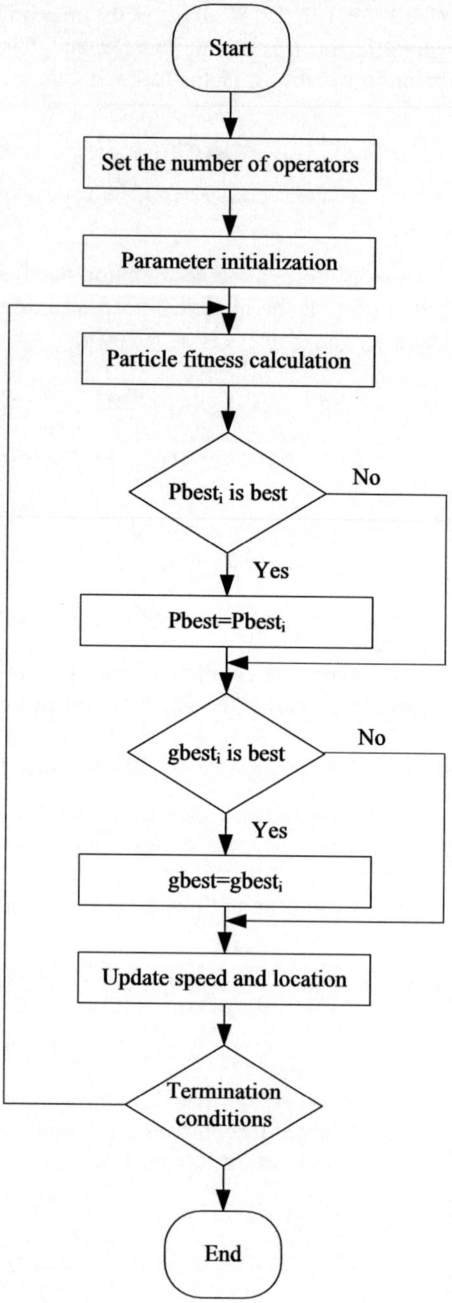

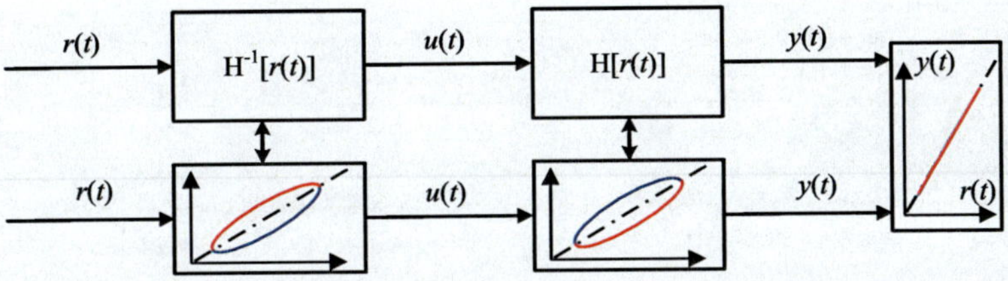

Fig. 4 Feedforward inverse model compensation

Table 1 Model parameters

Parameter	Value
k	4.983×10^{-6}
α	0.4
A	-3.376×10^{-6}
β	0.21
λ	−0.0157

through the inverse model. Here, there is an inverse hysteresis relationship between the expected input and the control signal. It is this relationship that the control signal $u(t)$, the positive hysteresis effect of the piezoelectric actuator is counteracted, and the actual output $y(t)$ and expected input $r(t)$ can be finally achieved linear relationship.

The relationship between the control signal $u(t)$ and the input signal $r(t)$ can be described as:

$$\begin{cases} u(t) = \alpha k r(t) + (1-\alpha) k z(t) \\ \dot{z} = A\dot{r} - \beta|\dot{r}||z|^{n-1}z - \lambda\dot{r}|z|^{n} \end{cases} \tag{10}$$

Similarly, the improved PSO method proposed in the previous paper is used for parameter identification in the model, and the parameter values are shown in Table 1.

5 Experimental Result

In order to verify the effectiveness of the method in this paper, a fast steering mirror experimental system is designed, as shown in Fig. 5. The experimental platform includes that PC, laser, power supply, control circuit, driving circuit, piezoelectric actuator and mirror body. The piezoelectric ceramic actuator adopts S-330 actuator made in Germany, the driving voltage range is 0–100V, the maximum deflection angle is 2mrad. The special adhesive is adsorbed on the piezoelectric actuator platform in a certain way to form a fast steering mirror.

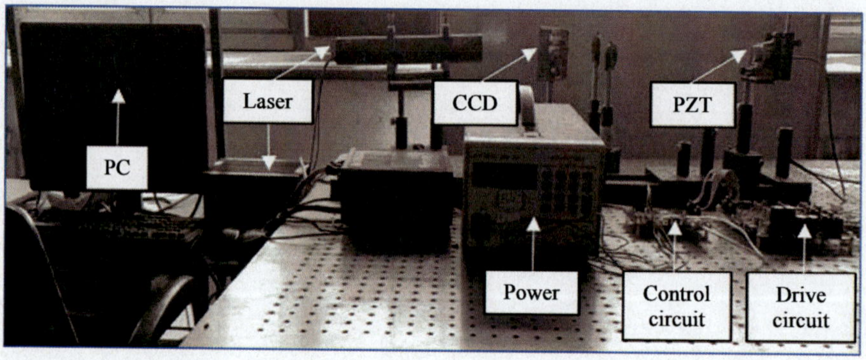

Fig. 5 FSM experimental platform

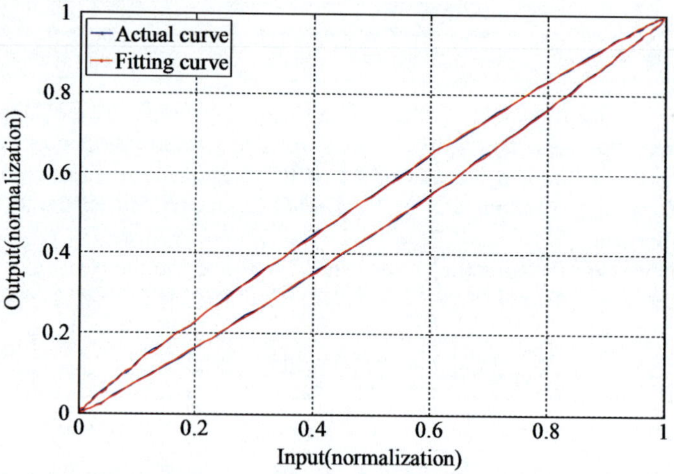

Fig. 6 Actual and fitting curves

Experiment 1: model fitting

For the sinusoidal voltage with input frequency of 10Hz and amplitude of 80V, the actual feedback value is measured, and the hysteresis characteristics are fitted according to the input and output values. The fitting results are shown in Fig. 6 and the fitting error is shown in Fig. 7, The model fitting error is shown in Fig. 7.

It can be seen from Figs. 6 and 7 that Bouc-Wen model can fit the hysteresis of fast steering mirror, and the fitting error is less than 1%, which indicates that the improved PSO model parameter identification method proposed in this paper has high identification accuracy, therefore, the fitting accuracy of the model is high.

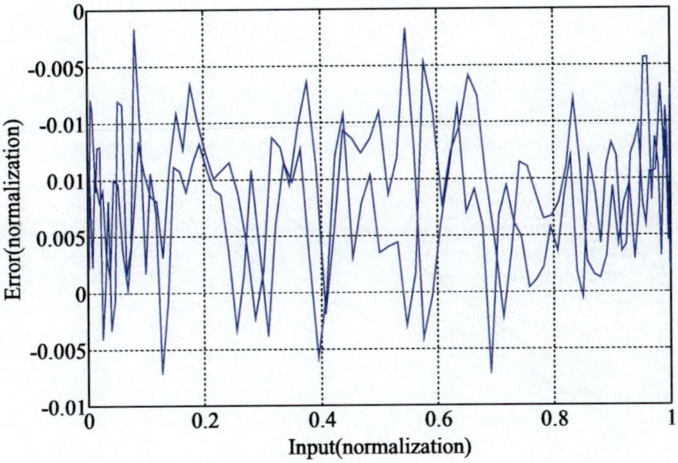

Fig. 7 Fitting error

Experiment 2: feedforward linearization

Similarly, for the sinusoidal voltage with 10Hz input frequency and 80V amplitude, the actual feedback value after linearization is measured. The original hysteresis curve is shown in Fig. 8, the linearization compensation curve is shown in Fig. 9 and the compensation error is shown in Fig. 10. All curves were normalized.

It can be seen from Figs. 8 and 9 that before the feedforward linearization compensation, the average linearity is 8.9%, and the hysteresis error is less than 7.3%, which has obvious hysteresis effect. After the feedforward linearization compensation, the average linearity is 1.9%. From the feedforward linearization error in Fig. 10, the hysteresis error after compensation is less than 0.6%, the linearity is greatly improved, and the hysteresis effect is obviously improved.

Experiment 3: tracking performance

The hysteresis of the fast steering mirror is caused by the inherent hysteresis characteristics of its actuator piezoelectric ceramics. The purpose of compensation is to improve the linearity and improve the accuracy of the fast steering mirror in the process of motion. Therefore, it is necessary to test the tracking performance of the system with the input voltage amplitude of 60V and frequency of 100Hz, the tracking results and error curves are shown in Figs. 11 and 12.

It can be seen from Figs. 11 and 12 that the compensated actuator has a high tracking accuracy. From the tracking error curve in Fig. 12, the maximum tracking error is 0.41v, accounting for 0.68% of the total voltage. There is a high linearity between the expected input and the actual feedback output, and the hysteresis effect should be significantly reduced to meet the practical application requirements. It shows that the method proposed in this paper is feasible and effective.

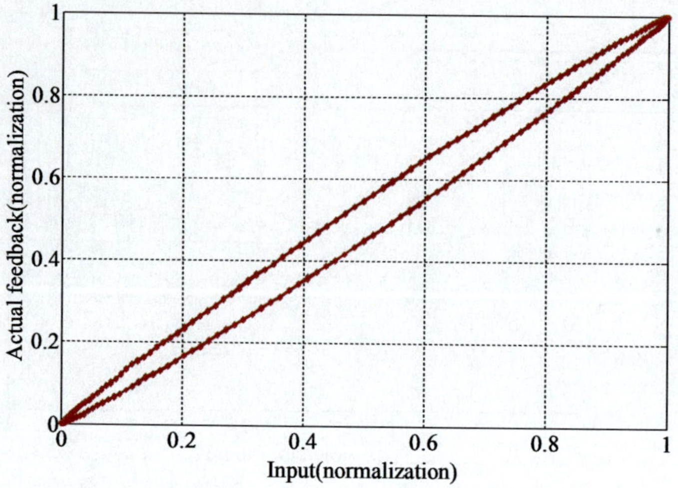

Fig. 8 Original hysteresis curve

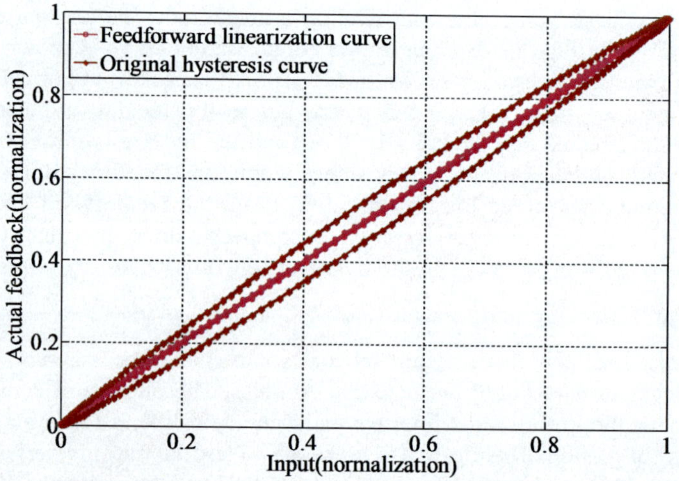

Fig. 9 Feed forward linearization

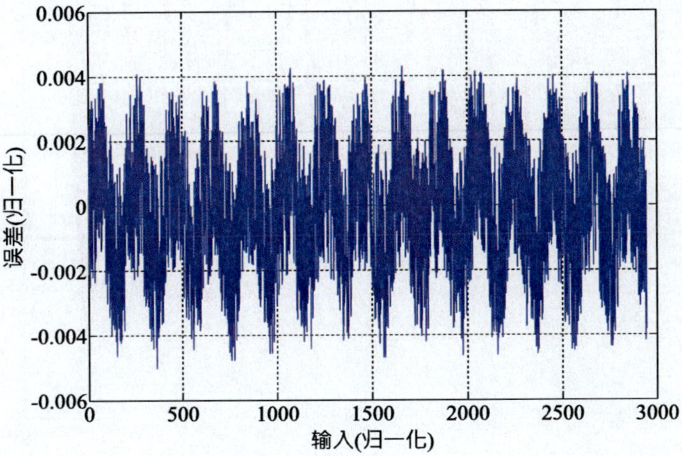

Fig. 10 Error of feed forward linearization

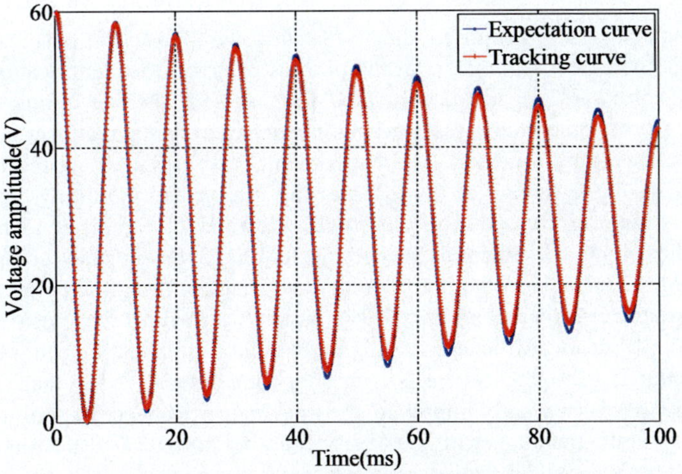

Fig. 11 100 Hz verification curve with damped sinusoid input

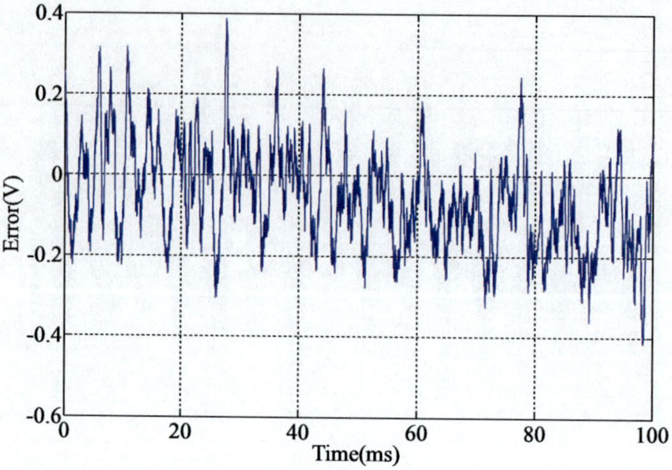

Fig. 12 Tracking error with damped sinusoid input

6 Conclusion

The advantage of the proposed method is that the piezoelectric actuator can't identify the nonlinear parameters in the process of nonlinear identification on this basis, a feedforward linearization method for inverse hysteresis compensation is designed. At the same time, a fast steering mirror experimental platform is designed to verify the proposed method. The experimental results show that the hysteresis model can fit the hysteresis of the fast steering mirror, and the fitting error is less than 1%, which indicates that the improved PSO model parameter identification method has high identification accuracy; through the inverse compensation experiment, it can be seen that the average compensation frontline degree is 8.9%, and the hysteresis error is less than 7.3%, which has obvious hysteresis effect.

After the feedforward linearization compensation, the linearity is improved the average value is 1.9%, the hysteresis error after compensation is less than 0.6%, and the linearity has been greatly improved; the tracking performance experiment shows that the maximum tracking error is 0.41v, accounting for 0.68% of the total voltage, which meets the practical application requirements.

The method proposed in this paper can effectively solve the nonlinear hysteresis problem of fast steering mirror in satellite laser communication. The method is effective and feasible, and has important practical reference significance for the realization of satellite laser communication link.

References

1. van der Geer J, Hanraads JAJ, Lupton RA (2000) The art of writing a scientific article. J Sci Commun 163:51–59
2. Chaudhary S, Sharma A, Singh V (2019) Optimization of high speed and long-haul inter-satellite communication link by incorporating differential phase shift key and orthogonal frequency division multiplexing scheme. J Optics 176:185–190
3. Gao T, Guo Q, Wang H et al (2017) A new static routing algorithm of GEO/LEO hybrid satellite networks. In: International conference in communications, signal processing, and systems. Springer, Singapore, pp 84–92
4. Ramasamy Y, Densmore AC (2014) Technology trends and challenges of antennas for satellite communication systems. J IEEE Trans Antennas Propag 63(4):1191–1204
5. Abe Y, Tsuji H, Miura A et al (2018) Frequency resource management based on model predictive control for satellite communications system. J IEICE Trans Fundam Electron Commun Comput Sci 101(12):2434–2445
6. Sravani KG, Guha K, Rao KS (2018) Analysis on selection of beam material for novel step structured RF-MEMS switch used for satellite communication applications. J Trans Electric Electron Mater 19(6):467–474
7. Xu Z, Xu J, Meng H et al (2018) A balanced sub-harmonic mixer for EHF satellite communications. J IEICE Electron Express 15:20180931
8. Narimani R, Farhoudi L (2017) Design of a self-phased quadrivial helix antenna for satellite communication. J Eng Technol Appl Sci Res 7(6):2273–2276
9. Wang J, Wang G, Bai R et al (2017) Ground simulation method for arbitrary distance optical transmission of a free-space laser communication system based on an optical fiber nanoprobe. J Opt Commun Netw 9(12):1136–1144
10. Jiles DC, Atherton DL (1986) Theory of ferromagnetic hysteresis. J Magn Magn Mater 61(1–2): 48–60
11. Preisach F (1935) Über die magnetische Nachwirkung. J. Zeitschriftfür Physika Hadrons and Nuclei 94(5):277–302
12. Macki JW, Nistri P, Zecca P (1993) Mathematical models for hysteresis. J. SIAM review 35(1):94–123
13. Bouc R (1967) Forced vibration of mechanical systems with hysteresis. In: Proceedings of the fourth conference on non-linear oscillation, Prague, Czechoslovakia
14. Ikhouane FA, MañOsa VC, Rodellar J (2005) Adaptive control of a hysteretic structural system. J Automatica 41(2):225–231
15. Coleman BD, Hodgdon ML (1987) On a class of constitutive relations for ferromagnetic hysteresis. J Arch Ration Mech Anal 99(4):375–396
16. Liu S, Su CY (2013) A modified generalized Prandtl-Ishlinskii model and its inverse for hysteresis compensation. In: American control conference (ACC), 2013. IEEE, pp 4759–4764
17. Wei Q, Zhang CJ, Zhang D, Wang CL (2012) Neural network control for piezo-actuator using sliding-mode technique. J Opt Precis Eng 5(20):1055–1063
18. Cheng L, Liu W, Hou ZG et al (2015) Neural-network-based nonlinear model predictive control for piezoelectric ceramic actuators. J IEEE Trans Ind Electron 62(12):7717–7727
19. Wang DH, Zhu W (2010) Hysteretic modeling and experimental verification for WTYD type piezoceramic micro-actuators. J Opt Precis Eng 18(1):205–211
20. Liu W, Chen M, Yang W (2009) Modeling of micro-piezoelectric motion platform for compensation and neuro-PID controller design. In: International conference on intelligent computing. Springer, Berlin, Heidelberg, pp 784–793
21. Wang XH, Guo SX, Ye XF, Ru CH (2009) Modeling and feed-forward control based on piezoelectric ceramic hysteretic. J Electric Mach Control 13(5):766–771

22. Feng Y, Hu YM, Su CY (2006) Adaptive control of a class of nonlinear systems with Prandtl-Ishlinskii hysteresis. J Acta automatica sinica 32(3):450–455
23. Lai ZL, Liu XD, Chen Z (2012) Multi-mode time-varying sliding mode control of piezoceramic actuator based on inverse compensation. J Electric Mach Control 16(1):92–103

Construction of Remote Sensing User Growth System Based on Hooked Mode

Caiping Li, Xiaoming Zhou, Jisheng Zhang, and Lei Chang

Abstract Remote sensing information has been widely used in various industries of national economy. In order to promote the benign ecological development of remote sensing information producers and users, it is necessary to build a growth system of remote sensing users. This paper analyzes the common methods of typical Internet user growth system, and summarizes the essence of user growth system. Aiming at the construction of remote sensing user growth system, firstly, the target users, application scenarios, product concepts, service platforms and strategic directions of remote sensing information are defined. Then, the driving force based on hooked model is designed, and the RFA model suitable for the growth of remote sensing users is constructed. Finally, a closed-loop remote sensing user growth system is constructed by taking the integral method as an example, so that users can continuously apply remote sensing information, Promote the continuous development of remote sensing means and form a virtuous circle.

Keywords Hooked model · Remote sensing · Growth system

1 Introduction

User growth system is a kind of operation means of Internet businesses and enterprises, that is, to judge user behavior through quantitative models and standards. It is a method to distinguish users from different values and implement differentiated incentive and operation strategies. Almost all Internet websites have user growth system, the difference is that they have their own targeted system design, expecting to form a stable ecosystem for users through the intervention of operators, so as to achieve the strategic objectives of operation. With the rapid development of remote sensing satellite, remote sensing information has been widely used in forest resources inventory, crop yield estimation, land and resources

C. Li · X. Zhou (✉) · J. Zhang · L. Chang
Beijing Institute of Remote Sensing Information, Beijing, China
e-mail: zxm2913@163.com

L. Wang et al. (eds.), *Proceedings of the 7th China High Resolution Earth Observation Conference (CHREOC 2020)*, Lecture Notes in Electrical Engineering 757, https://doi.org/10.1007/978-981-16-5735-1_37

survey, weather forecast, marine environment monitoring, urban planning and so on. In order to give full play to the application benefits of satellites, the manufacturers of remote sensing information have built service systems for related products. For example, luojia-01 satellite has built a product data service subsystem, a web-based query and display subsystem and a mobile query and display subsystem [1]. Users can query, preview and download data through the web or intelligent mobile terminal, and their essence is mutual Promotion of remote sensing products in the network mode. In order to promote the benign ecological development of remote sensing information producers and users, it is necessary to learn from the successful experience of typical websites and construct the growth system of remote sensing users around the application characteristics of remote sensing users.

2 Common Methods of User Growth System

2.1 *Internet User Growth System Model*

The system of user growth is based on the user data model, through finding the key path and core driving force of user growth, building an incentive channel for user growth and a touch channel connecting user behavior, forming a set of operation mechanism to drive user growth.

2.2 *Internet User Growth System Model*

Growth system is an evaluation standard and value orientation that reflects the users' use or participation in the website. There are many forms of growth system, whether it is the most common points, growth value, virtual currency, membership level, achievements, medals, and even hidden conditions and restrictions, the purpose is to realize the business interests of the website and meet the core needs of users. For the website operators, the growth system can ensure that high-quality resources or value-added services are used by the core target users, reasonably guide the user behavior, ensure the unity of user interests and website interests, and build a healthy and stable website ecological environment; for users, the growth system can make high-quality users gain a sense of superiority and achieve the use value of the website In the process, they can get positive and negative incentives in time and enjoy the privileges brought by the promotion of the level. Quantitative analysis of website user behavior is one of the most important tasks for all successful websites. The common user growth system of several domestic successful websites is shown in Table 1. It can be seen that the growth system of users of these relatively successful websites, from media to pan SNS, from social map to interest

Table 1 Typical website user system

Website name	Evaluation relation quantity	Evaluation core
Sina Weibo	Vermicelli consumption	The core idea is to create cattle expert (vermicelli, certifying cattle)
	Certification level	
	Number of medals	
	…	
Tencent QQ	Growth value	Embodying user's virtual "satisfaction of self value"
	Q coin	
	…	
Renren.com	Growth value	Renren's users are more concerned about the relationship between the circle of friends and the content around them
	Membership level	
	…	
Douban.com	Bean yield	"Reflect the value of users" through various interests and participation of users
	…	
Baidu Encyclopedia	Empirical value	Constructing virtual identity between strangers based on the hierarchy of content/answer authors
	Wealth value	
	…	
zhihu.com	Amount of attention	Different from Baidu Encyclopedia, Zhihu's questions and answers need more practical experience and deep understanding
	Invitation code	
	…	
TaoBao	Credit value	Focusing on "trading", on the one hand, we should pay attention to the accumulation of trading experience, which is reflected by the membership level; on the other hand, we should give the actual interest concession, which is reflected by points
	Membership level	
	Integral	
	…	
JD.com	Credit evaluation	Around the "transaction", gradually strengthen the attention to the experience and evaluation of commodity credit
	User level	
	integral	
	…	

map, from Knowledge Q & A to experience Q & A, and from C2C to B2C, their user growth system is closely related to the strategic positioning of the website. As an operator, "what do we want the website to get?" and "what do our users do with the website?" are two basic problems that are related to how to establish the user growth system of the website.

Table 2 Common methods of user growth system

Method	Content
Number of fans	That is, how many people pay attention to you, which can also reflect your influence and voice right in the community or the website to a certain extent (microblog, Zhihu)
authentication	Real name authentication, enterprise certification, etc. (micro blog)
Grade	Different user levels match different rights and requirements (microblog);
Integral/growth value/ experience value	Quantitative index of evaluation grade (QQ, Baidu know)
Medals/medals/ achievements	Incentive method to evaluate participation (microblog)
Evaluate	Self evaluation is personal recommendation, others' evaluation is social satisfaction (QQ)
Virtual currency	The virtual currency circulating in a community or specific group can obtain and realize resource exchange(QQ)
Label	It can be added by yourself or others, similar to evaluation, and more concise (QQ, microblog)
Ranking	Rank users according to different dimensions and display them accordingly (microblog)
Sign in/task/share	Increase user stickiness and social recommendation (QQ, microblog)

Summarizing the growth system of typical website users, it can be seen that the methods involved in user growth system are shown in Table 2.

Of course, in fact, there are more than the above methods, but the change is not separated from its original, and the operation goal is the same: adopt differentiated strategy to screen users, retain valuable users, and cultivate high-value users.

3 Remote Sensing User Growth System Model

Before designing the user growth system, it is necessary to define the application scope of remote sensing information, as shown in Fig. 1. For the application of remote sensing information, the target users are mainly research institutes/colleges and national ministries and commissions, etc., which use remote sensing information to carry out scientific and technological research and verification of forestry, agriculture and urban construction. The demand of target users for remote sensing information is more focused on whether the product quality can meet its application. From the user's point of view, we should pay more attention to the imaging quality, positioning accuracy and time resolution of remote sensing products. Its application scenarios include forest resources inventory, crop yield estimation, land and resources survey, weather forecast, marine environment monitoring, urban

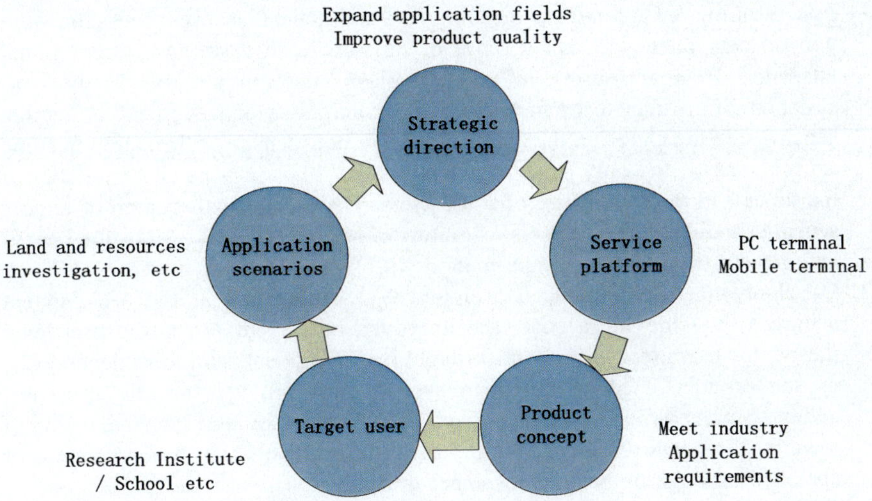

Fig. 1 Application activities of remote sensing information

planning, etc. different industries have different requirements for the application of remote sensing information. As a producer of remote sensing products, its strategic direction is to continuously expand the application fields and improve the quality of remote sensing products. Product concept is more for the national economic construction services, so the design and production of its products are around the improvement of product quality, the design of product concept is consistent with the needs of target users. Considering that users have the application requirements of PC and mobile for remote sensing information, the service platform should have the application ability of PC and mobile. This is also a general requirement for regular Internet sites. The construction of user growth system needs to focus on five aspects of remote sensing information activities.

3.1 Design of Driving Force Based on Hooked Model

Do not use any macros for the figures and tables (We will not be able to convert such papers into our system).

The essence of the growth system is to encourage users to operate spontaneously or even dependently in the product, and the user growth system is just as a fulcrum to pry users into the feeling of "addiction". Eight core driving forces [2], such as sense of mission, achievement and progress, and creativity, are designed in the hooked model to "attract" users. For remote sensing users, the eight core driving forces are still applicable, but the particularity of remote sensing information application needs to be considered in the design.

1. Epic meaning & Calling/Epic meaning & Emotional call, sense of mission: what you do can make you feel a sense of mission, with a strong era, emotional attributes. Remote sensing information plays an important role in the construction of national economy. With the continuous enhancement of remote sensing satellite capacity, the application of remote sensing information in some aspects may be the only solution, which makes it become the characteristics of the times and the superiority of some users. The user growth system of remote sensing should strengthen users' cognition of remote sensing, application benefit and growth through the design method with "sense of ceremony".
2. Development & compliance/Development & Achievement: the achievement achieved by achieving a goal. The improvement of remote sensing user level means that users feel the benefits brought by the products and form dependence on the products. The user growth system must provide rewards after user actions, among which the exclusive privileges opened by user level are a kind of reward. The application of remote sensing information and the completion of specified tasks is the best achievement in itself.
3. Empowerment of creativity & feedback: it can provide users with DIY, reconstruction and timely feedback. Exclusive privileges should be designed so that users can choose their preferred application mode according to their preferences. Users can get timely feedback after action, which is helpful to quickly develop user habits.
4. Ownership & possession/ownership and possession: satisfy users' desire for control and possession. Exclusive privileges belong to the things that users can actually possess and dominate.
5. Social influence & relatedness/social influence & connection: the embodiment of a certain status in society and the relationship with others. It mainly refers to the application groups with common tasks, and the groups undertaking key tasks will enjoy exclusive privileges, and there will be certain competition or cooperation among the groups.
6. Scarcity & impatience/scarcity & impatience: the rarer you are, the more you cherish it. What you can't get is always in turmoil. The user system of most products will take advantage of scarcity to influence users, such as the common senior members. These scarce users in the product have already obtained certain resources, and they are more likely to rely on the product.
7. Unpredictability & Culture/unpredictability & curiosity: curiosity and the unknown will drive human beings to explore continuously. What problems can remote sensing information solve? To what extent? Users who want to solve the application practice will have such curiosity and imagination. Therefore, in the process of continuous exploration, remote sensing users have explored the application fields of remote sensing information. In addition, exclusive privileges have certain unpredictability, because they can not be tried, and there is no detailed introduction, which will inevitably bring certain attraction.
8. Loss & avoidance/loss, loss & avoidance: a successful user system will give users a sense of crisis. When the user does not insist on the application of remote

sensing information, the persistence index will be reduced or even cleared, so that the user will feel lost, so as to stimulate the user's sense of success.

3.2 Growth Quantitative Model

1. RFA model

The growth of conventional Internet users is quantified by using RFM user consumption behavior model [3], which is an important method to measure user value and profit. Among them, R (revenue) is the most recent consumption from a certain point in time, and users who have recently purchased tend to buy again; f (frequency) is the consumption frequency in a period of time, and the frequent users have a high probability of purchasing again; m (monetary) is the consumption amount corresponding to this period of time, and customers with more consumption amount are more likely to re purchase. Since the application benefits of remote sensing information are more reflected in the national economic construction, the consumption amount is not the main factor. Therefore, in order to quantify the growth of remote sensing users, the RFM model is improved to RFA model, as shown in Fig. 2. Among them, a (amount) is the number of applications of remote sensing products in the corresponding time, and users with more applications are more likely to apply again.

In RFA model, the time, frequency and product quantity of remote sensing information are managed in the RFA model, and different incentive mechanisms are adopted for users in different periods. The LTV formula of users' lifetime value is as follows:

$$\mathrm{LTV} = \frac{F_r \times \sum_{i=1}^{n}(A_i \times B_i)}{T_r - T_0} M_i \tag{1}$$

Among them, F_r is the monthly application frequency of remote sensing information of a user, A_i is the number of applications of class I products by users, B_i is the aging characteristics of class I products, T_0 is the current time to calculate

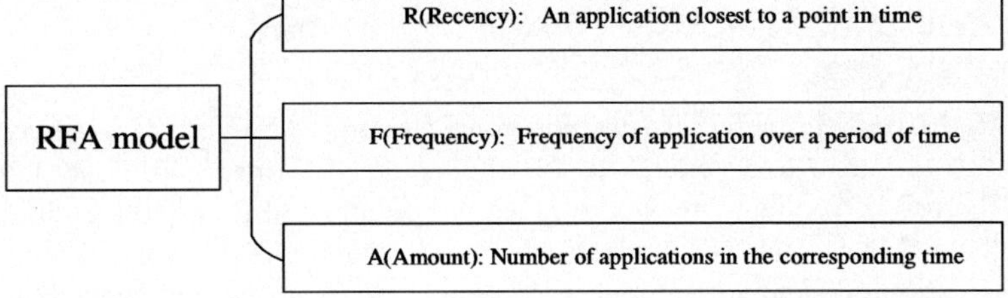

Fig. 2 RFA model

the lifetime value of users, T_r is from t_ The latest application time at T_0. Where M_i is the identification number of remote sensing products.

2. Set the grade

Based on the quantitative analysis of online, download and search activity of remote sensing user behavior data, the user level is divided into four categories: VIP users, active users, general users and zombie users [4]. When the evaluation parameter 80% $<\Sigma(k) \leq 100\%$, it is set as VIP user; when 50% $<\Sigma(k) \leq 80\%$, it is set as active user; when 0% $<\Sigma(k) \leq 50\%$, it is set as general user; when $\Sigma(k) = 0\%$, it is set as Zombie user. Where $\Sigma(k) = a\xi(k) + b\varphi(k) + c\lambda(k)$, $\xi(k)$ is the online rate, $\varphi(k)$ is the download rate, $\lambda(k)$ is the search rate, $a + b + c = 1$. According to different user levels, different user rights are given.

4 Construction of Remote Sensing User Growth System

4.1 Remote Sensing User LTV Model

According to the system model construction method in Sect. 3, the architecture of remote sensing users is designed. The lifetime value LTV of remote sensing users is closely related to the remote sensing products used by users. The specific relevant values are shown in Table 3.

Among them, download rate φ (k) and search rate λ (k) can be calculated by calculation, and other formula values correspond to the corresponding items in the table, then the user growth quantitative model can be calculated.

4.2 Integral Incentive Design for Remote Sensing Users

According to the above ideas, referring to the Internet e-commerce platform, taking the score acquisition as an example, the background of remote sensing user growth system is designed, and its framework is shown in Fig. 3.

4.3 Remote Sensing User Growth Architecture

Through the growth model and user incentive strategy, the interaction between the two forms the growth system. With the interaction of elements in the system, the better the use of products, the more incentive points, and then the recommended products tend to be rich and accurate, the user's sense of experience is improved, the application efficiency of products increases, and its social and economic value increases. The relationship can be described in Fig. 4.

Table 3 Description of LTV and product and user information weight

Image product information

Target name	Satellite payload	position	resolving power	accuracy
** chemical plant	visible light	A place in the North	1 m	20 m
M_i				

Image content information

Feature type	Target properties	Changes	Research and judgment results	Characteristic state
Building	Civil facilities	New plant	Capacity up * * tons	There are changes
				$T_r - T_0$

Image usage information

Last download time	Last query time	amount of downloads	Number of queries	Application direction
May 4, 2020	May 1, 2020	2	5	Environmental monitoring
		A_i	B_i	

User information

User name	Online rate	User rights
** Environmental law enforcement team	30%	Whole
F_r	$\xi(k)$	

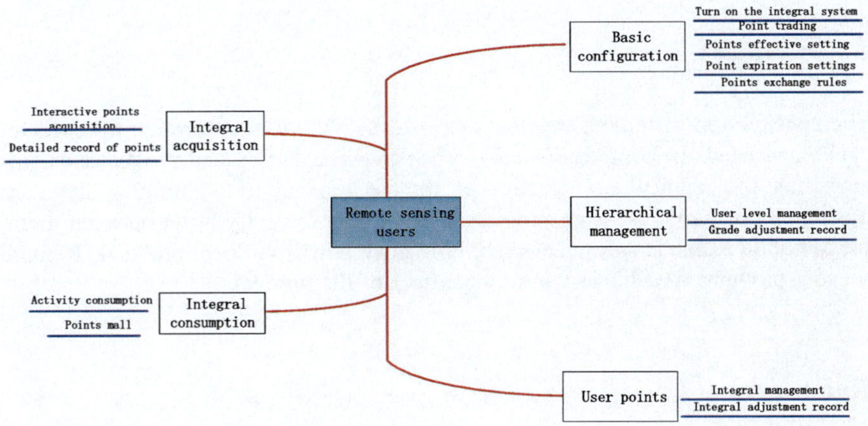

Fig. 3 Background framework of integral driven user growth system

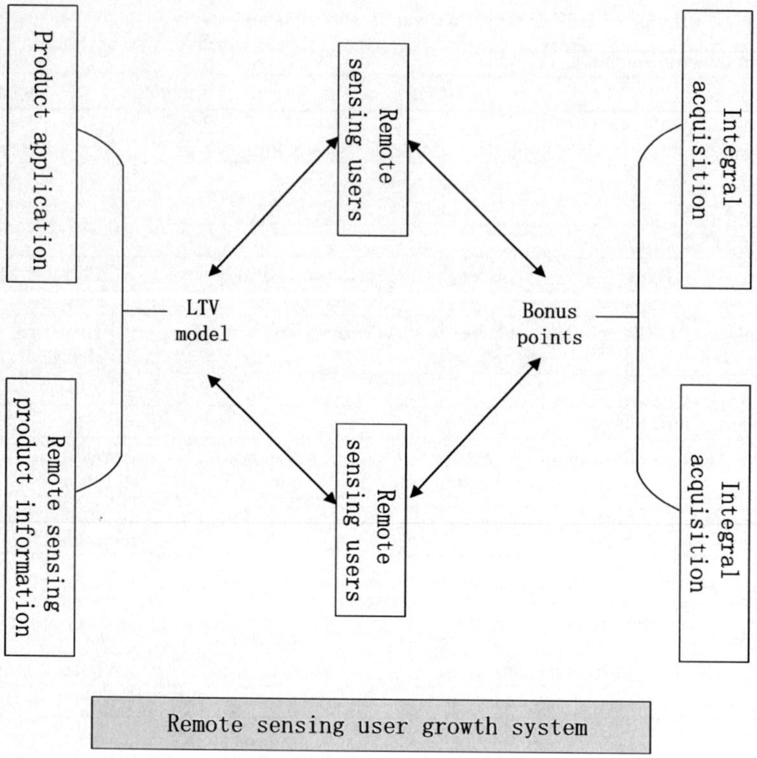

Fig. 4 Remote sensing user growth architecture

5 Summary

The construction of remote sensing user growth system runs through the product cycle, and needs to comprehensively consider the user side and product side. By analyzing the application behavior of remote sensing users, refining the core positioning of remote sensing products, and finding the entry point between them, we can build a unique sustainable user growth system of our own products. Remote sensing products to achieve the ultimate goal of the product.

References

1. van der Geer J, Hanraads JAJ, Lupton RA (2000) The art of writing a scientific article. J Sci Commun 163:51–59
2. Zhang G et al (2019) Design and processing of night light remote sensing of luojia-1 satellite 01. M Science Press

3. Sun R (2019) Research on interaction design of user growth system in Internet Environment. J Art Design
4. Wang J et al (2018) Research on user growth strategy of scientific data service platform. J Inf Technol Informatization

Ship Azimuth Velocity Estimation in Spaceborne SAR Based on Minimum-Entropy Criterion and Newton's Method

Yating Huang, Dexin Li, and Zhen Dong

Abstract Spaceborne synthetic aperture radar is of great significance in ship monitoring. Estimation of motion parameters of moving ships has a high application value in civil and military fields. In this paper, we propose a method to estimate azimuth velocity of moving ships in spaceborne SAR as well as improve the image quality. This method is applied directly on the single-look complex (SLC) image data, and does not demand extra information or special configurations. Applying Newton's method to search the calibration phase corresponding to the minimum of the entropy, the Doppler frequency modulation (FM) rate and azimuth velocity of ships can be obtained by fitting the quadratic coefficient of the correction phase, and the image blur caused by the movement of ships is effectively alleviated by means of phase compensation. Finally, the proposed method is validated by the Gaofen-3 (GF-3) complex data, and the experimental results perfectly fit the data recorded by the international automatic identification system, which explains the effectiveness of the proposed approach.

Keywords Azimuth velocity estimation · Moving ships · Minimum-entropy criterion · Newton's method · AIS

1 Introduction

Spaceborne synthetic aperture radar (SAR) is a kind of active detection sensor, which can provide two-dimensional high-resolution image of illuminated scene. It plays a significant role in civil and military fields related to remote sensing detection. As a hot issue, ship monitoring, including the detection, positioning and motion parameters estimation of moving ships, has received extensive attention.

Y. Huang · D. Li · Z. Dong (✉)
College of Electronic Science and Technology, National University of Defense Technology, Changsha, Hunan, People's Republic of China
e-mail: dongzhen@vip.sina.com

L. Wang et al. (eds.), *Proceedings of the 7th China High Resolution Earth Observation Conference (CHREOC 2020)*, Lecture Notes in Electrical Engineering 757, https://doi.org/10.1007/978-981-16-5735-1_38

To obtain the velocity of ships, scholars at home and abroad have proposed a variety of algorithms for estimation of ship velocity [1–10]. Most of these algorithms are based on extra information or require special configurations which bring restrictions for traditional method in real applications [11]. Little attention has been paid to the motion information contained in the SAR data itself [12]. In this paper, we estimate the azimuth velocity by measuring the defocusing effect of moving ships in SAR images.

This paper is organized as follows. First, the signal model of moving ships are introduced in Sect. 2. In Sect. 3, an autofocus algorithm based on Minimum-entropy criterion and Newton's method is proposed to estimate azimuth velocity and improve the image quality of moving ships. AIS data and Gaofen-3 (GF-3) complex data is used to validate the proposed method and the experimental results are shown in Sect. 4. Finally, conclusions are drawn in Sect. 5.

2 Signal Model

In this paper, only strip map SLC SAR data are concerned and the squint angle as well the beam center time being zero for all the following derivations. The geometry model between spaceborne SAR and moving ships is shown in Fig. 1.

After demodulation and range compression, the azimuth time domain echo [13] of the ideal moving ship scattering points can be expressed as

$$s_a(t_a) = A_0 \times w_a(t_a - t_c) \times \exp\left(-j\frac{4\pi R(t_a)}{\lambda}\right) \tag{1}$$

where A_0 is an arbitrary complex constant proportional to the target reflection coefficient, and t_a and t_c respectively represent the slow time and the beam centre deviation time.$w_a(\cdot)$ is the sinc square type azimuth envelope and λ represents radar

Fig. 1 SAR geometry between satellite and ships

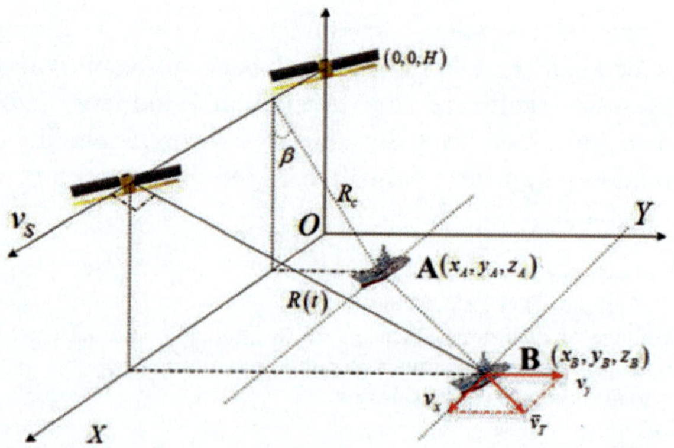

wavelength. At t_a, the $R(t_a)$ is the distance between radar and target can be approximately expressed as follow

$$R(t_a) \approx R_c + v_r(t_a - t_c) + \frac{(v_S - v_x)^2}{2R_c}(t_a - t_c)^2 \tag{2}$$

where v_S is speed of the radar platform which moves along the azimuth direction. The $v_r = v_y \sin\beta$ is the radial velocity, β is the look angle. v_x and v_y respectively represent the azimuth velocity and ground range velocity of the ship. R_c and y_A respectively stand for the range and ground range between the target and the platform when $t_a = 0$.

The Doppler centroid is

$$f_{dc} = -\frac{2v_r}{\lambda} \tag{3}$$

In the azimuth time domain, the Doppler FM rate of the moving ship can be expressed as

$$f_{dr} = -\frac{2}{\lambda R_c}(v_S - v_x)^2 \tag{4}$$

The Doppler FM rate of the stationary target can be expressed as

$$f_{dr}^{*} = -\frac{2v_S^2}{\lambda R_c} \tag{5}$$

It can be found that the moving ships' Doppler FM rate changes comparing with the stationary target, and the difference due to ship's motion is

$$\Delta f_{dr} = f_{dr} - f_{dr}^{*} = -\frac{2}{\lambda}\frac{v_x^2 - 2v_S v_x}{R_c} \tag{6}$$

Assuming the matched filtering is performed according to the stationary scene Doppler FM rate, there will be a mismatch and result in an azimuth blur. Therefore, autofocus algorithms will be applied to make the images clear again. The next section will introduce an Autofocus algorithm based on Minimum-entropy criterion and Newton's method, which can estimate azimuth speed of the moving ship and refocus the image simultaneously.

3 Methodology

As the analysis in the previous section, autofocus algorithms can remove the influence of ships' movements, and the autofocus technologies have been pretty mature [14]. The pixel value after the autofocus processing can be expressed as [15]

$$s_m(n) = \frac{1}{N}\sum_{k=0}^{N-1} S_m(k)\exp[j\varphi(k)]\exp\left[j\frac{2\pi}{N}kn\right] \quad (7)$$

where m represents the mth range unit, n is the azimuth position, N is azimuth sample number, and $S_m(k) = DFT[s_m(k)]$ represents the SAR image of the m-th range unit in the Doppler domain after the azimuth Fourier transform. k is azimuth frequency, and $\varphi(k)$ is the correction phase of the azimuth frequency domain, which is the opposite of the error phase, namely $\varphi(k) = -\hat{\varphi}_e(k)$. Therefore, to refocus the moving ship is to find proper $\varphi(k)$.

3.1 Minimum-Entropy Criterion

Information entropy is to measure the average amount of information produced by a probabilistic stochastic source of data mathematically. As for SAR images, if it has higher entropy, that means there are more changes and noise in pixel values. The SAR images with smaller entropy have been better focused. Thus, SAR images autofocus is executed based on the global minimum-entropy criterion to find the optimum $\varphi(k)$.

The image entropy is formulated as

$$E(m,n) = -\frac{1}{P}\sum_{m=0}^{M-1}\sum_{n=0}^{N-1}|s_m(n)|^2\sum_{i=1}^{n}\ln|s_m(n)|^2 + \ln P \quad (8)$$

where

$$P = \sum_{m=0}^{M-1}\sum_{n=0}^{N-1}|s_m(n)|^2 \quad (9)$$

Therefore, the estimation of $\varphi(k)$ based on the minimum-entropy criterion can be formulated as

$$\varphi = \arg\min_{\varphi}[s_m(n);\varphi(k)] \quad (10)$$

Considering that Newton's method is the most widely used algorithm with better optimization and high efficiency at present, the following will introduce finding the optimum $\varphi(k)$ based on Newton's method.

3.2 *The Optimum Correction Phase Estimation Based on Newton's Method*

The basic notion of Newton's method is to find the next estimated minimum by means of conducting a second-order Taylor expansion of the objective function in the vicinity of the existing estimated minimum. Setting the image entropy of the moving ship as the objective function, the correction phase of the moving ships can be acquired, and the phase value corresponding to the minimum entropy is what we want.

Assuming that φ_l is the estimated correction phase in the lth iteration, the second-order Taylor approximation for the image entropy $E(s|\varphi)$ can be formulated as

$$E(s|\varphi) = E(s|\varphi_l) + \nabla E(s|\varphi_l)(\varphi - \varphi_l) + \frac{1}{2}\nabla^2 E(s|\varphi_l)(\varphi - \varphi_l) + o\left(|\varphi - \varphi_l|^2\right) \tag{11}$$

Ignore the high-order infinitesimals, using $T(s|\varphi)$ to represent it. At the minimum, the first derivative of $T(s|\varphi)$ is zero that is

$$\nabla T(s|\varphi) = \nabla T(s|\varphi_l) + \nabla^2 E(s|\varphi_l)(\varphi - \varphi_l) = 0 \tag{12}$$

Taking the stagnation point of $T(s|\varphi)$ as $E(s|\varphi)$'s to derive the estimated correction phase in the $l+1$th iteration, it can be expressed as

$$\varphi_{l+1} = \varphi_l - \frac{\nabla E(s|\varphi_l)}{\nabla^2 E(s|\varphi_l)} \tag{13}$$

As shown in Eq. (13), that the second-order derivative of objective function is continuous is a necessary condition to execute optimization with Newton's method.

The first-order derivative of the ship image entropy with respect to the correction phase can be formulated as

$$\nabla E(s|\varphi) = \frac{\partial E(s|\varphi)}{\partial \varphi(k)} = \frac{2}{N}\mathrm{Im}\left\{\sum_{k=0}^{N-1} S_m(k)\exp(j\varphi(k)) \cdot S_m^*(k)\right\} \tag{14}$$

where

$$S_m(k) = \sum_{n=0}^{N-1} \left(1 + \ln|s_m(n)|^2\right) \cdot s_m(n) \exp\left[j\frac{2\pi}{N}kn\right] \tag{15}$$

The second-order derivative of the ship image entropy with respect to the correction phase can be formulated as

$$\begin{aligned} \nabla^2 E(s|\varphi) = \frac{\partial^2 E(s|\varphi)}{\partial\varphi^2(k)} \approx -\frac{2}{N^2}\sum_{m=1}^{M}\sum_{n=1}^{N}\left[|S_m(k)|^2 \cdot \left(2 + \ln|s_m(n)|^2\right)\right] \\ + \frac{2}{N}\mathrm{Re}\left\{\sum_{k=1}^{N} S_m(k)\exp(j\varphi(k)) \cdot S_m^*(k)\right\} \end{aligned} \tag{16}$$

Substituting Eq. (14) Eq. (15) and Eq. (16) into Eq. (13) we can obtain the estimated correction phase in the $l+1$th iteration. Finding the calibration phase corresponding to the minimum-entropy through the loop iteration, and the estimation corrected phase $\hat{\varphi}(k)$ of the moving ships in the Range-Doppler domain can be obtained. Then, phase compensation will be performed to alleviate the image blur caused by the motions of the ships.

3.3 Ship Azimuth Velocity Estimation

The phase error in the Range-Doppler domain caused by azimuth velocity is [13]

$$\varphi_e(k) = \frac{\Delta f_{dr}}{(f_{dr}^* + \Delta f_{dr})f_{dr}^*}\frac{k^2}{\pi} \approx \frac{\Delta K_a}{\pi f_{dr}^{*2}}k^2 \tag{17}$$

where k is azimuth frequency, assuming that $\hat{\varphi}(k)$ can be fitted and formulated as:

$$\hat{\varphi}_e(f_a) = a_1 + b_1 f_a + c_1 f_a^2 \tag{18}$$

Considering $\varphi(k) = -\hat{\varphi}_e(k)$, according to Eq. (17), there be

$$\Delta\hat{f}_{dr} = -c_1 \cdot \pi f_{dr}^{*2} \tag{19}$$

According to Eq. (6), we can get

$$2v_x^2 - 4v_S v_x + \lambda R_c \Delta f_{dr} = 0 \tag{20}$$

From the root-finding formula of the quadratic equation in one variable, and considering the fact, the speed of ship cannot be greater than the radar', the azimuth velocity of the ship can be formulated as

$$v_x = v_S - \sqrt{v_S^2 - \lambda R_c \Delta f_{dr}/2} \quad (21)$$

Substituting Eq. (19) into Eq. (21), we can obtain

$$\hat{v}_x = v_S - \sqrt{v_S^2 + \lambda R_c \cdot c_1 \cdot \pi f_{dr}^{*2}/2} \quad (22)$$

When $\hat{v}_x$ is negative, it means that the azimuth velocity direction of the ship is opposite to the radar.

4 Result

Experiments were executed utilizing GF-3 satellite SLC image data. Automatic Identification System (AIS) data was considered as reference in this paper, and several ship targets were analyzed to compare the velocity estimated by proposed approach with AIS data. AIS is a system to manage vessel traffic [16] and provide the ground truth data about ships. Besides, we made comparison between proposed method and the algorithm in literature [10].

The basic satellite parameters are listed in Table 1.

Considering both SAR and AIS data, four ships were selected from strip-map images of GF-3 satellite in the sea area near Changxing Island of China. The red box was the selected ships, as shown in Fig. 2.

The chose ships were processed with the proposed method and the approach in literature [10]. Figure 3 shows the experimental results. where the negative sign indicates the velocity direction is opposite to *x*-axis direction. It is apparent that the proposed approach has higher precision. Quantificationally, the mean absolute error (MAE) and the standard deviation of proposed method respectively are 0.3925 m/s and 0.4093m^2/s^2, while those of the approach in literature [10] respectively are 0.9350 m/s and 0.9565m^2/s^2.Therefore, the proposed approach in this paper has higher precision to estimate azimuth velocity.

On the other hand, while obtaining the azimuth velocity of the ships, the proposed method also makes the image clear. In general, visual effects, the image entropy, contrast and peak values are the parameters to evaluate the refocusing performance and defocusing extent. Figure 4 shows the visual effects of four ships,

Table 1 GF-3 satellite parameters

Parameters	Value	Parameters	Value
PRF(Hz)	4055.36	R_c(m)	1,050,710.56
B_r(MHz)	80	$S_{azimuth}$(m)	1.66
B_a(Hz)	2469.43	$\lambda_{(m)}$	0.055517
R_{ref}(m)	1,065,442.01	v_S(m/s)	6734.80
R_{near}(m)	1,053,446.56	SquintAngle θ(°)	0

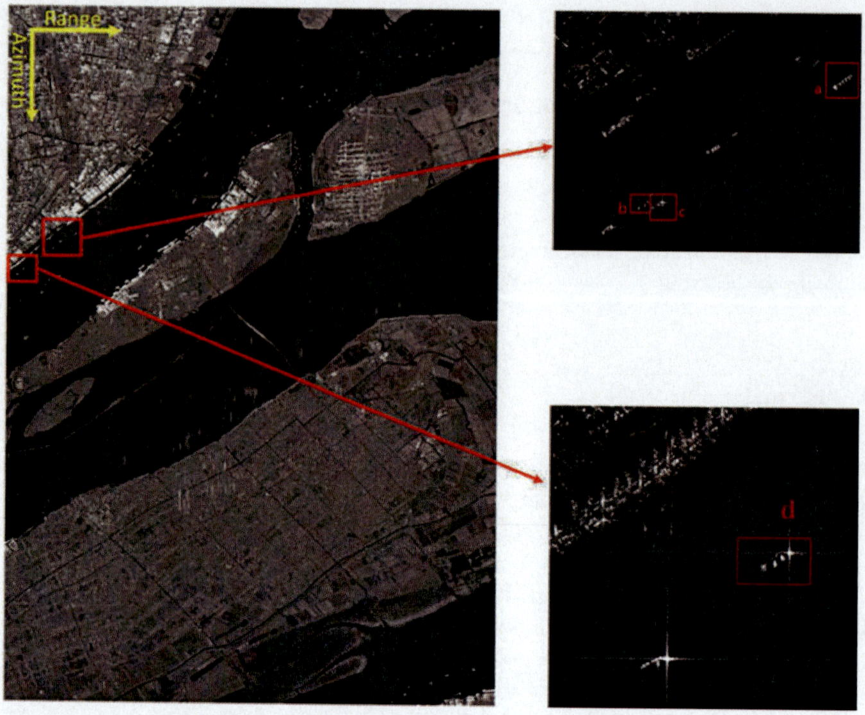

Fig. 2 Ships in GF-3 SAR image near Changxing Island

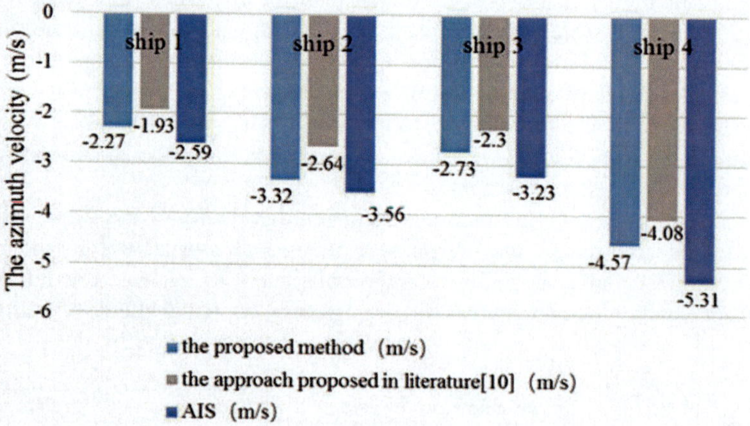

Fig. 3 The comparison of two algorithm

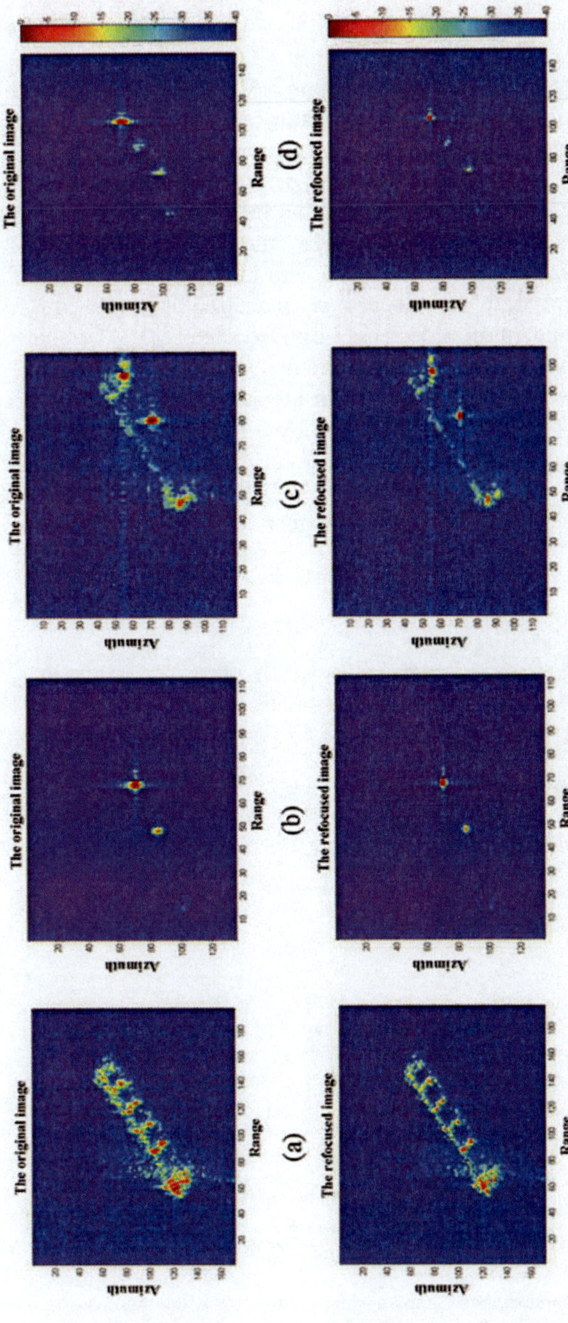

Fig. 4 Ships near Changxing Island **a–d** are unfocused ships **a1–d1** are refocused ships

it is obvious that the outline of the focused target is clearer, and the image quality is improved.

Besides, from the experiments results, the ship 1 and 4 are 100-m-class ship targets, the targets 2 and 3 are 10-m-class ship targets, and they occupy different pixels in the image; the proposed algorithm has fairy behavior for all the four targets. To conclude, the proposed algorithm can obtain high-precision speed estimation and splendid refocusing performance for ships with different size.

As for the contrast, Image entropy, and peak value, the changes between the original and the focused four ships are shown in Fig. 5. Figure 5a–c respectively display the changes of the contrast, image entropy, and peak value. From the experimental results, the images refocused with the proposed method, the contrast of the four ship targets 1, 2, 3, 4 respectively increased by 0.1698, 1.7804, 0.2272, 1.9882, and the image entropy decreased by respectively 0.3119, 0.6917, 0.5129, 0.9995. The peak values increased by respectively 345.4824, 3925.1116, 358.6191, 4882.6403. Processed with the approach in literature [10], the contrast of the four ship targets 1, 2, 3, 4 respectively increased by 0.132, 1.2071, 0.1829, 1.0877, and the image entropy decreased by respectively 0.2302, 0.4547, 0.3719, 0.5528. The peak values increased by respectively 300.3188, 2962.7730, 276.5770, 3057.7224. And it can be found that the defocusing extent of the ship 4 is the most serious,

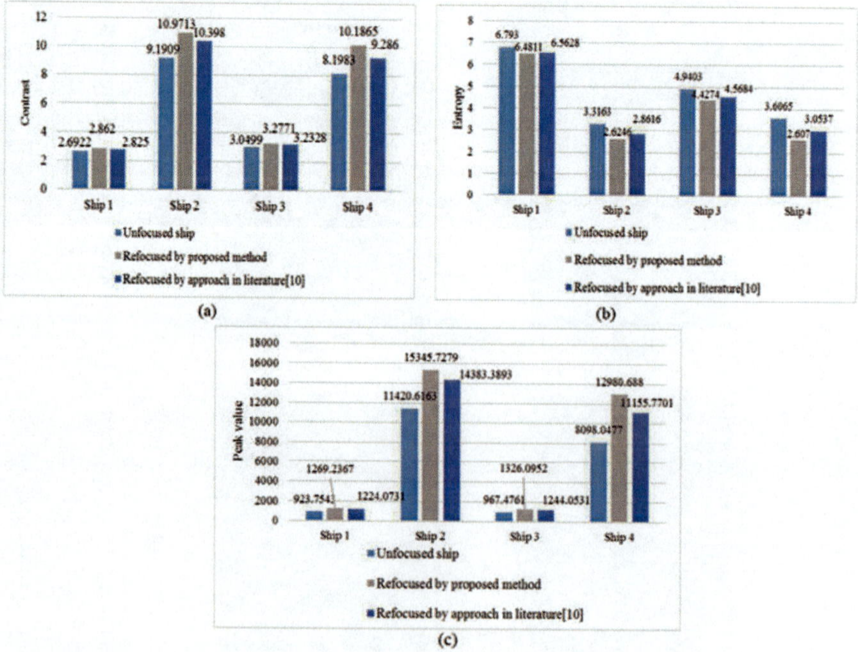

Fig. 5 Image quality parameters of ships **a–c** Respectively display the changes of the contrast, Image entropy, and peak value between the original and the focused with two methods

which can be account for that its' the azimuth velocity is biggest among four ships. Comparing the refocusing performance of the two algorithms, the four ship targets 1, 2, 3, 4 processed with the method in this paper, the contrast increased by 1.38%, 6.24%, 1.45%, 10.98% more than the other method did, and the image entropy decreased by respectively 1.2%,7.14%, 2.85%, 12.38% more. The peak values increased by respectively 4.89%, 8.43%, 8.48%, and 22.53% more. Apparently, the proposed method performed better in refocusing ships.

In summary, the proposed method has better performance than the approach in literature [10].

5 Conclusion

In this paper, a novel approach to estimate azimuth velocity of moving ships in spaceborne SAR is proposed, which is capable of improving the image quality as well, and has better performance than some present methods. The proposed algorithm is based on Newton's method to search the calibration phase corresponding to the minimum-entropy. And phase compensation effectively alleviates the image blur. Furthermore, the results of processing the GF-3 complex data in the sea area near Changxing Island of China perfectly fit the data recorded by the international automatic identification system (AIS). In conclusion, the proposed approach can estimate the ship velocity with high precision, and improve the image quality with magnificent performance as well. However, this paper obtained the Doppler FM rate and azimuth velocity of ships by fitting the quadratic coefficient of the corrected phase, which may bring great errors when there is a jump value. In addition, selecting the proper section and removing the effects of edge value is essential to the precision of fitting. Decreasing the biases of fitting is a new task to fulfill.

References

1. Ouchi K, Iehara M, Morimura K et al (2002) Nonuniform azimuth image shift observed in the Radarsat images of ships in motion. IEEE Trans Geosci Remote Sens 40(10):2188–2195
2. Dragosevic MV, Vachon PW (2008) Estimation of ship radial speed by adaptive processing of RADARSAT-1 fine mode data. IEEE Geosci Remote Sens Lett 5(4):678–682
3. Zilman G, Zapolski A, Marom M (2004) The speed and beam of a ship from its wake's SAR images. IEEE Trans Geosci Remote Sens 42(10):2335–2343
4. Eldhuset K (2002) An automatic ship and ship wake detection system for spaceborne SAR images in coastal regions. IEEE Trans Geosci Remote Sens 34(4):1010–1019
5. Jin-song C, Yue O, Min-hui Z (2006) Detection of ocean target in synthetic aperture radar imagery. Ocean Press, Beijing, pp 96–105
6. Renga A, Moccia A (2014) Ship velocity estimation by Doppler centroid analysis of focused SAR data. In: Proc. IEEE international geoscience and remote sensing symposium, Quebec City, Canada, pp 1809–1812

7. Sun H, Wang X, Chong J (2012) Ocean dynamic information acquisition based on matching SAR ocean sub-aperture sequence images. J Electron Inf Technol 34(1):179–186
8. Renga A, Moccia A (2016) Use of doppler parameters for ship velocity computation in SAR images. IEEE Trans Geosci Remote Sens 54(7):1–17
9. Fei-xiang W, Xiao-qing W, Jin-song C (2018) Ship azimuthal speed estimation method based on local region Doppler centroid in SAR images. J Electron Inf Technol 40(9):2242–2249
10. Ao D, Datcu M, Schwarz G et al (2018) Moving ship velocity estimation using TanDEM-X data based on subaperture decomposition. IEEE Geosci Remote Sens Lett 15(10):1560–1564
11. Gierull CH, Cerutti-Maori D, Ender JHG (2008) Ground moving target indication with tandem satellite constellations. IEEE Geosci Remote Sens Lett 5(4):710–714
12. Lv G, Wang J, Liu X (2013) Ground moving target indication in SAR images by symmetric de-focusing. J. IEEE Geosci Remote Sens Lett 10:241–245
13. Cumming IG, Wong FH (2005) Digital processing of synthetic aperture radar data: algorithms and implementation. Artech House Inc., Norwood
14. Li X, Liu G, Ni J (1999) Autofocusing of ISAR images based on entropy minimization. IEEE Trans Aerosp Electron Syst 35(4):1240–1252
15. Zeng T, Wang R, Li F (2013) SAR image autofocus utilizing minimum-entropy criterion. IEEE Geosci Remote Sens Lett 10:1552–1556
16. Information on http://en.wikipedia.org/wiki/Automatic_identification_system. Accessed 19 Jun 2018

Mapping 30-m Resolution Land Cover of China Based on Full Convolutional Neural Network

Yinhe Liu and Yanfei Zhong

Abstract Large-scale land cover maps are essential geoscience information for many applications. However, the cost of the current mapping process greatly limits the update frequency of large-scale land cover products. Moreover, the inconsistency among land cover maps also hindered further long time-series land cover change analysis and research. Therefore, we proposed a novel framework to efficiently generate new land cover maps using historical land cover products. The framework mainly includes the following three innovations: (1) To solve the influence of clouds and seasonal inconsistency, use Google Earth Engine to composite the multi-temporal remote sensing images to generate consistent cloud-free images; (2) To avoid the repeated collection of training samples, training data were generated by integrating multiple historical products; (3) To process massive generated training data and improve mapping accuracy, the data-driven deep fully convolutional network model is used to achieve end-to-end land cover mapping. Based on the proposed mapping approach, the 30-m resolution land cover map of China in 2015 was automatically completed with improved accuracy, which shows the potential for frequent large-scale land cover product integration and updating.

Keywords Land cover mapping · Google earth engine · Deep learning · Full convolutional network

1 Introduction

Land cover refers to the physical and biological characteristics of the earth's surface, including cropland, forest, shrubs, grass, wetland, water, and artificial surface. Large-scale land cover products can provide accurate national or even global land

---

Y. Liu · Y. Zhong (✉)
State Key Laboratory of Information Engineering in Surveying, Mapping, and Remote Sensing, Wuhan University, Wuhan 430079, Hubei, China
e-mail: zhongyanfei@whu.edu.cn

L. Wang et al. (eds.), *Proceedings of the 7th China High Resolution Earth Observation Conference (CHREOC 2020)*, Lecture Notes in Electrical Engineering 757, https://doi.org/10.1007/978-981-16-5735-1_39

cover information to address issues such as biogeochemical, ecosystem, biodiversity, and climate modeling.

Primitive global land cover products were often mapped at coarser resolutions (300 m–1 km). Finer than 100-m resolution global land cover mapping has been possible since Landsat archive became freely available in 2008. The first 30 m resolution global land cover production Finer Resolution Observation and Monitoring of Global Land Cover (FROM-GLC) was mapped with pixel-based supervised classification methods [1]. Nevertheless, the accuracy and consistency of FROMGLC are not sufficient. Soon afterward, two 30m global land cover maps for the years 2000 and 2010 entitled GlobeLand30 have been produced [2]. The products achieved impressive accuracy with a pixel-object-knowledge-based (POK) classification approach. However, the process of expert interpretation and manual interaction in POK approach makes mapping extremely expensive and time costing.

Moreover, some applications such as deforestation, wetland loss, and urban planning, require not only higher resolutions but also more frequent global land cover maps. Most of the current large-scale land cover mapping methods require great efforts are needed in data acquisition, processing, and training samples collection, which limits the updating frequency of land cover products. In addition, it has also been found that the inconsistency among the land cover maps is more significant than the actual land cover change. These unfavorable factors significantly reduce the usability of existing products for long-term serial land cover change detection and analysis.

In fact, the annual land cover changes only make up only a tiny fraction of the land surface. Which make the existing historical land cover products can be reused as training data. To facility the land cover mapping and integrate the existing land cover products, a novel land cover mapping framework is proposed to classify new land cover maps with finer accuracy and consistency from historical maps. In this framework, the existing land cover datasets are treated as labels to train a new classification network with the annual composite image. Annual composite images and synthetic samples derived from existing land cover data are used to train a deep neural network. The contribution of this paper can be summarized as follows.

1. Google Earth Engine is employed to facilitate image access and preprocessing; a multi-temporal composite method is carried to mitigate the influence of cloud occlusion and seasonal differences.
2. To avoid manually collecting training samples, multisource existing land cover maps and auxiliary crowdsourced data are fused to generate training dataset.
3. A modified semantic segmentation network is used to learn from massive generated training labels; refined 30 m resolution land cover map of China in 2015 is completed.

2 Land Cover Mapping Framework

The proposed land cover framework is shown as Fig. 1. The mapping process contains two parts: Composite image and synthetic training samples were prepared on Google Earth Engine (GEE) at first [3]. Afterward, using the synthetic data to train a modified semantic segmentation network, and the land cover map was driven from the trained model.

2.1 *Multi-temporal Image Composition*

Because Landsat images are often covered by clouds, it is difficult to obtain high-quality cloudless images in some areas. Moreover, even if cloudless images are acquired, there are considerable differences between the images due to the different acquisition times, which are not conducive to large-scale land cover mapping.

Therefore, in order to solve the impact of the problems mentioned above on the classification of land cover, first, pixel-by-pixel synthesis of Landsat 8 whole-year images: First, calculate the cloud index for each pixel for all available images, and filter cloudy pixels according to a certain threshold. The median values of the remaining pixels are calculated per band. Finally, a composite image is obtained from all the images throughout the year, excluding the influence of clouds and seasons.

The multi-temporal image composite is implemented by GEE [3], which is a PB-scale geospatial data cloud computing platform. GEE provides convenient access and process for worldwide coverage of satellite imagery like MODIS, Landsat, Sentinel, and other geoscience data. GEE has therefore been widely used for large-scale geological data processing and analysis.

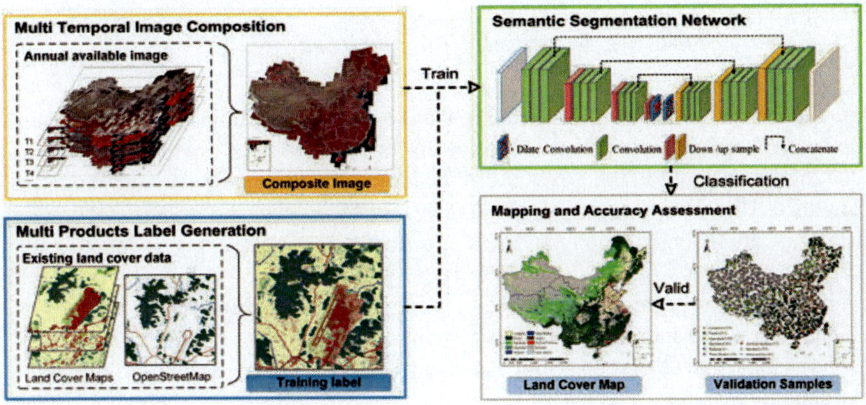

Fig. 1 Proposed land cover mapping framework

2.2 Multi-products Based Label Generation

When performing automatic land cover classification, manually labeled training samples are needed, but this process is costly. At larger scales, such as national and global land cover mapping, changes in land cover due to human activities or natural forces often only account for a small part of the overall scope, which makes land cover data Reusable. So we treat existing land cover products as samples. However, it is easy to introduce system errors that are difficult to eliminate for the classifier using only a single data source as a sample. Therefore, multi-source sample data is fused to provide more reliable sample information.

This paper refers to the classification system of GlobeLand30 and FROM-GLC products, which also use Landsat as the primary data source. The classification system used includes ten classes, namely cropland, forest, grassland, shrubland, wetland, water bodies, tundra, artificial surfaces, bare land, and snow (ice). All the used land cover data are mapped into this classification system. Afterward, the land cover maps and vector features were first reprojected to the same geographic coordinate system (WGS-1984). Then the data were resampled into 30m resolution by interpolation and rasterization. The land cover maps are fused based on corresponding confidence:

$$S_{i,j} = \sum_{1}^{L} \left(p_{i,j}^{L} * w^{L}\right) / Z$$

where $Z = \sum_{1}^{L}(w^{L})$ for normalization, $S_{i,j}$ denotes the generated result in location i, j, L denotes the land cover maps to be fused, $p_{i,j}^{L}$ and w^{L} denotes the land cover class probability and confidence of land cover map L. Since the confidence of the product is unknown, we use corresponding user precision instead.

2.3 Deep Semantic Segmentation Network

For a given image and training samples, the classification methods play a vital role in the final mapping results. Support vector machine (SVM) and random forest are the most commonly used land cover classification methods because of their stable performance. However, these methods are difficult to process the millions of training data after multi-source sample synthesis. In recent years, data-driven deep learning methods have been proven to achieve better performance in surface cover classification, and are suitable for incremental learning from massive synthetic samples. An efficient deep semantic segmentation network is modified from U-Net [4]:

The input images first pass into a sequence of layer blocks, namely "encoder", to extract the semantic features of the input images. Meanwhile, adding a residual connection unit in each resolution level to achieve the fusion of low-level features and high-level features facilitates the propagation of gradients, which can significantly improve the classification accuracy without increasing model parameters. In addition, the batch normalization (BN) [5] layer is now always used after convolutional layer to facilitate model training. However, since semantic segmentation models often have a large number of trainable parameters, and the size of remote sensing images is significantly larger than that of conventional natural images, the batch size during training is usually small, which will significantly decrease the performance of BN and even increase the error of the model. Different from BN, group normalization (GN) [6] is not normalized in the dimension of the batch, but grouped in the dimension of the band, and then normalized each group, thus avoiding the batch size dependence. This makes the GN normalization method significantly better than BN in the case of small batches, so the normalization method of GN instead of BN is selected in this network to speed up the training process of the model, and replace the Dropout layer to prevent the model from overfitting occurs.

After passing the encoder, because the size of the geo-objects in remote sensing images is various, it is difficult to extract enough texture information from a limited receptive field. To detect large geo-objects and complex global spatial relationships, excessive pooling will significantly reduce the resolution of remote sensing images, making it difficult to restore the original spatial details. At the same time, simply increasing the size of the convolution kernel will further increase the amount of network parameters that are already relatively complex. Therefore, dilated convolution kernels under different dilate rates are used in the middle of the network to extract multi-scale features while maintaining the resolution.

Finally in the part symmetrical to the encoder, namely "decoder", the feature maps in the same level of resolution are concatenated to restore more spatial details. In order to avoid the grid effect brought by deconvolution, we use simple bilinear interpolation to restore the resolution of the image[7]. Finally, through the Softmax layer, the output of the model is converted into the probability of each pixel's land cover type, and outputs the classification result (Fig. 2).

3 Experimental Analysis

China covers all the land cover classes except tundra and has made significant development in recent years. China was selected therefore as the area of the experiment because of its vast area and rich ecological environment.

3.1 Experiment Design

All available Landsat8 top of atmosphere (TOA) images were acquired and composited on GEE. All bands except the thermal bands were used for classification.

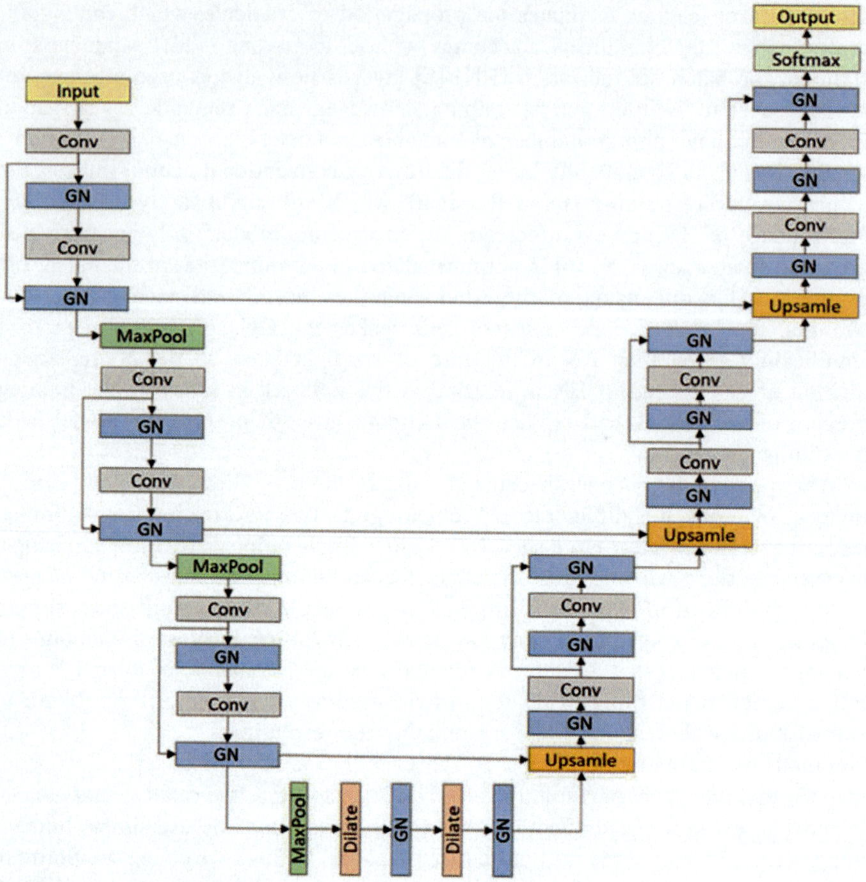

Fig. 2 Structure of the modified U-Net semantic segmentation network

GlobeLand30 [2], FROMGLC version2 (http://data.ess.tsinghua.edu.cn), and FROMGLC10 [8] were used to synthesize samples, and the OSM features were downloaded as auxiliary data.

The image composite and sample generation steps are carried on GEE with JavaScript, and the classification model training and testing are implemented by Python with PyTorch framework and run on a server with four NVIDIA Titan X Pascal GPU accelerators. The main land of China is divided uniformly into 115 map sheets of dimensions 4° in longitude and 3° in latitude. All annual composite Landsat images and synthetic samples are used to train the proposed network. The training image and corresponding masks are cropped randomly at a $256 * 256$ patch size, and the batch size is 12. Adam optimizer is used to avoid more parameters, and the learning rate is 0.001 as default. The model parameters were randomly initialized without pre-training.

In order to make a more reasonable assessment of the large-scale research area in this paper, a set of independent and uniformly distributed verification sample data sets are used [9]. Most of the samples in this sample set have been verified by Google Earth's high-resolution images, with very high confidence. The samples were checked and updated with the help of Google Earth's high-resolution image in 2015. The spatial distribution of the map sheets and the valid samples are shown in Fig. 3.

3.2 *Mapping Result and Validation*

Figure 3 illustrates the 30-m land cover map for 2015 of China derived from annual composite image and synthesis samples. The result is consistent with the true distribution of land cover of China and the main water bodies, cities, grasslands, forests, and bare land can be distinguished on the map. Besides, the mosaic effect caused by using single date images for classification is slight with our annual composite image.

Table 1 shows the confusion matrix validated against the validation samples. The overall accuracy of the derived land cover map reaches 70.33%, which is higher than the FROMGLC version2 of 60.93% under the same validation samples. However, there are also some problems with this result. Because the number of

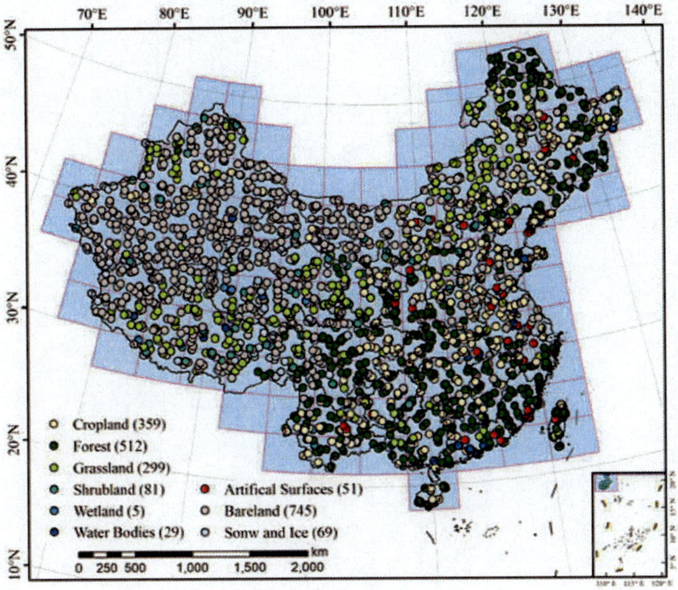

Fig. 3 Distribution of the map sheets and valid samples

Table 1 Confusion matrix of land cover mapping result

	Cropland	Forest	Grass	Shrubland	Wetland	Water	Impervious	Bareland	Ice/ Snow	Total
Cropland	258	63	23	0	0	3	11	1	0	359
Forest	28	469	10	0	0	0	5	0	0	512
Grass	18	58	192	0	0	1	1	29	0	299
Shrubland	1	21	41	0	0	2	0	16	0	81
Wetland	0	0	2	0	1	0	0	2	0	5
Water	4	0	1	0	0	20	3	1	0	29
Impervious	11	4	2	0	0	1	32	1	0	51
Bare land	71	14	130	0	0	5	5	520	0	745
Snow/ice	0	0	26	0	0	1	0	22	20	69
Total	391	629	427	0	1	33	57	592	20	

Overall accuracy = 70.33% Kappa coefficient = 0.6053

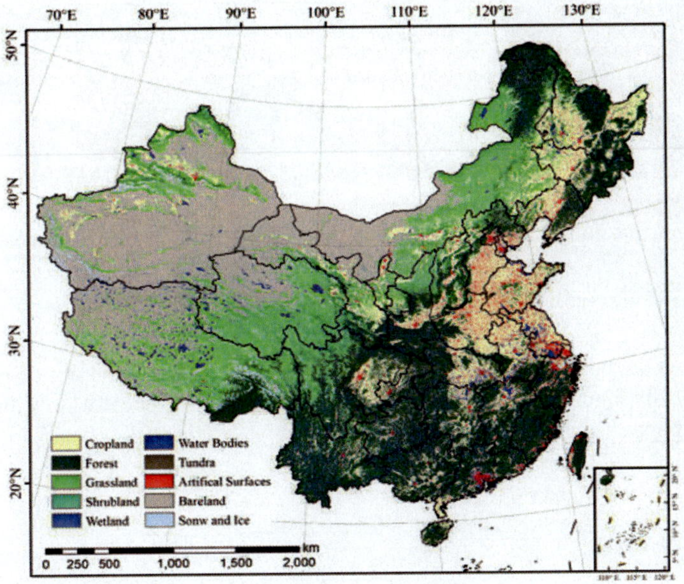

Fig. 4 The 30-m resolution land cover map of China in 2015

shrubs and wetlands in the synthetic sample accounted for a tiny proportion, the two classes were significantly misclassified. Besides, the confusion between grassland and bareland is also obvious (Fig. 4).

3.3 *Comparison with Existing Product*

Figure 5 gives an example of the comparison between the existing product and our result. The land cover map of this paper compares with the FROM-GLC surface cover products in the same year in 2015. It can be seen that FROM-GLC has an obvious band on the left side of figure, which corresponds to the edge of the two row and column images of Landsat images 124-37 and 123-37, which shows that in FROM-GLC land cover products are prone to appear inconsistency at the edges of the images. However, due to the multi-temporal image composite method, seamless stitching can be achieved in the overlapping areas of different images in our result. At the same time, since FROM-GLC uses a random forest pixel-by-pixel classification method, it is susceptible to interference from spectral information and confusion between categories. It can be seen that FROM-GLC has significantly underestimated the impervious surface category, and classified the villages around the city as cultivated land, and there is a large amount of "salt and pepper noise"

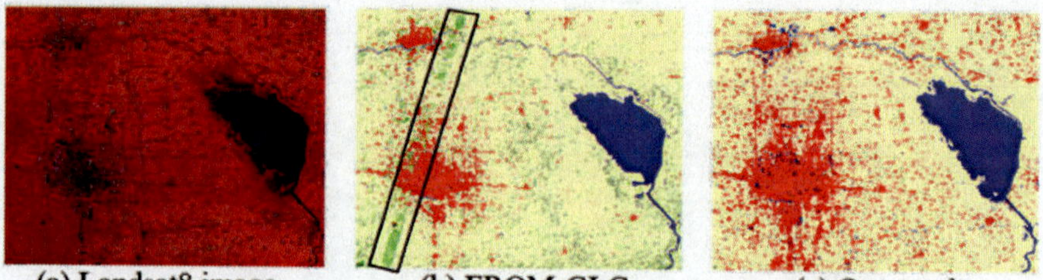

Fig. 5 Comparison between the existing land cover products and our result

in the forest land among the cultivated land. However, the method in this paper can effectively distinguish the cities and towns, and there is basically no influence of "salt and pepper noise".

4 Conclusion

In this paper, we proposed a novel large-scale land cover mapping framework. Annual composite images and synthetic samples derived from existing land cover data are used to train a semantic segmentation network. The result in China shows that the proposed framework can efficiently complete largescale land cover mapping with refined accuracy, which shows the potential for frequent global-scale land cover product integration and updating of the proposed mapping approach.

Acknowledgements This work was supported by National Natural Science Foundation of China under Grant Nos. 41771385.

References

1. Gong P, Wang J, Yu L et al (2013) Finer resolution observation and monitoring of global land cover: First mapping results with Landsat TM and ETM+ data. Int J Remote Sens 34(7):2607–2654
2. Chen J, Chen J, Liao A et al (2015) Global land cover mapping at 30 m resolution: a POK-based operational approach. ISPRS J Photogrammetry Remote Sens 103:7–27
3. Gorelick N, Hancher M, Dixon M et al (2017) Google earth engine: planetary-scale geospatial analysis for everyone. Remote Sens Environ 202:18–27
4. Ronneberger O, Fischer P, Brox T (2015) U-Net: convolutional networks for biomedical image segmentation. In: Medical image computing and computer-assisted intervention, pp 234–241
5. Ioffe S, Szegedy C (2015) Batch normalization: accelerating deep network training by reducing internal covariate shift. arXiv preprint arXiv:1502.03167
6. Wu Y, He K (2018) Group normalization. In: Proceedings of the European conference on computer vision (ECCV), pp 3–19

7. Odena A, Dumoulin V, Olah C (2016) Deconvolution and checkerboard artifacts. Distill 1(10): e3–e3
8. Gong P, Liu H, Zhang M et al (2019) Stable classification with limited sample: transferring a 30-m resolution sample set collected in 2015 to mapping 10-m resolution global land cover in 2017. Sci Bull 64(6):370–373
9. Zhao Y, Gong P, Yu L et al (2014) Towards a common validation sample set for global land-cover mapping. J Remote Sens 35(13):4795–4814

Research on Brightness Compensation Algorithm for GF-3 SAR Images

Zengguo Sun, Xiaopeng Yan, Qianfu Chen, and Weirong Chen

Abstract GF-3, the first C-Band full-polarimetric synthetic aperture radar (SAR) satellite with a space resolution up to 1 m, has multiple strip and scan imaging modes. In this paper, a brightness compensation model is proposed to correct the unbalanced radiation image of GF-3 satellite in some areas based on mathematical morphology principle, SAR imaging model, and histogram mapping model, so that the image radiation is relatively consistent and has moderate contrast. The proposed algorithm is tested on the GF-3 SAR image with radiation difference in Taklimakan desert area of Xinjiang, and is compared with the MASK dodging algorithm in terms of brightness mean, standard deviation, information entropy, and average gradient. Results of brightness compensation show that, the proposed algorithm makes the radiation differences within images smaller, and the image quality is obviously improved, which provides a non-radiation difference image for subsequent applications.

Keywords GF-3 synthetic aperture radar images · Brightness compensation · Mathematical morphology · Histogram mapping

Z. Sun (✉)
Key Laboratory of Modern Teaching Technology, Ministry of Education, Xi'an, China
e-mail: sunzg@snnu.edu.cn

Z. Sun · X. Yan
School of Computer Science, Shaanxi Normal University, Xi'an, China

Q. Chen
Land Satellite Remote Sensing Application Center, MNR, Beijing, China

W. Chen
China Centre for Resources Satellite Data and Application, Beijing, China

L. Wang et al. (eds.), *Proceedings of the 7th China High Resolution Earth Observation Conference (CHREOC 2020)*, Lecture Notes in Electrical Engineering 757, https://doi.org/10.1007/978-981-16-5735-1_40

1 Introduction

When using synthetic aperture radar (SAR) satellite image data for digital orthophoto map (DOM) product mosaic production and multi-source image fusion, it is required that the brightness of SAR image is balanced, the contrast is moderate and the color is consistent [1]. Since SAR is a microwave slant range imaging radar, the amplitude information (or brightness information) of SAR images reflects the backscattering intensity of ground targets to radar waves [2, 3]. Complex texture features of terrain objects and imaging geometry make the amplitude of the SAR image have a large dynamic range, therefore, the radiation of SAR image data acquired by different time phases and imaging modes is quite different, the main manifestation is that the radiation (or brightness) of SAR image is not balanced. GF-3 satellite is the first civil SAR radar in China. The uneven brightness of SAR images is a problem when using GF-3 images, which makes it difficult to use and produce SAR images. We focus on the uneven brightness of the GF-3 image. Firstly, we introduce the problems of SAR image and the current brightness compensation algorithm in this paper. Then the flowchart and principle of the proposed algorithm are elaborated in detail. Finally, the experiment and result analysis prove that the proposed algorithm can achieve better brightness compensation effect.

2 Current Compensation Algorithm Analysis

GF-3 satellite is a multi-imaging mode and multi-polarization C-band SAR satellite independently developed in China. It can monitor global ocean and land information all day and all weather [4]. At present, with the continuous and stable operation of GF-3 in orbit, the application of massive SAR image data has been more and more in-depth research [5]. During the period of making National Orthophoto Map of Radar Remote Sensing with GF-3 data [6], we found that in the data collected in Taklimakan Desert area have the phenomenon of uneven brightness in close range and dark in long range. 25 of 76 images have the phenomenon of uneven brightness, which need radiometric correction. Therefore, it is necessary to study the brightness compensation processing algorithm for GF-3 SAR images. The example of GF-3 uneven brightness image is shown in Fig. 1.

Up to now, domestic and foreign scholars have done a lot of research on the algorithm of brightness compensation for remote sensing images. According to the principle and model of brightness compensation algorithm, the main algorithms can be divided into three categories [7]: based on the additive noise model algorithm, the illumination-reflectance model algorithm and the statistical method of the brightness compensation algorithm. The additive noise model algorithm considers that the remote sensing image with uneven brightness distribution result from the remote sensing image with uneven brightness distribution and the noise image with

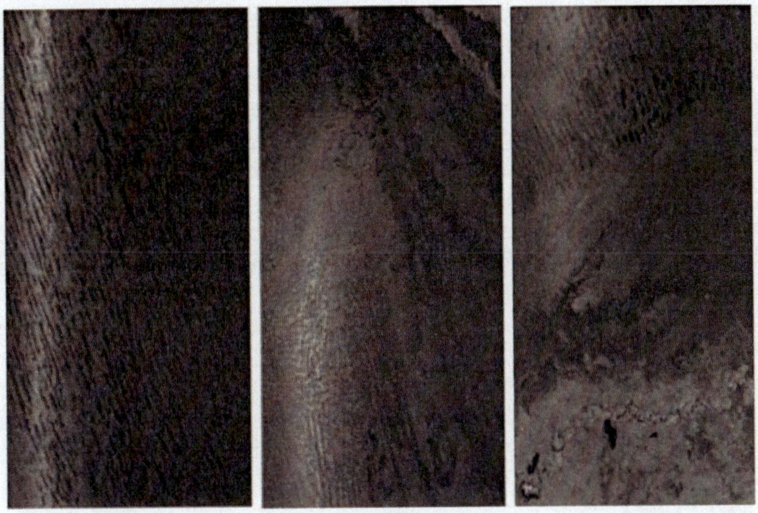

Fig. 1 Example of uneven brightness in GF-3 image

uneven brightness distribution. Its representative algorithm is Mask brightness compensation method based on photo printing technology [8]. This algorithm has a wide range of applications and can filter out most of the image details to obtain a better uniformity effect, but it is easy to produce halo around the image, resulting in color cast in the brightness compensation image. An improved algorithm is proposed for parameter selection of background image filter based on adaptive determination in [9]. According to the image information and optimizing the adaptive filter parameters, the stretching parameters of the image are automatically adjusted to solve the problem of unbalanced brightness of the image. The algorithm based on illuminance-reflectance model considers that remote sensing image is composed of the product of illumination and scene reflectance. The classical algorithm is Retinex brightness compensation method [10]. The theory of this algorithm is derived from the Retinex model proposed by Edwin [11], which is used by human visual system to perceive the color and brightness of the scene. This algorithm is mainly used to correct the uneven brightness in a single image, and the effect of dealing with the uneven of illumination is better, but the contrast and clarity of the processed image are reduced to a certain extent. By replacing the logarithmic transformation with the fast alternating direction optimization algorithm, the existing Retinex algorithm is improved in [12]. The results show that the improved method achieves good results in eliminating the creases and uneven brightness of the scanned map. The algorithm based on statistical method considers that the hue and brightness of remote sensing image are expressed by its mean value, and the clarity and contrast of image are expressed by its variance [13]. The classical algorithm is Wallis filtering algorithm. The principle of this algorithm is to map the mean and variance of image to the given mean and variance of gray level through Wallis filter, so that the mean and variance of different regions are approximately equal [14]. The algorithm has an obvious effect on the brightness compensation of multiple images, but its effect on single image is not obvious, and it will produce "block phenomenon" [15]. An improved Wallis algorithm is proposed in [16], which calculates the mean and variance in a fixed range centered on each point to be processed in turn, compares with the statistical parameters of the standard image, and effectively avoids the "block phenomenon". However, some experiments are needed to determine the parameters according to different images, which is not universal.

Since different brightness compensation algorithms are different in scope and situation, it is difficult to solve all the problems in a single algorithm, because the corresponding algorithms need to be selected according to the actual situation. In order to solve the above problems, a systematic brightness compensation process is proposed in this paper. Through experimental analysis, good compensation effect is achieved after the data radiation correction of uneven brightness of GF-3. On this basis, a comparative experiment is carried out with the traditional MASK method. The evaluation of spatial detail information can be reflected by gray mean, standard deviation, average gradient and information entropy, so they are used as the statistics of the evaluation of brightness compensation results, and the evaluation index results obtained are better than the traditional MASK brightness compensation method.

3 GF-3 SAR Image Brightness Compensation Algorithm

Aiming at the problems of GF-3 SAR image data, we propose a systematic brightness compensation algorithm, which is processed according to the process of morphological operation, brightness compensation and gray enhancement. Firstly, we use the top-hat transform to reduce the bright target in the image. Then, we use the SAR imaging model to further compensate the brightness of the image block. Finally, we choose the image with moderate contrast and no radiation difference as the reference image, and use histogram mapping method to enhance the image. The flowchart of the brightness compensation method for GF-3 SAR image is presented in this paper as shown in Fig. 2.

3.1 Top-Hat Transform

Top-hat transform is an important algorithm in mathematical morphology. Its principle is the difference between the original image and the open result of the image. An important purpose of top-hat transform is to correct the influence of uneven illumination for bright objects on dark background [17]. Mathematical

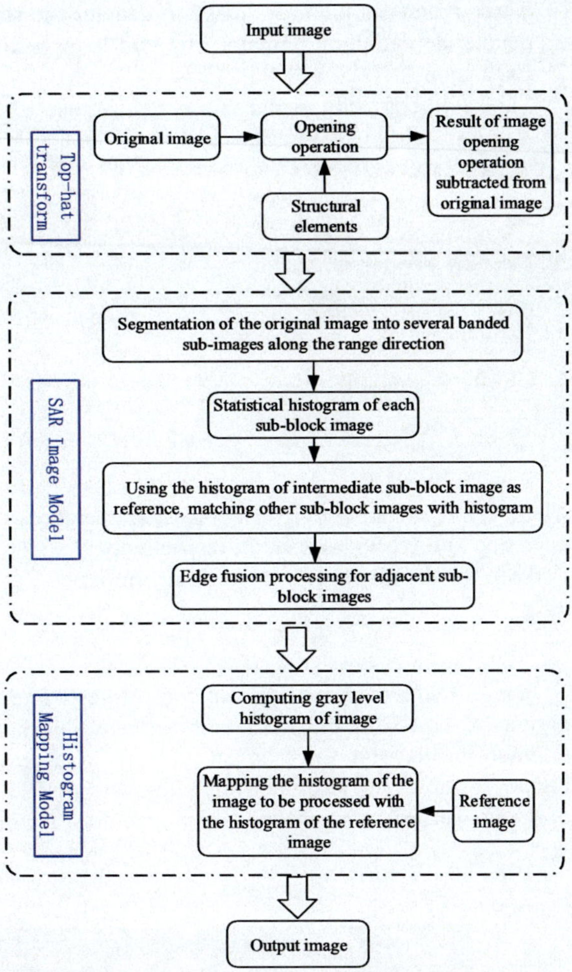

Fig. 2 Flowchart of brightness compensation method for SAR image

morphology is a subject based on set theory and integral geometry. Its basic operations include corrosion, expansion, opening and closing operations [18]. Through these basic operations, it can be deduced and combined into various practical mathematical morphology algorithms. The four basic operations of morphology are as follows.

1. Corrosion
   Corrosion is defined as:

$$(f\Theta b)(s,t) = \min\{f(s+x,t+y) - b(x,y)|(s+x),(t+y) \in D_f; (x,y) \in D_b\} \tag{1}$$

where f denotes the gray image, b denotes the structural elements, D_f and D_b denote the definition domains of f and b, respectively.

2. Expansion
Similar to corrosion, expansion is defined as:

$$(f \oplus b)(s,t) = \max\{f(s-x, t-y) + b(x,y) | (s-x), (t-y) \in D_f; (x,y) \in D_b\} \quad (2)$$

3. Opening operation

$$f \circ b = (f \Theta b) \oplus b \quad (3)$$

4. Closing operation

$$f \bullet b = (f \oplus b) \Theta b \quad (4)$$

where Θ, $\oplus$, $\circ$, and $\bullet$ represent corrosion operation, expansion operation, opening operation and closing operation respectively.

Then the definition of top-hat transform is:

$$T(f) = f - (f \circ b) \quad (5)$$

Figure 3 shows an example of the image before and after top-hat transform. Figure 3a shows that there is a contrast between the upper and lower parts of the original image, and there is an obvious phenomenon of uneven brightness. Figure 3b shows the results of top-hat transform. It can be seen that the gray level of the image results is uniform and the contrast is consistent.

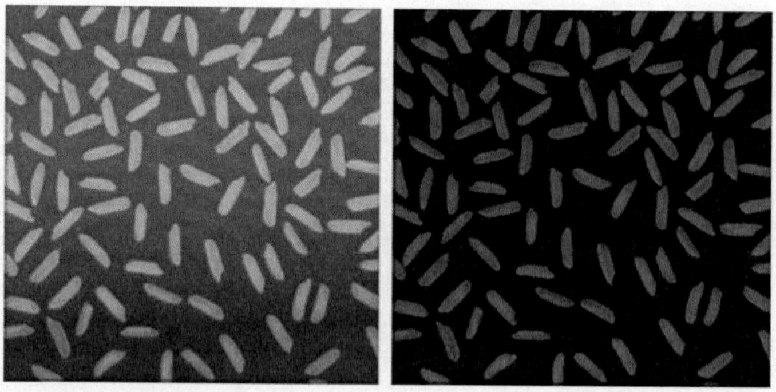

(a) (b)

Fig. 3 Examples of top-hat transform: **a** original image; **b** image after top-hat transform

3.2 SAR Imaging Model

Since SAR satellite system is a slant range imaging system, SAR satellite images not only have geometric deformation such as layover and foreshortening, but also radiation distortion along the range direction. SAR imaging model method is based on radar signal [19]. For a ground area with constant backscattering, the energy distribution of radar signal is shown in Fig. 4a. When the side lobe signal irradiates the ground target, the backscattering intensity of the ground object is relatively weak, while the signal emitted by the main lobe of the radar is strong, and the corresponding echo signal intensity is also very strong. Figure 4b shows the radar imaging schematic, reflecting the relationship between range and azimuth directions.

The steps of SAR imaging model method are as follows:

1. SAR image is segmented into several blocks with overlapping edges along the distance direction, and the number of pixels in each block is counted.
2. In each sub-block, the gray histogram is counted, and the histogram is normalized. Then, the histogram of one sub-block in the middle of the original image is selected as the reference histogram, and the histogram of all other sub-blocks is matched with the reference histogram.
3. The overlapping edges of adjacent sub-blocks are fused.

Figure 5 shows an example of SAR imaging model before and after block transformation.

Figure 6 shows the histogram distribution before and after sub-block bright-ness compensation.

Fig. 4 Energy distribution of the radar signal and radar imaging: **a** energy distribution of the radar signal; **b** radar imaging

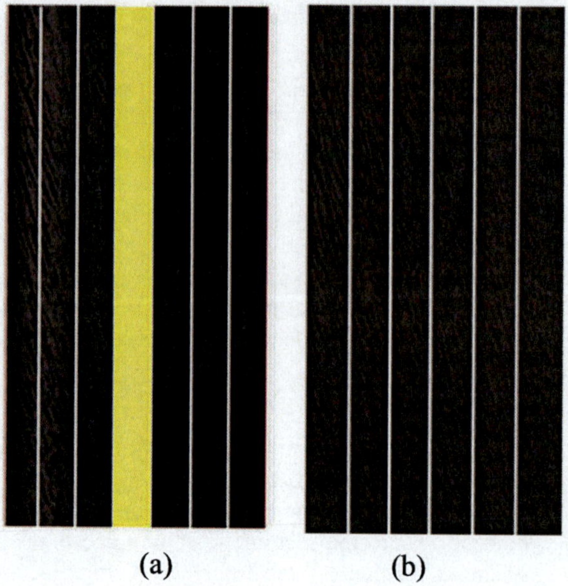

Fig. 5 Example of the sub-block transform of SAR imaging model: **a** before sub-block transform; **b** after sub-block transform

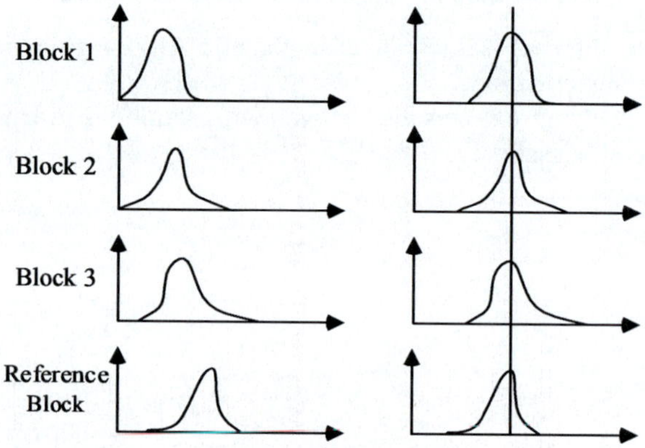

Fig. 6 Histogram distribution of the sub-block transform

3.3 *Histogram Mapping Model*

In digital image processing, histogram is the basis of many spatial domain processing, and histogram operation can be used for image enhancement. Histogram mapping model, also known as histogram matching or histogram specification [20–22], is to make the histogram of the original image refer to an image histogram, and establish the relationship between the original image and the reference image. The gray value of the original image is adjusted according to the histogram of the prescribed shape, and the overall color tone and local contrast of the original image can be changed to perform image enhancement.

The gray level of the unprocessed image is represented by r, and $[0, L-1]$ is the value range of r. The following gray level mapping is established.

$$s = T(r), 0 \leq r \leq L-1 \tag{6}$$

The probability of gray level r_k is as follows:

$$p_r(r_k) = \frac{n_k}{MN}, k = 0, 1, 2, 3, \cdots, L-1 \tag{7}$$

where n_k denotes the number of pixels of gray r_k, and L denotes the gray level of an image, therefore,

$$s_k = T(r_k) = (L-1)\sum_{j=0}^{k} p_r(r_j) = \frac{(L-1)}{MN}\sum_{j=0}^{k} n_j k = 0, 1, 2, 3, \cdots, L-1 \tag{8}$$

where MN is the total number of pixels in an image, and n is the number of pixels in the gray value r.

The following transformation function $G(z)$ is established.

$$G(z_q) = (L-1)\sum_{j=0}^{q} p(z_j) \tag{9}$$

where $p(z)$ is the discrete probability density function of $G(z)$. $q = 0, 1, 2, 3, \cdots, L-1$, and for q,

$$G(z_q) = s_k \tag{10}$$

Based on the inverse transformation,

$$z_q = G^{-1}(s_k) \tag{11}$$

This process gives a z value for each s value; thus, a mapping from s to z is formed. In actual processing, the gray level processed is an integer level, such as 8-bit gray image, each q can get a transformation value, and then after rounding these values are stored into a table. Given a s_k value, we can find the most matching values stored in the table.

4 Experiment and Result Analysis

4.1 Experiment Data

The research area of this paper is located in Taklimakan desert area of Xinjiang. Among 76 image data of desert area, 25 image data have uneven brightness. We elaborate the processing process and evaluation results based on two image data in this paper. In the first part of the experiment, the GF-3 in the fine strip II mode of October 11, 2017 is selected. The area shown in the green frame in Fig. 7 has a central longitude and latitude of 83° 52′ E and 38° 38′ N. The polarization mode is HV and the nominal resolution is 10 m. In the second part of the experiment, the GF-3 in the fine strip II mode of November 11, 2017 is selected. The area shown in the yellow frame in Fig. 7 has a central longitude and latitude of 79° 35′ E and 38° 6′ N. The polarization mode is HH and the nominal resolution is 10 m. All the data selected above have been pre-processed.

4.2 Experiment 1: Result and Analysis of the Proposed Method

In this section, SAR images of GF-3 desert area are tested according to the proposed method steps, as shown in Fig. 8. Figure 8d with moderate brightness and consistent contrast is selected as the reference image for brightness enhancement. Figure 9 is the profile line of the image along the range, and the sub-graphs (a), (b), (c), (d) of Fig. 9 correspond to the profile lines of (a), (b), (c), (e) of the sub-graphs of Fig. 8, respectively. Table 1 gives the corresponding quantitative analysis results. We use the quantitative index brightness mean and standard deviation to

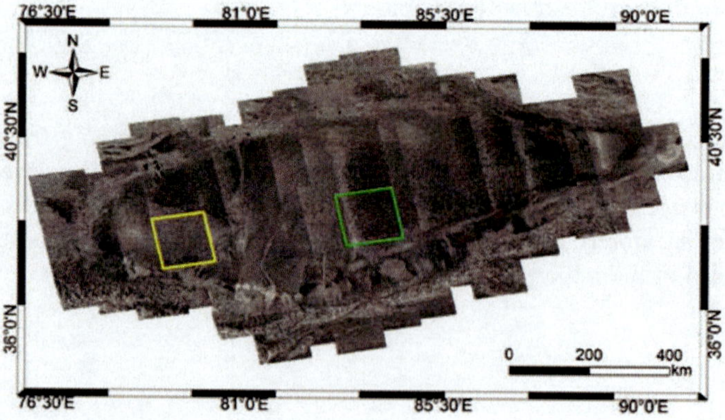

Fig. 7 Mosaic of desert areas

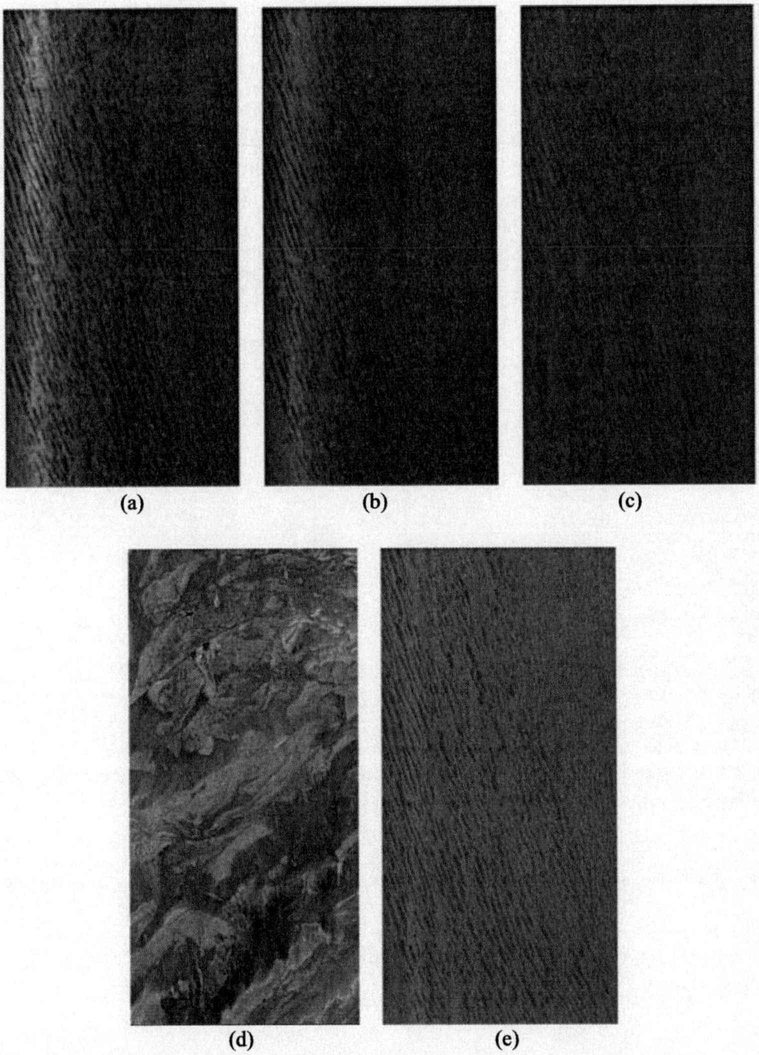

Fig. 8 Processing procedure of the proposed method: **a** original image; **b** the result of Top-hat transform based on (**a**); **c** the result of SAR imaging model based on (**b**); **d** reference image; **e** the result of histogram mapping model based on (**c**)

evaluate the image quality. The brightness mean reflects the brightness information of the image. If the brightness mean is moderate, the overall visual effect of the image will be good, and the image will not be brighter or darker. The standard deviation reflects the discrete gray distribution of the image. The larger the standard deviation, the larger the dynamic range of the gray value of the image.

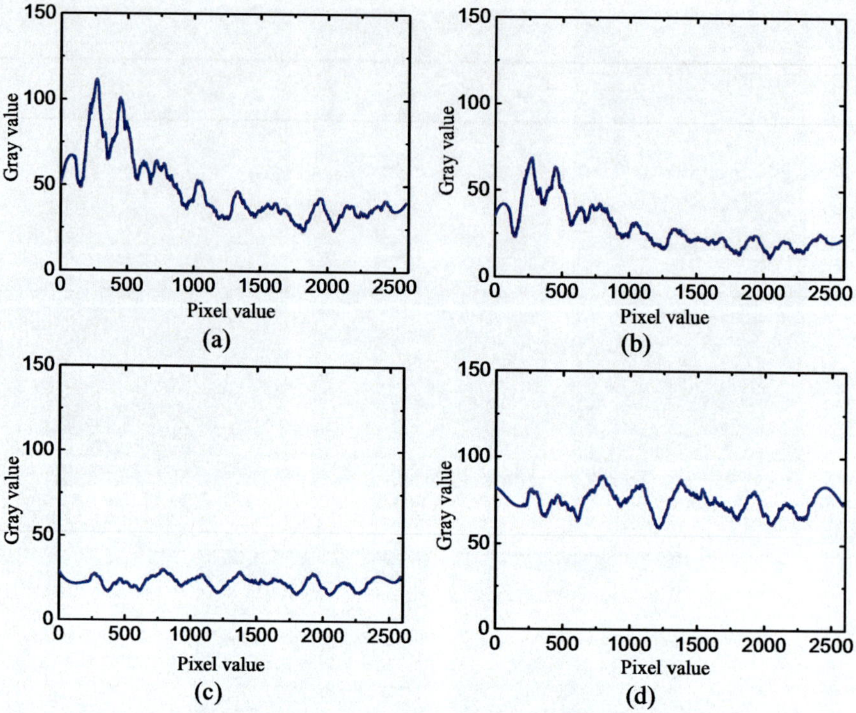

Fig. 9 Profile line of image gray: **a** Profile line of original image; **b** profile line of Top-hat transform result; **c** profile line of SAR imaging model result; **d** profile line of histogram mapping model result

Table 1 Brightness mean and standard deviation of study image before and after compensation

Image	Original image	Reference image	The proposed method
Brightness mean	42.36	84.21	108.65
Standard deviation	49.43	55.18	49.42

For the GF-3 image brightness compensation method proposed in this paper, it can be seen from Fig. 8a that the left part of the original image has a larger brightness, while the right part is darker, the overall brightness of the image is not balanced, and the contrast is not appropriate. Combining with the corresponding profile line of Fig. 9a of the original image, the curve undulates obviously, which further indicates that the original image has a serious brightness uneven phenomenon. After top-hat transform, SAR imaging model transform and histogram mapping model transform combined with reference image, the processing result of the proposed method is shown in Fig. 8e, and the following conclusions are drawn:

1. From Fig. 8, it can be seen that the brightness of the original image in Fig. 8a and the reference image in Fig. 8d are quite different. The overall gray level of the image after Fig. 8e processed is basically the same as that of the reference image in Fig. 8d, and the brightness is moderate. Moreover, the brightness problem of the original image has been well solved, which can better compensate the radiation difference within the image.
2. From Fig. 9, it can be seen that the profile line curve of the original image in Fig. 9a changes dramatically. After the systematic processing of proposed method, the change of the profile line curve of the image results is gradually smooth. The final result of the profile line curve is shown in Fig. 9d. The profile line curve of the processed image is smooth, and the range of gray value is reasonably improved.
3. From Table 1, it can be seen that the brightness mean and standard deviation of the image output after the proposed method are 66.29 and 0.01 different from the original image, respectively. The statistical results show that the brightness of the processed image is enhanced, and the relative relationship between the image pixel values remains good.

4.3 Experiment 2: Comparison Between the Proposed Method and MASK Brightness Compensation Method

In order to fully verify the effectiveness of proposed method, the proposed method is compared with the MASK brightness compensation method. The experimental results are shown in Fig. 10.

From Fig. 10, it can be seen that MASK brightness compensation method and the proposed method achieves a certain compensation purpose. The overall brightness of the image processed by MASK brightness compensation method decreases, the recognition rate of terrain and ground objects is low, and the left side of the image still has local gray-scale uneven, which has a certain contrast with the rest of the image. In the proposed method, the brightness of the image has been eliminated, the overall brightness remains the same, the image is clear, and the brightness has been enhanced.

In order to further quantitatively evaluate the quality of two kinds of brightness compensation image, we use four evaluation indicators of brightness mean, standard deviation, information entropy, and average gradient for quality evaluation, to evaluate the compensation results quality. The brightness mean and standard deviation has been explained in the previous section, and will not be explained in detail here. Information entropy reflects the amount of information contained in an image. The bigger the value of information entropy, the more information contained in an image. Average gradient reflects the richness of image details. The bigger the average gradient, the bigger the contrast of image details and the clearer the image.

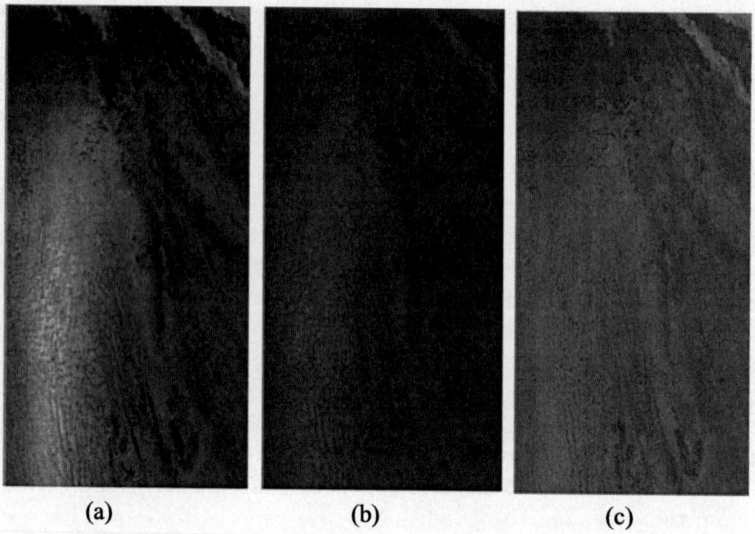

Fig. 10 Comparison of the results of two methods: **a** original image; **b** the result of MASK brightness compensation method; **c** the result of proposed method

In order to evaluate the image accurately, the original image and two results image compensated by the two algorithms are divided into 9 image blocks of 3×3. The image blocks are numbered from left to right and from top to bottom (as shown in Fig. 11). The brightness mean and standard deviations of each image block are counted separately, and the statistical values are plotted in the order of image block numbering. The distribution curves of brightness mean and standard deviation are shown in Figs. 12 and 13, respectively. The information entropy and average gradient of the image processed by two methods are shown in Table 2.

Through the analysis of the above figures and table, the following conclusions are drawn.

1. From Fig. 12, it can be seen that the brightness distribution curve of the original image changes dramatically, as shown by the black curve, which indicates that the original image has a serious uneven brightness distribution. The image curve processed by MASK brightness compensation method is smoother than the

Fig. 11 Schematic sketch of image division and numbering

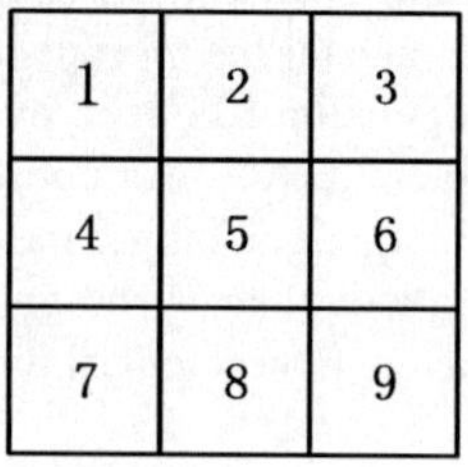

Fig. 12 Contrast chart of brightness mean

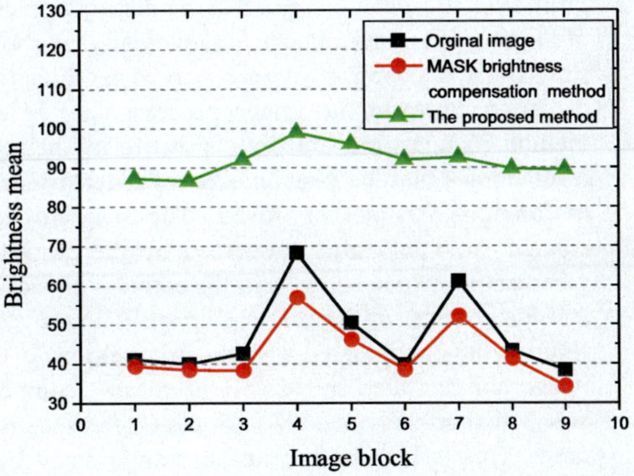

Fig. 13 Contrast chart of standard deviation

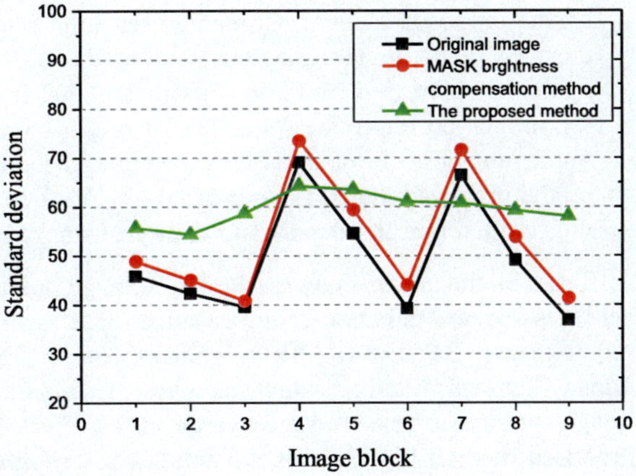

Table 2 Information entropy and average gradient of original image and two algorithms

Image	Original image	MASK brightness compensation method	The proposed method
Brightness mean	6.70	5.21	6.12
Standard deviation	22.36	23.25	38.35

original image curve, but the curve change is still obvious, as shown by the red curve in the image, and the result shows that the method does not effectively improve the original image brightness inhomogeneity. The image curve changes more smoothly after processed by the proposed method, as shown by the green curve, and the result shows that the brightness distribution of the processed image is uniform and the radiation consistency of the image is improved.

2. From Fig. 13, it can be seen that the distribution curve of the standard deviation of the original image changes dramatically, as shown by the black curve, which shows that the contrast of each part of the image is inconsistent. The standard deviation curve of the image processed by MASK brightness compensation method changes dramatically, as shown by the red curve in the figure, and the result shows that the phenomenon of inconsistency of the contrast of each part of the image is not improved. The standard deviation curve of the image changes smoothly after processed by the proposed method, as shown in the green curve, which shows that the contrast of each part of the image is moderate and the contrast of the original image is improved.
3. From Table 2, compared with the original image, the information entropy of the image compensated by MASK brightness compensation method and the proposed method decreases by 1.49 and 0.58 respectively, and the former decreases more. This is because the image compensated by MASK brightness compensation method is still affected by the uneven brightness phenomenon, and the overall quality of the processed image is lower than that of the image processed by the proposed method. Compared with the original image, the average gradient of result image compensated by the MASK brightness compensation method is increased by 0.89. This means that the MASK brightness compensation method improves the details of original image at some degree, but such improvement is not obvious. The average gradient of result image compensated by the proposed method is increased by 15.99 compared to that of the original image, therefore the details and clarity of image are improved obviously.

Based on the above analysis, the GF-3 image processed by the proposed method achieves better radiation compensation and satisfies the image application requirements. Because MASK brightness compensation method is based on frequency low-pass filtering, which removes the low-frequency information results in image contrast decline, and the overall visual effect of the image is not good. The proposed method has obvious advantages in brightness compensation. Figure 14

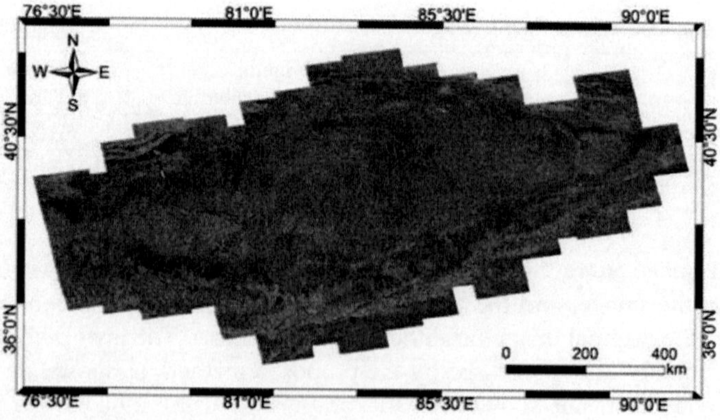

Fig. 14 Mosaic of desert areas after processed

shows the mosaic fusion results of 76 images processed in GF-3 desert area. Among them, 25 images have uneven brightness. The proposed method compensates the brightness of each image separately, and then fuses with other image. The image results show that the brightness is balanced and the transition is even, and the expected effect is achieved.

5 Conclusions

In order to solve the problem of uneven brightness of GF-3 images, we propose a method of brightness compensation for GF-3 image in this paper. Firstly, top-hat transform is used to reduce the bright targets in the images. Secondly, SAR imaging model method is used to further compensate the brightness of the whole image. Finally, with reference image, histogram mapping method is used to enhance the image. The experimental results show that the proposed method has a good compensation effect for GF-3 images with uneven brightness, and the image quality of the processed image remains good. The overall visual effect of the image has been greatly improved. It effectively solves the problem of uneven brightness of GF-3 image, and has a great practical value in data processed with GF-3.

Acknowledgments This work is supported by the National Natural Science Foundation of China (no. 61102163), the Fundamental Research Funds for the Central Universities (no. GK201903085), and the Key Laboratory of Land Satellite Remote Sensing Application Center, Ministry of Natural Resources of the People's Republic of China (no. KLSMNR-202004).

References

1. De-ren L, Mi W, Jun P (2006) Auto dodging processing and its application for optical RS images. Geomatics Inf Sci Wuhan Univ 31(9):753–756
2. Oliver C, Quegan S (1998) Understanding synthetic aperture radar images. Artech House, Boston, MA
3. Guo-xiang L (2004) Principles of imaging SAR and characteristics of SAR image. Si-chuan Bur Surv 27(3):141–143
4. Qing-jun Z (2017) System design and key technologies of the GF-3 satellite. Acta Geodaetica et Cartographica Sinica 46(3):269–277
5. Ze-yu L, Bin L, Wei-wei G et al (2017) Ship detection in GF-3 NSC mode SAR image. J Radars 6(5):473–482
6. Tao L, Jing L (2018) National orthophoto map of radar remote sensing with GF-3. [EB/OL] http://www.sasmac.cn/xwfb/zxdt/201809/t20180920_408025.html 2018/9/20
7. Peng F (2013) Dodging research for remote sensing image. University of Chinese Academy of Science
8. Mi W, Jun P (2004) A method of removing the uneven illumination for digital aerial image. J Image Graphics 9(6):744–748
9. Peng-cheng L, Dong W, Ji-guo Z (2018) Improved adaptive mask dodging algorithm. Remote Sens Inf 33(4):86–90

10. Orsini G, Ramponi G, Carrai P, Federico RD (2003) A modified retinex for image contrast enhancement and dynamic control. In: IEEE international conference on image processing, Sept 14–17 Barcelona, Spain, p 393
11. Edwin H (1971) Lightness and retinex. J Opt Soc Am 1:1–11
12. Fu XY, Sun Y, Li MH et al (2014) A novel retinex based approach for image enhancement with illumination adjustment. In: The 39th international conference on acoustics, speech, and signal processing May 4–9 Florence, Italy, p 1190
13. Wen-jie Y, Radiometric dodging for remote sensing image M.S. dissertation. PLA Information Engineering University
14. Fan C, Xu-shuai C, Zhong L et al (2017) Improved Wallis dodging algorithm for large-scale super-resolution reconstruction remote sensing images. Sensors 17(3):623
15. Zhi-jun W, De-ren L, Qing-quan L (2000) Application of Wallis transform in wavelet based image fusion. J Wuhan Tech Univ Surv Map 25(4):338–342
16. Bin-cai C, Bao-shan Z, Run-sheng L et al (2012) Wallis algorithm for single image dodging. J Geomatics Sci Technol 29(5):373–377
17. Prasertsakul P, Toshiaki K (2017) Video shot classification using 2D motion histogram. In: 14th international conference on electrical engineering/electronics, computer, tele-communications and information technology (IEEE) May 14–17 Nakhon Ratchas, Thailand, p 5
18. Morale S, Naranjo V, Jesus A (2013) Automatic detetion of optic disc based on PCA and mathematical morphology. IEEE Trans Med Imaging 32(4):786–796
19. Bang-song W (2015) Method of automatic registration and mosaic for SAR images. Wuhan University
20. Long-tao Z, Yu-qiu S (2016) Improved histogram matching algorithm based on fuzzy entropy. J Southwest Univ (Natural Science Edition) 38(4):124–129
21. Guang-sheng C, Chao L, Wei W et al (2019) Fully convolutional neural network with augmented atrous spatial pyramid pool and fully connected fusion path for high resolution remote sensing image segmentation. Appl Sci 9(9)
22. Bin Z, Xue-mei D, Dong-jun Y et al (2019) Multi-level features extraction for discontinuous target tracking in remote sensing image monitoring. Sensors 19(22)
23. Han YT (2014) Research on key technology of color consistency processing for digital ortho map mosaicing. Wuhan University
24. Bang-song W, Hai-bin A, Hong A (2011) Research on color consistency processing of aerial images. Remote Sens Inf (1):45–49

The Spaceborne SAR 3D Imaging Method Based on 2D Compressed Sensing

He Tian, Chunzhu Dong, and Hongcheng Yin

Abstract Synthetic Aperture Radar (SAR) three-dimensional (3D) imaging technology uses multiple flight observations in the cross-track direction, to obtain three-dimensional spatial distribution of observation scenes. In order to shorten the revisit time of the satellite and increase the number of orbits, this paper proposes a sparse 3D imaging method of two-dimensional (2D) Compressed Sensing (CS). Under the large orbit distribution and sparse flight, the linear measurement matrix in range and cross-track directions is established, which is conducive to the joint solution of CS theory for images under sparse representation and to avoid the coupling of echo signals in the orbit and distance directions. The simulation results of imaging resolution of 0.5 m for single satellite and double satellites are given, verifying the effectiveness of the method.

Keywords Sparse flight · SAR · 3D imaging · compressed sensing · Spaceborne

1 Introduction

Three-dimensional (3D) Synthetic Aperture Radar (SAR) [1] applying the azimuthal synthetic aperture principle to the cross-track direction, to obtain the cross-track imaging resolution, which has important application value in terrain mapping, battlefield reconnaissance and other fields [2]. For the spaceborne SAR platform, due to the limitation of the satellite platform, the cross-track flights are non-uniform and under-sampled. Therefore, the traditional imaging method based on matched filtering will make the image seriously blurring. The Compressive Sensing (CS) [3–7] theory points out that for sparse signal, the low-dimensional observations obtained far below the Nyquist sampling rate can be used to accurately recover the signal through an appropriate sparse signal reconstruction algorithm. Based on CS theory, the Tomographic SAR (TomoSAR) imaging technique has the

H. Tian (✉) · C. Dong · H. Yin
Science and Technology on Electromagnetic Scattering Laboratory, Beijing 100854, China
e-mail: tianhe0407@126.com

L. Wang et al. (eds.), *Proceedings of the 7th China High Resolution Earth Observation Conference (CHREOC 2020)*, Lecture Notes in Electrical Engineering 757, https://doi.org/10.1007/978-981-16-5735-1_41

ability to estimate the positions and scattering intensity of multiple scattered points from the sparse observation. This kind of method can solve the SAR reconstruction problem under sparse sampling in cross-track direction, but it ignores the data coupling of signal in cross-track and range directions, and the cross-track aperture length is limited by the optimal vertical baseline length [7]. For example, Zhu [8–10] showed that in terraSAR-X satellite data imaging test, the equivalent aperture length of the cross-track direction was about 270 m, and the resolution was 33 m. Ge [11] conducted Differential SAR Tomography (D-TomosAR) using the tandem-X data, and the equivalent aperture length in cross-track direction formed by the sparse flight is about 750 m. Therefore, it is necessary to study the 3D imaging method of spaceborne SAR in the large aperture range.

With the development of small satellite technology, the use of formation flying satellites to achieve topographic mapping and 3D imaging has become a research focus in spaceborne SAR technology [12]. The existence of several satellites provides the possibility of cross-track synthesis of large aperture. At present, several European teams have carried out in-depth studies on formation satellite SAR imaging technology, and typical schemes include the Cartwheel project from France [13] and the tandem-X mission from Germany [14, 15]. Since the formation satellites can obtain multiple baselines only through one flight, the combination of formation satellite technology and sparse flight can effectively shorten the satellite revisit time, greatly increase the number of observing flights, and expand the length of synthetic aperture in cross-track direction. Therefore, this paper combines with the actual orbit parameters of the satellite, and gives the sparse flight distribution of both single satellite and double satellites, with the equivalent aperture length in cross-track direction of 20 km. In the large range of orbit distribution, the range and cross-track linear measurement matrix is established, and CS theory is used to jointly solve the problem, so as to avoid the coupling problem of echo signals in the cross-track and range direction.

2 Imaging Model

Figure 1 shows the 3D imaging geometry for spaceborne SAR with sparse flight. In imaging coordination system, the x, y, z axis represents the ground direction, azimuth direction and height direction, respectively; r represents radar range direction, s is the cross-track direction; θ is the angle of incidence, H is the platform height and R_0 is the referential distance between radar and scene center.

Assume that the number of Effective Phase Centers (EPCs) formed by satellite motion is M, the number of sparse flights aimed at same scene is N, and the effective aperture length is L, $L = (MN\text{-}1)\ d_a$, d_a is the average interval length. Therefore, we can define the sparse sampling rate in spaceborne SAR as

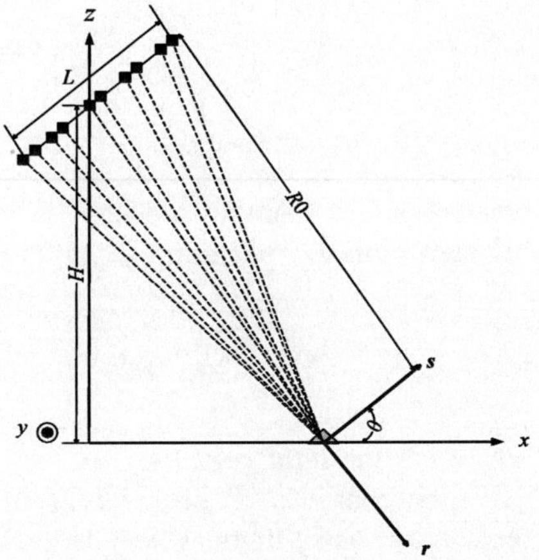

Fig. 1 3D imaging geometry for spaceborne SAR with sparse flight

$$\eta = \left[1 - \frac{MN - 1}{\left(\frac{L}{d_{\min}} \right)} \right] \times 100\% \tag{1}$$

where d_{min} is the minimum interval length. Assume that the radar working band is X, the center frequency is 10 GHz, the platform height is 550 km, the angle of incidence at broadside mode is 45°. When the length of effective aperture in cross-track direction is 20 km, the EPCs number M is 2, and the flight number N is 23, therefore we can calculate the sparse sampling rate is 74.4%.

3 Signal Procession

Assume that SAR transmit a linear frequency modulation signal

$$p(t) = \mathrm{rect}(\frac{t}{T_p}) \exp(j2\pi f_c t + j\pi K_r t^2) \tag{2}$$

where T_{p} is the pulse width, f_c is center frequency and K_r is the chirp-rate. For a point target q, the spatial position in s-y-r coordination is (s_q, y_q, r_q), and the back scattering coefficient is σ_q. Assume the satellite motion forms M EPCs and N sparse flights. Therefore, the cross-track position of mth EPC in nth sparse flight is s_{mn}, $s_{mn} = -L/2 + (mn - 1)d_a$. Then the SAR received echo can be written as

$$g(t,y,s_{mn}) = \sigma_q \mathrm{rect}\left[\frac{t-\tau_q(y,s_{mn})}{T_p}\right]\exp\left\{-j2\pi f_c\tau_q(y,s_{mn}) + j\pi K_r\left[t-\tau_q(y,s_{mn})\right]^2\right\} \tag{3}$$

where y is the sampling position in azimuth direction, $\tau_q(y, s_{mn})$ the time delay between sampling position (s_{mn}, y, R_0) and target position (s_q, y_q, r_q)

$$\tau_q(y,s_{mn}) = \frac{2}{c}\sqrt{(R_0-r_q)^2 + (y-y_q)^2 + (s_{mn}-s_q)^2} \tag{4}$$

where c is the light velocity.

In spaceborne SAR observation, the effective aperture in cross-track direction formed by sparse flight is very large. To avoid complex phase compensation and Range Migration Correlation (RMC), we use CS-based method to jointly process SAR echo in range and cross-track directions [16, 17]. Assume that after azimuthal compression, the SAR echo can be written as

$$\begin{aligned} g_{ac}(t,y,s_{mn}) = \sigma_q \mathrm{sinc}(y-y_q)\mathrm{rect}\left[\frac{t-2R_{s_{mn}}/c}{T_p}\right]\exp\left(-j\frac{4\pi R_{s_{mn}}}{\lambda}\right) \\ \cdot\exp\left\{j\pi K_r[t-2R_{s_{mn}}/c]^2\right\} \end{aligned} \tag{5}$$

where $R_{s_{mn}} = \sqrt{(R_0-r_q)^2 + (s_{mn}-s_q)^2}$.

We divide the range and cross-track imaging space into $N_s \times N_r$ units, and in this grid, the position of resolution cell P_{ij} at ith range sample and jth cross-track sample is (s_{Pij}, r_{Pij}), $1 \le i \le N_s$, $1 \le j \le N_r$. For mth resolution cell in azimuth direction, we establish the 2D measuring matrix in range and cross-track directions as $\mathbf{\Phi}$

$$\mathbf{\Phi} = \begin{bmatrix} \varphi[R_{s_{11}}(P_{11})] & \varphi[R_{s_{11}}(P_{12})] & \ldots & \varphi[R_{s_{11}}(P_{ij})] & \ldots & \varphi[R_{s_{11}}(P_{N_sN_r})] \\ \varphi[R_{s_{12}}(P_{11})] & \varphi[R_{s_{12}}(P_{12})] & \ldots & \varphi[R_{s_{12}}(P_{ij})] & \ldots & \varphi[R_{s_{12}}(P_{N_sN_r})] \\ & & & \vdots & & \vdots \\ \varphi[R_{s_{mn}}(P_{11})] & \varphi[R_{s_{mn}}(P_{12})] & \ldots & \varphi[R_{s_{mn}}(P_{ij})] & \ldots & \varphi[R_{s_{mn}}(P_{N_sN_r})] \\ & & & \vdots & & \vdots \\ \varphi[R_{s_{MN}}(P_{11})] & \varphi[R_{s_{MN}}(P_{12})] & \ldots & \varphi[R_{s_{MN}}(P_{ij})] & \ldots & \varphi[R_{s_{MN}}(P_{N_sN_r})] \end{bmatrix} \tag{6}$$

where $\varphi\left[R_{s_{mn}}(P_{ij})\right] = \exp\left\{j\pi K_r\left[t - 2R_{s_n}(P_{ij})/c\right]^2\right\}$,$t = \left[t_1\ t_2\ \cdots\ t_{N_{rg}}\right]^{\mathrm{T}}$ is the fast time sequence, $R_{s_{mn}}(P_{ij}) = \sqrt{(R_0 - r_{P_{ij}})^2 + (s_{mn} - s_{P_{ij}})^2}$, N_{rg} is the number of sampling points in range direction.

Assume the back scattering coefficient at imaging cell P_{ij} is σ_{ij}, then the restored image can be represented as vector $\boldsymbol{\sigma}$

$$\boldsymbol{\sigma} = \left[\sigma_{11} \cdots \sigma_{1N_r}\ \sigma_{21}\ \cdots\ \sigma_{2N_r} \cdots \sigma_{N_s1} \cdots \sigma_{N_sN_r}\right]^{\mathrm{T}} \tag{7}$$

The 2D echo in range and cross-track directions can be represented as vector $\boldsymbol{y}$

$$\boldsymbol{y} = \left[g_{ac}(\boldsymbol{t}, y, s_{11})^{\mathrm{T}}\ \cdots\ g_{ac}(\boldsymbol{t}, y, s_{mn})^{\mathrm{T}}\ \cdots\ g_{ac}(\boldsymbol{t}, y, s_{MN})\right]^{\mathrm{T}} \tag{8}$$

Then the liner measuring model is

$$\boldsymbol{y} = \boldsymbol{\Phi\sigma} \tag{9}$$

We use CS technique to establish the optimization function based on $\ell_{1/2}$ as

$$\min_{\boldsymbol{\sigma}}\{\|\boldsymbol{y} - \boldsymbol{\Phi\sigma}\|_2^2 + \xi\|\boldsymbol{\sigma}\|_{1/2}^{1/2}\} \tag{10}$$

where ξ represents the regularization parameter. By solve the Eq. (10), the scattering coefficient vector can be restored. Furthermore, by implementing the 3D coordination transformation, the reconstructed image of observed scene with high resolution is obtained. The flow chart of spaceborne SAR 3D imaging algorithm based on 2D CS and sparse flight sampling method is as follow (Fig. 2).

4 Sparse Flight Sampling Analysis

In this paper we take a set of parameters for a real satellite with platform height of 550 km. Aiming at a fixed observing scene, the observation time of this satellite is 120 days, and the orbital distribution length in cross-track direction is 20 km. The orbital distribution for single satellite is shown is Fig. 3. From Fig. 3 we can see that the number of sparse flights (tracks) is 23, and the minimum interval between each track are 111 m. In X band and broadside mode with angle of 45°, the imaging range without blurring in cross-track direction is 105 m, which basically meets the need of 3D imaging for surface features.

We calculate the pattern of sparse flight for single satellite and antenna with size of 10 m, as shown in Fig. 4. In range of [–0.003°, 0.003°], the sparse flight pattern has one main lope and avoids the influence of side-lopes. Therefore we can find that the

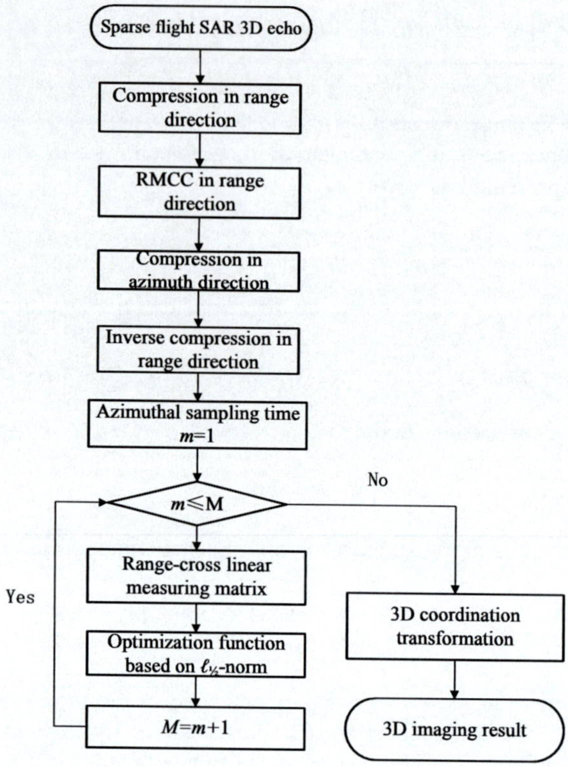

Fig. 2 3D imaging method based on 2D compressed sensing for spaceborne SAR with sparse flight

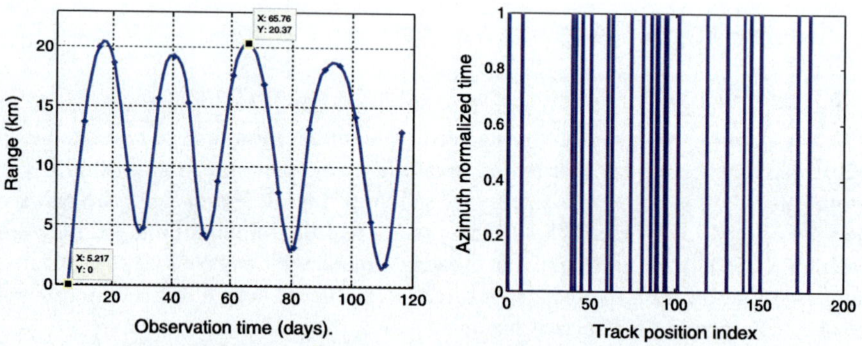

(a) The range migration of orbits (b) the spatial distribution of orbits

Fig. 3 Orbital distribution for single satellite

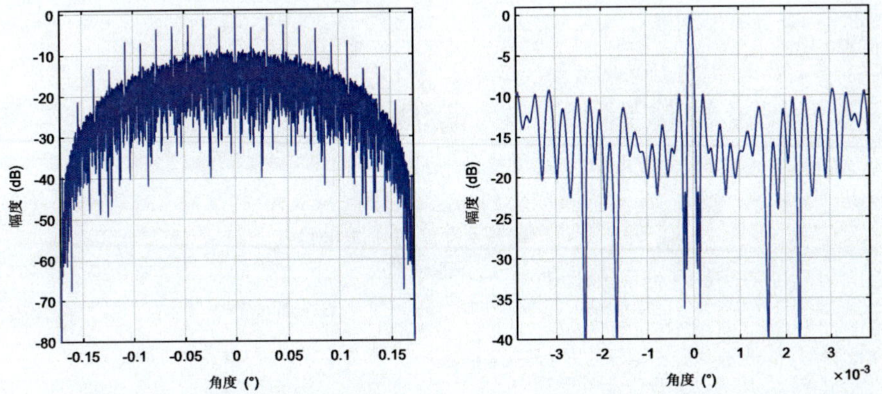

(a) Pattern in antenna main-lope width (b) Pattern in imaging width (without blurring)

Fig. 4 Sparse flight pattern for single satellite with 10 m antenna

imaging range without blurring is 105 m in cross-track direction and 74 m in height direction.

5 Simulation Analysis

The simulation parameters of spaceborne SAR 3D imaging is shown in Table 1. The minimum interval of orbital track is 111 m, and the effective aperture in cross-track direction is 20 km. Therefore, the range resolution is 0.50 m, the azimuthal resolution is 1.00 m, and the cross-track resolution is 0.58 m. The size of the observed scene is 40 m (*x* direction) × 40 m (*y* direction) × 30 m (*z* direction), which including a cone with diameter of 7 m and height of 30 m, as shown in Fig. 5.

Table 1 Simulating parameters

Parameters	Values
Working wavelength (λ)	0.03 m
Bandwidth (B_s)	300 MHz
Platform height (H)	550 km
Angle of incidence	45°
Resolution in azimuth direction	1 m
Velocity of platform	7 km/s
Minimum orbital interval of single satellite	111.3 m
Minimum orbital interval of double satellites	222.6 m
Effective aperture length in cross-track direction	20 km
Orbit number	23

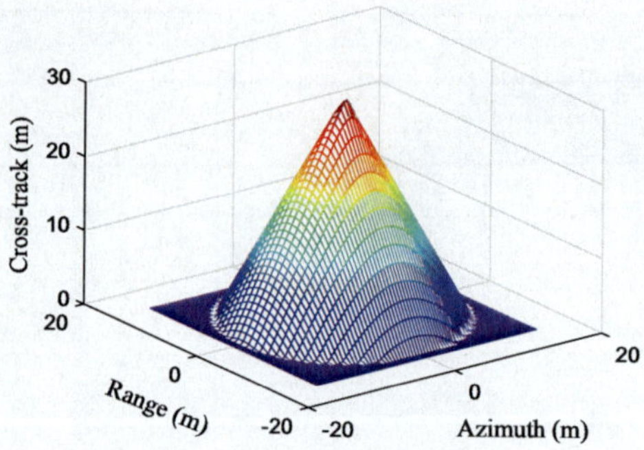

Fig. 5 The observed scene in simulation analysis

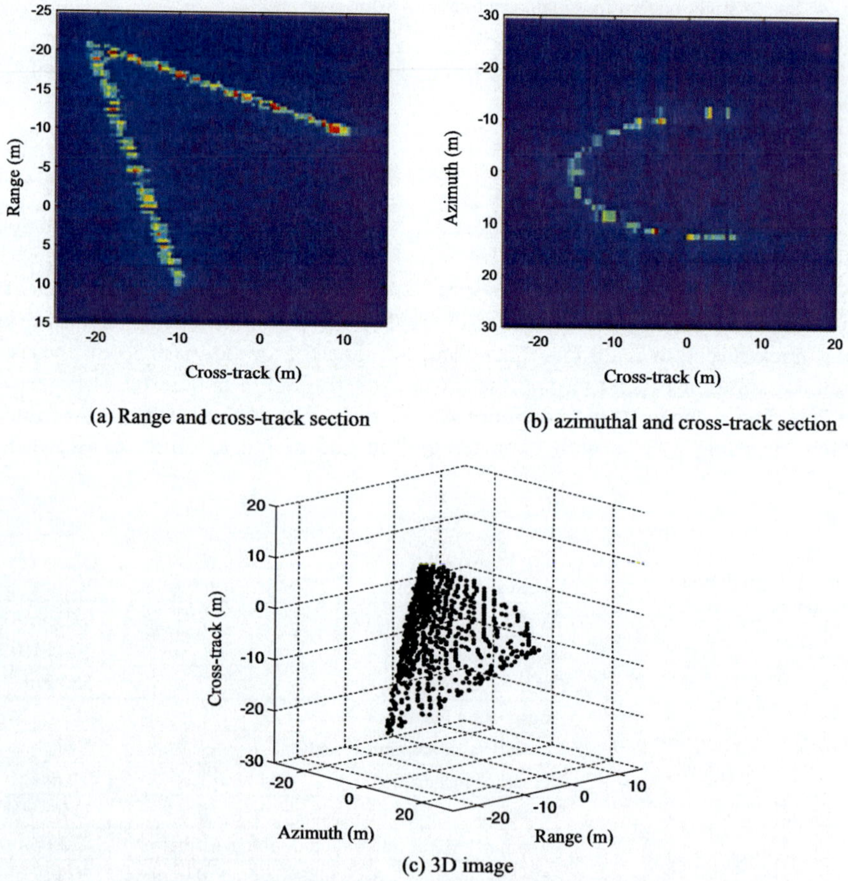

Fig. 6 3D imaging results based on BP algorithm and 180 tracks

According to the minimum interval the effective aperture in cross-track direction, the number of flights in dense sampling mode is 180. Figure 6 is the SAR imaging result base on 180 tracks and 3D back projection (BP) algorithm. In subsequent processing, we use this imaging result to implement the reconstruction performance evaluation.

5.1 *Simulation with Single Satellite and Sparse Flight*

For the single satellite, under the condition of 23 sparse flights, the corresponding 3D BP imaging results are shown in Fig. 7. It can be seen from the figure, under the

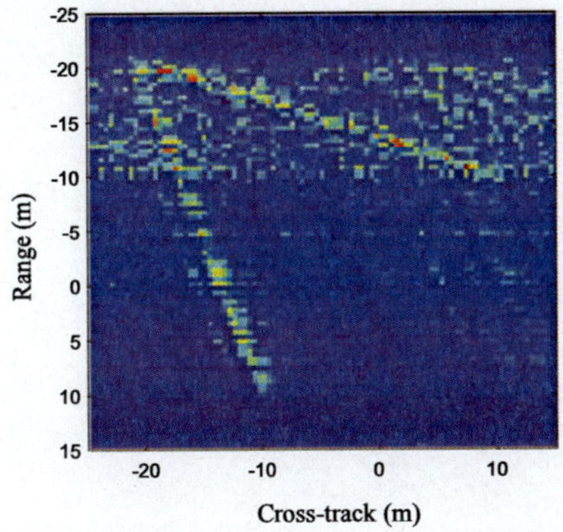

(a) Range and cross-track section

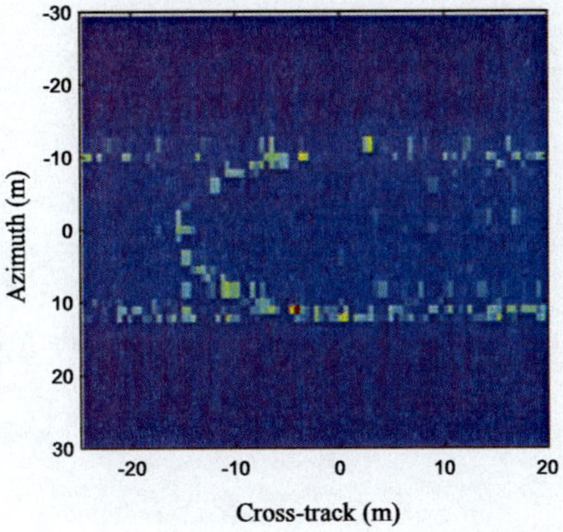

(b) azimuthal and cross-track section

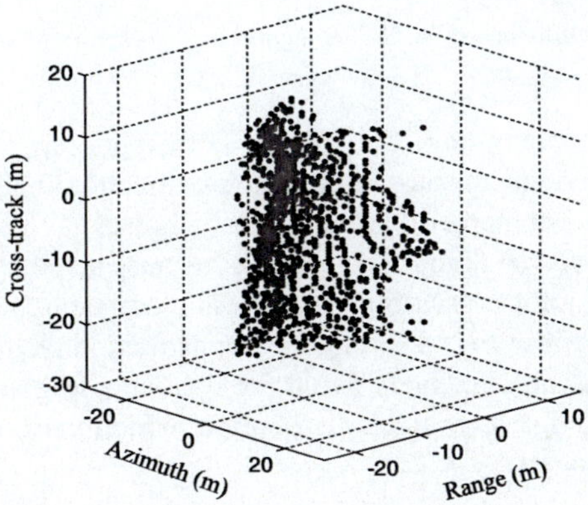

(c) 3D image

Fig. 7 3D imaging results based on BP algorithm, 23 tracks and single satellite

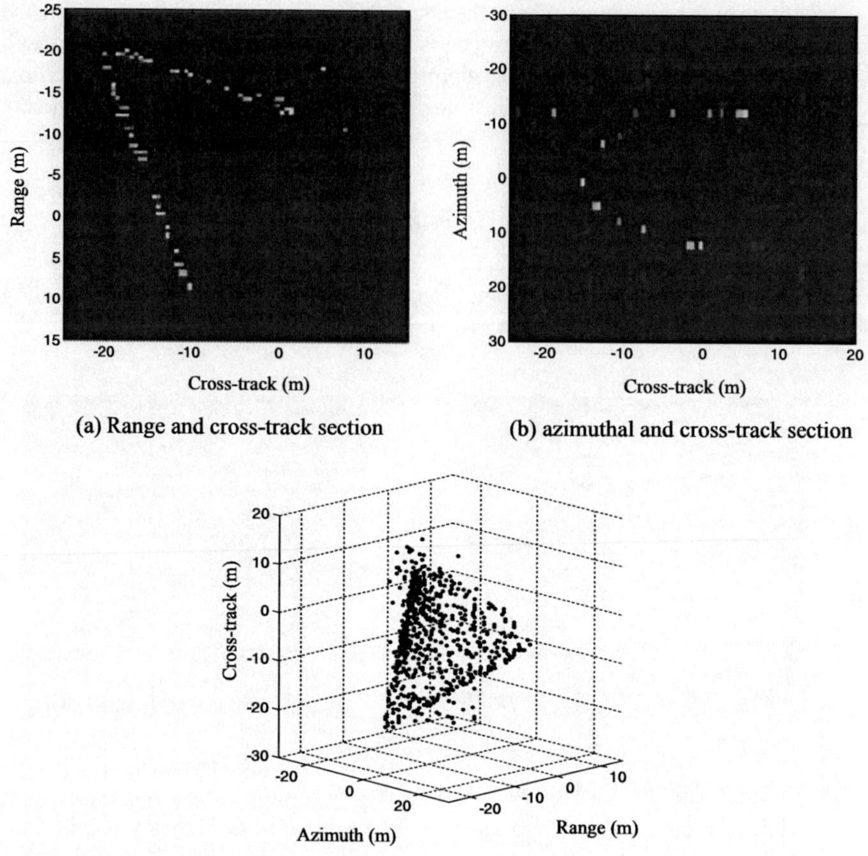

(a) Range and cross-track section

(b) azimuthal and cross-track section

(c) 3D image

Fig. 8 3D imaging results based on 2D CS algorithm, 23 tracks and single satellite

condition of sparse sampling, the images corresponding to the direct imaging method generate serious aliasing in cross-track direction, and the scattering intensity of the target is submerged by the side-lobe noise.

On the basis of sparse flight, the echo was reconstructed by using the range and cross-track 2D CS joint processing method. The corresponding image reconstruction results are shown in Fig. 8. From the reconstruction results, it can be seen that the 2D CS method can effectively avoid the data coupling and distance migration problems caused by the large track distribution, which verifies the effectiveness of the proposed method.

5.2 Simulation with Double Satellites and Sparse Flight

In this paper, a formation structure based on two small satellites is designed, with an interval of 222.6 m for small satellites, which generates two EPCs with an interval of 111.3 m. The sparse flight method corresponding to the two small satellites is shown in Fig. 9a, where 46 orbits are formed and the sparse sampling rate is 74.4%. The Fig. 9b shows the self-correlation function of the sparse flight distribution of the double satellites, from which it can be seen that the distribution mode has a good random property, which is conducive to gate lobe suppression.

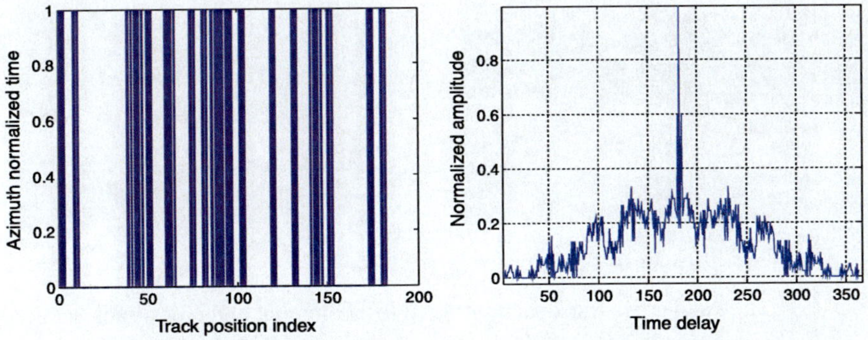

(a) the spatial distribution of orbits (b) the self-correlation plot of orbits

Fig. 9 The self-correlation plot on orbital distribution for double satellite

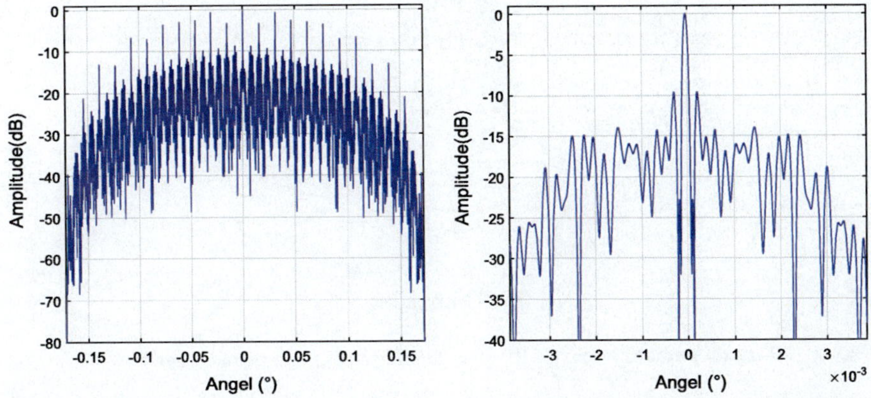

(a) Pattern in antenna main-lope width (b) Pattern in imaging width (without blurring)

Fig. 10 Sparse flight pattern for double satellite with 10 m antenna

The cross-track sparse flight pattern formed by the formation double satellites is shown in Fig. 10. It can be seen from the figure that the sparse flight pattern does not exist the impact of grate lobes within [–0.003°, 0.003°]. Additionally, the side lobe in in the far area of the pattern is lower than that of single satellite, which is conducive to imaging. Based on this sampling method, combined with further image reconstruction, the observation scene can be accurately restored.

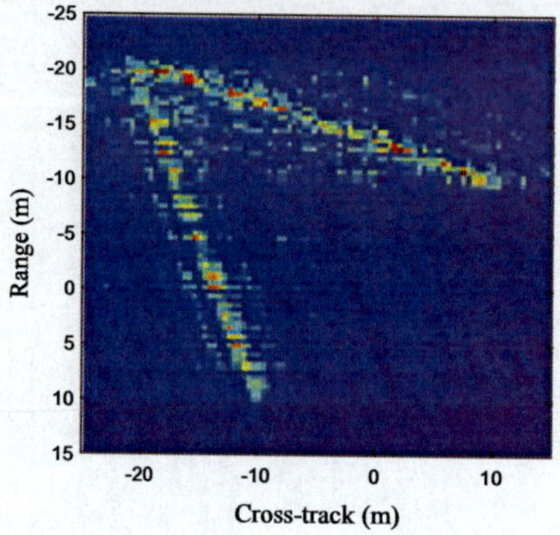

(a) Range and cross-track section

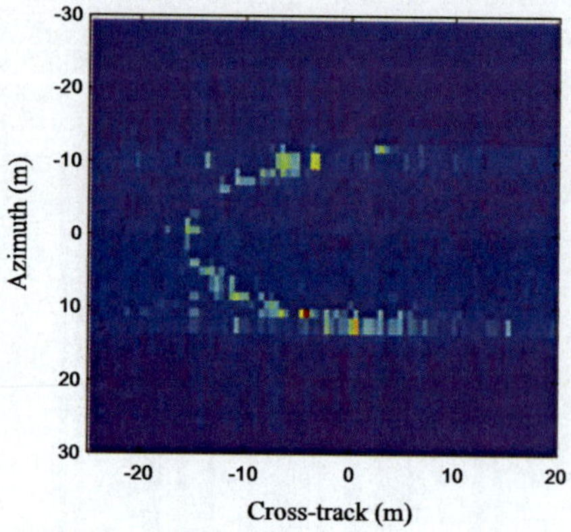

(b) azimuthal and cross-track section

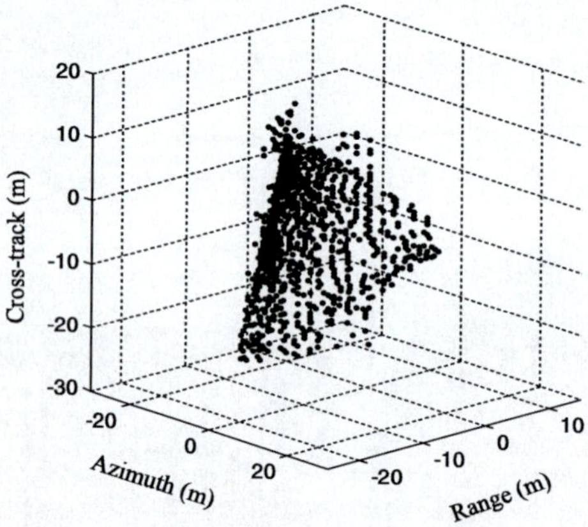

(c) 3D image

Fig. 11 3D imaging results based on BP algorithm, 23 tracks and double satellites

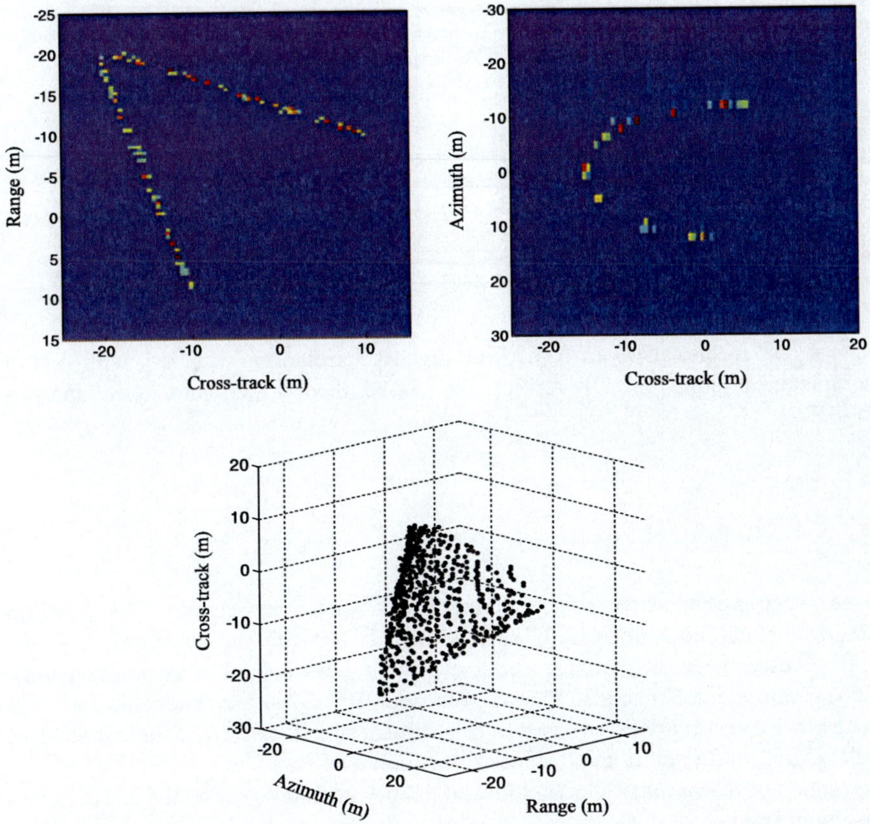

Fig. 12 D imaging results based on 2D CS algorithm, 23 tracks and double satellites

For the formation of double satellites, the 3D BP imaging results under the condition of 23 sparse flight are shown in Fig. 11. The image reconstruction results based on 2D CS method are depicted in Fig. 12. From the reconstruction results, it can be seen that the 3D imaging method of 2D CS can effectively suppress the side-lobe noise brought by sparse sampling, and correctly reconstruct the intensity and position of the scattered points in target.

5.3 Imaging Evaluation

The error analysis of image reconstruction results is carried out, and the evaluation indexes are image entropy, correlation coefficient and Root Mean Square Error (RMSE), respectively. It can be seen from the table that the imaging results obtained by the 2D CS proposed in this paper have the lowest image entropy, the

Table 2 Error analysis of image reconstruction

Observation methods	Algorithm	Evaluation index		
		Imaging entropy	Correlation coefficient	RMSE
180 tracks	3D BP	0.4532		
Single satellite, 23 tracks	3D BP	0.7214	0.6566	0.0604
Double satellites, 23 tracks	3D BP	0.6980	0.7903	0.0453
Single satellite, 23 tracks	2D CS	0.0095	0.8002	0.0432
Double satellites, 23 tracks	2D CS	0.0093	0.8994	0.0414

lowest root mean square error, and the highest correlation coefficient with that of complete sampling, which verifies the above theory and simulation analysis (Table 2).

6 Conclusion

This paper presented the sparse flight distribution of spaceborne SAR both on single satellite and formation flying satellites. In order to shorten the revisit time and increase the orbit number, a sparse 3D imaging method of two-dimensional (2D) Compressed Sensing (CS) was proposed. The linear measurement matrix in range and cross-track directions was established, which can avoid the coupling of echo signals in the orbit and distance directions. The simulation results of imaging resolution of 0.5 m for single satellite and double satellites verified the effectiveness of the method.

References

1. Giret R, Jeuland H, Enert P (2004) A study of 3D-SAR concept for a millimeter wave imaging radar onboard an UAV. In: European radar conference 2004, Amsterdam, Netherland, pp 201–204
2. Hong W (2019) Research progress on three-dimensional SAR imaging techniques. J Radars: 1–22[2019-08-06]
3. Reigber A, Moreira A, Papathanassiou KP (1999) First demonstration of airborne SAR tomography using multibaseline L-band data. In: IEEE 1999 international geoscience and remote sensing symposium. IGARSS'99 (Cat. No.99CH36293), Hamburg, Germany, pp 44–46, vol. 1. https://doi.org/10.1109/IGARSS.1999.773395
4. Candes EJ, Wakin MB (2008) An introduction to compressive sampling. IEEE Signal Process Mag 25(2):21–30
5. Candès E (2006) Compressive sampling. In: proceedings of the international congress of mathematicians, Madrid, Spain, pp 1–20
6. Donoho DL (2006) Compressed sensing. IEEE Trans Inf Theory 52(4):1289–1306

7. Xu HP, Zhou YQ, Li CS (2003) Baseline analysis and design for distributed spaceborne inteferometric SAR. J Electron Inf Technol (09):1194–1199
8. Zhu X, Bamler R (2010) Very high resolution spaceborne SAR tomography in urban environment. IEEE Trans Geosci Remote Sens 48(12):4296–4308
9. Zhu XX, Bamler R (2012) Sparse tomographic SAR reconstruction from mixed TerraSAR-X/TanDEM-X data stacks. In: IEEE international geoscience and remote sensing symposium, Munich, pp 7468–7471. https://doi.org/10.1109/IGARSS.2012.6351905
10. Zhu XX, Bamler R (2011) Let's do the time warp: multicomponent nonlinear motion estimation in differential SAR tomography. IEEE Geosci Remote Sens Lett 8(4):735–739. https://doi.org/10.1109/LGRS.2010.2103298
11. Ge N, Zhu XX (2019) Bistatic-like differential SAR tomography. IEEE Trans Geosci Remote Sens 57(8):5883–5893. https://doi.org/10.1109/TGRS.2019.2902814
12. Jianfeng Y, Qingjun Z, Jie L et al (2018) A review on development of formation flying interferometric SAR satellite system. Spacecraft Eng 27(1):116–122
13. Peterson EH, Fotopoulos G, Zee RE (2009) A feasibility assessment for low-cost InSAR formation-flying microsatellites. IEEE Trans Geosci Remote Sens 47(8):2847–2858. https://doi.org/10.1109/TGRS.2009.2017521
14. Zink M, Bartusch M, Miller D (2011) TanDEM-X mission status. IEEE international geoscience and remote sensing symposium, Vancouver, BC, pp 2290–2293. https://doi.org/10.1109/IGARSS.2011.6049666
15. Tridon DB et al (2015) TanDEM-X going for the DEM: acquisition, performance, and further activities. In: 2015 IEEE 5th Asia-pacific conference on synthetic aperture radar (APSAR), Singapore, pp 163–168. https://doi.org/10.1109/APSAR.2015.7306180
16. Zhenfang LI, Zheng BAO, Hongyang WANG et al (2006) Performance improvement for constellation SAR using signal processing techniques. IEEE Trans Aerosp Electron Syst 42(3):436–452
17. Li L, Li D, Huang P (2016) Airship sparse array antenna radar real aperture imaging based on compressed sensing and sparsity in transform domain. J Radars 5(1):109–117. https://doi.org/10.12000/JR14159